D1032806

Digital Control

Digital Control

Kannan M. Moudgalya
Indian Institute of Technology, Bombay

John Wiley & Sons, Ltd

Telephone (+44) 1243 779777

Email (for orders and customer service enquiries): cs-books@wiley.co.uk
Visit our Home Page on www.wiley.com

Other Wiley Editorial Offices

John Wiley & Sons Inc., 111 River Street, Hoboken, NJ 07030, USA

Jossey-Bass, 989 Market Street, San Francisco, CA 94103-1741, USA

Wiley-VCH Verlag GmbH, Boschstr. 12, D-69469 Weinheim, Germany

John Wiley & Sons Australia Ltd, 42 McDougall Street, Milton, Queensland 4064, Australia

John Wiley & Sons (Asia) Pte Ltd, 2 Clementi Loop #02-01, Jin Xing Distripark, Singapore 129809

John Wiley & Sons Canada Ltd, 6045 Freemont Blvd, Mississauga, ONT, Canada L5R 4J3

Wiley also publishes its books in a variety of electronic formats. Some content that appears in print may not be available in electronic books.

Anniversary Logo Design: Richard J. Pacifico

Library of Congress Cataloging-in-Publication Data

Moudgalya, Kannan M.
Digital control / Kannan M. Moudgalya.
p. cm.
Includes bibliographical references and index.
ISBN 978-0-470-03143-8 (cloth)
1. Digital control systems. I. Title.
TJ223.M53M689 2007
629.8'9–dc22

2007024694

British Library Cataloguing in Publication Data

A catalogue record for this book is available from the British Library

ISBN 978-0-470-03143-8 (HB)
ISBN 978-0-470-03144-5 (PB)

Typeset by the author using LaTeX software.
Printed and bound in Great Britain by Antony Rowe Ltd, Chippenham, Wiltshire
This book is printed on acid-free paper responsibly manufactured from sustainable forestry in which at least two trees are planted for each one used for paper production.

To my parents

Muthu Meenakshi

Kuppu Subramanian

Contents

Preface

This book has evolved over a period of several years of teaching the courses *Digital Control* and *Modelling and System Identification* at IIT Bombay. Several undergraduate and postgraduate students, some without a good analog control background, have successfully completed these courses. Several working professionals, some with a long gap in their education, have also gone through these courses successfully. This material has also been used in the year 2004 to teach the course *Intermediate Process Control* at the University of Alberta. There were students with a gap in their education, as they had enrolled in the academia–industry co-op programme.

Keeping the above mentioned requirement to cater for different kinds of students, we have designed this book to teach digital control from scratch. Apart from a few elementary concepts on calculus and differential equations, not much background is required to understand this book. In particular, the reader need not have a good analog control background to understand most of the topics presented. As a result, this book is suitable also for students whose domain of interest is made up of only discrete time systems, such as the field of computing systems.

In order to make the book self-contained, we have presented the topic of digital signal processing in reasonable detail. This is because most students, except those in electrical engineering, do not take a course in this important topic, exposure to which is useful in understanding the control design techniques presented in this book.

Because we focus on discrete time techniques, it is possible to present some advanced topics as well. For example, this approach allows us to present the important topic of system identification in some detail. Indeed, it is possible to use the material presented in this book to teach a first course in identification. Although the topic of identification is useful in all areas, it may be the only way to obtain realistic models in human–made systems, such as computing systems.

We have adopted a transfer function approach, more or less exclusively. We strongly believe that one gets good insight into the control problems through this approach. Practising engineers are generally more comfortable with transfer function based design techniques. They present an ideal platform to deal with a large number of basic facts about control design in an elementary fashion. Explaining some of these topics through the state space approach would have been more difficult. For completeness, however, we also present the state space approach to control design. We restrict our attention to single input single output systems.

We believe that one of the most important benefits that the reader of this book will acquire is to *think digital*. We strongly believe that learning to discretize analog controllers alone does not prepare the student to exploit the full potential of digital systems.

This book is useful also for students who may have a good analog control background. Comparison of the concepts presented in this book with those from analog control will only help strengthen these ideas. These students, as well as industry professionals, will find the coverage on PID controllers useful.

Working exclusively with discrete time domain has its own shortcomings. The most glaring one amongst them is the antialiasing filter, which is generally designed using continuous time techniques. Addressing this requirement through discrete time techniques may require fast sampling rates. We do not discuss the effects of quantization errors in this book. Nevertheless, because these are generally modelled with the help of random noise, the material developed in this book should help address these problems.

A first course on control can be taught using this book, especially in disciplines that deal with only discrete time models. It may also be used to teach a second course on control. Because it is more or less self-contained, this book could be useful to people who have had a break in their studies, such as practising professionals. It follows that a course using the book may also be useful for students who may not have taken a good analog control course.

A first course on identification may be taught using this book. The first six chapters could form the basis of a digital signal processing course. Finally, using the material presented in this book on identification and control, it is possible to formulate an adaptive control course.

We have provided an extensive set of Matlab[1] routines to explain the various ideas presented in this book. The heart of the routines is a solver for Aryabhatta's identity [39], which is also known as the Diophantine equation [59]. We have implemented the algorithm of [8] in Matlab to solve this polynomial equation. We have used the notation in [27] to represent polynomials in this book. We also make use of some basic routines from this reference.

An extensive index of Matlab routines is given to help locate the programs. The built–in functions, including the routines of Matlab toolboxes, appear at the beginning of this index, with the routines developed for this book appearing at the end. Most of the latter are listed in the book, at the end of every chapter. A few routines, although freely downloadable, are not listed in the book, owing to space constraints. Appendix A.2 explains the procedure for downloading and installing the software. We recommend that the reader traces through the calculations using the Matlab debugger.

The website `http://www.moudgalya.org/dc/` provides much useful material related to this book. All Matlab programs discussed in this book are available at this location. Instructional material, such as slides, based on this book are also available at this website. Finally, the reader will find at this website some other useful material, omitted in the book owing to space constraints.

Authors of previous books on the topics presented in this book have used titles such as Discrete Time Control [45]. This nomenclature presupposes that time is an independent variable and that the control effort is something different. We have shown, however, that it is possible to apply the control design principles to problems in which the sampling time itself is the control effort [35]. In view of this generality, we choose to use the title of *Digital Control* for this book.

---

[1] Matlab® and Simulink® are registered trademarks of The Mathworks, Inc. Other product or brand names are trademarks or registered trademarks of their respective holders.

Acknowledgements

This book has evolved over a period of several years of teaching the courses *CL 692, Digital Control* and *CL 625, Modelling and System Identification* at IIT Bombay, through class room instruction, as well as through the distance mode of education. Partial funding to write this book has been provided by the curriculum development programme at IIT Bombay. I would first like to acknowledge the freedom and support given by the education system at IIT Bombay to be the main factors that helped me conceive this project.

Many colleagues from IIT Bombay and other universities have given a lot of input in shaping this work. Notable amongst them are Preeti Rao, Arun Tangirala, K. P. Madhavan, A. Subramanyam, S. Sabnis, S. Narayana Iyer, Eric Ostertag and Inderpreet Arora.

I want to thank Sirish Shah for giving me an opportunity to spend one year at University of Alberta and work on this book. He also constantly urged me to develop Matlab based programs to explain the concepts. I wrote a large fraction of the Matlab codes reported in this work during my stay in Edmonton. I also want to thank Sirish for funding my visit to Taiwan to present an IEEE conference paper [41] that explains the approach followed in this work. I would like to thank Fraser Forbes for giving me an opportunity to teach the course *CHE 576, Intermediate Process Control* at UoA using the approach followed in this book. I also want to thank the researchers at UoA for giving several useful suggestions. Notable amongst them are Vinay Kariwala and Hari Raghavan. The input that I received at UoA helped sharpen the presentation.

I would like to especially mention the help that I received from Arun Tangirala in writing the material on identification. The many discussions that I had with him on this topic, at UoA as well as at IIT Bombay and at IIT Madras, helped clear my doubts in this subject. Arun has also been kind enough to let me use some examples that he has developed.

I would also like to thank Huibert Kwakernaak for his permission to use some of the basic polynomial manipulation Matlab programs reported in [27]. We have used his method of representing polynomial matrices.

I would like to thank S. Sudarshan, Soumen Chakrabarty, Inderpreet Arora and Neville Hankins for their help in preparing this manuscript: Sudarshan and Soumen patiently answered my LaTeX related questions; Inderpreet did the proofreading; Neville carried out copyediting.

I would like to thank my wife Kalpana and daughter Sukanya for bearing my absence from their lives for extended periods of time so that I would progress in this project. Finally, I would like to acknowledge my parents: without their sacrifices, I could not have written this book, let alone reach where I am today.

Kannan M. Moudgalya
IIT Bombay
17 May 2007

List of Matlab Code

List of Acronyms

1-DOF One degree of freedom
2-DOF Two degrees of freedom
A/D Analog to digital
ACF Auto covariance function
Auto correlation function
AR Auto regressive
ARIMAX Auto regressive integrated, moving average, exogeneous
ARIX Auto regressive, integrated, exogeneous
ARMA Auto regressive moving average
ARMAX Auto regressive moving average, exogeneous
ARX Auto regressive, exogeneous
AWC Anti windup control
BIBO Bounded input, bounded output
BJ Box–Jenkins
BLUE Best linear unbiased estimate
CCF Cross covariance function
Cross correlation function
D/A Digital to analog
DFT Discrete Fourier transform
DMC Dynamic matrix controller
DSP Digital signal processing
FIR Finite impulse response
GCD Greatest common divisor
GMVC Generalized minimum variance controller
GPC Generalized predictive controller
I/O LTI Input–output linear time invariant
IIR Infinite impulse response
IMC Internal model controller
LCM Least common multiple
LQG Linear quadratic Gaussian
LQR Linear quadratic regulator
LSE Least squares estimation
LTI Linear time invariant
MPC Model predictive control
MVC Minimum variance control

ODE	Ordinary differential equation
OE	Output error
PACF	Partial autocorrelation function
PID	Proportional, integral, derivative
PP	Pole placement
PRBS	Pseudo random binary sequence
ROC	Region of convergence
RPC	Remote procedure call
SISO	Single input single output
SNR	Signal to noise ratio
ZOH	Zero order hold

Chapter 1

Introduction

The concept of control is ubiquitous. We see the application of control in everyday appliances and equipment around us: washing machines, elevators, automobiles, satellites, aeroplanes, room heater, etc. It is no wonder that control is an important component in all engineering disciplines. Lately, the topic of control has become an important one in computing [22] and supply chain systems [14].

A typical feedback control configuration is given in Fig. 1.1. It consists of a plant or a process and a controller, along with the necessary control accessories, such as sensors and actuators. The objective of the controller is to maintain the plant output y close to the desired value r, known as the *reference signal* or *setpoint*. The role of controller is to provide stable and agile performance in the presence of any disturbance, even if we do not have an exact knowledge of the plant.

The requirements expected from a controller in different fields are not identical, because of the differences in plants/processes. For example, in the mechanical and aerospace industries, while stability is an issue, nonlinearity and delays are not that significant. Just the converse is true in chemical engineering. Because of these reasons, advanced control in these areas have become separate fields in their own right. Nevertheless, the basic principles of control are the same in all fields. We present some of these common basic principles in this book.

There are two broad classes of control design techniques: analog and digital. These work with continuous time and discrete time systems, respectively. Because many real world systems are described by differential equations, analog control design techniques have become popular. Because most controllers are made of digital systems, the latter is equally popular. Although it is possible to discretize analog controllers to

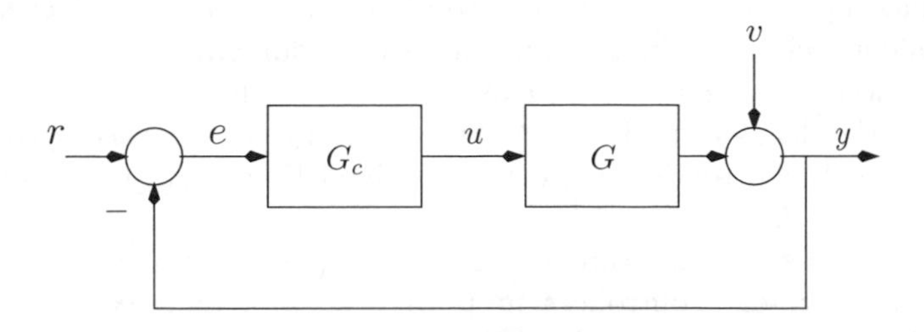

Figure 1.1: A schematic of feedback control loop

obtain digital controllers, an independent study of the latter could be useful [2, 20]. We present techniques for the design of digital controllers, from scratch, in this book.

Why is it that most controllers are now implemented using digital systems? In order to answer this, we need to first compare analog and digital circuits. In analog circuits, noise is a serious problem in all signal transmissions. In contrast, the digital signals refer to a range of values and hence it is possible to provide good noise margins. For example, in transistor transistor logic (TTL), there is a noise margin of 0.4 V in both high and low levels. The basic ability of digital systems to reject noise in a fundamental way has resulted in these systems becoming extremely popular.

We now list some of the advantages of digital systems over analog systems [38]. It is easy to implement and modify digital controllers – we just have to change the coefficients! The margins mentioned above can take care of unavoidable difficulties, such as the noise and drift. We can achieve the desired accuracy by using a sufficient number of bits. It is possible to implement error checking protocols in digital systems. Because of the advances in the manufacturing processes of digital circuits, the components can be produced in large volumes. Large digital circuits can be fully integrated through VLSI. Through a multiplexer, a single processor can handle a large number of digital signals. Digital circuits pose no loading problems, unlike analog circuits. Because of these reasons, digital devices became popular, which formed an important impetus for the advancement of digital systems. Digital devices have become rugged, compact, flexible and inexpensive. It is no wonder that most modern electronic devices, such as controllers, watches, computers, etc., are made of digital systems.

We will now present some of the advantages that digital control has over analog control. For some naturally occurring discrete time plants, a continuous time version may not even exist. For example, consider a scalar state space equation,

$$\dot{x}(t) = fx(t) + gu(t)$$

where f and g are constants. Under uniform sampling, the ZOH equivalent of this plant is

$$x(k+1) = ax(k) + bu(k)$$

where $a = e^{fT_s}$ and T_s is the sampling time. Note that $a > 0$. Thus, if a turns out to be negative in an identification exercise applied to plant data, no continuous control design technique may be suitable.

It is possible to design first-cut filters in the discrete time domain by the placement of poles and zeros at appropriate locations [49]. To amplify a signal at a particular frequency, place a pole at that frequency; to filter out a frequency, place a zero. There are no equivalent techniques in the continuous time domain.

There are design techniques that exist only in the discrete domain, the notable example being the dead-beat controller. Discrete time control theory comes up with unusual, but useful, control designs, with negative PID tuning parameters being one of them, see [30, pp. 58–62].

In several respects, discrete time control techniques are easier to understand. For example, we can present controllers in powers of z^{-1}, and realize them through recursive techniques. Most of the identification procedures known today determine discrete time models [32]. Identification of continuous time models is in its infancy.

In view of this, it is a lot easier to teach identification techniques in the discrete time domain. It is also a lot easier to teach control design techniques, such as model predictive and adaptive controllers, in the discrete time domain.

Handling of delays is a lot easier in the discrete time domain, whether they are fractional or multiples of the sampling period [2, 17]. The concept of a unit pulse is a lot easier in the discrete time domain: it is a signal of unit height with one sample width.

Although most real life systems are nonlinear, we generally use linear controllers with them. The reason is that a lot more is known about linear, as opposed to nonlinear, controllers. Moreover, if the controllers are well designed, the deviations from the operating point are expected to be small. When the deviations are small, nonlinear systems can be approximated reasonably well with first order Taylor approximations. Because of these reasons of ease and adequacy, we generally use linear controllers to regulate our systems. We restrict our attention to the design of linear controllers in this book.

We have organized this book into four parts and fourteen chapters. In the second chapter, we lay the framework for control system design. If the plant is described in continuous time, we outline a procedure to connect it with digital systems, through discretization and A/D and D/A converters.

We devote the first part to the topic of digital signal processing. In the third chapter, we present the basic concepts of linearity, time invariance and causality. While linear systems in the transfer function framework should have their initial state at zero, there is no such constraint in the state space framework. We define the former concept as input/output linearity and bring out its relation to the latter concept, which we refer to as simply linearity.

In signal processing books, the variable u is used to denote the unit step input. In control texts, however, this symbol is reserved for control effort. The variable x is used for input and state in signal processing and control books, respectively. In this book, we use u, x and y for input, state and output, respectively. We use $1(n)$ to denote the discrete time unit step signal, as practised in some popular control texts [44].

We present the topic of the Z-transform in the fourth chapter. We also present the concept of region of convergence in great detail, unlike most control books. This study helps derive the conditions for stability and causality of transfer functions and helps understand the limits of performance of control systems.

In order to understand the concept of spectrum, required in identification, we need to work with noncausal signals. This, in turn, necessitates the use of the two sided Z-transform, although most control books work only with the one sided Z-transform. Because the two sided Z-transform requires signals to be defined from the time instant of $-\infty$, we need to precisely define the concept of initial rest in state space systems.

We present frequency domain analysis in the fourth chapter. We explain the topics of Fourier transform, frequency response, sampling and reconstruction. We also explain filter design by placement of poles and zeros.

We present the topic of identification in the third part. We have given an exhaustive introduction to this topic, because this subject is becoming increasingly important, owing to the increased automation, and consequently increased availability of data. Most plants use digital devices for data acquisition and storage. As the price of storage devices has dropped, most plants also have a large amount of operating data.

Identification is concerned with the important topic of determining the plant transfer function from measurements, for the explicit purpose of achieving good control.

We devote the fourth part to transfer function based control design techniques. In the seventh chapter, we present some of the important principles required for control system design. We present a Z-transform based direct design method for lead–lag controllers. In this book, we introduce a procedure to translate the performance specifications directly to a desired region of the Z-transform complex plane, and use this to design controllers.

We devote Chapter 8 to PID controllers. Keeping in mind the popularity of PID controllers in industry, we discuss in detail different methods to discretize them, in case they have already been designed using continuous time techniques. This is the only chapter in the entire book that requires an analog control background. All other control techniques presented in this book do not require such a background.

In Chapter 9, we present pole placement controller design to arrive at two degrees of freedom controllers. We use this framework to present all controllers, including implementation of specific ones, such as anti windup controllers, in the rest of the book.

Because the coverage of this book is broad, there is a paucity of symbols. The two degrees of freedom discrete time controller, which is popular in the transfer function approach, is usually known as the RST controller, with three polynomials, R, S and T. We use the symbols R_c, S_c and T_c, respectively, for this purpose. We reserve R, S and T for reference signal, and sensitivity and complementary sensitivity functions, respectively.

We have used the phrases *good* and *bad* in place of the invertible and noninvertible factors of a polynomial [48]. Although somewhat unconventional, we believe that this usage helps remember the underlying concepts clearly. It is also less confusing than the use of notation, such as $+$ and $-$, borrowed from analog control.

It turns out that working with polynomials in powers of z^{-1} helps simplify controller design. We work with polynomials in powers of z as well as z^{-1}. We have avoided using the operator q to denote shift operation. The Z-transform variable z is used to define the shift as well.

We discuss the Smith predictor and internal model control technique in Chapter 10. In Chapter 11, we present minimum variance and generalized minimum variance controllers. In Chapter 12, we discuss model predictive controllers. In particular, we present generalized predictive control and dynamic matrix control design techniques. We also make an attempt to implement these controllers through a PID framework. We devote Chapter 13 to the design of linear quadratic Gaussian control system design.

We devote the last part to state space techniques for control design. In Chapter 14, we discuss pole placement controllers, linear quadratic regulators, observers, and the combined controller and observer. We touch upon topics such as Kalman filters.

An extensive set of Matlab routines have been provided along with the book. The heart of these is the program to solve Aryabhatta's identity [39], also known as Bezout identity or Diophantine equation. Links have been provided to download the programs from the web. A procedure to download the entire set of programs is given in Appendix A.2.

Chapter 2

Modelling of Sampled Data Systems

In this chapter, we present an overview of feedback control systems, consisting of a mix of analog and digital systems. We will refer to analog systems as continuous time systems and present a few examples. One way of connecting analog and digital systems is through sampling of the former or, equivalently, through discretization of continuous time systems. We will summarize analog to digital conversion and digital to analog conversion, through which the signals become compatible with the systems under consideration. We also present models of naturally occurring discrete time systems. We conclude this chapter with a brief discussion on different approaches to controller design and its validation.

2.1 Sampled Data System

There are many reasons why we use controllers. Some of these are: to stabilize unstable plants, to improve the performance of plants and to remove the effect of disturbances. There are two major types of controllers – feed forward and feedback. We will discuss these in more detail in Chapter 7. For the current discussion, we will consider the feedback control structure, a schematic of which is given in Fig. 1.1.

In this figure, the block G denotes the plant or the process that needs to be controlled. It could be a continuous system, described by differential equations. Examples of a plant are a room whose temperature has to be regulated, an RC circuit, a robot, a reactor, a vehicle or a distillation column. In this chapter, we will use models of the following form to represent these systems:

$$\dot{x}(t) = Fx(t) + Gu(t) \tag{2.1a}$$

$$y(t) = Cx(t) + Du(t) \tag{2.1b}$$

These are known as state space equations. The first one is the state equation and the second one is the output equation. The variable $x(t)$ is known as the state. State is defined as the set of variables required to completely understand the system. The state variables are supposed to represent the effect of all past inputs to the system.

Digital Control Kannan M. Moudgalya

For example, knowing the current state and the future inputs to the system, one can completely specify the future states and the future outputs of the system.

The input to the system can consist of two types of variables: disturbance variable and manipulated or control variable. The former is one over which we have no control. These are typically external to a system. The manipulated or control variables help make the system behave the way we want. The values of these variables are calculated by the controller.

Typically, one does not measure all the states, but only a smaller number of variables, which could be functions of states. We use the symbol y to refer to the measured vector and call it the output. Eq. 2.1b is known as the *output equation*. Here, C and D are constant matrices.

The plant could also be a discrete time system. Examples of these are computing and supply chain systems. These systems are modelled using the variables available at specific time instants only:

$$x(n+1) = Ax(n) + Bu(n) \tag{2.2a}$$
$$y(n) = Cx(n) + Du(n) \tag{2.2b}$$

As in the continuous time case, x, u and y refer to state, input to and output from the plant, respectively.

In the figure, G_c denotes the controller. We will restrict G_c to a digital controller, that is, a control algorithm implemented in a digital device. Examples of a digital device are personal computers, printers and calculators. Digital devices work with quantized data only. For example, a calculator can represent numbers only to a certain precision. This is because all the required information has to be represented with the help of a finite number of bits only.

There are many advantages in using digital controllers. Digital controllers are robust and flexible. It is possible to implement any complicated algorithm using digital devices. Because of the low prices of digital components, it is possible to achieve a sophisticated amount of automation without too much expenditure. In summary, digital controllers provide sophisticated yet rugged performance at affordable prices. Design of digital controllers is easier than that of analog controllers. The reason is that the analysis of difference equations is easier than that of differential equations.

If the plant is naturally described by a discrete time model, design of the digital controller is straightforward. In case the plant is described by a continuous time model, however, two different methods could be used: the first approach is to design a continuous time controller for the continuous time plant and discretize it. The second approach is to discretize the continuous time plant and to design a controller for it. We will use the second approach in this book.

In order to design digital controllers for continuous time systems, it is necessary to convert the continuous time models, given in Eq. 2.1, into discrete time models, given in Eq. 2.2. In addition, we should convert the signals so that the discrete and the continuous time systems that participate in the control structure can understand each other. We address these issues in this chapter.

We will next look at the variables that help the blocks communicate with each other. The output variable y is also known as the controlled variable. The reference variable for y is given by r, which is also known as the setpoint for y. The error, namely the difference $r - y$, is used by the controller to decide the control effort u,

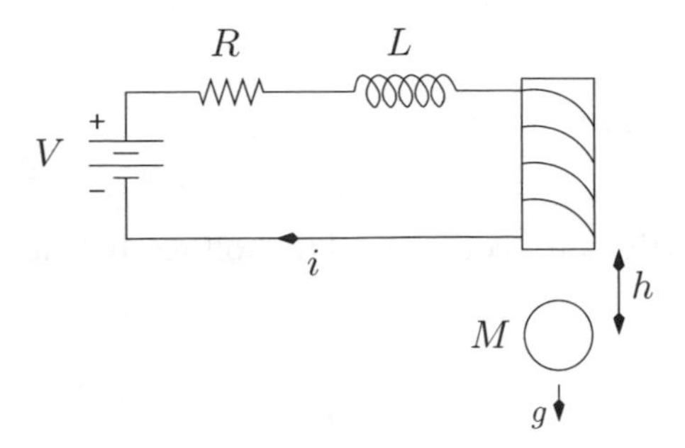

Figure 2.1: Magnetically suspended steel ball. The current through the coil creates a magnetic force, which counter balances the force due to gravity.

which is also known as the manipulated variable or the plant input. Finally, v denotes the disturbance variable.

The connections indicated in the diagram are not straightforward. The reason is that the controller G_c is a digital device that can understand only digital signals, represented through a string of binary numbers. The plant, on the other hand, could be a continuous time system described by a system of differential equations. How does one make these two, different, types of systems understand each other? We will answer this question in Sec. 2.5.

The next question is how does a digital device communicate the control action to be implemented to the plant? The answer is not obvious because the digital devices can work with numbers defined at discrete points in time only, whereas the plant could be a continuous time system described by variables that are defined at every instant of time. We will address this issue in Sec. 2.4. Once these questions are answered, we will be able to develop a unified view of all the devices present in the control structure of Fig. 1.1. In the next section, we present models of a few popular continuous time systems.

2.2 Models of Continuous Time Systems[1]

In this section, we will present the models of a magnetically suspended ball, a DC motor, an inverted pendulum, a flow system and a chemical reactor, and show how they can be represented using state space equations of the form given by Eq. 2.1.

2.2.1 Magnetically Suspended Ball

In this section, we will present the model of a *magnetically suspended ball* system, a schematic of which is shown in Fig. 2.1. The current passing through the wire wound around the armature creates a magnetic force, which attracts the steel ball and counter balances the force due to gravity [12].

The magnetic force is proportional to the square of the current and inversely proportional to the distance between the ball and the armature. The force balance

[1]If the reader does not have to deal with continuous time systems, they can skip this section.

can be written as

$$M\frac{d^2h}{dt^2} = Mg - \frac{Ki^2}{h} \tag{2.3}$$

where K is the proportionality constant. The voltage balance in the circuit can be written as

$$V = L\frac{di}{dt} + Ri \tag{2.4}$$

Suppose that the current i is such that the ball is stationary at a chosen distance h_s. We would like to derive a linear model that relates a deviation in h to a deviation in i. Let the force balance corresponding to the stationary point be modelled as

$$M\frac{d^2h_s}{dt^2} = Mg - \frac{Ki_s^2}{h_s} = 0 \tag{2.5}$$

Subtracting Eq. 2.5 from Eq. 2.3, we obtain

$$M\frac{d^2\Delta h}{dt^2} = -K\left[\frac{i^2}{h} - \frac{i_s^2}{h_s}\right] \tag{2.6}$$

Linearizing the right-hand side about the stationary point, we obtain

$$\begin{aligned}\frac{i^2}{h} &= \frac{i_s^2}{h_s} + 2\left.\frac{i}{h}\right|_{(i_s,h_s)}\Delta i - \left.\frac{i^2}{h^2}\right|_{(i_s,h_s)}\Delta h \\ &= \frac{i_s^2}{h_s} + 2\frac{i_s}{h_s}\Delta i - \frac{i_s^2}{h_s^2}\Delta h\end{aligned}$$

Substituting in Eq. 2.6, we obtain

$$M\frac{d^2\Delta h}{dt^2} = -K\left[\frac{i_s^2}{h_s^2} + 2\frac{i_s}{h_s}\Delta i - \frac{i_s^2}{h_s^2}\Delta h - \frac{i_s^2}{h}\right]$$

Simplifying, we obtain

$$\frac{d^2\Delta h}{dt^2} = \frac{K}{M}\frac{i_s^2}{h_s^2}\Delta h - 2\frac{K}{M}\frac{i_s}{h_s}\Delta i \tag{2.7}$$

We will next derive the voltage balance in deviational form. Using the variables corresponding to the force balance, we obtain

$$V_s = L\frac{di_s}{dt} + Ri_s \tag{2.8}$$

Subtracting this from Eq. 2.4, we obtain

$$\Delta V = L\frac{d\Delta i}{dt} + R\Delta i \tag{2.9}$$

We define the state variables as follows:

$$
\begin{aligned}
x_1 &\triangleq \Delta h \\
x_2 &\triangleq \Delta \dot{h} \\
x_3 &\triangleq \Delta i \\
u &\triangleq \Delta V
\end{aligned}
\tag{2.10}
$$

From the definition of x_1 and x_2, we obtain

$$
\frac{dx_1}{dt} = x_2
$$

Eq. 2.7 becomes

$$
\frac{dx_2}{dt} = \frac{K}{M}\frac{i_s^2}{h_s^2}x_1 - 2\frac{K}{M}\frac{i_s}{h_s}x_3
$$

Eq. 2.9 becomes

$$
\frac{dx_3}{dt} = -\frac{R}{L}x_3 + \frac{1}{L}u
$$

Combining these equations, we arrive at

$$
\frac{d}{dt}\begin{bmatrix} x_1 \\ x_2 \\ x_3 \end{bmatrix} = \begin{bmatrix} 0 & 1 & 0 \\ \frac{K}{M}\frac{i_s^2}{h_s^2} & 0 & -2\frac{K}{M}\frac{i_s}{h_s} \\ 0 & 0 & -\frac{R}{L} \end{bmatrix} \begin{bmatrix} x_1 \\ x_2 \\ x_3 \end{bmatrix} + \begin{bmatrix} 0 \\ 0 \\ \frac{1}{L} \end{bmatrix} u
\tag{2.11}
$$

The following are typical values of the parameters:

M	Mass of ball	0.05 kg
L	Inductance	0.01 H
R	Resistance	1 Ω
K	Coefficient	0.0001
g	Acceleration due to gravity	9.81 m/s^2
h_s	Distance	0.01 m

The current corresponding to the stationary point is obtained from Eq. 2.5 as

$$
\begin{aligned}
i_s^2 &= \frac{Mgh_s}{K} = \frac{0.05 \times 9.81 \times 0.01}{0.0001} \\
i_s &= 7.004 \text{ A}
\end{aligned}
$$

With these values, we arrive at the state space equation, given by Eq. 2.1a, with

$$
x = \begin{bmatrix} x_1 \\ x_2 \\ x_3 \end{bmatrix}, \quad F = \begin{bmatrix} 0 & 1 & 0 \\ 981 & 0 & -2.801 \\ 0 & 0 & -100 \end{bmatrix}, \quad G = \begin{bmatrix} 0 \\ 0 \\ 100 \end{bmatrix}
\tag{2.12}
$$

Let us now see by how much we should change the voltage if we want to move the ball by 1%. As this displacement is small, we expect the linear model to hold. Let the variables corresponding to the new position be indicated by a prime. The new distance is,

$$h'_s = 0.0101$$

From Eq. 2.5, we obtain

$$\begin{aligned} i'^2_s &= \frac{Mgh'_s}{K} = \frac{0.05 \times 9.81 \times 0.0101}{0.0001} \\ i'_s &= 7.039 \\ \Delta i &= i'_s - i_s = 0.035\,\text{A} \\ \Delta V &= R\Delta i = 0.035\,\text{V} \end{aligned} \tag{2.13}$$

In other words, we have to increase the voltage by 0.035 V to maintain the ball at the new stationary point, given by $h_s = 0.0101$ m.

We have converted a second order ordinary differential equation (*ODE*) into a set of two first order ODEs. This procedure can be used to convert higher order ODEs as well into systems of first order.

2.2.2 DC Motor

A *DC motor* is a popular rotary actuator in control systems. On application of an electrical voltage, the rotor rotates, as per the following Newton's law of motion:

$$\frac{J}{b}\ddot{\theta} = -\dot{\theta} + \frac{K}{b}V \tag{2.14}$$

where θ and J are the angular position and the moment of inertia of the shaft, respectively. V, b and K are the voltage, damping factor and a constant, respectively. The initial angular position and the angular velocity may be taken to be zero. Suppose that in an implementation, we have

$$\frac{K}{b} = 1 \tag{2.15}$$

If we define

$$\begin{aligned} x_1 &= \dot{\theta} \\ x_2 &= \theta \end{aligned} \tag{2.16}$$

it implies that $\ddot{\theta} = \dot{x}_1$, and the above system gets reduced to the state space equation given by Eq. 2.1a, with

$$x = \begin{bmatrix} x_1 \\ x_2 \end{bmatrix}, \quad F = \begin{bmatrix} -b/J & 0 \\ 1 & 0 \end{bmatrix}, \quad G = \begin{bmatrix} b/J \\ 0 \end{bmatrix} \tag{2.17}$$

Because the initial angular position and the angular velocity are given as zero, we obtain $x_1(0) = 0$ and $x_2(0) = 0$.

The same model can be used to describe a satellite tracking antenna system [17], as well as ships, if J/b is interpreted as the time constant [2].

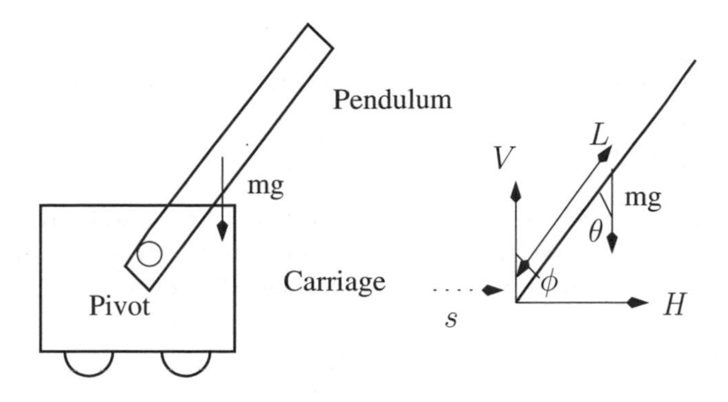

Figure 2.2: Inverted pendulum

2.2.3 Inverted Pendulum Positioning System

The next example is concerned with the positioning of a pendulum in an inverted position in a carriage. The pivot of the pendulum is mounted on a carriage, see Fig. 2.2. Although the natural position of the pendulum is to point downwards from the pivot, we are interested in positioning the pendulum upwards. This is proposed to be achieved by moving the carriage in a horizontal direction. The carriage is driven by a small motor that at time t exerts a force $\mu(t)$ on the carriage. This force is the input variable to the system.

We define $s(t)$, $\phi(t)$, m, L, J and M, respectively, as displacement of pivot at time t, angular rotation of pivot at time t, mass of pendulum, distance of pivot from the centre of gravity (CG) of the pendulum, moment of inertia about CG and mass of carriage. Let $H(t)$ and $V(t)$ denote the horizontal and vertical reaction forces at t.

The horizontal and vertical force balances and the angular momentum balance on the pendulum are given, respectively, by

$$m\frac{d^2}{dt^2}[s(t) + L\sin\phi(t)] = H(t)$$

$$m\frac{d^2}{dt^2}[L\cos\phi(t)] = V(t) - mg$$

$$J\frac{d^2\phi(t)}{dt^2} = LV(t)\sin\phi(t) - LH(t)\cos\phi(t)$$

The force balance on the carriage, with the assumption of zero frictional force, is given by

$$M\frac{d^2 s}{dt^2} = \mu(t) - H(t)$$

Linearizing the equations about the equilibrium point defined by $\mu = \phi = \dot{\phi} = \dot{s} = 0$, and letting x_1, x_2, x_3 and x_4 denote Δs, $\Delta\phi$, $\Delta\dot{s}$ and $\Delta\dot{\phi}$, respectively, we obtain

$$\frac{d}{dt}\begin{bmatrix} x_1 \\ x_2 \\ x_3 \\ x_4 \end{bmatrix} = \begin{bmatrix} 0 & 0 & 1 & 0 \\ 0 & 0 & 0 & 1 \\ 0 & -\gamma & 0 & 0 \\ 0 & \alpha & 0 & 0 \end{bmatrix} \begin{bmatrix} x_1 \\ x_2 \\ x_3 \\ x_4 \end{bmatrix} + \begin{bmatrix} 0 \\ 0 \\ \delta \\ -\beta \end{bmatrix} \Delta\mu \tag{2.18}$$

where

$$\alpha = \frac{mgL(M+m)}{\Gamma}, \quad \beta = \frac{mL}{\Gamma}, \quad \gamma = \frac{m^2L^2g}{\Gamma}, \quad \delta = \frac{J+mL^2}{\Gamma}$$
$$\Gamma = (J+mL^2)(M+m) - m^2L^2$$

The control effort in the above equation is in terms of horizontal force, which is not measured. We will now show how it can be represented in terms of the measurable variable, voltage. The expression for applied voltage E is given as

$$E = I_m R_m + K_m K_g \omega_g = I_m R_m + K_m K_g \frac{\dot{s}}{r}$$

where E, I_m, K_m, K_g, ω_g and r are voltage applied to motor in volts, current in motor in amperes, back EMF constant in V s/rad, gear ratio in motor gear box, motor output angular velocity in rad/s and radius of motor pinion that meshes with the track in m.

The torque generated by the motor is given by $T = K_m K_g I_m$, which is transmitted as a force to the carriage through the pinion by $\mu = T/r$. Substituting the expression for T and then for I_m from the voltage equation, we obtain

$$\mu = \frac{K_k K_g}{R_m r} E - \frac{K_m^2 K_g^2}{R_m r^2} \dot{s}$$

As $\dot{s} = \Delta\dot{s}$, we can write this as $\mu = \alpha_1 E - \alpha_2 \Delta\dot{s}$. With $\mu = \Delta\mu$, Eq. 2.18 becomes

$$\frac{d}{dt}\begin{bmatrix} x_1 \\ x_2 \\ x_3 \\ x_4 \end{bmatrix} = \begin{bmatrix} 0 & 0 & 1 & 0 \\ 0 & 0 & 0 & 1 \\ 0 & -\gamma & -\alpha_2\delta & 0 \\ 0 & \alpha & \alpha_2\beta & 0 \end{bmatrix} \begin{bmatrix} x_1 \\ x_2 \\ x_3 \\ x_4 \end{bmatrix} + \begin{bmatrix} 0 \\ 0 \\ \alpha_1\delta \\ -\alpha_1\beta \end{bmatrix} E$$

With the following values,

$$\begin{aligned} K_m &= 0.00767 \text{ V s/rad} \\ K_g &= 3.7 \\ R_m &= 2.6\ \Omega \\ r &= 0.00635 \text{ m} \\ M &= 0.522 \text{ kg} \\ m &= 0.231 \text{ kg} \\ g &= 9.81 \text{ m/s}^2 \\ L &= 0.305 \text{ m} \\ J &= 0 \end{aligned}$$

we arrive at

$$\frac{d}{dt}\begin{bmatrix} x_1 \\ x_2 \\ x_3 \\ x_4 \end{bmatrix} = \begin{bmatrix} 0 & 0 & 1 & 0 \\ 0 & 0 & 0 & 1 \\ 0 & -4.3412 & -14.7164 & 0 \\ 0 & 46.3974 & 48.2506 & 0 \end{bmatrix} \begin{bmatrix} x_1 \\ x_2 \\ x_3 \\ x_4 \end{bmatrix} + \begin{bmatrix} 0 \\ 0 \\ 3.2929 \\ -10.7964 \end{bmatrix} E \tag{2.19}$$

which is in the form of Eq. 2.1a. M 2.1 shows how to construct the state space equation for this system.

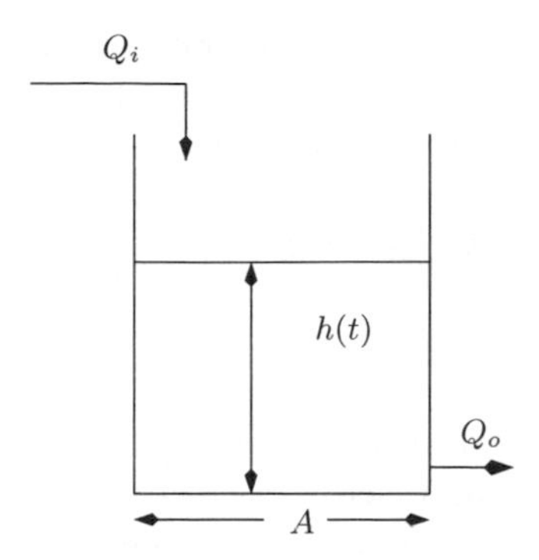

Figure 2.3: Liquid flow system

2.2.4 Liquid Flow Systems

We frequently come across flow systems in everyday life. We present a model of these in this section. Let a liquid flow into a tank of uniform cross-sectional area A at a flow rate of $Q_i(t)$. Let the output flow rate be $Q_o(t)$. Let the height of the liquid in the tank be $h(t)$. Let us assume that the density is constant in the tank. A schematic of this system is given in Fig. 2.3. The mass balance for this system is

$$A\frac{dh(t)}{dt} = Q_i(t) - Q_o(t) \tag{2.20}$$

Suppose that the outflow rate is proportional to the square root of height in the tank, that is, $Q_o(t) = k\sqrt{h(t)}$. Then the linearized model of this system about the operating point (Q_{is}, h_s) can be derived as

$$\frac{d\Delta h(t)}{dt} = -\frac{k}{2A\sqrt{h_s}}\Delta h(t) + \frac{1}{A}\Delta Q_i(t) \tag{2.21}$$

where

$$\begin{aligned} \Delta h(t) &\triangleq h(t) - h_s \\ \Delta Q_i(t) &\triangleq Q_i(t) - Q_{is} \end{aligned} \tag{2.22}$$

with the initial condition at $t = 0$ given as

$$\Delta h(t) = 0 \tag{2.23}$$

This is in the standard state space equation form, given by Eq. 2.1a.

2.2.5 van de Vusse Reactor

In this section, we will present a chemical reaction system, studied by [3]. Consider the following two reactions carried out at a constant temperature,

$$\begin{aligned} A &\xrightarrow{k_1} B \xrightarrow{k_2} C \\ 2A &\xrightarrow{k_3} D \end{aligned} \tag{2.24}$$

that describe how the reactant A is converted into a desired product B and undesired chemicals C and D. In order to understand how the concentration of B varies with time, we need to study the following mass balance equations:

$$\begin{aligned}\frac{dC_A}{dt} &= \frac{F}{V}(C_{Af} - C_A) - k_1 C_A - k_3 C_A^2 \\ \frac{dC_B}{dt} &= -\frac{F}{V} C_B + k_1 C_A - k_2 C_B\end{aligned} \tag{2.25}$$

Here, C_{Af} denotes the concentration of A in the feed. C_A and C_B denote the concentration of A and B in the reactor, respectively. All of these have the units of gmol/l. F denotes the flow rate and V denotes the volume in the reactor. We define deviation variables

$$\begin{aligned}x &= \begin{bmatrix} x_1 \\ x_2 \end{bmatrix} = \begin{bmatrix} C_A - C_{As} \\ C_B - C_{Bs} \end{bmatrix} \\ m &= \frac{F}{V} - \frac{F_s}{V} \\ d &= C_{Af} - C_{Afs}\end{aligned} \tag{2.26}$$

where x, m and d, respectively, refer to state, manipulated and disturbance variables. $C_{Af} - C_{Afs}$ is termed the disturbance variable, because variations in it cannot be avoided. On the other hand, the deleterious consequences due to this variation can be countered by adjusting m – hence the name, manipulated variable. Linearizing about the operating point defined by $C_{As} = 3$ gmol/l, $C_{Bs} = 1.117$ gmol/l and $F_s/V = 0.5714$ min^{-1}, we obtain the following state space model:

$$\dot{x} = Fx + G_1 m + G_2 d \tag{2.27}$$

where

$$F = \begin{bmatrix} -2.4048 & 0 \\ 0.8833 & -2.2381 \end{bmatrix}, \quad G_1 = \begin{bmatrix} 7 \\ -1.117 \end{bmatrix}, \quad G_2 = \begin{bmatrix} 0.5714 \\ 0 \end{bmatrix} \tag{2.28}$$

An objective of developing this model is to explore the possibility of regulating the concentration of B, namely C_B, by manipulating the flow rate F, despite the variations in C_{Af}. If we define $u = \begin{bmatrix} m & d \end{bmatrix}^T$, Eq. 2.27 gets reduced to the standard form given by Eq. 2.1a with $G = \begin{bmatrix} G_1 & G_2 \end{bmatrix}$. We will refer to this system as the *van de Vusse reactor*.

In this section, we have seen several continuous time systems and methods to discretize them. There are also naturally occurring discrete time systems, some of which will be presented in the next section.

2.3 Naturally Occurring Discrete Time Systems

In this section, we consider plants that have naturally discrete time models. That is, these systems may not even have a differential equation model. Computing systems are ideal examples of this category. Indeed, a digital control text book devoted to the control of only computing systems has recently appeared [22]. In this section, we present the example of an IBM Lotus Domino server and an example from the supply chain area.

2.3.1 IBM Lotus Domino Server

IBM Lotus Domino server is an email server. In this discussion, we will abbreviate it as simply the server. Clients may be considered as the conduits through which the end users access the server. Clients access the database of emails maintained by the server through Remote Procedure Calls (RPCs). The number of RPCs, denoted as RIS, has to be controlled. If the number of RIS becomes large, the server will be overloaded, with a consequent degradation of performance. RIS should not be made small either. If RIS is less than what can be handled, the server is not being used optimally.

Unfortunately, it is not possible to regulate RIS directly. Although RIS is closely related to the number of users, the correspondence could be off at times, because some users could just be idling. Regulation of RIS may be achieved by limiting the maximum number of users (MaxUsers) who can simultaneously use the system.

Based on experience in operating the server, it is possible to come up with an optimal RIS to achieve acceptable performance. If the actual RIS is smaller than the reference value of RIS, MaxUsers may be increased and vice versa.

It is difficult to come up with an exact analytical model that relates MaxUsers and the actual RIS in the server. This is because the server could have different types of administrative loads, such as memory swapping and garbage collection at different times. Moreover, it is difficult to predict how many users will idle at a given time.

A possible way to arrive at a model is through identification, which consists of the following steps: experimentation, data collection and curve fitting, to be explained in detail in Chapter 6. Hellerstein *et al.* [22] report an experiment in which the parameter MaxUsers is varied about the operating level of $\overline{\text{MaxUsers}} = 165$ and the corresponding variation of RIS about the operating level of $\overline{\text{RIS}} = 135$. Defining

$$\begin{aligned} x(k) &= \text{RIS}(k) - \overline{\text{RIS}} \\ u(k) &= \text{MaxUsers}(k) - \overline{\text{MaxUsers}} \end{aligned} \tag{2.29}$$

and carrying out an identification exercise, they arrive at the following relationship:

$$x(k+1) = 0.43x(k) + 0.47u(k) \tag{2.30}$$

which is in the form of Eq. 2.2a. Discrete time models are a natural choice in the field of computing systems.

2.3.2 Supply Chain Control

Another field in which discrete time modelling is on the upswing is the area of supply chains. The main reasons for the increased activity in this field are increasing competition and availability of data. We now present an example from this field.

If a manufacturer produces less than what is required, they may lose market share. On the other hand, if they produce more than required, they will incur losses through inventory costs, interest, etc.

We will now present an example from the general area of supply chain management [58]. The idea is that if we have a good estimate of the inventory and the demand for goods, we can decide how much to manufacture. The system inventory is made up of the inventories at the manufacturer and the distributor, as well as the goods in transit, corrected for estimated arrival times.

The functions of the production ordering and inventory management system include demand forecasting, customer order fulfilment, production ordering (determining the production release quantities), and the production process. The forecasted demand FD of the products is based on first order exponential smoothing of the customer sales rate SALES, with a smoothing constant ρ and a sampling interval δ:

$$\mathsf{FD}(n) = \mathsf{FD}(n-1) + \rho\delta(\mathsf{SALES}(n-1) - \mathsf{FD}(n-1)) \tag{2.31}$$

The sampling interval δ or the integration time step is said to correspond to the frequency at which the information is updated within the system. The inventory level (INV) accumulates the difference in the production rate (PRATE) and the sales rate:

$$\mathsf{INV}(n) = \mathsf{INV}(n-1) + \delta(\mathsf{PRATE}(n) - \mathsf{SALES}(n)) \tag{2.32}$$

The term production release refers to the quantity ordered for production. The work in process (WIP) level accumulates the difference in the production release (PREL) and production rate (PRATE):

$$\mathsf{WIP}(n) = \mathsf{WIP}(n-1) + \delta(\mathsf{PREL}(n) - \mathsf{PRATE}(n)) \tag{2.33}$$

The production release quantities (PREL) are determined using the ordering rule in Eq. 2.33, based upon the forecasted demand, the difference between the desired level of WIP and the current WIP level, and the difference between the desired level of inventory and the current inventory level:

$$\begin{aligned}\mathsf{PREL}(n) = {} & \mathsf{FD}(n-1) + \alpha(L \times \mathsf{FD}(n-1) - \mathsf{WIP}(n-1)) \\ & + \beta(\mathsf{FD}(n-1) - \mathsf{INV}(n-1))\end{aligned} \tag{2.34}$$

where α is the fractional adjustment rate for WIP and it describes how much of the discrepancy between the desired and current levels of WIP are to be added to the production release order. Similarly, β is the fractional adjustment rate for inventory and it describes how much of the discrepancy between the desired and current levels of inventory are to be added to the production release order. These variables are to be chosen so as to achieve a good performance. This topic is discussed in more detail in Example 4.20 on page 88.

Based on the Little's law, the desired WIP in the system is set to yield the desired throughput (set equal to forecasted demand), given the lead time, L [58]. To provide adequate coverage of inventory, the manufacturer seeks to maintain a desired level of inventory set equal to the forecasted demand. In the following equation, the production process is typically modelled as a fixed pipeline delay L:

$$\mathsf{PRATE}(n) = \mathsf{PREL}(n-L) \tag{2.35}$$

An objective of this model is to determine production release as a function of sales.

In this section, we have presented two models that have been posed directly in discrete time. For similar models in diverse fields, such as banking and the criminal justice system, the reader is referred to [4].

Figure 2.4: Analog to digital converter

2.4 Establishing Connections

In this section, we will discuss the issues in connecting up the feedback loop as given in Fig. 1.1 on page 1. As mentioned earlier, the connections are not straightforward, because the plant may be a continuous time system, while the controller is digital.[2]

Digital systems can understand only binary numbers. They can also produce only binary numbers, at discrete time intervals, synchronized by a clock. The real life systems, on the other hand, could be continuous with their properties specified by variables that are defined at all times. In this section, we explain how these two types of devices communicate with each other.

2.4.1 Continuous Time to Digital Systems – A/D Converter

We will first look at the transformation to be carried out on the signal from the plant, before it can be communicated to the digital device: it has to be quantized for the digital device to understand it.

For example, suppose that the variable y in Fig. 1.1 denotes a voltage in the range of 0 to 1 volt. This could be the conditioned signal that comes from a thermocouple, used in a room temperature control problem. Suppose also that we use a digital device with four bits to process this information. Because there are four bits, it can represent sixteen distinct numbers, 0000 to 1111, with each binary digit taking either zero or one. The resolution we can achieve is 1/15 volt, because, sixteen levels enclose fifteen intervals. The continuous variable y that lies in the range of 0 to 1 volt is known as the analog variable.

Representation of an analog value using binary numbers is known as analog to digital conversion. The circuit that is used for this purpose is known as an analog to digital converter, abbreviated as *A/D converter*. It reads the analog variable at a time instant, known as *sampling*, and calculates its binary equivalent, see Fig. 2.4.

This calculation cannot be done instantaneously; it requires a nonzero amount of time. This is the minimum time that should elapse before the converter attempts to quantize another number. Sampling is usually done at equal intervals, synchronized by a clock. In all digital devices, the clock is used also to synchronize the operations of different components. Because of this reason, the binary equivalents of an analog signal are evaluated at discrete time instants, known as the sampling instants, only. The binary signals produced by the A/D converter are known as the *digital signals*. These signals are quantized in value, because only specific quantization levels can be represented and discrete in time, and because these conversions take place only at sampling instants. This is the reason why the device used for this purpose is known as the A/D converter.

[2] If the reader does not have to deal with continuous time systems, they can now go straight to Sec. 2.4.3.

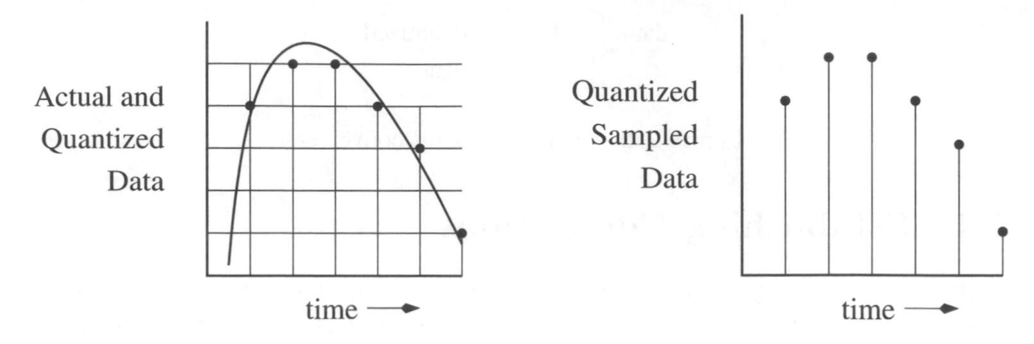

Figure 2.5: Sampling, quantization of analog signals

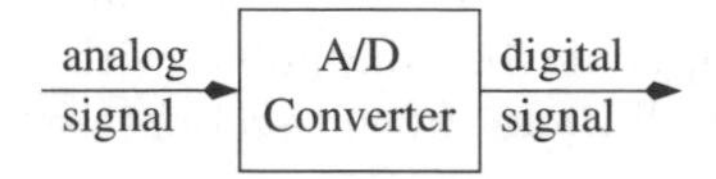

Figure 2.6: Quantization

The left diagram in Fig. 2.5 illustrates the fact that the A/D converter takes a continuous function of real values, known as an analog signal, and produces a sequence of quantized values. The data seen by the digital device for this example are given in the right diagram of the same figure. In view of the above discussion, Fig. 2.4 can be redrawn as in Fig. 2.6.

The error in quantization varies inversely with the number of bits used. The falling hardware prices have ensured that even low cost devices have large numbers of bits with small quantization errors. As a result, it is reasonable to assume that no major difficulties arise because of quantization.

It should be pointed out that the time required for A/D conversion usually depends on the value of the analog signal itself. Nevertheless, the A/D converters are designed so as to sample the analog signal at equal intervals, with the requirement that the interval is at least as large as the maximum conversion time required. Most A/D converters sample analog signals at equal time intervals only. This is referred to as uniform sampling. The digital processing devices that are connected to the output of A/D converters also read the digital signals at the same instants.

As it takes a finite but nonzero amount of time for A/D conversion, it can be assumed that the analog signal, sampled at a time instant, will be available for digital processing devices in quantized form, at the next time instant only.

The next question to ask is whether there is any loss in information because of sampling in the A/D converter. The answer is no, provided one is careful. Many variables of practical interest vary slowly with time – in any case, slower than the speeds at which the data acquisition systems work. As a result, the loss of information due to sampling is minimal. Moreover, it is possible to get high performance A/D converters with small sampling time at a low cost and, as a result, the sampling losses can be assumed to be minimal. Indeed it is possible to use a single A/D converter to digitize several analog signals simultaneously. That is, an A/D converter can sample several analog signals, one at a time, without affecting the performance.

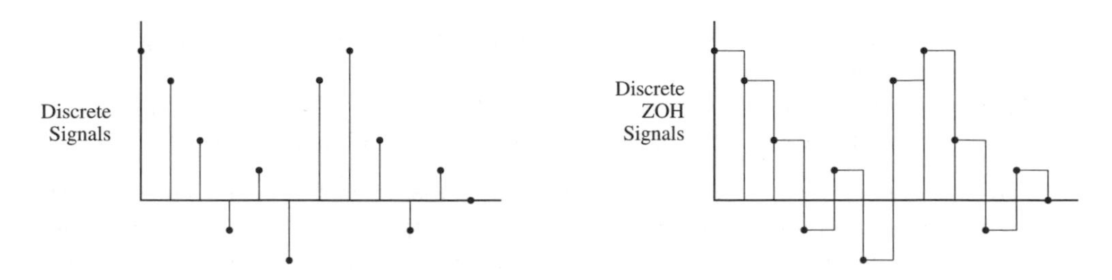

Figure 2.7: Zero order hold

We have talked about the procedure to follow to make the continuous time devices, such as the plant in Fig. 1.1 on page 1, communicate with digital devices. We will now present the issues in the reverse communication – now the digital devices want to communicate to continuous time devices.

2.4.2 Digital to Continuous Time Systems – D/A Converter

The outputs from the digital devices are also digital signals and they are available only at regular intervals. A typical signal is presented in the left hand diagram of Fig. 2.7.

The real world that has to deal with the digital device may be analog in nature. For example, suppose that the controller, through binary numbers, indicates the desired position of a valve position. The valve, being a continuous time device, can only understand analog values. This is where the digital to analog converter, abbreviated as D/A converter, comes in. It converts the binary vector into a decimal number.

There is one more issue to be resolved. The valve cannot work with the numbers arriving intermittently. It should know the value of the signal during the period that separates two samples. The most popular way to do this is to hold the value constant until the next sampling instant, which results in a *staircase approximation*, as shown in the right diagram of Fig. 2.7. Because a constant value is a polynomial of zero degree, this is known as the *zero order hold (ZOH)* scheme. Although more complicated types of hold operations are possible, the ZOH is most widely used and it is usually sufficient. The ZOH operation usually comes bundled as a part of the D/A converter. In the rest of this book, we will use only ZOH.

We have talked about the procedure to convert the signals so that they are suitable to the receiving systems. Now let us discuss some issues in connecting the plant to the loop.

2.4.3 Input–Output View of Plant Models

We have shown that it is possible to express several continuous time systems in the form of Eq. 2.1a on page 5, namely $\dot{x}(t) = Fx(t) + Gu(t)$. Here, x, known as the state vector, denotes the variables that characterize the state of the system. The variable $u(t)$, which denotes the input to the system, can consist of disturbance and manipulated or control variables. As most systems are nonlinear, linearization is required to arrive at such state space models.

We will show in the next section that it is possible to discretize continuous time systems to arrive at an equation of the form of Eq. 2.2a on page 6, namely

$x(n+1) = Ax(n) + Bu(n)$. Such equations also arise in naturally occurring discrete time systems, as we have seen in Sec. 2.3.

Although the states contain all the information about a system, it may not be possible to measure all of them, unfortunately. We give below two reasons:

1. It may be very expensive to provide the necessary instruments. For example, consider a distillation column with 50 trays. Suppose the temperature in each tray is the state variable. It may be expensive to provide 50 thermocouples. Moreover the column may not have the provision to insert 50 thermocouples. Usually, however, a limited number of openings, say in the top and the bottom tray, will be available for thermocouple insertions.

2. There may not be any sensors, to measure some of the states. Suppose for instance the rate of change of viscosity is a state vector. There is no sensor that can measure this state directly.

Because of these reasons, only a subset of the state vector is usually measured. Sometimes a function of states also may be measured. The following example illustrates this idea.

Example 2.1 Consider a system in which there are two immiscible liquids. Suppose the height of the two liquids forms the state vector. Construct the output equation for the following two cases:

1. Only the level of the second fluid is measured.
2. Only the sum of two levels is measured.

Let x_1 and x_2 denote the heights of liquids 1 and 2, respectively. We obtain

$$x = \begin{bmatrix} x_1 \\ x_2 \end{bmatrix}$$

Let y denote the measured variable. When the level of the second fluid is measured, we obtain

$$y = \begin{bmatrix} 0 & 1 \end{bmatrix} \begin{bmatrix} x_1 \\ x_2 \end{bmatrix} = x_2$$

Note that the above expression can be written as $y = Cx + Du$, where C is given by $\begin{bmatrix} 0 & 1 \end{bmatrix}$ and $D = 0$. When the sum of levels is measured, we obtain

$$y = \begin{bmatrix} 1 & 1 \end{bmatrix} \begin{bmatrix} x_1 \\ x_2 \end{bmatrix} = x_1 + x_2$$

This can once again be written as $y = Cx + Du$, where C is given by $\begin{bmatrix} 1 & 1 \end{bmatrix}$ and $D = 0$.

∎

In view of the reasons explained above, complete state space models are given by Eq. 2.1 on page 5 or Eq. 2.2 on page 6.

The control effort or the manipulated variable $u(n)$ becomes the input to the plant, with $y(n)$ being the output from it. These two variables connect to the external world.

2.5 Discretization of Continuous Time Systems

Recall that the objective of this chapter is to establish connections between the building blocks of Fig. 1.1 and to study them in a uniform framework. We demonstrated that A/D and D/A converters can be used to establish connections. The models of the plant and the controller are quite different from each other, though. The former is often made up of differential equations with continuous time variables. The latter, on the other hand, can understand only the transitions between the sampling intervals. If the controller has to take meaningful control decisions, it better understand how the plant works. One way to do this is to transform the continuous time plant model into a discrete time model that relates the parameters of the system at the sampling intervals. Through this mechanism, we can bring the plant and the controller models to a uniform framework. It is then possible to design the control law. This section is devoted to the discretization of continuous time models.

2.5.1 Solution to the State Space Equation

In this section, we will discuss how to discretize a continuous time model of the form Eq. 2.1a on page 5, reproduced here for convenience,

$$\dot{x}(t) = Fx(t) + Gu(t) \tag{2.36}$$

One of the popular methods of discretization is to solve the system and choose appropriate time values, as we will explain shortly. With this objective in mind, we define the exponential of a square matrix. Through its derivative, we construct an integrating factor, using which we obtain an explicit solution to the state space equation.

The exponential of a square matrix F, namely e^{Ft}, is defined as follows:

$$e^{Ft} \triangleq I + Ft + \frac{1}{2!}F^2t^2 + \cdots \tag{2.37}$$

We will present an example to illustrate this idea.

Example 2.2 Determine the exponential of the matrix F for the DC motor system, defined in Sec. 2.2.2, with $J/b = 1$.

It is given that

$$F = \begin{bmatrix} -1 & 0 \\ 1 & 0 \end{bmatrix}$$

We will first calculate powers of F:

$$F^2 = \begin{bmatrix} -1 & 0 \\ 1 & 0 \end{bmatrix} \begin{bmatrix} -1 & 0 \\ 1 & 0 \end{bmatrix} = \begin{bmatrix} 1 & 0 \\ -1 & 0 \end{bmatrix}$$

$$F^3 = \begin{bmatrix} -1 & 0 \\ 1 & 0 \end{bmatrix} \begin{bmatrix} 1 & 0 \\ -1 & 0 \end{bmatrix} = \begin{bmatrix} -1 & 0 \\ 1 & 0 \end{bmatrix}$$

Using the definition given in Eq. 2.37, we obtain the exponential of Ft as

$$e^{Ft} = \begin{bmatrix} 1 & 0 \\ 0 & 1 \end{bmatrix} + \begin{bmatrix} -1 & 0 \\ 1 & 0 \end{bmatrix} t + \frac{1}{2} \begin{bmatrix} 1 & 0 \\ -1 & 0 \end{bmatrix} t^2 + \frac{1}{3!} \begin{bmatrix} -1 & 0 \\ 1 & 0 \end{bmatrix} t^3 + \cdots$$

Carrying out a term by term summation, we obtain

$$e^{Ft} = \begin{bmatrix} 1 - t + t^2/2 - t^3/3! + \cdots & 0 \\ t - t^2/2 + t^3/3! - \cdots & 1 \end{bmatrix} = \begin{bmatrix} e^{-t} & 0 \\ 1 - e^{-t} & 1 \end{bmatrix}$$

There are many ways to calculate a matrix exponential. Matlab calculates it using a numerically stable procedure. M 2.2 shows how this calculation is done for $t = 1$. We obtain

$$e^{F} = \begin{bmatrix} 0.3679 & 0 \\ 0.6321 & 1 \end{bmatrix}$$

∎

We also need the concept of derivative of the exponential of a matrix. Differentiating both sides of Eq. 2.37, we obtain

$$\begin{aligned} \frac{d}{dt}\left(e^{Ft}\right) &= \frac{d}{dt}\left(I + Ft + \frac{1}{2!}F^2t^2 + \frac{1}{3!}F^3t^3 + \cdots\right) \\ &= 0 + F + \frac{1}{2!}F^2 2t + \frac{1}{3!}F^3 3t^2 + \cdots \\ &= F + F^2 t + \frac{1}{2!}F^3t^2 + \cdots \\ &= (I + Ft + \frac{1}{2!}F^2t^2 + \cdots)F = e^{Ft}F \end{aligned}$$

Note that we could have factored F on the left-hand side as well. Thus, we obtain

$$\frac{d}{dt}(e^{Ft}) = Fe^{Ft} = e^{Ft}F \tag{2.38}$$

Now consider solving the state space equation of Eq. 2.36. First we rewrite it as follows:

$$\dot{x}(t) - Fx(t) = Gu(t)$$

Premultiply both sides by e^{-Ft} to obtain

$$e^{-Ft}\dot{x}(t) - e^{-Ft}Fx(t) = e^{-Ft}Gu(t)$$

Using Eq. 2.38, the left-hand side can be simplified as

$$\frac{d}{dt}(e^{-Ft}x(t)) = e^{-Ft}Gu(t)$$

Integrating both sides with respect to time from t_0 to t, we obtain

$$\begin{aligned} e^{-Ft}x(t) - e^{-Ft_0}x(t_0) &= \int_{t_0}^{t} e^{-F\tau}Gu(\tau)d\tau \\ e^{-Ft}x(t) &= e^{-Ft_0}x(t_0) + \int_{t_0}^{t} e^{-F\tau}Gu(\tau)d\tau \end{aligned}$$

Premultiplying both sides by e^{Ft}, we obtain

$$x(t) = e^{F(t-t_0)}x(t_0) + \int_{t_0}^{t} e^{F(t-\tau)}Gu(\tau)d\tau \tag{2.39}$$

This is the solution to the state space equation, given in Eq. 2.36.

2.5.2 Zero Order Hold Equivalent of the State Space Equation

We will now make an assumption that all inputs to analog devices that describe continuous processes are outputs of a ZOH device. This assumption allows us to develop a systematic theory of systems consisting of both analog and digital devices. If the sampling rate is sufficiently high, this assumption poses no major problems even when the staircase approximation is used for smoothly varying analog signals.

The easiest way to study both analog and digital devices in a single framework is to observe their behaviour at sampling instants only. Let us denote by t_n and t_{n+1} two successive sampling instants. Let us substitute t_n and t_{n+1}, respectively, for t_0 and t in Eq. 2.39. We obtain

$$x(t_{n+1}) = e^{F(t_{n+1}-t_n)}x(t_n) + \int_{t_n}^{t_{n+1}} e^{F(t_{n+1}-\tau)}Gu(\tau)d\tau \tag{2.40}$$

We will next make the ZOH assumption, that is, assume that u is held constant in a sampling interval, as in Fig. 2.7:

$$u(\tau) = u(t_n), \quad t_n \leq \tau < t_{n+1} \tag{2.41}$$

We will also assume that we use uniform sampling intervals, T_s, defined by

$$T_s \triangleq t_{n+1} - t_n, \ \forall n \tag{2.42}$$

We obtain

$$x(t_{n+1}) = e^{FT_s}x(t_n) + \left[\int_{t_n}^{t_{n+1}} e^{F(t_{n+1}-\tau)}Gd\tau\right] u(t_n).$$

We define

$$A \triangleq e^{FT_s} \tag{2.43}$$

$$B \triangleq \int_{t_n}^{t_{n+1}} e^{F(t_{n+1}-\tau)}Gd\tau \tag{2.44}$$

Note that A and B are constants as are F and T_s. We obtain

$$x(t_{n+1}) = Ax(t_n) + Bu(t_n)$$

We now assume that we begin the sampling operations at $t_0 = 0$. Making use of the uniform sampling assumption of Eq. 2.42, we obtain

$$x((n+1)T_s) = Ax(nT_s) + Bu(nT_s)$$

Because the sampling period is constant, we don't have to explicitly state the presence of T_s. It is sufficient to use only the *sampling number*, namely n and $n+1$. We obtain

$$x(n+1) = Ax(n) + Bu(n) \tag{2.45}$$

This is known as the ZOH equivalent of the continuous time state space model, given by Eq. 2.36. We observe the following:

1. There is no approximation in discretization, so long as $u(t)$ is kept constant during one interval. Thus both discrete and continuous time models predict identical behaviour at the sampling instants.

2. Consider the case of F being a scalar. As it is defined as the exponential of a real number, A cannot be negative. Thus the ZOH equivalent of continuous models will give rise to positive A.

3. There could be systems with $A(n)$ negative,[3] for example, when we use plant data for identification, the topic of discussion in Chapter 6. Because of the previous observation, such systems cannot be represented by the usual continuous time state space models. As a result, the discrete time model equations given above can accommodate a larger class of systems.

4. The discrete time model derived above cannot explain what happens between sampling instants. If the sampling interval is sufficiently small and if there are no hidden oscillations,[4] it is possible to make a reasonable guess of what happens between the samples.

5. By a change of variable, the expression for $B(n)$ can be written as

$$B(n) \triangleq \left[\int_0^{T_s} e^{Ft} dt \right] G \tag{2.46}$$

   see Problem 2.4. As a result, $B(n)$ also is a constant matrix.

We now illustrate this idea with an example.

Example 2.3 Calculate the ZOH equivalent of the DC motor, presented in Sec. 2.2.2, for $J/b = 1$, with the sampling time $T_s = 1$ s.

From Eq. 2.17, the continuous time matrices are given by

$$F = \begin{bmatrix} -1 & 0 \\ 1 & 0 \end{bmatrix}, \quad G = \begin{bmatrix} 1 \\ 0 \end{bmatrix}$$

The exponential of this matrix has been calculated in Example 2.2 as

$$e^{Ft} = \begin{bmatrix} e^{-t} & 0 \\ 1 - e^{-t} & 1 \end{bmatrix}$$

[3]An example of this is presented in Sec. 9.8.

[4]A procedure to avoid intra sample oscillations is presented in Example 9.10 on page 343.

A and B matrices are calculated next. Using Eq. 2.43 and Eq. 2.46, with $T_s = 1$, we obtain

$$A = e^{FT_s} = \begin{bmatrix} e^{-1} & 0 \\ 1 - e^{-1} & 1 \end{bmatrix} = \begin{bmatrix} 0.3679 & 0 \\ 0.6321 & 1 \end{bmatrix}$$

$$\begin{aligned} B &= \int_0^1 e^{F\tau} d\tau G \\ &= \int_0^1 \begin{bmatrix} e^{-\tau} & 0 \\ 1 - e^{-\tau} & 1 \end{bmatrix} d\tau \begin{bmatrix} 1 \\ 0 \end{bmatrix} = \begin{bmatrix} -e^{-\tau} & 0 \\ \tau + e^{-\tau} & \tau \end{bmatrix}_0^1 \begin{bmatrix} 1 \\ 0 \end{bmatrix} \\ &= \begin{bmatrix} -e^{-\tau} \\ \tau + e^{-\tau} \end{bmatrix}_0^1 = \begin{bmatrix} -e^{-1} + 1 \\ 1 + e^{-1} - 1 \end{bmatrix} = \begin{bmatrix} 0.6321 \\ 0.3679 \end{bmatrix} \end{aligned}$$

M 2.3 shows how to carry out the above calculation in Matlab. ∎

Now we will present another approach to find the ZOH equivalent of continuous time systems. In this, we will first diagonalize the system matrix F.

Example 2.4 Using the diagonalization procedure of Appendix A.1.2, determine the ZOH equivalent model of the antenna control system described in Sec. 2.2.2, with $J/b = 10$ and $T_s = 0.2$ s.

The system has a state space model

$$\begin{aligned} \dot{x} &= Fx + Gu \\ y &= Cx \end{aligned}$$

with

$$F = \begin{bmatrix} -0.1 & 0 \\ 1 & 0 \end{bmatrix}, \quad G = \begin{bmatrix} 0.1 \\ 0 \end{bmatrix}, \quad C = \begin{bmatrix} 0 & 1 \end{bmatrix}$$

and $x_1(0) = 0$, $x_2(0) = 0$. It is easy to verify that the eigenvalues of F are given by $\lambda = 0$ and -0.1. The corresponding eigenvectors are given by

$$\begin{bmatrix} 0 \\ 1 \end{bmatrix} \text{ and } \begin{bmatrix} 1 \\ -10 \end{bmatrix}$$

Thus the matrix F can be diagonalized as $F = S\Lambda S^{-1}$, where

$$S = \begin{bmatrix} 0 & 1 \\ 1 & -10 \end{bmatrix}, \quad \Lambda = \begin{bmatrix} 0 & 0 \\ 0 & -0.1 \end{bmatrix}$$

Finally we obtain the discrete time state space matrices as

$$A = e^{FT_s} = Se^{\Lambda T_s}S^{-1}$$

where

$$e^{\Lambda T_s} = \begin{bmatrix} 1 & 0 \\ 0 & e^{-0.1T_s} \end{bmatrix}$$

and hence

$$A = \begin{bmatrix} 0 & 1 \\ 1 & -10 \end{bmatrix} \begin{bmatrix} 1 & 0 \\ 0 & e^{-0.1T_s} \end{bmatrix} \begin{bmatrix} 10 & 1 \\ 1 & 0 \end{bmatrix} = \begin{bmatrix} e^{-0.1T_s} & 0 \\ 10\left(1 - e^{-0.1T_s}\right) & 1 \end{bmatrix}$$

Using the procedure described in Example 2.3, we obtain

$$B = \begin{bmatrix} 1 - e^{-0.1T_s} \\ T_s + 10(e^{-0.1T_s - 1}) \end{bmatrix}$$

When we use a sampling interval of 0.2 s, *i.e.*, $T_s = 0.2$ s, we obtain

$$A = \begin{bmatrix} 0.9802 & 0 \\ 0.19801 & 1 \end{bmatrix}, \quad B = \begin{bmatrix} 0.01980 \\ 0.001987 \end{bmatrix}$$

∎

2.5.3 Approximate Method of Discretization

The method presented in the previous section is not the only one available to discretize continuous time models. In this section, we present a simple method of discretization. More methods will be discussed in Sec. 8.2.

If we replace the derivative of x with a rectangular approximation, Eq. 2.36 becomes

$$\frac{x(n+1) - x(n)}{T_s} = Fx(n) + Gu(n)$$

from which it follows that

$$x(n+1) = (I + FT_s)x(n) + T_sGu(n) \tag{2.47}$$

We will achieve the same result if we substitute the definition of e^{FT_s} given in Eq. 2.37 and expand Eq. 2.43 and Eq. 2.46:

$$\begin{aligned} A &= I + FT_s + \text{ higher degree terms in } T_s \\ B &= \int_0^{T_s} e^{F\tau} d\tau G = F^{-1} e^{F\tau}\Big|_0^{T_s} G \\ &= F^{-1}[e^{FT_s} - I]G \\ &= F^{-1}[I + FT_s + \text{ higher degree terms in } T_s - I]G \\ &= T_sG + \text{ higher degree terms in } T_s \end{aligned}$$

If T_s is small, the higher degree terms in T_s may be dropped in the above expressions for A and B, which, when substituted in Eq. 2.45, become identical to Eq. 2.47.

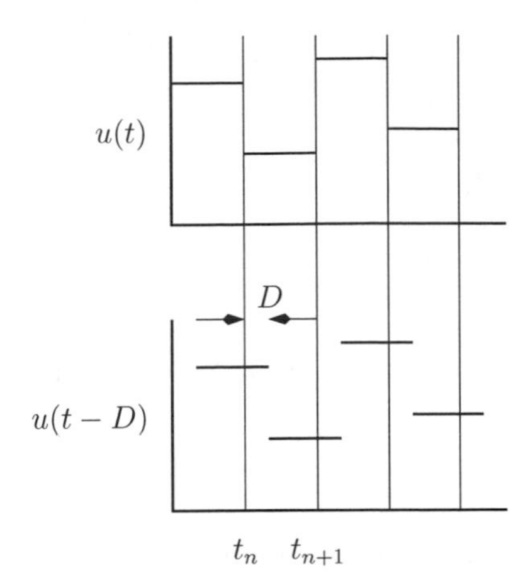

Figure 2.8: Original and time shifted versions of a signal

2.5.4 Discretization of Systems with Delay

Often, changes in the input u do not affect the state immediately. Suppose that the control effort u is the flow rate of some liquid. If a long pipe were to separate the control valve and the place where the liquid enters the plant, it would take some time for any change in the flow rate to be felt by the plant.

Oil and process industries have plants of extremely large orders. These are often sluggish systems. It is conventional to approximate these systems with low order models and a *dead time* [33], also known as *time delay*. If a system has a large time delay or dead time, it will take a long time to respond to any stimulus.

Suppose that any change in u is felt at the plant only after D time units, as expressed by the following state space equation:

$$\dot{x}(t) = Fx(t) + Gu(t - D), \quad 0 \le D < T_s \tag{2.48}$$

This can be graphically illustrated as in Fig. 2.8 where $u(t)$ and $u(t - D)$ are drawn. The solution to this equation is identical to Eq. 2.40, but for the delay D in u. We obtain

$$x(t_{n+1}) = e^{F(t_{n+1}-t_n)}x(t_n) + \int_{t_n}^{t_{n+1}} e^{F(t_{n+1}-\tau)}Gu(\tau - D)d\tau \tag{2.49}$$

From Fig. 2.8, we see that the effect of the delay is to shift the constant values of u by D units to the right. If there had been no delay, we would have used $u(t_n)$ during the interval (t_n, t_{n+1}), see Eq. 2.41. Because of the delay, $u(t_{n-1})$ will be used during the period $(t_n, t_n + D)$ and $u(t_n)$ during the rest of the interval. As a result, the last term in the above equation can be written in two parts:

$$\begin{aligned}\int_{t_n}^{t_{n+1}} e^{F(t_{n+1}-\tau)}Gu(\tau - D)d\tau &= \int_{t_n}^{t_n+D} e^{F(t_{n+1}-\tau)}Gd\tau u(t_{n-1}) \\ &\quad + \int_{t_n+D}^{t_{n+1}} e^{F(t_{n+1}-\tau)}Gd\tau u(t_n) \\ &\triangleq B_1u(t_{n-1}) + B_0u(t_n)\end{aligned}$$

where the symbol $\triangleq$ has to be read as *defined as*. The above expressions for B_1 and B_0 can be simplified to arrive at the following relations, see Problem 2.5:

$$B_1 = e^{F(T_s-D)} \int_0^D e^{Ft} G dt \tag{2.50a}$$

$$B_0 = \int_0^{T_s-D} e^{Ft} G dt \tag{2.50b}$$

Substituting the above expressions in Eq. 2.49 and making use of Eq. 2.42 and Eq. 2.43, we obtain

$$x(n+1) = Ax(n) + B_1 u(n-1) + B_0 u(n) \tag{2.51}$$

which can be written using an augmented state vector:

$$\begin{bmatrix} x(n+1) \\ u(n) \end{bmatrix} = \begin{bmatrix} A & B_1 \\ 0 & 0 \end{bmatrix} \begin{bmatrix} x(n) \\ u(n-1) \end{bmatrix} + \begin{bmatrix} B_0 \\ I \end{bmatrix} u(n) \tag{2.52}$$

where, as before, A is given by Eq. 2.43 and I is an identity matrix of the same size as u. If we define the vector $(x(n), u(n-1))$ as an augmented state at the sampling instant n, Eq. 2.52 is reduced to the standard state space model of the form Eq. 2.45.

Recall that the concept of state has been introduced to remember the effects of the past. The presence of time delay requires us to remember the previous control effort as well. What better place to store this information than the state? We present an example below.

Example 2.5 Discretize the state space equation

$$\dot{x}(t) = -\frac{1}{\tau} x(t) + \frac{1}{\tau} u(t-D) \tag{2.53}$$

where τ and D are constants and the delay is less than one sampling period, *i.e.*, $D < T_s$.

Using Eq. 2.50a,

$$B_1 = e^{-(T_s-D)/\tau} \int_0^D e^{-t/\tau} \frac{1}{\tau} dt = e^{-(T_s-D)/\tau} \left[1 - e^{-D/\tau}\right] \tag{2.54a}$$

Using Eq. 2.50b,

$$B_0 = \int_0^{T_s-D} e^{-t/\tau} \frac{1}{\tau} dt = \left[1 - e^{-(T_s-D)/\tau}\right] \tag{2.54b}$$

From Eq. 2.52, we arrive at the discrete time equivalent of Eq. 2.53 as

$$\begin{bmatrix} x(n+1) \\ u(n) \end{bmatrix} = \begin{bmatrix} \Phi & B_1 \\ 0 & 0 \end{bmatrix} \begin{bmatrix} x(n) \\ u(n-1) \end{bmatrix} + \begin{bmatrix} B_0 \\ 1 \end{bmatrix} u(n) \tag{2.54c}$$

where

$$\Phi = e^{-T_s/\tau} \tag{2.54d}$$

∎

What do we do when the delay is greater than the sampling time, *i.e.*, $D > T_s$? A procedure to deal with this situation is given in Problem 2.5.

We have so far discussed how to discretize the state equation only. What about discretization of the output equation, discussed in detail in Sec. 2.4.3? At the sampling instant, the relation for y is unchanged, provided we synchronize the sampling of $y(t)$ with that of $x(t)$ and $u(t)$. Thus, it is possible to arrive at the discrete time state space model, given by Eq. 2.2 on page 6. The discrete time state equation also arises naturally, see Sec. 2.3. In these systems also, it is difficult to measure all the states; the reasons are same as that for continuous time systems. Thus, once again, we arrive at Eq. 2.2.

2.6 Approaches to Controller Design and Testing

In this chapter, we have discussed in detail how to connect the blocks given in Fig. 1.1. If the plant G denotes a discrete time model, the blocks can be directly connected. If, on the other hand, the plant is described through a continuous time model, we discretize it, as explained in Sec. 2.5, and use it. In this case, the signals also have to be converted, as explained above.

There are two broad categories of control design techniques. The first one is based on a transfer function approach. In this approach, one gets an algebraic relation between the input u and the output y with the help of the Z-transform, known as the transfer function, to be explained in Sec. 4.3. The control design technique based on transfer functions is known as the transfer function approach. The transfer function methods are easy to explain and implement. We present several control design techniques that use this approach.

The second category of control design methods uses the state space models directly. This method is useful, especially when one has to deal with several input variables u. In Chapter 14, we present a few state space model based control techniques as well.

Validation of the controller is an important issue, especially in hybrid systems that consist of a continuous time plant and a digital controller. What is the guarantee that we did not lose information while discretizing the plant? What is the guarantee that the plant behaves properly in between the sampling instants? The minimum rate at which we have to sample a system is given by Shannon's sampling theorem, to be derived in Sec. 5.3.2. In reality, we have to sample the continuous time much faster. Unfortunately, however, there are only heuristic guidelines for this, some of which are presented in Sec. 8.1.

If one takes some precautions while designing the controller, one does not have to worry about how the system will behave in between the sampling instants. This topic is discussed in Chapter 9. The performance of the final digital controller is tested on the continuous time system through simulations, with tools such as Simulink. The only difference between evaluating the performance of the digital controller on plant that is described by a continuous time or a discrete time model is in the usage of the zero order hold element. It is present in the former only. There is no need to explicitly state the presence of an A/D converter in these simulations.

A large number of pre-defined Simulink programs are given in Appendix A.2 on page 524, the usage of which will be studied in great detail in Parts III and IV.

2.7 Matlab Code

Matlab Code 2.1 Model of inverted pendulum, presented in Sec. 2.2.3. This code is available at `HOME/ss/matlab/pend_model.m`[5]

```
1  Km = 0.00767; Kg = 3.7; Rm = 2.6; r = 0.00635;
2  M = 0.522; m = 0.231; g = 9.81; L = 0.305; J = 0;
3
4  D1 = (J+m*L^2)*(M+m)-m^2*L^2;
5  alpha = m*g*L*(M+m)/D1;
6  beta = m*L/D1;
7  gamma = m^2*g*L^2/D1;
8  delta = (J+m*L^2)/D1;
9  alpha1 = Km*Kg/Rm/r;
10 alpha2 = Km^2*Kg^2/Rm/r^2;
11
12 A = zeros(4); A(1,3) = 1; A(2,4) = 1;
13 A(3,2) = -gamma; A(3,3) = -alpha2*delta;
14 A(4,2) = alpha; A(4,3) = alpha2*beta;
15 B = zeros(4,1); B(3) = alpha1*delta; B(4) = -alpha1*beta;
```

Matlab Code 2.2 Exponential of the matrix presented in Example 2.2 on page 21. Available at `HOME/chap2/matlab/mat_exp.m`

```
1  F = [-1 0;1 0];
2  expm(F)
```

Matlab Code 2.3 ZOH equivalent state space system, as discussed in Example 2.2 on page 21. Available at `HOME/chap2/matlab/ZOH1.m`

```
1  F = [-1 0;1 0]; G = [1; 0];
2  C = [0 1]; D = 0; Ts=1;
3  sys = ss(F,G,C,D);
4  sysd = c2d(sys,Ts,'zoh')
```

[5] `HOME` stands for `http://www.moudgalya.org/dc/` – first see the software installation directions, given in Appendix A.2.

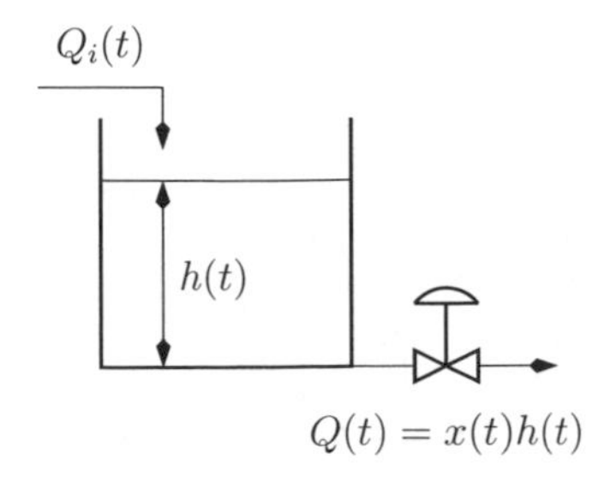

Figure 2.9: Water flow in a tank

2.8 Problems

2.1. It is desired to model the flow of water into and out of a tank of uniform cross-section, see Fig. 2.9. The inflow rate $Q_i(t)$ need not be steady. The opening of the outflow valve can be adjusted so as to reduce the variations in the outflow. The outflow rate $Q(t)$ is a multiple of the height of water in the tank and a variable $x(t)$ that is a function of valve opening, *i.e.*,

$$Q(t) = x(t)h(t) \tag{2.55}$$

Derive a linearized model of this system in deviation variables.

2.2. Repeat Problem 2.1 with the difference that the outflow now is proportional to square root of the height of water in the tank:

$$Q(t) = x(t)\sqrt{h(t)} \tag{2.56}$$

2.3. The motion of a unit mass in an inverse square law force field is governed by a pair of second order equations

$$\begin{aligned}
\frac{d^2r(t)}{dt^2} &= r(t)\left(\frac{d\theta(t)}{dt}\right)^2 - \frac{K}{r^2(t)} + u_1(t) \\
\frac{d^2\theta(t)}{dt^2} &= -\frac{2}{r(t)}\frac{d\theta(t)}{dt}\frac{dr(t)}{dt} + \frac{1}{r(t)}u_2(t)
\end{aligned}$$

where $r(t)$ and $\theta(t)$ are defined in Fig. 2.10. The radial thrust of the unit mass (say a satellite) is denoted by $u_1(t)$ and that in the tangential direction is denoted by $u_2(t)$. If $u_1(t) = u_2(t) = 0$, the equations admit the solution

$$\begin{aligned}
r(t) &= \sigma \text{ constant} \\
\theta(t) &= \omega t, \text{ with } \omega \text{ constant}
\end{aligned}$$

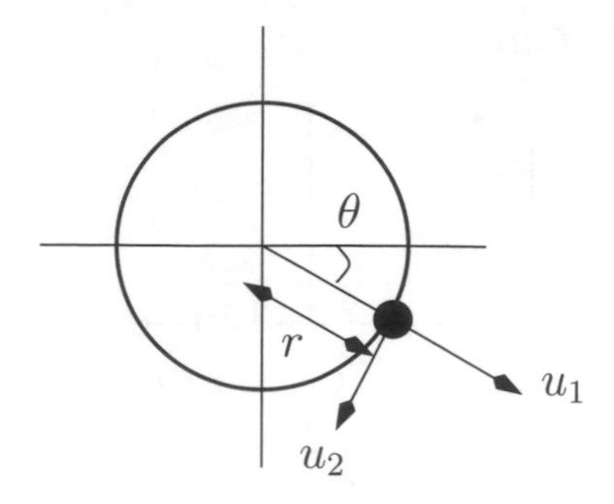

Figure 2.10: Satellite in a circular orbit

where $\sigma^3\omega^2 = K$. This means that circular orbits are possible. Define $x_1 = r - \sigma$, $x_2 = \frac{dr}{dt}$, $x_3 = \sigma(\theta - \omega t)$, $x_4 = \sigma(\frac{d\theta}{dt} - \omega)$ and write down the linearized differential equations which describe the resulting motion for small deviations from a circular orbit. Show that the F and G matrices are given by

$$F = \begin{bmatrix} 0 & 1 & 0 & 0 \\ 3w^2 & 0 & 0 & 2w \\ 0 & 0 & 0 & 1 \\ 0 & -2w & 0 & 0 \end{bmatrix}, \quad G = \begin{bmatrix} 0 & 0 \\ 1 & 0 \\ 0 & 0 \\ 0 & 1 \end{bmatrix}$$

2.4. Derive Eq. 2.46 on page 24. Also, show that B can be written as $B = A_1 T_s G$, with A_1 given by $A = I + FT_s A_1$, where A is given by Eq. 2.43.

2.5. This problem is concerned with discretization of time delay systems, addressed in Sec. 2.5.4.

(a) Derive the expressions for B_0 and B_1, derived in Eq. 2.50.

(b) Derive the ZOH equivalent discrete time state space model for $d = 2$.

(c) Is there any relation between the matrices B_0 and B_1 that we derived for $d = 1$ and those for $d = 2$?

(d) Can you repeat the above two questions for a general integer $d > 0$?

2.6. The following equations model a thermal system where the states denote temperature and the control effort $u(t)$ can stand for either steam or electric current:

$$\frac{d}{dt}\begin{bmatrix} x_1(t) \\ x_2(t) \end{bmatrix} = \begin{bmatrix} 1 & 1 \\ 1 & 1 \end{bmatrix}\begin{bmatrix} x_1(t) \\ x_2(t) \end{bmatrix} + \begin{bmatrix} 0 \\ 1 \end{bmatrix} u(t)$$

Find a ZOH equivalent discrete system. Let the sample time be T_s in consistent units. Use the diagonalization procedure described in Sec. A.1.2 and check the results with that obtained through the direct method.

Part I

Digital Signal Processing

Chapter 3

Linear System

The approach followed in this book is to design linear controllers using linear models. We also have to ensure that the controllers are realizable, or equivalently, implementable. We need the concept of causality for this purpose. In this chapter, we present the concepts of linearity, causality, as well as time invariance and stability.

We will be concerned with discrete time signals that are defined at discrete time instants, from $-\infty$ to ∞. An example of such a sequence is given by

$$\{u(n)\} \triangleq \{\ldots, u(-2), u(-1), u(0), u(1), u(2), \ldots\} \tag{3.1}$$

where $u(i)$ on the right-hand side of the equation denotes the scalar value of u taken at time instant i. For example, $u(-1)$, $u(0)$ refer to the values of u taken at time instants of -1 and 0. In general, we will denote such a sequence u as $\{u(n)\}$ with double braces. We will denote the value of the sequence u taken at time instant n as $u(n)$. When there is no confusion, we will drop the braces, even when talking about sequences.

We work with finite energy signals in this book. The energy of a discrete time signal $\{u(n)\}$ is defined as

$$E_u = \sum_{n=-\infty}^{\infty} |u(n)|^2 \tag{3.2}$$

where, u can be complex. In case of real signals, the energy is the sum of the squares of u at al time instants. We would also be interested in signals with finite power. Recall that power is defined as energy per unit time. Periodic sequences are an example of signals with finite power.

In this chapter, we will show the importance of convolution operation in the modelling of linear time invariant systems: for any arbitrary input, the plant output can be obtained by a convolution of the input and the impulse response.

3.1 Basic Concepts

In this section, we introduce the concepts of linearity, time (or shift) invariance and causality. One that has all these properties is known as a linear time invariant (LTI) causal system. In this book, we will deal mostly with LTI causal systems. We begin the discussion with the property of linearity.

3.1.1 Linearity

Linear systems are easier to understand compared to nonlinear systems. Often, controllers designed with the assumption of the underlying process being linear are sufficient. In view of this observation, the concept of linearity, to be discussed in this section, becomes extremely important.

Let us first recall the definition of linearity of functions. A function $f(x)$ is said to be linear if

$$f(\alpha x_1 + \beta x_2) = \alpha f(x_1) + \beta f(x_2) \tag{3.3}$$

where α and β are arbitrary scalars. For example, $f_1(x) = 2x$ is linear, but $f_2(x) = \sin x$ is nonlinear. We now ask the question, is

$$f_3(x) = 2x + 3 \tag{3.4}$$

linear? Let us see whether Eq. 3.3 is satisfied with $\alpha = \beta = 1$. The left-hand side becomes

$$f_3(x_1 + x_2) = 2(x_1 + x_2) + 3$$

while the right-hand side is

$$f_3(x_1) + f_3(x_2) = (2x_1 + 3) + (2x_2 + 3) = 2(x_1 + x_2) + 6$$

We see that the two sides are unequal. Because Eq. 3.3 has to be satisfied for all α and β in order that the function be linear, we see that f_3 is not linear. Thus, addressing functions of the form of Eq. 3.4 as linear should at best be considered as colloquial. The presence of the nonzero constant in Eq. 3.4 is the reason why f_3 is nonlinear. Thus, we arrive at the following necessary condition for linearity:

$$f(0) = 0 \tag{3.5}$$

Now we will present the concept of a linear *system*. Consider a system at the state $x(n_0)$ at time instant n_0. Suppose that we apply an input sequence of the form Eq. 3.1, starting at time n_0, *i.e.*,

$$\{u(n)\} = \{u(n_0), u(n_0 + 1), u(n_0 + 2), \ldots\}, \quad n \geq n_0$$

Suppose that the future states and the output of the system are defined as functions of the input and the initial state:

$$\begin{aligned} x(n) &= f_n(x(n_0), \{u(n)\}) \\ y(n) &= g_n(x(n_0), \{u(n)\}) \end{aligned}$$

This can be represented as in Fig. 3.1, where the inputs, shown on the left-hand side of the block diagram, are

1. $\{u(n)\}$ – input sequence
2. $x(n_0)$ – initial state

and the outputs, shown on the right-hand side of the block diagram, are

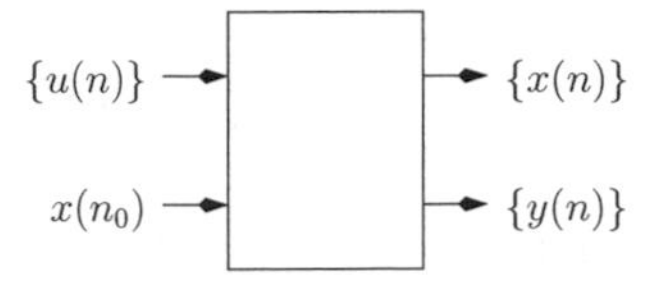

Figure 3.1: Diagrammatic representation of a system

1. $x(n),\ n > n_0$, the future states
2. $y(n),\ n > n_0$, the future outputs

If f_n and g_n are linear functions of $x(n_0)$ and $\{u(n)\}$, $n \geq n_0$, the system is said to be linear. In other words, a linear combination of $(x(n_0), \{u(n)\})$ should give the same combination of $(x(n), y(n))$ for all future n. In other words, suppose that under the action of f_n, we obtain the following two mappings:

$$\begin{bmatrix} x_1(n_0) \\ \{u_1(n)\} \end{bmatrix} \xrightarrow{f_n} \{x_1(n)\} \tag{3.6a}$$

$$\begin{bmatrix} x_2(n_0) \\ \{u_2(n)\} \end{bmatrix} \xrightarrow{f_n} \{x_2(n)\} \tag{3.6b}$$

If the system is linear, a linear combination of the left-hand sides should give the same combination of the right hand sides:

$$\alpha \begin{bmatrix} x_1(n_0) \\ \{u_1(n)\} \end{bmatrix} + \beta \begin{bmatrix} x_2(n_0) \\ \{u_2(n)\} \end{bmatrix} \xrightarrow{f_n} \alpha\{x_1(n)\} + \beta\{x_2(n)\} \tag{3.6c}$$

where, as before, α and β are arbitrary scalars. The output function g_n should also satisfy this property. That is, under the condition

$$\begin{bmatrix} x_1(n_0) \\ \{u_1(n)\} \end{bmatrix} \xrightarrow{g_n} \{y_1(n)\} \tag{3.7a}$$

$$\begin{bmatrix} x_2(n_0) \\ \{u_2(n)\} \end{bmatrix} \xrightarrow{g_n} \{y_2(n)\} \tag{3.7b}$$

the following also should hold true:

$$\alpha \begin{bmatrix} x_1(n_0) \\ \{u_1(n)\} \end{bmatrix} + \beta \begin{bmatrix} x_2(n_0) \\ \{u_2(n)\} \end{bmatrix} \xrightarrow{g_n} \alpha\{y_1(n)\} + \beta\{y_2(n)\} \tag{3.7c}$$

This concept can be illustrated as in Fig. 3.2. We refer to this property as the *superposition* principle when $\alpha = \beta = 1$. We will present an example below.

Example 3.1 The system given by the following equations is linear:

$$x(n+1) = A(n)x(n) + B(n)u(n) \tag{3.8}$$

$$y(n) = C(n)x(n) + D(n)u(n) \tag{3.9}$$

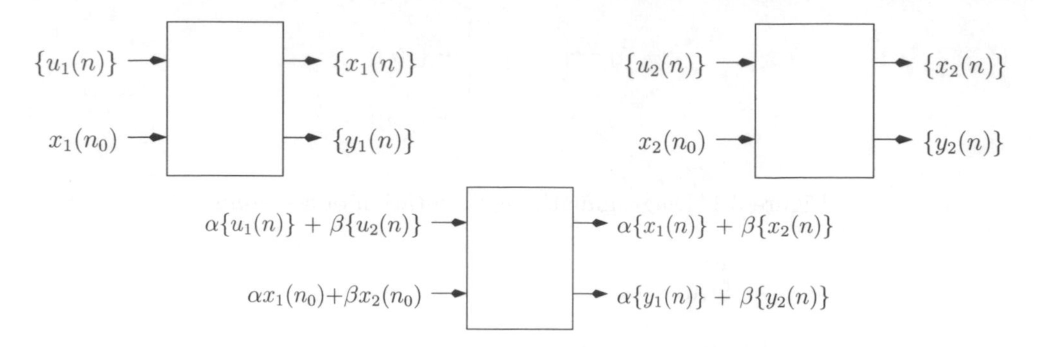

Figure 3.2: Linearity of a system defined

We will first show that Eq. 3.8 is linear. Let us suppose that the future states for $(x_1(n_0), \{u_1(n)\})$ and $(x_2(n_0), \{u_2(n)\})$ are $\{x_1(n)\}$ and $\{x_2(n)\}$, respectively. In particular, at the time instant $n_0 + 1$, we obtain the following relations:

$$
\begin{aligned}
x_1(n_0 + 1) &= A(n_0)x_1(n_0) + B(n_0)u_1(n_0) \\
x_2(n_0 + 1) &= A(n_0)x_2(n_0) + B(n_0)u_2(n_0)
\end{aligned}
$$

Now, suppose that we use a linear combination of $(x_i(n_0), \{u_i(n)\})$, $i = 1, 2$, as the starting point. At $n_0 + 1$, we obtain the following relation:

$$x(n_0 + 1) = A(n_0)(\alpha x_1(n_0) + \beta x_2(n_0)) + B(n_0)(\alpha u_1(n_0) + \beta u_2(n_0))$$

Note that we have used the same linear combination for both the input sequence and the initial state. This can be written as

$$
\begin{aligned}
x(n_0 + 1) = {} & \alpha(A(n_0)x_1(n_0) + B(n_0)u_1(n_0)) \\
& + \beta(A(n_0)x_2(n_0) + B(n_0)u_2(n_0))
\end{aligned}
$$

This is nothing but $\alpha x_1(n_0 + 1) + \beta x_2(n_0 + 1)$. That is, a linear combination of $(x_i(n_0), \{u_i(n)\})$, $i = 1, 2$, has given rise to the next state also being of the same combination. Thus, the linearity of f_n is verified for $n = n_0 + 1$. By induction, this can be shown to be true $\forall n \geq n_0$. Therefore f_n is linear. In a similar way, g_n can also be shown to be linear. Thus, it is a linear system. ∎

The next example is obvious.

Example 3.2 If the outflow from a tank y is proportional to the square root of the height of the liquid x in the tank, *i.e.*,

$$y = k\sqrt{x}$$

the system is not linear. ∎

y_u: Due to only input y_x: Due to only initial state

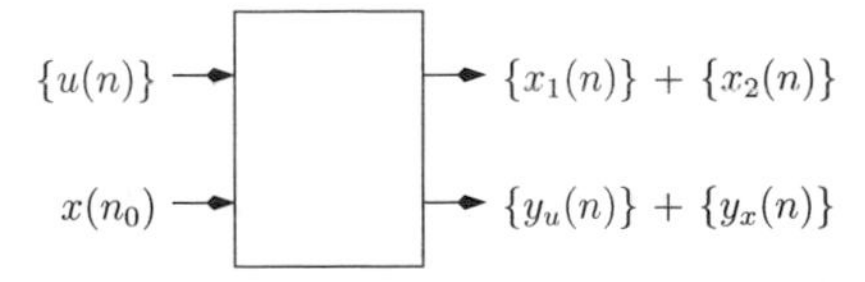

Figure 3.3: Output (y) is the sum of output due to input (y_u) and output due to initial state (y_x)

The superposition principle allows us to study the system with zero and nonzero values of the initial state. In particular, we have the following important property: In a linear system, the output $y(n)$ is the sum of output with zero input and output with zero initial state. Suppose that under the action of the function g_n, $(x(n_0), \{0\})$ is mapped to $\{y_x(n)\}$.[1] Also, $(0, \{u(n)\})$ gets mapped to $\{y_u(n)\}$. Pictorially, this can be seen as follows:

$$\begin{bmatrix} x(n_0) \\ \{0\} \end{bmatrix} \xrightarrow{g_n} \{y_x(n)\}$$

$$\begin{bmatrix} 0 \\ \{u(n)\} \end{bmatrix} \xrightarrow{g_n} \{y_u(n)\}$$

Using the property of linearity given by Eq. 3.7c, with $\alpha = \beta = 1$, we obtain

$$\begin{bmatrix} x(n_0) \\ \{u(n)\} \end{bmatrix} = \begin{bmatrix} x(n_0) \\ \{0\} \end{bmatrix} + \begin{bmatrix} 0 \\ \{u(n)\} \end{bmatrix} \xrightarrow{g_n} y_x(n) + y_u(n)$$

This superposition property can be summarized by the following equation:

$$y(n) = y_x(n) + y_u(n) \tag{3.10}$$

This idea is illustrated in Fig. 3.3.

3.1.2 Time Invariance

Suppose that an input $\{u\} = \{\alpha, \beta, \ldots\}$ is applied to a system at some initial state x_0 at time n_0. Thus, $\{u\}$ is just a sequence of numbers. Let the state and the output evolve as $\{x_1(n)\}$ and $\{y_1(n)\}$, $n \geq n_0$. That is, as given in Eq. 3.6a and Eq. 3.7a, respectively. Suppose also that the same input sequence $\{\alpha, \beta, \ldots\}$ is applied to the system at the same initial state x_0 but at some other time $n_0 - k_0$. Even though the

[1]The subscript x does not denote a partial derivative. It says that the output is due to nonzero value of initial state only.

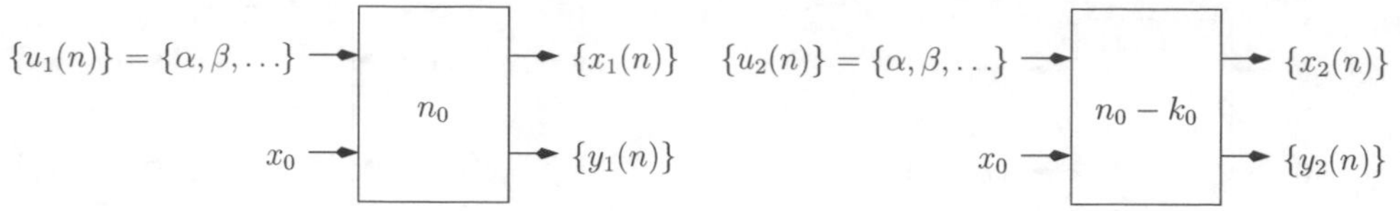

Figure 3.4: Time invariance of linear systems. Identical input sequence u is applied at $n = n_0$ (left) and at $n = n_0 - k_0$ (right).

same sequence of numbers is used, because these start at two different time instants, they will be represented by two different functions, u_2 and u_1, respectively. We have

$$\{u_2(n)\} = \{u_1(n - k_0)\}$$

Let the state and output corresponding to u_2 evolve, respectively, as $\{x_2(n)\}$ and $\{y_2(n)\}$, $n \geq n_0 - k_0$, *i.e.*, as given in Eq. 3.6b and Eq. 3.7b, respectively. This is illustrated in Fig. 3.4. Now if $x_2(n)$ and $y_2(n)$ are time shifted versions of $x_1(n)$ and $y_1(n)$, *i.e.*,

$$\begin{aligned} x_2(n) &= x_1(n - k_0) \\ y_2(n) &= y_1(n - k_0) \end{aligned}$$

for all $n \geq n_0$ and for all k_0, then the system is said to be time invariant. In this case, we can replace f_n by f and g_n by g. That is, f_n and g_n are not functions of time.

Example 3.3 It is easy to verify that the system described by

$$\begin{aligned} x(n+1) &= Ax(n) + Bu(n) \\ y(n) &= Cx(n) + Du(n) \end{aligned}$$

where A, B, C and D are constants, is a time invariant system. ∎

Now we will give an example of a system that does not satisfy the time invariance property.

Example 3.4 Determine whether the system described by

$$y(n) = nu(n) \tag{3.11}$$

is time invariant.

Since x does not appear in the equation, it is sufficient to consider u and y only. Suppose that when an input u_i is applied, we obtain the output y_i, where, i can take 1 or 2. That is,

$$y_i(n) = nu_i(n), \ i = 1, 2$$

Let us take $u_2(n) = u_1(n - k)$, $k \neq 0$. We would like to check whether under the application of $u_2(n)$, we obtain the output $y_2(n)$ equal to $y_1(n - k)$. We see that

$$y_2(n) = nu_2(n) = nu_1(n - k)$$

But, from Eq. 3.11, we obtain

$$y_1(n-k) = (n-k)u_1(n-k)$$

We see that $y_2(n) \neq y_1(n-k)$, even though $u_2(n) = u_1(n-k)$. Hence, the system is not time invariant. ∎

3.1.3 Causality and Initial Rest

We will now discuss the important topic of causality. A system is causal if its output depends only on the current and the past inputs. As a result, if two inputs that are identical up to some time are applied to a causal system, the corresponding outputs will also be identical up to that time. Such a system is known as nonanticipative as it does not anticipate the future values of the input. In contrast, systems whose output depends on future inputs are known as *noncausal*. An example of such a system has the following input–output mapping: $y(n) = u(n+1)$.

Noncausal systems cannot be implemented, because no physical system can predict the future. In view of this, we have to ensure that our designs produce causal systems. In turns out that improper specification of initial conditions can result in the loss of causality. One way to achieve causality is through initial conditions that correspond to the system being at rest, also known as *initial rest*. A system that is initially at rest will have all its past inputs and outputs zero. Equivalently, the initial state of the system is zero. We illustrate these ideas in the following example [47].

Example 3.5 Solve the system

$$y(k) - ay(k-1) = u(k) \tag{3.12}$$

for initial conditions given at different time instances.

We will first consider giving an initial condition to determine the solution. Multiply both sides of Eq. 3.12 by a^{-k} and sum the result from $N+1$ to n, with $N < n$. The left-hand side is

$$\begin{array}{l} \quad\quad\; + \dfrac{y(N+1)}{a^{N+1}} + \dfrac{y(N+2)}{a^{N+2}} + \cdots + \dfrac{y(n-1)}{a^{n-1}} + \dfrac{y(n)}{a^n} \\ -\dfrac{y(N)}{a^N} - \dfrac{y(N+1)}{a^{N+1}} - \dfrac{y(N+2)}{a^{N+2}} - \cdots - \dfrac{y(n-1)}{a^{n-1}} \end{array}$$

Summing, all terms except the first and the last vanish. We obtain

$$\frac{y(n)}{a^n} - \frac{y(N)}{a^N} = \sum_{k=N+1}^{n} \frac{u(k)}{a^k}$$

Solving this, we obtain

$$y(n) = a^{n-N}y(N) + \sum_{k=N+1}^{n} a^{n-k}u(k) \tag{3.13}$$

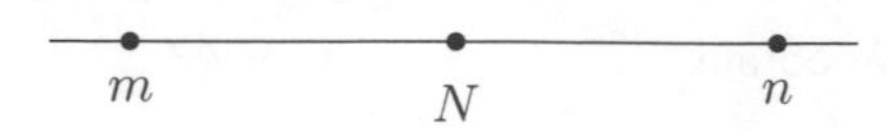

Figure 3.5: Specifying condition at a finite point, as discussed in Example 3.5

This shows that the output $y(n)$ depends on $y(N)$ and the values of the input between the interval N and n, see Fig. 3.5. If we use the initial condition, $\lim_{N\to-\infty} a^{-N}y(N) = 0$, we obtain

$$y(n) = \sum_{k=-\infty}^{n} a^{n-k}u(k) \tag{3.14}$$

Eq. 3.13 suggests that the output $y(m)$ could depend on $u(N)$ even when $m < N$. To see this, we carry out the following calculation. Multiplying both sides of Eq. 3.12 by $1/a^k$ and summing the result from $m+1$ to N, the left-hand side becomes

$$\begin{aligned} &+\frac{y(m+1)}{a^{m+1}} + \frac{y(m+2)}{a^{m+2}} + \cdots + \frac{y(N-1)}{a^{N-1}} + \frac{y(N)}{a^N} \\ -\frac{y(m)}{a^m} &- \frac{y(m+1)}{a^{m+1}} - \frac{y(m+2)}{a^{m+2}} - \cdots - \frac{y(N-1)}{a^{N-1}} \end{aligned}$$

Summing, we obtain

$$-\frac{y(m)}{a^m} + \frac{y(N)}{a^N} = \sum_{k=m+1}^{N} \frac{u(k)}{a^k}$$

Solving this, we obtain

$$y(m) = a^{m-N}y(N) - \sum_{k=m+1}^{N} a^{m-k}u(k) \tag{3.15}$$

This shows that the output $y(m)$ depends on $y(N)$ and the values of the input between the interval m and N, see Fig. 3.5. We see that the value of y is specified at a finite point N; the system loses causality.

Let us consider the extreme case of $\lim_{N\to\infty} a^{-N}y(N) = 0$. We obtain

$$y(m) = -\sum_{k=m+1}^{\infty} a^{m-k}u(k) \tag{3.16}$$

making it clear that the current output $y(m)$ depends on only future values of input, $u(k)$, $k > m$.

∎

From this example, we see that if the value of output y is specified at a finite point N, the system is not causal. There are two ways of handling this difficulty:

Figure 3.6: Unit step (left) and unit impulse (right) signals

- The first solution is to choose N to be $-\infty$. In this case, there is no question of even worrying about what happens before N. On the flip side, we will have to remember all the input values applied from $-\infty$ onwards.
- The second solution to this problem is to choose N to be finite, but let all values of u in the interval $(-\infty, N)$ be zero. If the system is I/O LTI, the output y will also be zero during this interval. This condition is known as initial rest. Thus, the condition of initial rest helps achieve causality.

We generally prefer the second solution. A way to accommodate nonzero initial conditions is explained in Sec. 3.4. The meaning of a nonzero initial condition is explained in Sec. 4.3.3. Although we can assign any finite value to it, we generally choose N to be zero.

If noncausal systems cannot be implemented, why should we study them? While identifying a plant from input–output data, we need the concept of past inputs (see Sec. 6.3.5), which calls for a general framework that can accommodate noncausal systems as well.

3.2 Basic Discrete Time Signals

In this section, we will discuss a few discrete time signals that we will encounter many times, their use and the relation between them. A unit step sequence $\{1(n)\} = \{\ldots, 1(-2), 1(-1), 1(0), 1(1), 1(2), \ldots\}$ is now defined:

$$1(n) = \begin{cases} 1 & n \geq 0 \\ 0 & n < 0 \end{cases} \tag{3.17}$$

As a result, $\{1(n)\} = \{\ldots, 0, 0, 1, 1, \ldots\}$, where the first one appears at the location corresponding to $n = 0$. The left-hand side of Fig. 3.6 shows the unit step sequence. We will often refer to it simply as a *step sequence*.

Unit impulse sequence or unit sample sequence is defined as $\{\delta(n)\} = \{\ldots, \delta(-2), \delta(-1), \delta(0), \delta(1), \delta(2), \ldots\}$, where

$$\delta(n) = \begin{cases} 1 & n = 0 \\ 0 & n \neq 0 \end{cases} \tag{3.18}$$

Thus $\{\delta(n)\} = \{\ldots, 0, 0, 1, 0, 0, \ldots\}$, where 1 appears at the location corresponding to $n = 0$. The right-hand side of Fig. 3.6 shows the unit impulse signal. We will often refer to it simply as an *impulse sequence.*

It is easy to verify the following relation between unit step and unit impulse signals:

$$\{\delta(n)\} = \{1(n)\} - \{1(n-1)\} \tag{3.19}$$

Note that $\{1(n)\}$ and $\{1(n-1)\}$ are unit steps that start at $n = 0$ and $n = 1$, respectively.

We will now present an important property of impulse sequences: it is possible to represent arbitrary sequences as linear combinations of impulse sequences. Let us consider an arbitrary sequence $\{u(n)\}$, defined in Eq. 3.1 on page 35, reproduced here for convenience:

$$\{u(n)\} = \{\ldots, u(-2), u(-1), u(0), u(1), u(2), \ldots\}$$

Writing this as a sum of an infinite number of sequences, each with only one nonzero term, we obtain

$$\begin{aligned}\{u(n)\} = \cdots &+ \{\ldots, 0, 0, u(-2), 0, 0, 0, 0, 0, \ldots\} \\ &+ \{\ldots, 0, 0, 0, u(-1), 0, 0, 0, 0, \ldots\} + \{\ldots, 0, 0, 0, 0, u(0), 0, 0, 0, \ldots\} \\ &+ \{\ldots, 0, 0, 0, 0, 0, u(1), 0, 0, \ldots\} + \{\ldots, 0, 0, 0, 0, 0, 0, u(2), 0, \ldots\} + \cdots\end{aligned}$$

where the summation sign indicates component wise addition. The first term on the right-hand side can be written as $u(-2)\{\delta(n+2)\}$, because the nonzero term occurs at $n = -2$. Note that this is a product of a scalar, $u(-2)$, and a shifted impulse sequence, $\{\delta(n+2)\}$. Similarly, all other terms can also be written as products of a scalar and a shifted sequence. We arrive at

$$\begin{aligned}\{u(n)\} = \cdots &+ u(-2)\{\delta(n+2)\} + u(-1)\{\delta(n+1)\} \\ &+ u(0)\{\delta(n)\} + u(1)\{\delta(n-1)\} + u(2)\{\delta(n-2)\} + \cdots\end{aligned}$$

Thus the sequences $\{\delta(n+2)\}$, $\{\delta(n-1)\}$, etc., can be thought of as unit vectors and $u(-2)$, $u(1)$, etc., can be thought of as components along these directions. The above equation can be written as

$$\{u(n)\} = \sum_{k=-\infty}^{\infty} u(k)\{\delta(n-k)\} \tag{3.20}$$

If $u(n) = 0$ for $n < 0$, Eq. 3.20 becomes

$$\{u(n)\} = \sum_{k=0}^{\infty} u(k)\{\delta(n-k)\} \tag{3.21}$$

3.3 Input–Output Convolution Models

Transfer function based control design techniques are popular in industry. A starting point for these models is the input–output convolution models. This section is devoted to a study of such models. We begin with the concept of input–output linear models.

3.3.1 Input–Output Linearity

There are times when we would be interested in systems that are linear from an *input-output perspective* only, that is, without worrying about the state. To be precise, we would want these systems to satisfy the following condition. If

$$\{u_1(n)\} \xrightarrow{g'_n} \{y_1(n)\} \tag{3.22a}$$

$$\{u_2(n)\} \xrightarrow{g'_n} \{y_2(n)\} \tag{3.22b}$$

then we would want

$$\alpha\{u_1(n)\} + \beta\{u_2(n)\} \xrightarrow{g'_n} \alpha\{y_1(n)\} + \beta\{y_2(n)\} \tag{3.22c}$$

In order to fulfil this requirement, we need to study the general result so as to arrive at the required condition on the initial state. Using the superposition principle of Eq. 3.10, Eq. 3.7c can be written as

$$\alpha \begin{bmatrix} x_1(n_0) \\ \{u_1(n)\} \end{bmatrix} + \beta \begin{bmatrix} x_2(n_0) \\ \{u_2(n)\} \end{bmatrix} \xrightarrow{g_n} \alpha\{y_{1x}(n)\} + \alpha\{y_{1u}(n)\} \\ + \beta\{y_{2x}(n)\} + \beta\{y_{2u}(n)\}$$

Comparing the outputs of g_n and g'_n, we see that the latter has the additional component of response due to initial state, y_x. We see that y_x has to be zero if the outputs of g'_n and g_n are to be equal. In other words, the right-hand sides of the above two mappings g'_n and g_n are equal if and only if $\{y_{1x}(n)\} \equiv 0$ and $\{y_{2x}(n)\} \equiv 0$ and $\{y_1(n)\} \equiv \{y_{1u}(n)\}$ and $\{y_2(n)\} \equiv \{y_{2u}(n)\}$. Using the definition that the output with the subscript x refers to the response due to initial state only, this observation is equivalent to the requirement

$$\begin{bmatrix} x_1(n_0) \\ \{0\} \end{bmatrix} \xrightarrow{g_n} \{0\} \quad \text{and} \quad \begin{bmatrix} x_2(n_0) \\ \{0\} \end{bmatrix} \xrightarrow{g_n} \{0\}$$

From Eq. 3.5, we know that this is possible if $x_1(n_0) = x_2(n_0) = 0$. As a result, we conclude that systems that are linear from the input–output perspective can be produced by enforcing a zero initial condition.

When this property of input–output linearity holds, we can easily handle a combination of two or more inputs. For example, suppose that a combination of m inputs is given as below:

$$u = \alpha_1 u_1 + \alpha_2 u_2 + \cdots + \alpha_m u_m$$

If the output corresponding to u_1 alone is y_{u1}, that for u_2 alone is y_{u2}, etc., then using input-output linearity, we arrive at

$$y_u = \alpha_1 y_{u_1} + \alpha_2 y_{u_2} + \alpha_3 y_{u_3} + \cdots + \alpha_m y_{u_m} \tag{3.23}$$

This idea is illustrated in Fig. 3.7.

An input–output model does not consist of state information and it can be constructed from input–output data[2] alone. In Sec. 3.4, we will show how it can be

[2]This topic is discussed in detail in Chapter 6.

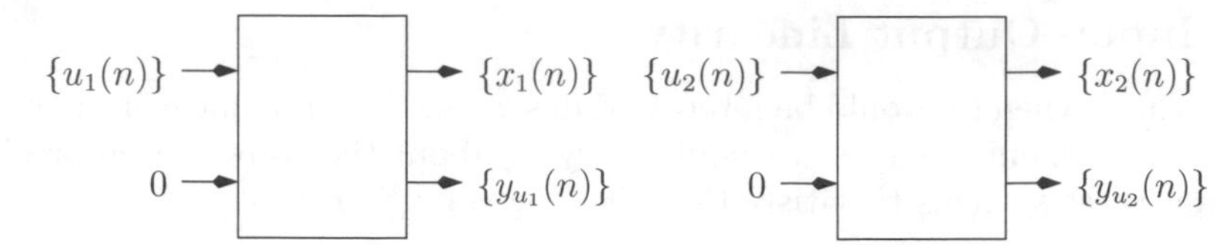

(a) Input/Output linear because the initial state is zero in both the experiments

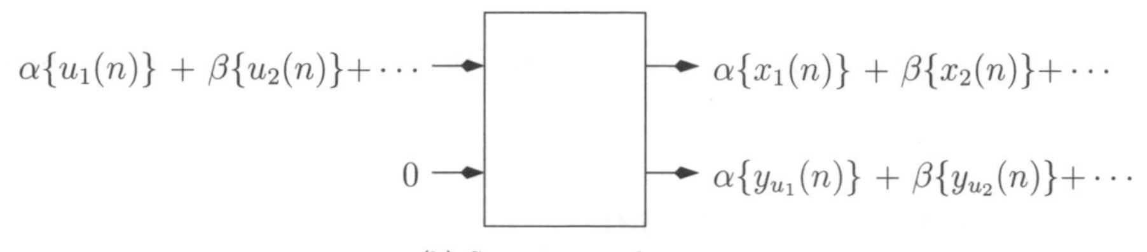

(b) Superimposed system

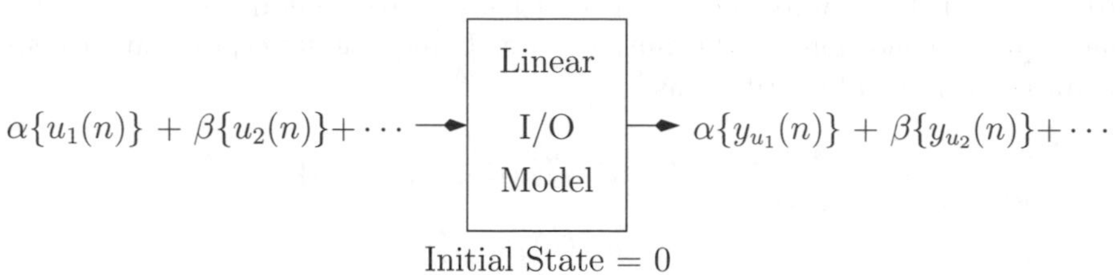

(c) Can drop the initial state information as the system is input/output linear

Figure 3.7: Superposition principle for zero initial state

generated from a state space model. In view of this, we can say that the input–output model is more general than the state space model.

Because of all of the above benefits, we will assume the initial state to be zero, *i.e.*, $x(n_0) = 0$, in the rest of the book, unless otherwise stated. We will refer to input–output linear systems as *I/O linear systems* and when no confusion arises as simply linear systems.

Now we will address the question of what happens if the initial state $x(n_0)$ and hence $\{y_x(n)\}$ are not zero. Recall that for linear systems, the output y can be written as

$$\begin{aligned} y = &\text{ Output at zero initial state and actual input } (y_u) \\ &+ \text{Output at actual initial state and zero input } (y_{x_0}) \end{aligned}$$

Substituting for y_u from Eq. 3.9, this equation becomes

$$y = \alpha_1 y_{u_1} + \alpha_2 y_{u_2} + \cdots + \alpha_m y_{u_m} + y_{x_0}$$

It is clear from this discussion that if we want I/O linearity (*i.e.*, linear combination of inputs giving the same linear combination of corresponding outputs), y_{x_0} must be zero. This is satisfied if $x_0 = 0$.

We will next study the important concept of impulse response models, which one can use to derive convolution models.

Figure 3.8: Impulse response (left) and response to a shifted impulse (right) for a time invariant system

3.3.2 Impulse Response Models

The *impulse response* of an LTI system is the output measured as a function of time for an impulse $\{\delta(n)\}$ when the initial state value is zero. Similarly, a step response model is the response due to a unit step $\{1(n)\}$ with zero initial state. The impulse response will be denoted by $\{g(n)\}$ and step response by $\{s(n)\}$. We will now derive expressions for these responses.

Suppose the input to an LTI system is $\{\delta(n)\}$. We obtain the response as in the left-hand side of Fig. 3.8. For a time invariant system, the response to a time shift in input will be a corresponding shift in the output also. As a result, we obtain the input–output behaviour as in the right-hand side of Fig. 3.8. Notice that if the system is not time invariant, we cannot use the same symbol g in both the figures.

We will now develop a formula to express the response to an arbitrary input $\{u(n)\}$ resolved in terms of impulses. Recall Eq. 3.21, reproduced here for convenience:

$$\{u(n)\} = \sum_{k=-\infty}^{\infty} u(k)\ \{\delta(n-k)\}$$

The term $u(k)$ on the right-hand side is a scalar. It refers to the value of $\{u\}$ taken at time k. Thus, u is a linear combination of sequences of the form $\{\delta(n-k)\}$. The output corresponding to this input u is the same linear combination of responses due to $\{\delta(n-k)\}$ – this follows from the I/O linearity of the system. Also note that if the response of $\{\delta(n)\}$ is $g(n)$, the response due to $\{\delta(n-k)\}$ is $g(n-k)$. This follows from time invariance, as illustrated in Fig. 3.8. In view of these observations, we obtain the following expression for y:

$$\{y(n)\} = \sum_{k=-\infty}^{\infty} u(k)\{g(n-k)\} \tag{3.24}$$

We have made three assumptions to arrive at the above equation: the initial state is zero, the system is linear and time invariant. In other words, the above equation holds for I/O LTI systems. The right-hand side of Eq. 3.24 is known as the *convolution sum* or *superposition sum* or *convolution*of the sequences $\{u(n)\}$ and $\{g(n)\}$, which will be denoted symbolically as

$$\{y(n)\} = \{u(n)\} * \{g(n)\} \tag{3.25}$$

The implication of Eq. 3.25 is that to find the output due to an arbitrary input, all we have to do is to convolve the latter with the impulse response function of the system. Hence it follows that all information about an LTI system at zero initial state is contained in its impulse response function.

We will now look at a simpler way of writing the above two equations. If we write Eq. 3.24 term by term, we obtain the following relation:

$$\{\ldots, y(-1), y(0), y(1), \ldots\} = \sum_{k=-\infty}^{\infty} u(k)\{\ldots, g(-1-k), g(-k), g(1-k), \ldots\}$$

On equating the two sides term by term, we obtain the following relations:

$$\begin{aligned} y(-1) &= \sum_{k=-\infty}^{\infty} u(k)g(-1-k) \\ y(0) &= \sum_{k=-\infty}^{\infty} u(k)g(-k) \\ y(1) &= \sum_{k=-\infty}^{\infty} u(k)g(1-k) \end{aligned}$$

This can be generalized as

$$y(n) = \sum_{k=-\infty}^{\infty} u(k)g(n-k) \tag{3.26}$$

In other words, the output at sampling instant n is obtained by summing the terms of the form $u(k)g(n-k)$ for all k values. Note that the sum of the arguments of u and g is n, which is the same as the argument of y. This equation is also written as

$$y(n) = u(n) * g(n) \tag{3.27}$$

Comparing the above two equations with Eq. 3.24–3.25, we see that the only difference is the presence or absence of the braces.[3] In case the initial state is not zero, we obtain

$$y(n) = y_x(n) + u(n) * g(n) \tag{3.28}$$

Now we present an example to calculate the convolution sum.

Example 3.6 If a system with impulse response $\{g(n)\} = \{1, 2, 3\}$ is subjected to an input sequence $\{u(n)\} = \{4, 5, 6\}$, determine the output sequence $\{y(n)\}$. It can be assumed that both g and u start at $n = 0$ and that they are zero for $n < 0$.

It is given that $g(0) = 1$, $g(1) = 2$ and $g(2) = 3$. Similarly the values of u at $n = 1, 2$ and 3 are $4, 5$ and 6, respectively. Evaluating Eq. 3.26 for different values of n, we obtain

$$\begin{aligned} y(0) &= u(0)g(0) = 4 \\ y(1) &= u(0)g(1) + u(1)g(0) = 13 \\ y(2) &= u(0)g(2) + u(1)g(1) + u(2)g(0) = 28 \\ y(3) &= u(1)g(2) + u(2)g(1) = 27 \\ y(4) &= u(2)g(0) = 18 \end{aligned}$$

[3]In view of this observation, we will write the convolution sum with or without braces.

All possible u and g combinations whose arguments sum up to that of y have appeared in every line of the above set of equations. All terms that don't appear in these equations are zero. M 3.2 shows how to do these computations in Matlab. ∎

In case the state at $k = 0$ is not zero, then Eq. 3.24 should be modified as

$$\{y(n)\} = \sum_{k=-\infty}^{\infty} u(k)\{g(n-k)\} + y_{x_0} \tag{3.29}$$

where y_{x_0} is the zero input response due to nonzero initial state. We will now present an example that requires the calculation of the impulse response.

Example 3.7 Through convolution sum calculation, verify that the impulse response of an LTI system is the response one gets for an impulse input.

Let the impulse response be $\{g(n)\}$. The output sequence $\{y(n)\}$ is calculated using Eq. 3.27:

$$y(n) = u(n) * g(n) = \sum_{k=-\infty}^{\infty} u(k)g(n-k)$$

We want to check whether we will obtain $y(n)$ as $g(n)$ if we substitute $\delta(n)$ for $u(n)$. After this substitution, we obtain

$$y(n) = \sum_{k=-\infty}^{\infty} \delta(k)g(n-k)$$

All the terms in the above sum are zero, except when k is zero, by the property of the impulse sequence, see Eq. 3.18. As a result, we obtain

$$y(n) = g(n)$$

This is in agreement with the definition of the impulse response. ∎

Systems that have finite numbers of terms in their impulse response are known as *finite impulse response* systems and these are abbreviated as *FIR* systems. The definition of *infinite impulse response* systems is obvious. These are abbreviated as *IIR* systems.

3.3.3 Properties of Convolution

As the response of a system with zero initial state to an arbitrary input is given in terms of this operation, convolution is fundamental to I/O LTI systems. In this section, we present some of the well known and useful properties of convolution.

Figure 3.9: Explanation of associative property of convolution

Commutativity: Convolution is a commutative operation, *i.e.*,

$$u(n) * g(n) = g(n) * u(n) \tag{3.30}$$

We start with the left-hand side. Using the definition of convolution sum, we obtain

$$u(n) * g(n) = \sum_{k=-\infty}^{\infty} u(k)g(n-k)$$

Substituting $r = n - k$, this becomes

$$u(n) * g(n) = \sum_{r=-\infty}^{\infty} u(n-r)g(r) = \sum_{r=-\infty}^{\infty} g(r)u(n-r)$$

which is the same as the right-hand side of Eq. 3.30. This property does not hold even if one of the sequences is either nonlinear or time varying, or both, see Problem 3.2.

Associativity: Convolution is associative, *i.e.*,

$$(u(n) * g_1(n)) * g_2(n) = u(n) * (g_1(n) * g_2(n)) \tag{3.31}$$

Because convolution has been defined with only two sequences at a time, brackets in the above equation are necessary to indicate the order of the calculations. Let y_1 denote the term on the left-hand side. We have

$$y_1 = \left(\sum_{r=-\infty}^{\infty} u(r)g_1(n-r) \right) * g_2(n) = \sum_{k=-\infty}^{\infty} \sum_{r=-\infty}^{\infty} u(r)g_1(n-r-k)g_2(k)$$

Let y_2 be the term on the right-hand side of Eq. 3.31. We have

$$y_2 = u(n) * \left(\sum_{k=-\infty}^{\infty} g_1(n-k)g_2(k) \right) = \sum_{r=-\infty}^{\infty} u(r) \sum_{k=-\infty}^{\infty} g_1(n-k-r)g_2(k)$$

which is the same as y_1. An interpretation is given in Fig. 3.9. We obtain y_1 as the final output while sending u into the first system and then sending the resulting output to the second system. The output y_2 is computed by sending the input u to the convolved system $g_1 * g_2$. We have shown that $y_1 = y_2$.

This property says that it does not matter in what order we carry out the convolution operation. In other words, it does not matter where we put the brackets in a convolution sum involving more than two terms. As a result, the expressions given in Eq. 3.31 can be written as $u(n) * g_1(n) * g_2(n)$. If the associativity property does not hold, however, we cannot remove the brackets in Eq. 3.31.

Figure 3.10: Explanation of distributive property of convolution

Distributivity: Convolution distributes over addition, *i.e.*,

$$u(n) * (g_1(n) + g_2(n)) = u(n) * g_1(n) + u(n) * g_2(n) \tag{3.32}$$

Let the left-hand side be equal to y_1. We have

$$y_1 = \sum_{r=-\infty}^{\infty} u(n-r)(g_1(r) + g_2(r))$$

Expanding the brackets in the second term, we obtain

$$y_1 = \sum_{r=-\infty}^{\infty} u(n-r)g_1(r) + \sum_{r=-\infty}^{\infty} u(n-r)g_2(r)$$

which is equal to the right-hand side of Eq. 3.32; call it y_2. The usefulness of this property is illustrated in Fig. 3.10. The output y_1 can be thought of as the response of a system with impulse response $g_1 + g_2$ for an input u. The output y_2, on the other hand, is the sum of responses of two systems with impulse responses g_1 and g_2, for the same input u. This property says that these two outputs y_1 and y_2 are equal. By induction, we can extend this to any number of systems in parallel. Thus, a parallel combination of LTI systems can be replaced by a single LTI system whose unit sample response is the sum of the individual unit sample responses in the parallel combination.

3.3.4 Step Response Models

In Sec. 3.3.2, We have presented the usefulness of impulse response models. The problem with these models, though, is that it is difficult to generate the required impulse signal at the input. For example, imagine opening a valve fully for only one instant and then closing it back fully at the next instant. In comparison, step signals are somewhat easier to generate. The response of an I/O LTI system to a unit step signal $\{1(n)\}$ is known as the *step response* and it will be denoted by $\{s(n)\}$.

In Sec. 3.3.2, we have shown that all information about an I/O LTI system is contained in its impulse response model. In this section, we will show that it is possible to characterize an I/O LTI system in terms of its unit step response also. That is, all the information about this system is contained in its unit step response. We will begin the discussion by calculating the step response. By substituting $1(k)$ in the place of

$u(k)$ in Eq. 3.24, we obtain

$$\{s(n)\} = \sum_{k=-\infty}^{\infty} 1(k)\{g(n-k)\}$$

Using the definition of $\{1(n)\}$ from Eq. 3.17, this is simplified as

$$\{s(n)\} = \sum_{k=0}^{\infty} \{g(n-k)\} \tag{3.33}$$

This shows that the step response is the sum of the impulse response.

We can also obtain the impulse response from the step response. Because the response to $\{1(n)\}$ is denoted as $\{s(n)\}$, the response to $\{s(n-1)\}$ is $\{1(n-1)\}$. This follows from the time invariance property of the system. Recall the relation between step and impulse signals, given by Eq. 3.19:

$$\{\delta(n)\} = \{1(n)\} - \{1(n-1)\}$$

Using the property of linearity, we obtain

$$\{g(n)\} = \{s(n)\} - \{s(n-1)\} \tag{3.34}$$

In Sec. 3.3.2, we have shown that the impulse response contains all information. The above equation, however, shows that given the step response $s(n)$ of an I/O LTI system, one can easily calculate its impulse response. This implies that the step response also contains all information about an I/O LTI system.

Since $\{s(n)\}$ also contains all the information about an I/O LTI system, it must be possible to express the response of the system to an arbitrary input $\{u(n)\}$ as a function of $\{s(n)\}$ only. To arrive at this relationship, we start with the convolution relation involving impulse response:

$$\{y(n)\} = \{u(n)\} * \{g(n)\} = \{g(n)\} * \{u(n)\}$$

Using Eq. 3.34, we obtain

$$\{y(n)\} = [\{s(n)\} - \{s(n-1)\}] * \{u(n)\}$$

Using the distributivity property of convolution, this becomes

$$\{y(n)\} = \{s(n)\} * \{u(n)\} - \{s(n-1)\} * \{u(n)\} \tag{3.35}$$

The second term on the right-hand side can be simplified:

$$\{s(n-1)\} * \{u(n)\} = \sum_{k=-\infty}^{\infty} s(k-1)\{u(n-k)\}$$

With $j = k - 1$, the above equation becomes

$$\{s(n-1)\} * \{u(n)\} = \sum_{j=-\infty}^{\infty} s(j)\{u(n-j-1)\} = \{s(n)\} * \{u(n-1)\}$$

Substituting this in Eq. 3.35, we obtain

$$\{y(n)\} = \{s(n)\} * [\{u(n)\} - \{u(n-1)\}]$$

where we have used the distributivity property of convolution. We finally arrive at

$$\{y(n)\} = \{s(n)\} * \{\triangle u(n)\} \tag{3.36}$$

where

$$\Delta u(n) \stackrel{\triangle}{=} u(n) - u(n-1) \tag{3.37}$$

This gives a procedure to calculate the response to an arbitrary input, given that the step response is known. Recalling the discussion to arrive at Eq. 3.28, we obtain

$$y(n) = y_x(n) + s(n) * \Delta u(n) \tag{3.38}$$

3.3.5 Impulse Response of Causal Systems

In Sec. 3.1.3, we have shown that one popular way to achieve causality in LTI systems is through initial-rest. Impulse response implies application of zero input everywhere, except at $n = 0$. For I/O LTI systems, this is nothing but initial rest. It follows that the output $y(n) = 0$ for $n < 0$. By the definition of impulse response, this is nothing but $\{g(n)\}$. Thus we have $g(n) = 0$ for $n < 0$ in case of causal, I/O LTI systems. Problem 3.5 examines this issue further.

The above observation that $g(n) = 0$ for $n < 0$ results in a simplification of the impulse response expression. For example, Eq. 3.26 becomes

$$y(n) = \sum_{k=-\infty}^{n} u(k)g(n-k)$$

Most signals that we deal with take nonzero values only from the zeroth sampling instant. For such systems, this equation becomes

$$y(n) = \sum_{k=0}^{n} u(k)g(n-k) \tag{3.39}$$

When expanded, the above equation becomes

$$y(n) = u(0)g(n) + u(1)g(n-1) + \cdots + u(n)g(0)$$

which can be rewritten as

$$y(n) = \sum_{m=0}^{n} g(m)u(n-m) \tag{3.40}$$

Note that Eq. 3.40 can also be obtained by starting with $y(n) = g(n) * u(n)$ and simplifying.

We extend the above discussed property of causality of systems to sequences: a sequence $\{u(n)\}$ is said to be causal if $u(k) = 0$ for $k < 0$. This definition is extremely useful because, in control applications, we often work with signals that are nonzero from the time instant zero onwards.

3.3.6 Parametric and Nonparametric Models

The step and the impulse response models are known as nonparametric models. These are characterized by a large number of variables. For example, we need to specify $\{g(n)\}$ for all n values in the case of impulse response models. Although an infinite number of parameters are implied, we need to specify only a finite number of them in most cases. In this case, the expression for the output given by Eq. 3.40 becomes

$$y(n) = \sum_{0}^{M} g(m)u(n-m) \tag{3.41}$$

where M is some large positive integer. Note that when $n \to \infty$, an infinite number of g values are required in Eq. 3.40, while only $M + 1$ values are required in the above equation. In view of the fact that only a finite number of impulse response coefficients are required, Eq. 3.41 is known as the finite impulse response (FIR) model, as mentioned earlier.

Although M is finite, it can still be large. For example, M is known to take values even as large as 100. In view of this large number, the FIR model also can be classified as nonparametric.

In contrast to the above, models that are characterized by a small number of parameters are known as parametric models. For example, the following model

$$y(n) + a_1 y(n-1) = b_1 u(n-1) \tag{3.42}$$

between the input u and the output y is classified as parametric. In Sec. 6.6, we will study such models from a probabilistic view point.

There is a tradeoff between the number of parameters and the ease of their determination. For example, although the FIR models have more parameters, it is easier to estimate them. In contrast, the identification of the nonparametric models is more involved. We discuss some of these issues in Sec. 6.7 and 6.8.

3.3.7 BIBO Stability of LTI Systems

One of the important objectives of control system design is to ensure stability. There are many notions of stability. Some are: stability in the sense of Lyapunov, asymptotic stability and asymptotic stability at large [29]. For linear systems, these notions are equivalent. In this section, we will present the concept of external stability. In Sec. 7.4.2, we present the concept of internal stability.

A system is said to be externally stable if every bounded input produces bounded output. This is also known as BIBO stability. We now characterize the condition under which this property is fulfilled.

Let an LTI system have an impulse response $g(k)$. Then it is BIBO stable if and only if

$$\sum_{k=-\infty}^{\infty} |g(k)| < \infty \tag{3.43}$$

First we will show that if Eq. 3.43 is satisfied, the system is BIBO stable. Consider an input sequence $\{u(n)\}$ such that $|u(n)| < M \ \forall n$. Using Eq. 3.26, we obtain the response to this input as

$$y(n) = \sum_{k=-\infty}^{\infty} g(k)u(n-k)$$

Taking absolute values, we obtain

$$|y(n)| \leq \sum_{k=-\infty}^{\infty} |g(k)||u(n-k)|$$

If we replace $|u|$ by its bound M, we obtain

$$|y(n)| \leq M \sum_{k=-\infty}^{\infty} |g(k)|$$

Because $\sum_{k=-\infty}^{\infty} |g(k)| < \infty$, we see that $y(n)$ is bounded. As this is true for all n, we find that the output signal is bounded at all the sampling instants.

Now we will show that if the system is BIBO stable, Eq. 3.43 will be satisfied. We will prove this by contradiction. That is, we will show that if Eq. 3.43 is violated, *i.e.*, $\sum_{k=-\infty}^{\infty} |g(k)| \to \infty$, the system will not be BIBO stable. We produce the following input signal that can take either 1 or -1

$$u(k) = \begin{cases} g(-k)/|g(-k)| & g(-k) \neq 0 \\ 0 & g(-k) = 0 \end{cases}$$

Because it can take only 1 or -1, this input is bounded. We will now calculate an expression for the output signal at time zero:

$$y(0) = \sum_{k=-\infty}^{\infty} u(k)g(n-k)|_{n=0} = \sum_{k=-\infty}^{\infty} u(k)g(-k)$$

Substituting the specific choice of $u(k)$ constructed above, we obtain

$$y(0) = \sum_{k=-\infty}^{\infty} |g(-k)|$$

Because of the assumption on g, $y(0)$ is unbounded. Note that an unbounded output has been obtained with a bounded input. As a result, we have shown that the system is not BIBO stable.

BIBO stability for causal LTI systems is equivalent to the following condition:

$$\sum_{k=0}^{\infty} |g(k)| < \infty \tag{3.44}$$

as $g(k) = 0$ for $k < 0$. Notice that if this violated, using arguments similar to the ones used above, one can show that a particular choice of bounded inputs applied from $n = 0$ onwards will produce an unbounded output at $n = \infty$. We can do this because in time invariant systems shifting the time makes no difference to the behaviour of the system. The fact that this condition implies stability can be shown using arguments identical to the ones used earlier. Note that Eq. 3.44 can be satisfied only if

$$g(k) = 0, \ \forall k > M \tag{3.45}$$

where M is some large positive integer.

3.4 State Space Models Revisited

We conclude this chapter with an input–output study of state space models. Consider the state space system

$$\begin{aligned} x(n+1) &= Ax(n) + Bu(n) \\ y(n) &= Cx(n) + Du(n) \end{aligned}$$

Suppose that the initial state is $x(0)$ and that the sequence $\{u(n)\} = \{u(0), u(1), u(2), \ldots\}$ is applied. We obtain

$$\begin{aligned} x(1) &= Ax(0) + Bu(0) \\ x(2) &= Ax(1) + Bu(1) = A[Ax(0) + Bu(0)] + Bu(1) \\ &= A^2x(0) + ABu(0) + Bu(1) \\ x(3) &= Ax(2) + Bu(2) = A^3x(0) + A^2Bu(0) + ABu(1) + Bu(2) \end{aligned}$$

Continuing, we get

$$x(n) = A^n x(0) + \sum_{i=0}^{n-1} A^{n-(i+1)} Bu(i), \qquad A^0 = I \tag{3.46}$$

Substituting this in the output equation, we obtain

$$y(n) = \underbrace{CA^n x(0)}_{\text{zero input response}} + \underbrace{\sum_{i=0}^{n-1} CA^{n-(i+1)} Bu(i) + Du(n)}_{\text{zero state response}} \tag{3.47}$$

Recall that in the input–output setting, we got the following relation for causal systems:

$$\begin{aligned} y(n) &= y_x + y_u \\ &= y_x + \sum_{i=0}^{n} u(i)g(n-i) = y_x + \sum_{i=0}^{n-1} u(i)g(n-i) + u(n)g(0) \end{aligned} \tag{3.48}$$

Comparing terms, we obtain

$$\begin{aligned} y_x &= CA^n x(0) \\ g(n) &= CA^{n-1}B, \; n > 0 \\ g(0) &= D \end{aligned} \tag{3.49}$$

Most real life systems have at least one time delay in the sense that the output depends only on the past inputs and not on the current one. That is, $y(n)$ can depend only on $u(n-k)$, $k > 0$. The reason is that realistic systems require a nonzero time to respond to external stimuli. As the input and output are synchronized by a clock, even a small delay in the response can reflect only in the next sample of the output. In view of this, D and hence $g(0)$ are zero.

3.5 Matlab Code

Matlab Code 3.1 Energy of a signal. This code is available at
`HOME/system/matlab/energy.m`[4]

```
1 u = [4 5 6];
2 Eu = norm(u)^2;
3 ruu = xcorr(u);
4 Lu = length(ruu);
5 Eu = ruu(ceil(Lu/2));
```

Matlab Code 3.2 Convolution of two sequences. This code is available at
`HOME/system/matlab/conv2.m`

```
1 h = [1 2 3];
2 u = [4 5 6];
3 y = conv(u,h)
```

[4]`HOME` stands for `http://www.moudgalya.org/dc/` – first see the software installation directions, given in Appendix A.2.

3.6 Problems

3.1. Suppose that a discrete LTI system has input $u(n)$, impulse response $g(n)$ and output $y(n)$. Suppose also that $g(n)$ is zero everywhere outside the interval $N_0 \le n \le N_1$ and that $u(n)$ is zero everywhere outside the interval $N_2 \le n \le N_3$. Determine over what maximum interval $y(n)$ is not zero.

3.2. Commutativity of an I/O LTI system, discussed in Sec. 3.3.3, depends on both linearity as well as time invariance of both systems. This is brought out by the following two examples [49].

(a) Consider two systems A and B, where A is an LTI system with unit sample response $g(n) = \left(\frac{1}{2}\right)^n 1(n)$. If the input to system B is $w(n)$, its output is $z(n) = nw(n)$, *i.e.*, B is linear but time-varying. Show that the commutativity property does not hold for these two systems by computing the response of each of the systems below to the input $u(n) = \delta(n)$.

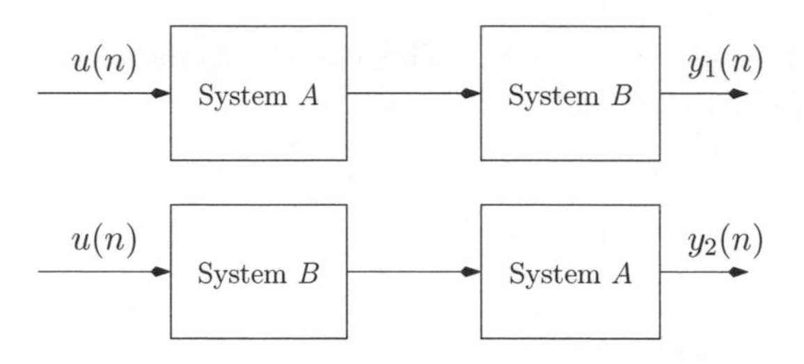

Figure 3.11: Checking commutativity

(b) Suppose that we replace system B in each of the interconnected systems of the above figure by the system with the following relationship between its input $w(n)$ and output $z(n)$: $z(n) = w(n) + 2$. Repeat the calculations of part (a) in this case.

3.3. Given $s(n) = \left(\frac{1}{2}\right)^n 1(n+1)$ and $u(n) = \left(-\frac{1}{2}\right)^n 1(n)$, find $y(n)$.

3.4. Let $\{s(n-1)\}$ denote the response of an I/O linear system to a unit step input given at $n = 1$, *i.e.*, response to $1(1)$. Comparing with Eq. 3.33, show that

$$\{s(n-1)\} = \sum_{k=1}^{\infty} \{g(n-k)\}$$

Subtracting the above from Eq. 3.33, show that

$$\{s(n)\} - \{s(n-1)\} = \{g(n)\}$$

the same as Eq. 3.34.

3.5. The objective of this problem is to show by direct calculations that the impulse response $\{g(n)\}$ of I/O LTI causal systems satisfies $g(n) = 0$ for $n < 0$. Do this as follows. Apply the following two inputs to this system:

$$u_1(n) \begin{cases} = u_2(n) & n \le n_0 \\ \ne u_2(n) & n > n_0 \end{cases}$$

Argue that the expressions for $y_1(n_0)$ and $y_2(n_0)$ should be equal. Comparing the convolution sum of these two outputs, show the following:

$$\sum_{-\infty}^{-1} g(k)u_1(n_0 - k) = \sum_{-\infty}^{-1} g(k)u_2(n_0 - k)$$

From this, conclude that $g(k) = 0$ for $k < 0$.

3.6. This problem is concerned with the solution of a discrete time linear equation.

(a) Consider the homogeneous difference equation

$$\sum_{k=0}^{N} a_k y(n-k) = 0 \tag{3.50}$$

Show that if z_0 is a solution of

$$\sum_{k=0}^{N} a_k z^{-k} = 0$$

Az_0^n is a solution of Eq. 3.50, where A is arbitrary.

(b) Show that if z_0 is a zero of the polynomial $p(z)$ defined as

$$p(z) \triangleq \sum_{k=0}^{N} a_k z^{N-k} \tag{3.51}$$

then also Az_0^n is a solution of Eq. 3.50.

(c) Show that if $y(n) = nz^{n-1}$ then

$$\sum_{k=0}^{N} a_k y(n-k) = \frac{dp(z)}{dz} z^{n-N} + (n-N)p(z)z^{n-N-1} \tag{3.52}$$

for $p(z)$ defined in Eq. 3.51.

(d) Suppose $p(z)$ defined in Eq. 3.51 is factored as

$$p(z) = a_0(z - z_1)^{m_1} \dots (z - z_r)^{m_r} \tag{3.53}$$

where $z_1, \dots, z_r$ are distinct zeros of $p(z)$ and $m_1, \dots, m_r$ are integers ≥ 1. If one of the zeros of $p(z)$, say z_i, has multiplicity 2, *i.e.*, $m_i = 2$, then show that at $z = z_i$, the right-hand side of Eq. 3.52 becomes zero.

(e) Using the fact from (d) above, show that nz_i^{n-1} and hence Bnz_i^{n-1} are solutions of Eq. 3.50, where B is arbitrary.

(f) Using the results from (b) and (e) above, show that if $m_i = 2$ in Eq. 3.53, Az_i^n and Bnz_i^{n-1} are solutions of Eq. 3.50.

(g) Qualitatively argue by extending the logic of (c), (d) and (e) above that if $m_i > 1$ in Eq. 3.53,

$$B\frac{n!}{j!(n-j)!} z^{n-j}$$

satisfies Eq. 3.50, where $j = 0, 1, \dots, m_i - 1$.

Chapter 4

Z-Transform

In the previous chapter, we have seen that convolution is an important operation in I/O LTI systems. In particular, the output at any instant is obtained as a convolution sum involving an infinite number of products of impulse response and input. If the output has to be evaluated at another time instant, the convolution sum has to be recalculated. We show in this chapter that the Z-transform is a tool to simplify convolution operations. The Z-transform makes easy the task of analysis and design of useful devices, such as filters and controllers. This aspect of the Z-transform will become clear as we proceed through the book.

We first present the conditions under which one can use the Z-transform. The conditions for stability and causality follow immediately. After presenting some facts about Z-transforms, we move on to the topic of transfer functions. The final topic is related to inversion of the Z-transform, a step to be carried out for implementation.

4.1 Motivation and Definition of Z-Transform

We begin this section with a motivational discussion of the Z-transform. We discuss the conditions under which the infinite sum converges. We formally define the Z-transform next. We conclude this section with a discussion on properties of the region where the Z-transform converges.

4.1.1 Motivation

Z-transforms are used to simplify the operations involving difference equations. For example, the convolution of two discrete time signals gets reduced to the product of two algebraic expressions with the use of Z-transforms. To see this, consider the convolution sum

$$y(n) = \sum_{k=-\infty}^{\infty} u(k)g(n-k)$$

and let $\{u(n)\} = \{u(0), u(1), u(2)\}$, $\{g(n)\} = \{g(0), g(1), g(2)\}$. Let these sequences be zero at all other times. We obtain the following y values:

$$\begin{aligned} y(0) &= u(0)g(0) \\ y(1) &= u(0)g(1) + u(1)g(0) \\ y(2) &= u(0)g(2) + u(1)g(1) + u(2)g(0) \\ y(3) &= u(1)g(2) + u(2)g(1) \\ y(4) &= u(2)g(2) \end{aligned} \tag{4.1}$$

The same result will be obtained from the coefficients of the resultant series, if we define two functions $U(z)$ and $G(z)$

$$\begin{aligned} U(z) &= u(0) + u(1)z^{-1} + u(2)z^{-2} \\ G(z) &= g(0) + g(1)z^{-1} + g(2)z^{-2} \end{aligned} \tag{4.2}$$

and multiply them. Let $V(z)$ denote the product of U and V:

$$V(z) = G(z)U(z) \tag{4.3}$$

Substituting for G and U from above and multiplying out, V is obtained as a fourth degree polynomial in z^{-1}, of the following form:

$$V(z) = y(0) + y(1)z^{-1} + y(2)z^{-2} + y(3)z^{-3} + y(4)z^{-4} \tag{4.4}$$

where, $y(i)$, $i = 1$ to 4, are identical to the ones obtained in Eq. 4.1. The functional form $U(z)$ and $G(z)$ will be defined as the Z-transform in Sec. 4.1.3. At this point, we may think of z as a position marker. For example, the term that multiplies z^{-i} occurs at the ith time instant in Eq. 4.2. Thus, the function $U(z)$, given in Eq. 4.2, can be thought of as a way of representing a sequence $\{u(n)\} = \{u(0), u(1), u(2)\}$. The advantage in using $U(z)$ in place of $\{u(n)\}$ is that the former can often be represented by a more compact expression, even if the original sequence has infinitely many nonzero terms.

With an example, we have shown that the product of the polynomial equivalent of two sequences gives the same result as their convolution. This approach is not of great use to short sequences, because the effort involved in evaluating the convolution and polynomial product is about the same. The main motivation for proposing this approach is that it could come in handy for sequences consisting of an infinite number of elements. The moment we have a sum consisting of an infinite number of elements, we have to ascertain that it converges.

The second issue we have to address is whether we can recover from the product polynomial the time domain elements, *i.e.*, the values that the sequence takes at different time instants. We would also like to know whether we can recover the sequence uniquely from the polynomial. Both of these issues will now be addressed.

4.1.2 Absolute Convergence

We claim that the two issues raised above, namely convergence of the sum and uniqueness of inversion, can be satisfactorily answered by the concept of absolute

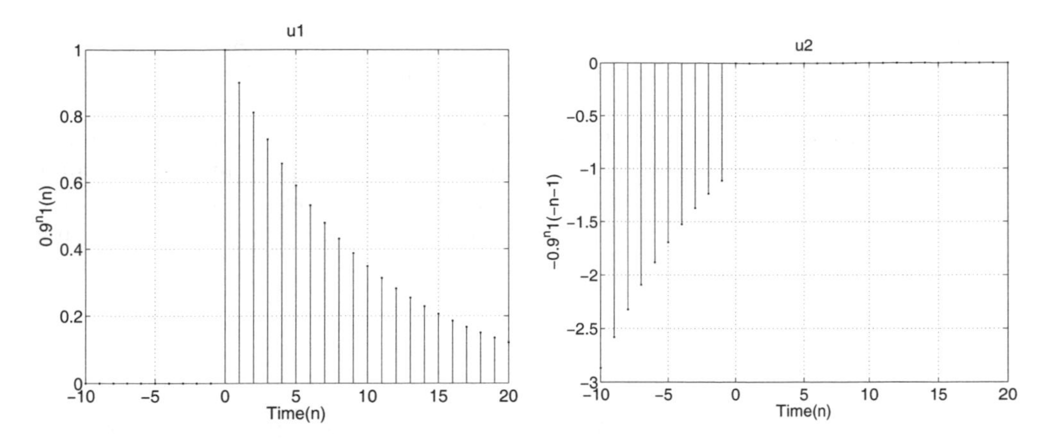

Figure 4.1: Plots of $0.9^n 1(n)$ (left) and $-(0.9)^n 1(-n-1)$ (right)

convergence. We begin the discussion with a motivational example. Let us study the following two sequences:

$$u_1(n) = a^n 1(n)$$
$$u_2(n) = -a^n 1(-n-1)$$

Fig. 4.1 shows the plots of these functions for $a = 0.9$. M 4.1 and M 4.2 have been used to generate these plots. We will use the position marker approach with z^{-1} and calculate the sum of these two sequences. We will assign upper case letters to denote this sum:

$$\begin{aligned} U_1(z) &= \sum_{n=0}^{\infty} a^n z^{-n} = \sum_{n=0}^{\infty} (az^{-1})^n \\ &= \frac{1}{1-az^{-1}} \quad \text{(if the sum exists)} \quad = \frac{z}{z-a} \\ U_2(z) &= -\sum_{n=-\infty}^{\infty} a^n 1(-n-1) z^{-n} = -\sum_{n=-1}^{-\infty} a^n z^{-n} \end{aligned}$$

Substituting $m = -n$, this becomes

$$\begin{aligned} &= -\sum_{m=1}^{\infty} a^{-m} z^m = 1 - \sum_{m=0}^{\infty} (a^{-1}z)^m \\ &= 1 - \frac{1}{1-a^{-1}z} \quad \text{(if the sum exists)} \quad = \frac{z}{z-a} \end{aligned}$$

Thus both $U_1(z)$ and $U_2(z)$ give $\frac{z}{z-a}$. We see that given the polynomial, such as $U(z)$, we cannot determine the underlying sequence $\{u(n)\}$ uniquely. We would like to explore whether this difficulty can be overcome. Note that we have not yet addressed the phrase *if the sum exists*. The conditions for the sums in u_1 and u_2 to exist, if a and z are real numbers, are

$$az^{-1} < 1 \text{ for } u_1 \text{ and}$$
$$a^{-1}z < 1 \text{ for } u_2$$

We see that these two conditions are completely different. We show below that this convergence condition can help overcome the difficulty in uniquely determining the time domain sequence.

In the above discussion, we have assumed z to be real. This can be restrictive. For example, we could be interested in the roots of the polynomial in z after Z-transformation. These roots could turn out to be complex. Moreover, in some problems, a can also be complex. Both situations are taken care of by modifying the above conditions as

$$\begin{aligned}
&|az^{-1}| < 1 \text{ for } u_1 \\
&|a^{-1}z| < 1 \text{ for } u_2
\end{aligned}$$

The above conditions result in *absolute convergence*. For example, the absolute sum $\sum_{n=0}^{\infty} |az^{-1}|^n$ converges. This implies that the original sum $\sum_{n=0}^{\infty}(az^{-1})^n$ also converges. Thus, convergence is not affected by the use of absolute values. The above two conditions result in the following pair:

$$\begin{aligned}
u_1(n) &\leftrightarrow \frac{z}{z-a}, \quad |az^{-1}| < 1 \text{ or } |a| < |z| \\
u_2(n) &\leftrightarrow \frac{z}{z-a}, \quad |a^{-1}z| < 1 \text{ or } |z| < |a|
\end{aligned}$$

Another advantage of using the absolute values is the restriction of the region of convergence to one side of a circle. For example, the u_1 sum converges for all z values outside the circle $z = |a|$ and the u_2 sum converges for all z values inside the same circle. Thus, the use of absolute values clearly demarcates the two regions of convergence. This fact helps us to arrive at the inverse uniquely. For example, if we know that the z value used is inside the circle of radius $|a|$, we can immediately say that the inverse of $z/(z-a)$ is $u_2(n)$. We will explain this idea further with the following calculations:

$$U_1(z) = \frac{z}{z-a} = \frac{1}{1-az^{-1}} = 1 + az^{-1} + (az^{-1})^2 + \cdots$$

since $|az^{-1}| < 1$. On inverting, we obtain

$$\{u_1(n)\} = \{1, a, a^2, \ldots\} = 1(n)a^n$$

with the sequence in the braces beginning at $n = 0$. As the region of convergence[1] of U_2 is different, it has to be factored differently:

$$\begin{aligned}
U_2(z) &= \frac{z}{z-a} = \frac{za^{-1}}{za^{-1}-1} = \frac{-za^{-1}}{1-za^{-1}} \\
&= (-za^{-1})[1 + (za^{-1}) + (za^{-1})^2 + \cdots], \text{ since } |za^{-1}| < 1 \\
&= -za^{-1} - (za^{-1})^2 - (za^{-1})^3 - \cdots \\
&\leftrightarrow u_2(n)
\end{aligned}$$

Thus ROC allows the sequences to be determined uniquely.

[1]We will abbreviate the region of convergence as ROC.

4.1.3 Definition of Z-Transform

The Z-transform of a sequence $\{u(n)\}$ is denoted by $U(z)$ and it is calculated using the formula

$$U(z) = \sum_{n=-\infty}^{\infty} u(n)z^{-n} \tag{4.5}$$

where z is chosen such that $\sum_{n=-\infty}^{\infty} |u(n)z^{-n}| < \infty$. As discussed earlier, the stronger condition of absolute convergence

1. allows complex quantities in u and z,
2. ensures that ROC lies on one side of a circle.

We will also use the following notation to represent a sequence $\{u(n)\}$ and its Z-transform $U(z)$:

$$\{u(n)\} \leftrightarrow U(z)$$

The main advantage in taking the Z-transform is that cumbersome convolution calculations can be done easier. Moreover, issues such as stability, causality and oscillatory behaviour of an infinite signal can easily be established by analyzing the closed form expression involving the Z-transform. This fact will come in handy while designing controllers.

Of course, there is an added burden of converting the polynomial to a closed form expression and inverting this procedure to arrive at the resultant polynomial. We will see later that it is possible to develop techniques to simplify these operations.

4.1.4 Region of Convergence

In Sec. 4.1.2, we have seen that $z/(z-a)$ is the Z-transform of

$$\begin{aligned} -a^n 1(-n-1) &\text{ for } |z| < |a| \\ a^n 1(n) &\text{ for } |z| > |a| \end{aligned}$$

From this observation it is clear that in specifying the Z-transform of a signal, both the algebraic expression and the range of values of z for which the expression is valid are required. In general, as mentioned earlier, the range of values of z for which the sum in the Z-transform converges absolutely is referred to as the ROC of the Z-transform.

It is convenient to describe ROC with respect to coordinate axes with the real part of the complex number being on the abscissa and the imaginary part on the ordinate, see Fig. 4.2.[2] Such a plane is referred to as the z plane. Since the condition $|z| > |a|$ refers to the region outside the circle of radius $|a|$ with centre at the origin, ROC for u_1 is shown on the left by shaded lines. Similarly, ROC for u_2 is shown on the right.

Example 4.1 Let us now consider a signal that is the sum of two real exponentials:

$$u(n) = \left(\frac{1}{2}\right)^n 1(n) + \left(\frac{1}{3}\right)^n 1(n)$$

[2] We will abbreviate imaginary as Im and real as Re throughout this book.

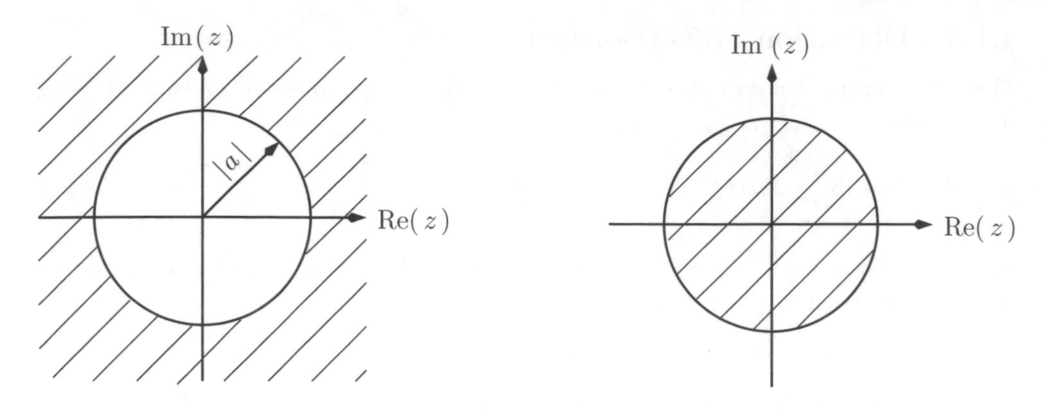

Figure 4.2: Region of convergence of $u_1(n) = a^n 1(n)$ (left) and $u_2(n) = -a^n 1(-n-1)$ (right), shown with shaded lines, on the z plane

The Z-transform is

$$\begin{aligned}
U(z) &= \sum_{n=-\infty}^{\infty} \left[\left(\frac{1}{2}\right)^n 1(n) + \left(\frac{1}{3}\right)^n 1(n) \right] z^{-n} \\
&= \sum_{n=-\infty}^{\infty} \left(\frac{1}{2}\right)^n 1(n) z^{-n} + \sum_{n=-\infty}^{\infty} \left(\frac{1}{3}\right)^n 1(n) z^{-n} \\
&= \sum_{n=0}^{\infty} \left(\frac{1}{2}\right)^n z^{-n} + \sum_{n=0}^{\infty} \left(\frac{1}{3}\right)^n z^{-n} \\
&= \sum_{n=0}^{\infty} \left(\frac{1}{2} z^{-1}\right)^n + \sum_{n=0}^{\infty} \left(\frac{1}{3} z^{-1}\right)^n
\end{aligned}$$

Notice that the first and the second sums are, respectively,

$$\frac{1}{1 - \frac{1}{2} z^{-1}} \text{ for } |z| > \frac{1}{2}, \qquad \frac{1}{1 - \frac{1}{3} z^{-1}} \text{ for } |z| > \frac{1}{3}$$

If we choose $|z| > 1/2$, both sums converge. We obtain

$$U(z) = \frac{1}{1 - \frac{1}{2} z^{-1}} + \frac{1}{1 - \frac{1}{3} z^{-1}} = \frac{z\left(2z - \frac{5}{6}\right)}{\left(z - \frac{1}{2}\right)\left(z - \frac{1}{3}\right)} \text{ for } |z| > \frac{1}{2}$$

We observe that U is a ratio of polynomials in z. ∎

The Z-transform of an exponential becomes a ratio of polynomials in z, because of the presence of an infinite number of terms. In view of this, suppose that we have the Z-transform of u as $U(z)$, with the following form:

$$U(z) = \frac{N(z)}{D(z)} \tag{4.6}$$

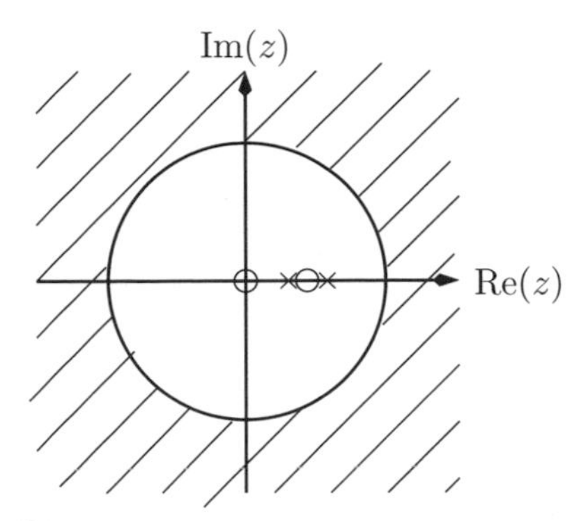

Figure 4.3: Pole–zero plot for $\left[\left(\frac{1}{2}\right)^n + \left(\frac{1}{3}\right)\right]^n 1(n)$, studied in Example 4.1. The poles are indicated by $\times$ and zeros by $\circ$.

where $N(z)$ is the numerator polynomial and $D(z)$ stands for the denominator polynomial. When $U(z)$ is of this form, it is referred to as a rational function of z. In the above example, we have seen that the sum of the Z-transform of two exponentials also is a rational. This property holds as long as we have a finite number of exponentials. In the above equation, if we assume that $N(z)$ is a polynomial of degree n, we can write it as

$$N(z) = K_1(z - z_1)(z - z_2)\dots(z - z_n)$$

$N(z)$ and hence $U(z)$ become zero at $z = z_i$, $i = 1, \dots, n$, which are the roots of the equation $N(z) = 0$. As at these values $U(z)$ becomes zero, z_i, $i = 1, \dots, n$, are known as the *zeros* of $U(z)$. Similarly if $D(z)$ is a polynomial of degree m, we can write it as

$$D(z) = K_2(z - p_1)(z - p_2)\dots(z - p_m)$$

At p_i, $i = 1, \dots, m$, the roots of $D(z) = 0$, $U(z)$ becomes infinite. These p_i, $i = 1, \dots, m$, are known as the *poles* of $U(z)$. The zeros in the above example are at zero and 5/12 while the poles are at 1/2 and 1/3. From the above discussion, the Z-transform of $u(n)$ consisting of a finite number of exponentials can be written as

$$U(z) = K\frac{(z - z_1)(z - z_2)\dots(z - z_n)}{(z - p_1)(z - p_2)\dots(z - p_m)}$$

where $K = K_1/K_2$. It is clear that but for a scale factor, the Z-transform of $\{u(n)\}$ can be specified by its poles, zeros and ROC. In view of this, we can say that the pole-zero plot in Fig. 4.3, along with ROC information, represents the function presented in Example 4.1. In this figure, we have marked the zeros by circles ($\circ$) and poles by crosses ($\times$). ROC is indicated by the shaded region. M 4.3 can also be used to produce a similar plot.

If the degree of the denominator polynomial is greater than that of the numerator polynomial, then $U(z)$ will become zero as z approaches infinity. Conversely, if the degree of the denominator polynomial is less than that of the numerator polynomial, then $U(z)$ will become unbounded as z approaches infinity. This behaviour can be interpreted as zeros or poles at infinity. If we include the zeros or poles at infinity in counting, the number of poles will be equal to that of zeros in any rational function.

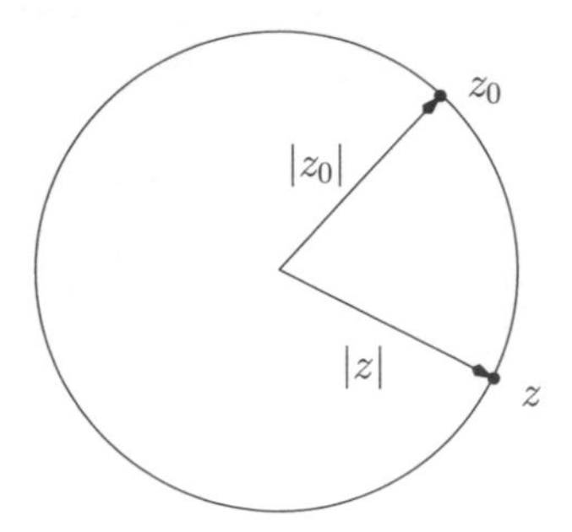

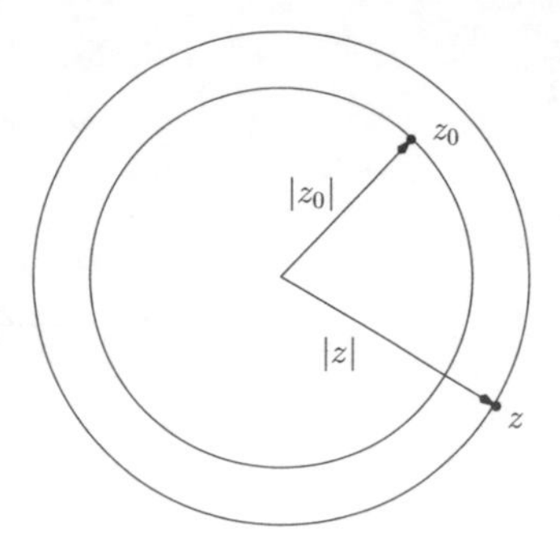

Figure 4.4: ROC has rings (left). Larger rings belong to ROC.

The *order* of a pole is the number of times it is repeated at a given location. Similarly, the order of a zero can be defined. A first order pole is known as a *simple pole*. In a similar way, a *simple zero* is defined. In the following transfer function,

$$U(z) = \frac{z^2(z-1)}{(z-0.6)^3}, \quad |z| > 0.6$$

there is a pole of order 3 at $z = 0.6$, a simple zero at $z = 1$ and a second order zero at $z = 0$.

When the degree of the numerator polynomial is less than or equal to that of the denominator polynomial, we describe the transfer function as *proper*. For example, if $\mathrm{d}N \leq \mathrm{d}D$ in Eq. 4.6, we describe U as proper. Here, the prefix d denotes *degree of*.[3] If, instead, the degree is strictly less, *i.e.*, $\mathrm{d}N < \mathrm{d}D$, we describe U as *strictly proper*. Finally, if the degree of N is higher than that of D, *i.e.*, $\mathrm{d}N > \mathrm{d}D$, it is *improper*.

We will refer to the sequence $\{u(n)\}$ and its Z-transform $U(z)$ as *Z-transform pair*. We will use the symbol $\leftrightarrow$ to indicate that one came from the other. We will indicate all of these by writing $u(n) \leftrightarrow U(z)$.

Before concluding this section, it is emphasized that the Z-transform is defined so as to represent an infinite sum in a compact way. If the sequence that is Z-transformed is a growing (unbounded) one, it is handled by choosing ROC appropriately, so as to make $|z|$ sufficiently large.[4]

4.1.5 Properties of Region of Convergence

In this section, some of the properties of ROC will be discussed. Suppose $\{u(n)\}$ is a sequence and $U(z)$ its transform.

ROC has rings: ROC of $U(z)$ consists of a ring in the z plane centred around the origin. The claim is that if a point $r_0 e^{j\phi}$ on a circle of radius r_0 belongs to ROC, then all points $r_0 e^{j\theta}$ where θ is arbitrary will also belong to ROC, see the diagram on the left-hand side of Fig. 4.4. Let us start with the definition of the Z-transform of $U(z)$:

[3] The prefix d will denote the degree of the polynomial that follows.

[4] Thus the compactness of the Z-transform has nothing to do with whether the sequence is growing or bounded.

$$U(z) = \sum_{n=-\infty}^{\infty} u(n)z^{-n}$$

By the requirement of absolute convergence of Z-transforms, we obtain

$$\begin{aligned}\infty > \sum_{n=-\infty}^{\infty} |u(n)z_0^{-n}| &= \sum_{n=-\infty}^{\infty} |u(n)||z_0^{-n}| \\ &= \sum_{n=-\infty}^{\infty} |u(n)||z_0|^{-n} \overset{|z|=|z_0|}{=} \sum_{n=-\infty}^{\infty} |u(n)||z|^{-n} = \sum_{n=-\infty}^{\infty} |u(n)z^{-n}|\end{aligned}$$

Thus, arbitrary z also belongs to ROC.

ROC has no pole: No pole can be present within ROC. This is because, at a pole, the Z-transform is infinite and hence, by definition, does not converge.

ROC for causal systems: Suppose $\{g(n)\}$ is causal. If the circle $|z| = r_0$ is in ROC then all z for which $|z| > r_0$ will also belong to ROC, see the diagram on the right-hand side of Fig. 4.4.

This is because the larger the radius of z, the smaller will be the sum, thanks to the negative powers of z. We will now demonstrate this. $G(z)$ is of the form $g(0) + g(1)z^{-1} + g(2)z^{-2} + \cdots$. This shows that whenever $z \in$ ROC, z cannot be zero. From this it follows that $r_0 > 0$. Because $z_0 \in$ ROC,

$$\sum_{n=0}^{\infty} |g(n)z_0^{-n}| < \infty$$

Rewriting this expression with obvious steps,

$$\begin{aligned}\infty > \sum_{n=0}^{\infty} |g(n)z_0^{-n}| &= \sum_{n=0}^{\infty} |g(n)||z_0^{-n}| = \sum_{n=0}^{\infty} |g(n)||z_0|^{-n} = \sum_{n=0}^{\infty} \frac{|g(n)|}{|z_0|^n} \\ &> \sum_{n=0}^{\infty} \frac{|g(n)|}{|z|^n}, \quad \because |z| > |z_0| \\ &= \sum_{n=0}^{\infty} |g(n)z^{-n}|\end{aligned}$$

We have shown that all z such that $|z| > |z_0|$ belongs to ROC. It is clear that ROC includes $\lim_{z\to\infty}$ as well. We say that larger circles belong to ROC.

Using the above discussed properties, we now proceed to prove two extremely useful properties.

Poles of stable systems: If a system with impulse response $g(n)$ is causal and stable, the poles of $G(z)$ will be inside the unit disc $|z| < 1$.

As $g(k)$ is causal and stable, we have $\sum_{k=0}^{\infty} |g(k)| < \infty$. From the expression for the Z-transform of $\{g(n)\}$,

$$G(z) = \sum_{k=0}^{\infty} g(k)z^{-k}$$

we obtain

$$\infty > \sum_{k=0}^{\infty} |g(k)| = \sum_{k=0}^{\infty} \left| g(k) z^{-k} \right|_{z=1}$$

Thus, there is absolute convergence at $|z| = 1$. In other words, the unit circle belongs to ROC. Because larger circles belong to ROC for causal sequences, all points outside the unit circle also belong to ROC. From the property of the poles, the unit circle or the region outside of it does not contain the poles and thus the poles should lie inside the circle.

Degrees of the Z-transform of causal systems: Let $u(k)$ be a causal sequence with its Z-transform $U(z) = N(z)/D(z)$, where $N(z)$ is a polynomial of degree n and $D(z)$ is a polynomial of degree m. Then $n \le m$.

Since $u(k)$ is causal, larger circles belong to ROC and hence $\lim_{z \to \infty}$ is in ROC. If $n > m$, $U(z)$ will diverge at ∞. Thus $n \le m$.

4.2 Z-Transform Theorems and Examples

In this section, we will present some properties of the Z-transform. We will demonstrate these with some examples. We will begin first with the Z-transform of some fundamental sequences.

Example 4.2 Find the Z-transform of the unit impulse sequence $\{\delta(n)\}$.

We obtain

$$\delta(n) \leftrightarrow \sum_{n=-\infty}^{\infty} \delta(n) z^{-n}$$

By the definition of δ, there is only one nonzero term in the above infinite summation. In view of this, we obtain

$$\delta(n) \leftrightarrow 1 \tag{4.7}$$

Because the above expression is true for all values of z, ROC includes all of the z plane. ∎

Example 4.3 Find the Z-transform of the unit step sequence $\{1(n)\}$.

Note that $1(n)$ is nothing but $u_1(n)$ with $a = 1$ in Sec. 4.1.2. As a result, we obtain

$$1(n) \leftrightarrow \frac{z}{z-1}, \quad |z| > 1 \tag{4.8}$$

Recall from Example 4.2 that there is no restriction on z for the Z-transform of the impulse function. But this is not true in the case of step functions. ∎

4.2.1 Linearity

The Z-transform of a linear combination of sequences is the same linear combination of the Z-transform of individual sequences:

$$Z\left[\alpha\{u_1(n)\} + \beta\{u_2(n)\}\right] = \alpha U_1(z) + \beta U_2(z)$$

where $u_1(n) \leftrightarrow U_1(z)$ and $u_2(n) \leftrightarrow U_2(z)$, with α and β as arbitrary scalars.

The left-hand side is given by

$$\begin{aligned}\sum_{n=-\infty}^{\infty}\left[\alpha u_1(n) + \beta u_2(n)\right] z^{-n} &= \sum_{n=-\infty}^{\infty} \alpha u_1(n) z^{-n} + \sum_{n=-\infty}^{\infty} \beta u_2(n) z^{-n} \\ &= \alpha U_1(z) + \beta U_2(z)\end{aligned}$$

which is equal to the right-hand side. We will use this result and calculate the Z-transform of a few signals.

Example 4.4 Find the Z-transform of

$$\begin{aligned}u_1(n) &= 2\delta(n) - 3\delta(n-2) + 4\delta(n-5) \\ U_1(z) &= 2\sum_{n=-\infty}^{\infty}\delta(n)z^{-n} - 3\sum_{n=-\infty}^{\infty}\delta(n-2)z^{-n} + 4\sum_{n=-\infty}^{\infty}\delta(n-5)z^{-n} \\ &= 2 - 3z^{-2} + 4z^{-5} \ \forall z^{-1} \text{ finite} \\ &= \frac{2z^5 - 3z^3 + 4}{z^5}, \quad |z| > 0\end{aligned}$$

That is, all points, except the origin of the z plane, are in ROC. ∎

Example 4.5 Find the Z-transform of a signal obtained by sampling the continuous function $e^{-t/\tau}$, $\tau > 0$, $t > 0$, with a period T_s.

$$u(n) = e^{-nT_s/\tau}1(n) = \left(e^{-T_s/\tau}\right)^n 1(n)$$

Taking the Z-transform, we obtain

$$U(z) = \frac{z}{z - e^{-T_s/\tau}}, \quad |z| > e^{-T_s/\tau}$$

This approach is useful in discretizing transfer functions that correspond to the continuous time domain, the topic to be discussed in detail in Sec. 8.2.2. ∎

Example 4.6 Find the Z-transform of

$$\begin{aligned}\{u_2(n)\} &= \left[2 + 4(-3)^n\right]\{1(n)\} \\ U_2(z) &= \sum_{n=0}^{\infty}\left[2 + 4(-3)^n\right] z^{-n} \\ &= \frac{2z}{z-1} + \frac{4z}{z+3}, \quad |z| > 1 \cap |z| > 3\end{aligned}$$

Simplifying this expression, we obtain

$$U_2(z) = \frac{6z^2 + 2z}{(z-1)(z+3)}, \quad |z| > 3$$

This shows that the region outside of the circle with radius 3 lies in ROC.

∎

Now we will demonstrate with an example that it is not always necessary for the Z-transform of a sequence to exist.

Example 4.7 Find the Z-transform of

$$\{u_3(n)\} = \{1(n)\} + \{0.5^n 1(-n)\}$$

The Z-transform of the first term is $z/(z-1)$ for $|z| > 1$. The Z-transform of the second term is now calculated:

$$\begin{aligned}\sum_{n=-\infty}^{\infty} 0.5^n\, 1(-n)\, z^{-n} &= \sum_{n=-\infty}^{0} 0.5^n z^{-n} = \sum_{m=0}^{\infty} (0.5^{-1}z)^m \\ &= \frac{1}{1 - 0.5^{-1}z}, \quad \left|0.5^{-1}z\right| < 1 \\ &= \frac{0.5}{0.5 - z}, \quad |z| < 0.5\end{aligned}$$

ROCs of the first and second terms are, respectively, $|z| > 1$ and $|z| < 0.5$, which are mutually exclusive. Thus the Z-transform for this example does not exist.

∎

This example is presented just to prove the point that we cannot take the existence of the Z-transform for granted. Fortunately, however, we will not come across this situation again in this book, as the signals of interest to us have Z-transforms.

Example 4.8 Find the Z-transform of $\cos\omega n\, 1(n)$ and $\sin\omega n\, 1(n)$. Since the Z-transform is easy to obtain for sequences of the form a^n, Euler's identity helps achieve precisely this:

$$\cos\omega n + j\sin\omega n = e^{j\omega n}$$

Take the Z-transform of both sides to obtain

$$Z\left[\cos\omega n + j\sin\omega n\right] = Z\left\{e^{j\omega n}\right\}$$

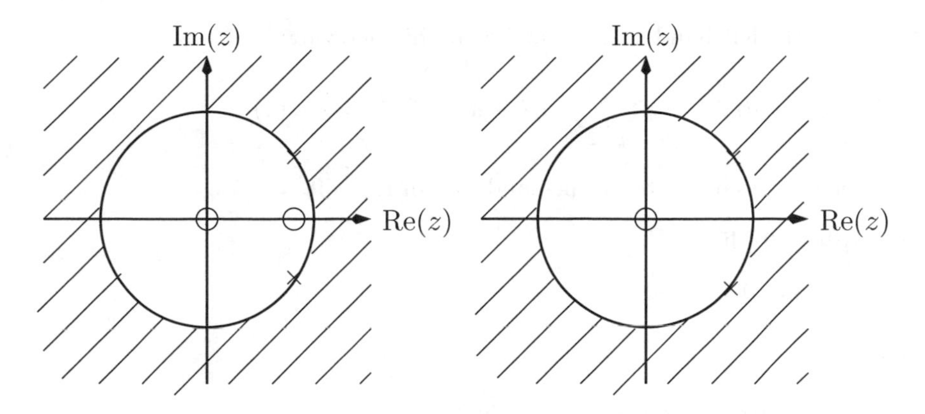

Figure 4.5: Poles and zeros of $\cos\omega n\, 1(n)$ (left) and $\sin\omega n\, 1(n)$ (right)

Comparing real and imaginary parts, we obtain $Z\left[\cos\omega n\right] = \text{Re}\left[Z\left\{e^{j\omega n}\right\}\right]$ and $Z\left[\sin\omega n\right] = \text{Im}\left[Z\left\{e^{j\omega n}\right\}\right]$. Let us now evaluate the Z-transform of $e^{j\omega n}$:

$$\begin{aligned}
Z\left[e^{j\omega n}\, 1(n)\right] &= \sum_{n=-\infty}^{\infty} e^{j\omega n}\, z^{-n}\, 1(n) = \sum_{n=0}^{\infty} e^{j\omega n}\, z^{-n} \\
&= \frac{z}{z - e^{j\omega}}, \quad |z| > \left|e^{j\omega}\right| = 1 \\
&= \frac{z(z - e^{-j\omega})}{(z - e^{j\omega})(z - e^{-j\omega})} \\
&= \frac{z(z - \cos\omega + j\sin\omega)}{(z - e^{j\omega})(z - e^{-j\omega})} \\
&= \frac{z(z - \cos\omega)}{(z - e^{j\omega})(z - e^{-j\omega})} + j\frac{z\sin\omega}{(z - e^{j\omega})(z - e^{-j\omega})}
\end{aligned}$$

Equating the real and imaginary terms, we obtain

$$\begin{aligned}
(\cos\omega n)1(n) &\leftrightarrow \frac{z(z - \cos\omega)}{(z - e^{j\omega})(z - e^{-j\omega})} \\
(\sin\omega n)1(n) &\leftrightarrow \frac{z\sin\omega}{(z - e^{j\omega})(z - e^{-j\omega})}
\end{aligned}$$

Notice that the zeros of $(\cos\omega n)1(n)$ are at 0 and $\cos\omega$ while the poles are at $e^{\pm j\omega}$. The zero of $(\sin\omega n)1(n)$ is at 0 and its poles are at $e^{\pm j\omega}$. Pole–zero plots of these two functions are given in Fig. 4.5.

∎

4.2.2 Shifting

The Z-transform of a shifted sequence is given as follows:

$$Z\left[u(k+d)\right] = z^d U(z) \tag{4.9}$$

Starting with the left-hand side, we arrive at the following:

$$\sum_{k=-\infty}^{\infty} u(k+d)z^{-k} = z^d \sum_{k=-\infty}^{\infty} u(k+d)z^{-(k+d)} = z^d U(z)$$

which is the expression on the right-hand side of Eq. 4.9.

Example 4.9 If

$$\{u(n)\} \leftrightarrow U(z)$$

then

$$\{u(n+3)\} \leftrightarrow z^3 U(z), \quad \{u(n-2)\} \leftrightarrow z^{-2} U(z)$$

∎

In Example 3.5 on page 41, we have explained that the time at which we specify the condition may determine the causality. We will continue this discussion here.

Example 4.10 Calculate the Z-transform of Eq. 3.14 on page 42 and Eq. 3.16.

Expanding the summation in Eq. 3.14, we obtain

$$y(n) = u(n) + au(n-1) + a^2 u(n-2) + \cdots$$

Making use of the linearity and the shifting property, we take the Z-transform of both sides to obtain

$$\begin{aligned} Y(z) &= U(z) + az^{-1}U(z) + a^2 z^{-2} U(z) + \cdots \\ &= \left[1 + \frac{a}{z} + \left(\frac{a}{z}\right)^2 + \cdots \right] U(z) \end{aligned}$$

Because there are an infinite number of entries within the square brackets, we need to enforce a convergence condition. We obtain

$$Y(z) = \frac{z}{z-a} U(z), \quad |z| > |a| \tag{4.10}$$

Note that ROC is the region outside a circle of radius $|a|$, as expected in causal systems.

Let us now consider taking the Z-transform of Eq. 3.16 on page 42. Expanding the summation term in this equation, we obtain

$$y(n) = -a^{-1}u(n+1) - a^{-2}u(n+2) - \cdots$$

Once again, we make use of the linearity and the shifting properties, take the Z-transform of both sides and obtain

$$\begin{aligned} Y(z) &= -a^{-1}zU(z) - a^{-2}z^2 U(z) - \cdots \\ &= -\left[\frac{z}{a} + \left(\frac{z}{a}\right)^2 + \cdots \right] U(z) = -\frac{z}{a}\left[1 + \frac{z}{a} + \left(\frac{z}{a}\right)^2 + \cdots \right] U(z) \end{aligned}$$

As before, we need to be concerned with ROC. We obtain

$$Y(z) = -\frac{z}{a}\frac{a}{a-z}, \quad |z| < |a| = \frac{z}{z-a}U(z), \quad |z| < |a| \tag{4.11}$$

Note that this ROC is inside the circle of radius $|a|$, as expected in noncausal systems. The expression for $Y(z)$ in Eq. 4.11 is identical to the one obtained in Eq. 4.10, but for ROC. The algebraic expressions are identical, because both are Z-transforms of Eq. 3.12 on page 41. ROCs are different because of the difference in the initial condition.

∎

4.2.3 Effect of Damping

The presence of an exponential in the time sequence results in the following:

$$Z\left[u(k)e^{-ak}\right] = U(ze^{a}) \tag{4.12}$$

i.e., U evaluated at ze^a. Starting from the left-hand side, we obtain

$$\sum_{k=-\infty}^{\infty} u(k)e^{-ak}z^{-k} = \sum_{k=-\infty}^{\infty} u(k)(e^{a}z)^{-k}$$

which is the right-hand side of Eq. 4.12.

4.2.4 Initial Value Theorem for Causal Signals

Provided the limit exists, the initial value of a causal signal can be obtained by its Z-transform,

$$u(0) = \lim_{z\to\infty} U(z) \tag{4.13}$$

From the definition of causal signals, we obtain

$$U(z) = \sum_{k=-\infty}^{\infty} u(k)z^{-k} = u(0) + u(1)z^{-1} + u(2)z^{-2} + \cdots$$

Taking the limit, we obtain

$$\lim_{z\to\infty} U(z) = u(0)$$

4.2.5 Final Value Theorem for Causal Signals

If $U(z)$ converges for all $|z| > 1$ and if all the poles of $U(z)(z-1)$ are inside the unit circle then

$$\lim_{k\to\infty} u(k) = \lim_{z\to 1}(1 - z^{-1})U(z) = \lim_{z\to 1}\frac{z-1}{z}U(z) \tag{4.14}$$

The condition on $U(z)$ assures that the only possible pole of $U(z)$ not strictly inside the unit circle is a simple pole at $z = 1$, which is removed in $(z-1)U(z)$. This allows the important situation of stable systems being excited by a unit step signal. Moreover, the fact that $U(z)$ is finite for arbitrarily large z implies that $u(k)$ is causal. As $u(\infty)$ is bounded we can evaluate the following expression, which has an extra $u(-1) = 0$:

$$\begin{aligned}
u(1) &= -u(-1) + u(0) - u(0) + u(1) \\
u(2) &= -u(-1) + u(0) - u(0) + u(1) - u(1) + u(2) \\
\lim_{k\to\infty} u(k) &= \underbrace{-u(-1) + u(0)}_{\Delta u(0)} \underbrace{-u(0) + u(1)}_{\Delta u(1)} \underbrace{-u(1) + u(2)}_{\Delta u(2)} - \cdots
\end{aligned}$$

With the definition $\Delta u(k) = u(k) - u(k-1)$, the above equation can be written as

$$\begin{aligned}
\lim_{k\to\infty} u(k) &= \Delta u(0) + \Delta u(1) + \Delta u(2) + \cdots \\
&= \lim_{z\to 1} \Delta u(0) + \Delta u(1)z^{-1} + \Delta u(2)z^{-2} + \cdots
\end{aligned}$$

Because it is a causal sequence, $\Delta u(k) = 0$ for all $k < 0$. Thus, we can extend the sum and obtain

$$\lim_{k\to\infty} u(k) = \lim_{z\to 1} \sum_{k=-\infty}^{\infty} \Delta u(k) z^{-k}$$

Invoking the meaning of $\Delta u(k)$, we obtain

$$\lim_{k\to\infty} u(k) = \lim_{z\to 1} \sum_{k=-\infty}^{\infty} [u(k) - u(k-1)] z^{-k}$$

Using linearity and shifting properties of the Z-transform, we obtain

$$\lim_{k\to\infty} u(k) = \lim_{z\to 1} \left[U(z) - z^{-1}U(z)\right] = \lim_{z\to 1} \left(1 - z^{-1}\right) U(z)$$

Note that we can write the right-hand side also as $\lim_{z\to 1}(z-1)U(z)$.

Example 4.11 Using the final value theorem, calculate the steady state value of $(0.5^n - 0.5)1(n)$ and verify that it agrees with the direct result.

We will start with the pair

$$(0.5^n - 0.5)1(n) \leftrightarrow \frac{z}{z-0.5} - \frac{0.5z}{z-1}, \quad |z| > 1$$

It is easy to see that as $n \to \infty$, the left-hand side becomes -0.5. We can also apply the final value theorem, by multiplying the right-hand side by $(z-1)/z$ and taking the limit as $z \to 1$. When we do this, only the second term is nonzero. We obtain

$$\lim_{z\to 1}(z-1)\mathrm{RHS} = -\lim_{z\to 1} \frac{z-1}{z}\frac{0.5z}{z-1} = -0.5$$

where RHS denotes right-hand side. We see that the two approaches give identical results.

∎

Example 4.12 Find the initial value and final value of a causal sequence whose Z-transform is given by

$$U(z) = \frac{0.792z^2}{(z-1)(z^2-0.416z+0.208)}$$

and verify.

$$\begin{aligned} \text{Initial Value} &= \lim_{z\to\infty} U(z) = \frac{0.792z^2}{z^3} = 0 \\ \text{Final Value} &= \lim_{z\to 1}(z-1)U(z) = \frac{0.792}{1-0.416+0.208} = 1 \end{aligned}$$

We will verify these results as follows. Expanding the expression for $U(z)$ by long division, we obtain

$$\begin{aligned} U(z) = {} & 0.792z^{-1} + 1.12z^{-2} + 1.091z^{-3} + 1.01z^{-4} \\ & + 0.983z^{-5} + 0.989z^{-6} + 0.99z^{-7} + \cdots \end{aligned}$$

It is easy to verify the initial and final values from this expression. ∎

The next example shows the importance of the preconditions required to be satisfied for the application of the final value theorem.

Example 4.13 Is it possible to use the final value theorem on $2^n 1(n)$?

$$2^n 1(n) \leftrightarrow \frac{z}{z-2}, \quad |z| > 2$$

Since the right-hand side is valid only for $|z| > 2$, the theorem cannot even be applied. On the left-hand side, we have a growing sequence without a limit; this once again violates the conditions of the theorem. Thus, the final value theorem cannot be used. ∎

We conclude this section with the observation that the final value theorem for causal signals cannot be used for growing sequences.

4.2.6 Convolution

Recall that while motivating the Z-transform, we showed with an example that the product of the Z-transform of two sequences gives the same result as the convolution of the said sequences. Now we will generalize this result. If $U(z)$ and $G(z)$ are Z-transforms of $\{u(n)\}$ and $\{g(n)\}$, respectively, then $U(z)G(z)$ is the Z-transform of $\{u(n)\} * \{g(n)\}$. In other words, if $u(n) \leftrightarrow U(z)$ and $g(n) \leftrightarrow G(z)$, then

$$g(n) * u(n) \leftrightarrow G(z)U(z) \tag{4.15}$$

Let $y(n) = g(n) * u(n)$. Using the definition of convolution, we obtain

$$y(n) = \sum_{k=-\infty}^{\infty} g(k)u(n-k)$$

Taking its Z-transform, we obtain

$$Y(z) = \sum_{n=-\infty}^{\infty} \sum_{k=-\infty}^{\infty} g(k)u(n-k)z^{-n} = \sum_{k=-\infty}^{\infty} g(k)z^{-k} \sum_{n=-\infty}^{\infty} u(n-k)z^{-(n-k)}$$

which is nothing but the right-hand side of Eq. 4.15.

The same result is applicable for causal g and u also, as we now show. Let

$$y(n) = \sum_{k=0}^{\infty} g(k)u(n-k)$$

Taking its Z-transform,

$$Y(z) = \sum_{n=0}^{\infty} \sum_{k=0}^{\infty} g(k)u(n-k)z^{-n}$$

With $m = n - k$,

$$Y(z) = \sum_{k=0}^{\infty} g(k)z^{-k} \sum_{n=0}^{\infty} u(n-k)z^{-(n-k)} = \sum_{k=0}^{\infty} g(k)z^{-k} \sum_{m=-k}^{\infty} u(m)z^{-m}$$

Because u is causal, the last term can be written as

$$Y(z) = \sum_{k=0}^{\infty} g(k)z^{-k} \sum_{m=0}^{\infty} u(m)z^{-m}$$

which is nothing but the right side of Eq. 4.15. It is possible to show the ROC condition for convolved sequences, see Problem 4.10.

Example 4.14 Determine the step response of a system with impulse response $g(n) = 0.5^n 1(n)$ using the convolution and Z-transform approaches.

First, we will evaluate the step response using the convolution approach:

$$\begin{aligned} y(n) &= \sum_{r=-\infty}^{\infty} g(r)u(n-r) = \sum_{r=0}^{\infty} 0.5^r u(n-r) \\ &= \sum_{r=0}^{\infty} 0.5^r 1(n-r) = 1(n) \sum_{r=0}^{n} 0.5^r \end{aligned}$$

Note that $1(n)$ appears in the last expression, because, for negative n, the previous summation is zero. As this is a geometric progression, we obtain

$$y(n) = 1(n) \frac{1 - 0.5^{(n+1)}}{1 - 0.5} = (2 - 0.5^n)1(n)$$

Next, we take the Z-transform of u and g and multiply them:

$$
\begin{aligned}
U(z) &= \sum_{j=-\infty}^{\infty} u(j)z^{-j} = \sum_{j=0}^{\infty} 0.5^j z^{-j} \\
&= [1 + 0.5z^{-1} + (0.5z^{-1})^2 + \cdots] = \frac{1}{1 - 0.5z^{-1}}
\end{aligned}
$$

where we have assumed $|z| > 0.5$. We also have

$$
G(z) = \sum_{j=-\infty}^{\infty} g(j)z^{-j} = \sum_{j=0}^{\infty} z^{-j} = [1 + z^{-1} + z^{-2} + \cdots] = \frac{1}{(1 - z^{-1})}
$$

where we have assumed $|z| > 1$. On multiplying these two expressions, we obtain

$$
Y(z) = G(z)U(z) = \frac{1}{1 - z^{-1}} \frac{1}{1 - 0.5z^{-1}}, \quad |z| > 1
$$

Note that we have taken ROC to be $|z| > 1$, which lies in ROC of both G and U. A procedure to handle expressions such as the one obtained above will be explained in Sec. 4.4. For the time being, it is easy to check that the above expression is equivalent to

$$
Y(z) = \frac{2}{1 - z^{-1}} - \frac{1}{1 - 0.5z^{-1}}, \quad |z| > 1
$$

On inverting this expression, we obtain

$$
y(n) = 2 \times 1(n) - 0.5^n 1(n) = (2 - 0.5^n)1(n)
$$

Thus the two expressions for $y(n)$ are identical. ∎

4.2.7 Differentiation

We have the following useful result that deals with derivatives:

$$
\begin{aligned}
&u(n) \leftrightarrow U(z) \text{ with ROC} = R_u \text{ then} \\
&nu(n) \leftrightarrow -z\frac{dU(z)}{dz} \text{ with ROC} = R_u
\end{aligned}
\tag{4.16}
$$

Differentiating the Z-transform of $u(n)$, we obtain

$$
\frac{dU(z)}{dz} = \frac{d}{dz} \sum_{n=-\infty}^{\infty} u(n)z^{-n} = -\sum_{n=-\infty}^{\infty} nu(n)z^{-n-1} = -z^{-1} \sum_{n=-\infty}^{\infty} nu(n)z^{-n}
$$

Therefore,

$$
-z\frac{dU(z)}{dz} = \sum_{n=-\infty}^{\infty} nu(n)z^{-n}
$$

which is what we have to show. We will present an example to illustrate this approach.

Example 4.15 Invert

$$U(z) = \log(1 + az^{-1}), \quad |z| > |a|$$

Differentiating both sides with respect to z, we obtain

$$-z\frac{dU(z)}{dz} = \frac{az^{-1}}{1 + az^{-1}}, \quad |z| > |a|$$

We need to find the inverse Z-transform of the right-hand side. But we know that

$$(-a)^n 1(n) \leftrightarrow \frac{1}{1 + az^{-1}}, \quad |z| > |a|$$

Multiplying both sides by a and applying the shifting theorem, we obtain

$$a(-a)^{n-1} 1(n-1) \leftrightarrow \frac{az^{-1}}{1 + az^{-1}}, \quad |z| > |a|$$

The left-hand side is equal to $-(-a)^n 1(n-1)$. By Eq. 4.16, it should be equal to $nu(n)$. Equating the two, we obtain

$$u(n) = -\frac{1}{n}(-a)^n 1(n-1)$$

∎

We can also carry out the differentiations with respect to a. By successively differentiating

$$a^n 1(n) \leftrightarrow \frac{z}{z-a} = \sum_{n=0}^{\infty} a^n z^{-n}, \quad \left|az^{-1}\right| < 1 \tag{4.17}$$

with respect to a, we obtain

$$\frac{(p-1)!z}{(z-a)^p} = \sum_{n=0}^{\infty} n(n-1)\dots(n-p+2)a^{n-p+1}z^{-n} \tag{4.18}$$

see Problem 4.9. By substituting $p = 2$, 3, respectively, we obtain

$$\begin{aligned} na^{n-1}1(n) &\leftrightarrow \frac{z}{(z-a)^2}, \quad |z| > |a| \\ n(n-1)a^{n-2}1(n) &\leftrightarrow \frac{2z}{(z-a)^3}, \quad |z| > |a| \end{aligned} \tag{4.19}$$

We will now illustrate the utility of this development with an example.

Example 4.16 Determine the Z-transform of $n^2 1(n)$.

We first split the given function into two terms:

$$n^2 1(n) = [n(n-1) + n]\, 1(n)$$

We find that Eq. 4.19 is applicable with $a = 1$. We obtain

$$n^2 1(n) \leftrightarrow \frac{2z}{(z-1)^3} + \frac{z}{(z-1)^2} = \frac{z^2+z}{(z-1)^3}, \quad |z| > 1$$

∎

If a were not 1, the above approach could still be used, with a little more work. This is illustrated in the next example.

Example 4.17 Find the Z-transform of $n^2 a^n 1(n)$.

First, we will split $n^2 a^n$ into a convenient form:

$$\begin{aligned} n^2 a^n 1(n) &= [n(n-1) + n]\, a^{n-2} a^2 \\ &= \left[n(n-1)a^{n-2} + n a^{n-1} a^{-1}\right] a^2 \end{aligned}$$

We will now use Eq. 4.19 and obtain

$$\begin{aligned} n^2 a^n 1(n) &\leftrightarrow \left[\frac{2z}{(z-a)^3} + \frac{z}{(z-a)^2} a^{-1}\right] a^2, \quad |z| > |a| \\ &= \frac{2za^2 + az(z-a)}{(z-a)^3} = \frac{az(z+a)}{(z-a)^3}, \quad |z| > |a| \end{aligned}$$

This reduces to the result of Example 4.16 for $a = 1$.

∎

4.2.8 Z-Transform of Folded or Time Reversed Functions

In control applications, signals defined in the time instant $n \geq 0$ only are used. On the other hand, identification techniques require the concept of signals being defined over negative n, see, for example, Sec. 6.3.5. This motivates the next result. If the Z-transform of $u(n)$ is $U(z)$, the Z-transform of $u(-n)$ is $U(z^{-1})$. If ROC of $U(z)$ is given by $|z| > |a|$, ROC of $U(z^{-1})$ is given by $|z| < |a^{-1}|$.

$$\begin{aligned} u(-n) \leftrightarrow &= \sum_{n=-\infty}^{\infty} u(-n) z^{-n} \\ &= \sum_{m=-\infty}^{\infty} u(m) z^{m}, \text{ where } m = -n \\ &= \sum_{m=-\infty}^{\infty} u(m)(z^{-1})^{-m} = U(z^{-1}) \end{aligned}$$

If ROC of u is given by $|z| > |a|$, it implies that $u(n) = a^n 1(n)$ and hence

$$U(z) = \sum_{n=0}^{\infty} a^n z^{-n} = 1 + az^{-1} + a^2 z^{-2} + a^3 z^{-3} + \cdots$$

Substituting z^{-1} in the place of z, we see that

$$U(z^{-1}) = 1 + az + a^2 z^2 + a^3 z^3 + \cdots$$

converges if $|az| < 1$ or $|z| < |1/a|$.

4.3 Transfer Function

The transfer function of a system is defined as the Z-transform of its impulse response. For example, if $g(n)$ is the impulse response, the transfer function is given by the Z-transform of g and it will be denoted by $G(z)$. Similarly, given the transfer function $G(z)$ of a system, its impulse response is denoted by $g(n)$. This is denoted by the following expression:

$$g(n) \leftrightarrow G(z)$$

Now suppose an arbitrary signal $\{u(n)\}$ is applied to such a system. The output $\{y(n)\}$ is given as

$$\{y(n)\} = \{g(n)\} * \{u(n)\}$$

Using the Z-transform of convolutions given by Eq. 4.15, we obtain

$$Y(z) = G(z)U(z)$$

In Sec. 3.3.2, we introduced the notion of FIR systems. Because FIR systems have a finite number of terms, the corresponding transfer function will be a polynomial in powers of z^{-1} with a finite number of terms. For example, if $\{g(n)\}$ has only three nonzero terms, say g_0, g_1 and g_2, the transfer function is given by Z-transform of $\{g(n)\}$, *i.e.*,

$$G(z) = g_0 + g_1 z^{-1} + g_2 z^{-2}$$

Because the above is a polynomial in z^{-1}, these systems are also known as *all zero systems*, even though, when written in powers of z, $G(z)$ has two poles at $z = 0$, as can be seen from the following:

$$G(z) = \frac{g_0 z^2 + g_1 z + g_2}{z^2}$$

In other words, the poles at $z = 0$ don't count. IIR systems, also defined in Sec. 3.3.2, have an infinite number of terms in the impulse response. The Z-transform of the impulse response of these systems generally results in a ratio of two polynomials. Indeed, all the impulse responses with infinite terms presented so far in this chapter have given rise to a ratio of two polynomials. Transfer functions of the following type,

$$G(z) = \frac{1}{A(z)}$$

where $A(z)$ is a polynomial in z^{-1}, are known as all pole systems. This is in spite of the fact that when written as powers of z, $G(z)$ will have zeros at $z = 0$.

4.3.1 Gain of a Transfer Function

The gain of a transfer function of a stable system to a unit step input is an important parameter, especially in control applications. Let the Z-transforms of input to and output from a system be $U(z)$ and $Y(z)$, respectively. Let $G(z)$, the Z-transform of the impulse response $g(n)$, be the transfer function of the system. Because $u(n)$ is a unit step sequence, $U(z) = z/(z-1)$. We obtain

$$Y(z) = G(z)U(z) = G(z)\frac{z}{z-1} \tag{4.20}$$

By applying the final value theorem,

$$\lim_{n\to\infty} y(n) = \lim_{z\to 1} \frac{z-1}{z} G(z) \frac{z}{z-1} = G(1) \tag{4.21}$$

where, as usual, $y(n)$ is the response y at the sampling instant n. As a result, the steady state gain of a stable system to a unit step input is simply $G(1)$.

4.3.2 Transfer Function of Connected Systems

We will often connect linear systems in series and in parallel. In the case of parallel interconnection of LTI systems, the impulse response is added. For example, we obtain from Fig. 3.10 on page 51,

$$g(n) = g_1(n) + g_2(n)$$

and from the linearity of the Z-transform,

$$G(z) = G_1(z) + G_2(z)$$

where G_1 and G_2 are respectively the Z-transforms of g_1 and g_2. In the case of LTI systems in series, their impulse responses are convolved. For example, from Fig. 3.9 on page 50, we obtain

$$g(n) = g_1(n) * g_2(n)$$

and the corresponding transfer functions satisfy the convolution property

$$G(z) = G_1(z)G_2(z)$$

The above properties help us to arrive at a transfer function of a feedback loop that arises in control applications. Consider the feedback loop given in Fig. 4.6. The different signals appearing in this system representation are given next:

$r(n)$	reference trajectory or setpoint or reference input
$e(n)$	error between reference and actual value
$u(n)$	control variable or manipulated variable
$y(n)$	output variable or controlled variable

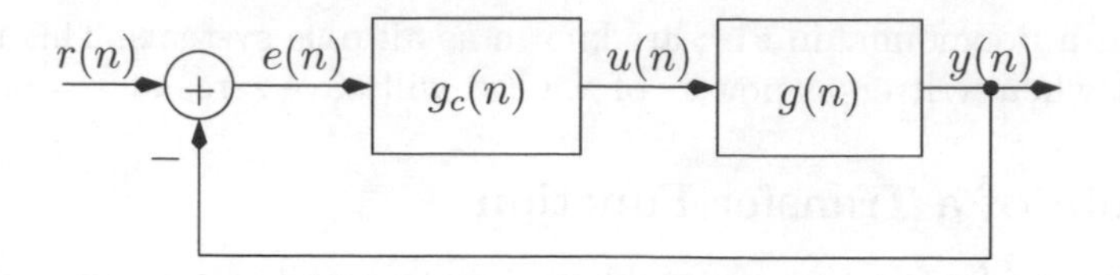

Figure 4.6: Feedback control

Analysis of this system in the Z-transform domain is straightforward. As usual, let the capitals indicate the Z-transform of the corresponding discrete time signal represented with lower case letters. We obtain

$$\begin{aligned} Y(z) &= G(z)U(z) \\ &= G(z)G_c(z)E(z) \\ &= G(z)G_c(z)(R(z) - Y(z)) \end{aligned}$$

Bringing all terms involving Y on one side, we obtain

$$Y(z) + G(z)G_c(z)Y(z) = G(z)G_c(z)R(z)$$

Solving this for Y, we obtain

$$Y(z) = \frac{G(z)G_c(z)}{1 + G(z)G_c(z)} R(z)$$

We define the closed loop transfer functions as

$$T(z) \triangleq \frac{G(z)G_c(z)}{1 + G(z)G_c(z)} \tag{4.22}$$

The closed loop transfer function maps the reference input $R(z)$ to the output variable $Y(z)$. An important objective in control design is to make $T(z)$ well behaved. Using similar arguments, the relation between the Z-transform of $e(n)$ and $r(n)$ is given by

$$E(z) = S(z)R(z) \tag{4.23}$$

where

$$S(z) = \frac{1}{1 + G(z)G_c(z)} \tag{4.24}$$

The use of Z-transforms allows us to replace time domain operations, such as convolution and time shifting with algebraic equations. The use of Z-transforms to convert system descriptions to algebraic equations is also useful in analyzing interconnections of LTI systems, such as series, parallel and feedback interconnections. This comes in handy while designing controllers. We conclude this section with a discussion of how Z-transforms can be applied to state space models.

4.3.3 Z-Transform of Discrete Time State Space Systems

In this section, we would like to study systems described by state space models. In addition to the advantages already explained, Z-transforming the state space system helps clearly state the implicit assumptions, including those in the nonzero initial conditions. Consider the standard state space system,

$$x(n+1) = Ax(n) + Bu(n) \tag{4.25}$$

$$y(n) = Cx(n) + Du(n) \tag{4.26}$$

with

$$x(0) = x_0 \tag{4.27}$$

The state space model is defined only for $n > 0$. For example, the state equation, given by Eq. 4.25, is not meant to be used to calculate $x(0)$ by setting $n = -1$. Because of this, the model does not clearly explain the state of the system before $n = 0$ and what happens at the time of transition, namely at $n = 0$. Nevertheless, our definition of Z-transform, given by Eq. 4.5 on page 65, requires values from $-\infty$. As a result, we are forced to explain what happens for $n \leq 0$. As a first step, we rewrite the state equation as suggested by [17]. We obtain

$$x(n+1) = Ax(n) + Bu(n) + \delta(n+1)x_0 \tag{4.28}$$

If, in addition, we assume that our system is initially at rest (see Sec. 3.1.3) and $u(n)$ is a causal sequence (see Sec. 3.3.5), the system is defined for all times, as we now describe.

Causal sequence implies that $u(n) = 0$ for all negative n. Initial rest implies that the state $x(n)$ is zero for negative n. We will now evaluate the validity of Eq. 4.28 for different values of n:

1. For all $n \leq -2$, both sides are zero.
2. For $n = -1$, it gets reduced to Eq. 4.27.
3. For $n \geq 0$, it is identical to Eq. 4.25.

Thus Eq. 4.28 and the conditions of initial rest and causal u present a clear picture: all the variables are zero prior to $n = 0$ and, somehow, x takes the value of x_0 at $n = 0$.

We are now in a position to take Z-transform of Eq. 4.28. We obtain

$$zX(z) = AX(z) + BU(z) + x_0 z$$

$$(zI - A)X(z) = BU(z) + x_0 z$$

As in the scalar case, we have to make a choice for z once again. We choose z in such a way that $zI - A$ is invertible. This selection procedure is explained in Problem 4.11.

$$X(z) = (zI - A)^{-1}BU(z) + z(zI - A)^{-1}x_0 \tag{4.29}$$

The Z-transform of Eq. 4.26 is

$$Y(z) = CX(z) + DU(z)$$

Substituting the expression for $X(z)$ from Eq. 4.29, we obtain

$$\begin{aligned} Y(z) &= C(zI - A)^{-1}BU(z) + DU(z) + C(zI - A)^{-1}zx(0) \\ &\triangleq \underbrace{G_u(z)U(z)}_{\text{zero state response}} + \underbrace{G_x(z)x_0}_{\text{zero input response}} \end{aligned}$$

The first term on the right-hand side is due to nonzero input. The second term is due to nonzero initial state. In Eq. 3.46 on page 56, we derived an expression for $y(n)$ in the time domain. Because $Y(z)$ is the Z-transform of $y(n)$, we see that the RHS of the above equation is the Z-transform of RHS of Eq. 3.46. Comparing the zero state response terms, we obtain

$$C(zI - A)^{-1}B \leftrightarrow CA^{n-1}B$$

Example 4.18 Find the transfer function of the antenna control system, discussed in Example 2.4 on page 25.

Because the initial conditions are zero, we obtain

$$\begin{aligned} G(z) &= C(zI - A)^{-1}B \\ &= \begin{bmatrix} 0 & 1 \end{bmatrix} \begin{bmatrix} z - 0.9802 & 0 \\ -0.19801 & z - 1 \end{bmatrix}^{-1} \begin{bmatrix} 0.0198 \\ 0.001987 \end{bmatrix} \\ &= \frac{\begin{bmatrix} 0 & 1 \end{bmatrix}}{(z-1)(z-0.9802)} \begin{bmatrix} z - 1 & 0 \\ 0.19801 & z - 0.9802 \end{bmatrix} \begin{bmatrix} 0.0198 \\ 0.001987 \end{bmatrix} \\ &= \frac{\begin{bmatrix} 0.19801 & z - 0.9802 \end{bmatrix}}{(z-1)(z-0.9802)} \begin{bmatrix} 0.0198 \\ 0.001987 \end{bmatrix} \\ &= \frac{0.001987z + 0.0019732}{(z-1)(z-0.9802)} \\ &= 0.001987 \frac{z + 0.9931}{(z-1)(z-0.9802)} \end{aligned}$$

M 4.4 illustrates how this calculation is carried out in Matlab.

∎

These calculations could be more involved in some applications. We illustrate this by taking the Z-transform of the supply chain problem, given in Sec. 2.3.2.

Example 4.19 Determine the transfer function between production release PREL and sales SALES, in the supply chain problem discussed in Sec. 2.3.2.

By taking the Z-transform of Eq. 2.31–2.35, we obtain

$$\mathsf{FD}(z) = \frac{\delta\,\rho\,\mathsf{SALES}(z)}{z-1+\delta\rho}$$

$$\mathsf{INV}(z) = \frac{z\delta(\mathsf{PRATE}(z)-\mathsf{SALES}(z))}{z-1}$$

$$\mathsf{WIP}(z) = \frac{z\delta(\mathsf{PREL}(z)-\mathsf{PRATE}(z))}{z-1}$$

$$\mathsf{PREL}(z) = \frac{(1+L\alpha+\beta)\mathsf{FD}(z)-\alpha\,\mathsf{WIP}(z)-\beta\,\mathsf{INV}(z)}{z}$$

$$\mathsf{PRATE}(z) = \frac{\mathsf{PREL}(z)}{z^L}$$

Solving the equations simultaneously, we obtain the following relation between PREL and SALES:

$$\frac{\mathsf{PREL}(z)}{\mathsf{SALES}(z)} = \frac{z^L\delta[(z-1)(1+L\alpha)\rho+\beta(z-1-\rho+z\rho+\delta\rho)]}{[z^{L+1}+(\beta-\alpha)\delta+z^L(\alpha\delta-1)](z-1+\delta\alpha)} \tag{4.30}$$

∎

In Matlab, the impulse response of a discrete time transfer function can be obtained by the function `dimpulse`.

4.3.4 Jury's Stability Rule

We have seen that for a causal LTI system to be stable, its poles should be inside the unit circle. While designing controllers, we would like to know the range of control parameters for which the closed loop system is stable. Jury's test provides an analytical procedure to determine this range. We present it in this section and give an example to illustrate the application.

Let the transfer function of a causal LTI system be given as

$$G(z) = \frac{B(z)}{A(z)}$$

with $\mathrm{d}B \leq \mathrm{d}A$, where d denotes the degree of the polynomial. Suppose that A is given by

$$A(z) = a_0 z^n + a_1 z^{n-1} + \cdots + a_n \tag{4.31}$$

Form the entries as in Table 4.1. Here, the first row consists of the coefficients from Eq. 4.31. The coefficients in the second row have been obtained by reversing those in the first row. The third row is obtained by multiplying the second row by $b_n = a_n/a_0$ and subtracting it from the first row. It is easy to see that the last coefficient in the third row will be zero. As a result, only the rest of the elements have been written in the third row. The fourth row is obtained by reversing the third row. The fifth row (not shown) is obtained by multiplying the fourth row by b_{n-1} and subtracting from the third row. Here, b_{n-1} is the ratio of the last entry of third row to the fourth row.

Table 4.1: Jury's table

a_0	a_1	$\cdots$	a_{n-1}	a_n	
a_n	a_{n-1}	$\cdots$	a_1	a_0	$b_n = a_n/a_0$
a_0^{n-1}	a_1^{n-1}	$\cdots$	a_{n-1}^{n-1}		
a_{n-1}^{n-1}	a_{n-2}^{n-1}	$\cdots$	a_0^{n-1}		$b_{n-1} = a_{n-1}^{n-1}/a_0^{n-1}$
$\vdots$					
a_0^0					

As a result, the fifth row is one term shorter than the third or fourth row. This process is continued until only the first entry in a row is nonzero.

Jury's stability rule can be stated as follows. If $a_0 > 0$, Eq. 4.31 has all its roots inside the unit circle if and only if $a_0^m > 0$, where $m = 0, 1, \ldots, n-1$. If no a_0^m is zero, the number of negative a_0^m is equal to the number of roots outside the unit circle.

We illustrate Jury's stability criterion with an example.

Example 4.20 Determine the stability region for the transfer function obtained in Example 4.19 with $L = 2$, $\rho = 1$ and $\delta = 1$.

With L, ρ and δ as chosen above, Eq. 4.30 becomes

$$\frac{\mathsf{PREL}(z)}{\mathsf{SALES}(z)} = \frac{z\left[(z-1)(1+2\alpha) + \beta(2z-1)\right]}{z^3 + (\alpha-1)z^2 + (\beta-\alpha)}$$

The characteristic polynomial is given by the denominator of the above transfer function:

$$\phi_{cl} = z^3 + (\alpha-1)z^2 + (\beta-\alpha) \tag{4.32}$$

We have the following values for variables:

$$n = 3,\ a_1 = \alpha - 1,\ a_2 = 0,\ a_3 = \beta - \alpha$$

Jury's table is constructed as in Table 4.2, where we have shown only the first six rows. Jury's table ends with one more row with the following entry:

$$1 - (\beta-\alpha)^2 - \frac{(\beta-\alpha)^2(\alpha-1)^2}{1-(\beta-\alpha)^2} - \frac{\left[(\alpha-1) + \frac{(\beta-\alpha)(\alpha-1)^2}{1-(\beta-\alpha)^2}\right]^2}{1-(\beta-\alpha)^2 - \frac{(\beta-\alpha)^2(\alpha-1)^2}{1-(\beta-\alpha)^2}} \tag{4.33}$$

We apply Jury's stability condition now. Because $a_0 = 1$, the conditions to be satisfied for stability are

$$a_0^2 = 1 - (\beta-\alpha)^2 > 0$$

$$a_0^1 = 1 - (\beta-\alpha)^2 - \frac{(\beta-\alpha)^2(\alpha-1)^2}{1-(\beta-\alpha)^2} > 0$$

Table 4.2: Jury's table for the problem discussed in Example 4.20

1	$\alpha - 1$	0	$\beta - \alpha$
$\beta - \alpha$	0	$\alpha - 1$	1
$1 - (\beta - \alpha)^2$	$\alpha - 1$	$-(\beta - \alpha)(\alpha - 1)$	
$-(\beta - \alpha)(\alpha - 1)$	$\alpha - 1$	$1 - (\beta - \alpha)^2$	
$1 - (\beta - \alpha)^2 - \frac{(\beta - \alpha)^2(\alpha - 1)^2}{1 - (\beta - \alpha)^2}$	$(\alpha - 1) + \frac{(\beta - \alpha)(\alpha - 1)^2}{1 - (\beta - \alpha)^2}$		
$(\alpha - 1) + \frac{(\beta - \alpha)(\alpha - 1)^2}{1 - (\beta - \alpha)^2}$	$1 - (\beta - \alpha)^2 - \frac{(\beta - \alpha)^2(\alpha - 1)^2}{1 - (\beta - \alpha)^2}$		

and that the expression given in Eq. 4.33 is greater than zero. We have mentioned in Sec. 2.3.2 that we have to assign the values of α and β so as to achieve a certain performance. One such performance indicator is stability: these parameters have to be chosen so as to make the system stable.

Although the conditions for stability look complicated, we can make some quick observations. For example, if we choose $\alpha = 0$ and $\beta = 1$, a_0^2 becomes 0, violating the condition for stability. This implies that even if we make a complete account of the shortcoming in the inventory ($\beta = 1$), if we ignore the work in process ($\alpha = 0$), the system will become unstable. Often it is difficult to get information on the work in process and hence one may like to ignore this factor. The above analysis suggests that it is suicidal to do so. For more details on how such an approach is used in supply chain problems, the reader should refer to studies such as [58] or [14].

A symbolic computing program is usually used for this purpose. An alternative approach is to assign values for the model parameters and to calculate the zeros of the characteristic equation, using a Matlab routine such as `roots`. In other words, one assigns values for α and β in Eq. 4.32, finds the roots, and repeats this procedure.

∎

4.4 Inverse of Z-Transform

As mentioned earlier, Z-transformation is carried out to simplify convolution. After completing all the calculations in the Z domain, we have to map the results back to the time domain so as to be of use. We will give a simple example next to explain this.

Example 4.21 How would you implement a system, whose transfer function is given by $G(z) = 1/(1 - 0.5z^{-1})$?

Let the input and the output be $e(n)$ and $y(n)$, respectively. We have

$$U(z) = G(z)E(z) = \frac{1}{1 - 0.5z^{-1}}E(z)$$

Cross multiplying,

$$(1 - 0.5z^{-1})U(z) = E(z)$$

Inverting,

$$u(n) - 0.5u(n-1) = e(n)$$

Thus, we obtain the system's output $u(n)$ as

$$u(n) = 0.5u(n-1) + e(n)$$

This equation says that the current output $u(n)$ is a sum of the previous output $u(n-1)$ and the current input $e(n)$. This is a popular way to implement transfer functions in real life.

▮

The above method is known as *inversion*. It is also known as *realization*, because it is through this methodology that we can realize transfer functions in real life.

Now we will see how to come up with methods to invert general transfer functions. In this section, we will present different techniques of inversion. We will present contour integration, partial fraction expansion, combined with table lookup and long division.

4.4.1 Contour Integration

We now discuss how to obtain the sequence $u(n)$ by contour integration, given its Z-transform. Recall that the Z-transform is defined by

$$U(z) = \sum_{k=-\infty}^{\infty} u(k)z^{-k} \tag{4.34}$$

Let us multiply both sides by z^{n-1} and integrate over a closed contour within ROC of $U(z)$; let the contour enclose the origin. We have

$$\oint_C U(z)z^{n-1}dz = \oint_C \sum_{k=-\infty}^{\infty} u(k)z^{n-1-k}dz$$

where C denotes the closed contour within ROC, taken in a counterclockwise direction. As the curve C is inside ROC, the sum converges on every part of C and, as a result, the integral and the sum on the right-hand side can be interchanged. The above equation becomes

$$\oint_C U(z)z^{n-1}dz = \sum_{k=-\infty}^{\infty} u(k) \oint_C z^{n-1-k}dz \tag{4.35}$$

Now we make use of the *Cauchy integral theorem*, according to which

$$\frac{1}{2\pi j}\oint_C z^{n-1-k} = \begin{cases} 1 & k = n \\ 0 & k \neq n \end{cases} \tag{4.36}$$

Here, C is any contour that encloses the origin. Using the above equation, the right-hand side of Eq. 4.35 becomes $2\pi j u(n)$ and hence we obtain the formula

$$u(n) = \frac{1}{2\pi j}\oint_C U(z) z^{n-1} dz \tag{4.37}$$

Depending on the nature of $U(z)$, the above integral can be simplified further. The *Cauchy residue theorem* can be used for this. Let $f(z)$ be a function of the complex variable z and let C be a closed contour in the z plane. If the derivative df/dz exists on and inside the contour C and if $f(z)$ has no pole at $z = z_0$, then

$$\frac{1}{2\pi j}\oint_C \frac{f(z)}{z - z_0} = \begin{cases} f(z_0) & \text{if } z_0 \text{ is inside } C \\ 0 & \text{if } z_0 \text{ is outside } C \end{cases} \tag{4.38}$$

If the $(m+1)$th order derivative of $f(z)$ exists and if $f(z)$ has no pole at $z = z_0$, then

$$\frac{1}{2\pi j}\oint \frac{f(z)}{(z - z_0)^m} dz = \begin{cases} \dfrac{1}{(m-1)!} \left. \dfrac{d^{m-1} f(z)}{dz^{m-1}} \right|_{z=z_0} & \text{if } z_0 \text{ is inside } C \\ 0 & \text{if } z_0 \text{ is outside } C \end{cases} \tag{4.39}$$

We can use Eq. 4.38 and Eq. 4.39 to determine the values of useful contour integrals. For example, suppose that the integrand of the contour integral is $G(z) = f(z)/A(z)$, where $f(z)$ has no pole inside the contour and $A(z)$ is a polynomial with simple zeros at z_k, $n \geq k \geq 1$. Then, the contour integral is given by

$$\begin{aligned} \frac{1}{2\pi j}\oint \frac{f(z)}{A(z)} dz &= \frac{1}{2\pi j}\oint \left[\sum_{k=1}^{n} \frac{f_k(z)}{z - z_k} \right] dz = \sum_{k=1}^{n} \frac{1}{2\pi j}\oint \frac{f_k(z)}{z - z_k} dz \\ &= \sum_{k=1}^{n} f_k(z_k) \end{aligned} \tag{4.40}$$

where

$$f_k(z) = (z - z_k) G(z) = (z - z_k) \left. \frac{f(z)}{A(z)} \right|_{z=z_k} \tag{4.41}$$

We call $f_k(z)$ the *residue* at the pole z_k. Thus, the contour integral is the sum of the residues of all the poles inside the contour. When higher order poles are present, one will have to make use of Eq. 4.39.

Example 4.22 Using the contour integration approach, calculate the inverse Z-transform of

$$U(z) = \frac{1}{1 - az^{-1}}, \quad |z| > |a|$$

We will now use Eq. 4.37 to arrive at

$$u(n) = \frac{1}{2\pi j} \oint \frac{z^{n-1}}{1 - az^{-1}} dz$$

where C is a circle of radius $> |a|$. We will first convert all the terms into polynomials in z:

$$u(n) = \frac{1}{2\pi j} \oint \frac{z^n}{z - a} dz \tag{4.42}$$

Let us evaluate this using the Cauchy residue theorem, stated in Eq. 4.38. Because $|z| > a$ is ROC, the closed contour, which is in ROC, will enclose the pole at $z = a$. We will have to consider two cases, one for $n \geq 0$ and another for $n < 0$. For $n < 0$, z^n will give rise to a simple or a multiple pole at $z = 0$.

$n \geq 0$: $f(z) = z^n$, which has no pole at $z = a$ ($z_0 = a$), and hence, using Eq. 4.38, we obtain

$$u(n) = f(a) = a^n$$

$n < 0$: z^n has an nth order pole at $z = 0$, which is also inside C. Thus we have poles at $z = 0$ and at $z = a$. In Eq. 4.37, we will substitute $n = -1, -2, \ldots$, evaluate each integral and thus obtain values of u at these negative n values. For $n = -1$,

$$u(-1) = \frac{1}{2\pi j} \oint_C \frac{1}{z(z-a)} dz$$

The right-hand side is equal to the sum of the residues at $z = a$ and at $z = 0$. Using Eq. 4.38, we obtain

$$u(-1) = \left.\frac{1}{z}\right|_{z=a} + \left.\frac{1}{z-a}\right|_{z=0} = 0$$

We do the same for $n = -2$:

$$u(-2) = \frac{1}{2\pi j} \oint \frac{1}{z^2(z-a)} dz$$

The right-hand side is once again equal to the sum of the residues at $z = a$ and at $z = 0$. For the first integral, we use Eq. 4.38 and for the second, we use Eq. 4.39 to obtain

$$u(-2) = \frac{1}{a^2} + \frac{1}{(2-1)} \frac{d}{dz} \left.\left(\frac{1}{z-a}\right)\right|_{z=0}$$

Differentiating and simplifying, we obtain

$$u(-2) = \frac{1}{a^2} - \left.\frac{1}{(z-a)^2}\right|_{z=a} = \frac{1}{a^2} - \frac{1}{a^2} = 0$$

We can continue this procedure and find that $u(n) = 0, \ \forall n < 0$. As a result, we obtain $u(n) = a^n 1(n)$, as expected. ∎

In the above example, we have taken ROC as $|z| > |a|$. From Sec. 4.1.2, we know that we should obtain a different answer if ROC is taken as $|z| < |a|$. We will illustrate this in the next example.

Example 4.23 Using the contour integration approach, calculate the inverse Z-transform of

$$U(z) = \frac{1}{1 - az^{-1}}, \quad |z| < |a|$$

The closed contour, which has to be inside ROC, will be inside the circle of radius $|a|$. As a result, the contour integral in Eq. 4.42 will have poles only at $z = 0$, and that too when $n < 0$. When $n \geq 0$, there will be no pole inside the closed contour and hence $u(n) = 0$. We will now carry out the calculations for $n < 0$.

For $n = -1$, we obtain

$$u(-1) = \frac{1}{2\pi j} \oint \frac{1}{z} \frac{1}{z - a}$$

Using the Cauchy residue theorem, we obtain

$$u(-1) = \left. \frac{1}{z - a} \right|_{z=0} = -\frac{1}{a}$$

For $n = -2$, we obtain

$$u(-2) = \frac{1}{2\pi j} \oint \frac{1}{z^2} \frac{1}{z - a}$$

Using the Cauchy residue theorem, we obtain

$$u(-2) = \left. \frac{d}{dz} \frac{1}{z - a} \right|_{z=0} = \left. \frac{-1}{(z - a)^2} \right|_{z=0} = -\frac{1}{a^2}$$

For $n = -3$, we obtain

$$u(-3) = \left. \frac{1}{2!} \frac{d^2}{dz^2} \frac{1}{z - a} \right|_{z=0} = \left. \frac{1}{(z - a)^3} \right|_{z=0} = -\frac{1}{a^3}$$

In summary, we obtain $u(n) = -a^n 1(-n - 1)$. Thus, the results of this example and Example 4.22 are in agreement with the results of Sec. 4.1.2. ∎

We would like to remark that depending on ROC and therefore the selection of the closed contour, we would obtain different inverse Z-transforms. Signal processing books, such as [49], state the condition on the contour in a general way, so as to accommodate noncausal transfer functions as well. Control texts, such as [44], on the other hand, make the contour enclose all poles of the transfer function, thus ensuring that the result of contour integration is a causal sequence.

The contour integration approach is especially suitable when we want to find the inverse at a few points only. In the next section, we will present a simpler and a more popular method of determining the inverse Z-transform.

Table 4.3: Z-transform of popular discrete time sequences

Time sequence	Z-transform
$\delta(n)$	1
$1(n)$	$\dfrac{z}{z-1}$, $\lvert z\rvert > 1$
$a^n 1(n)$	$\dfrac{z}{z-a}$, $\lvert z\rvert > \lvert a\rvert$
$na^{n-1}1(n)$	$\dfrac{z}{(z-a)^2}$, $\lvert z\rvert > \lvert a\rvert$
$\dfrac{1}{2!}n(n-1)a^{n-2}1(n)$	$\dfrac{z}{(z-a)^3}$, $\lvert z\rvert > \lvert a\rvert$
$\cos \omega n\, 1(n)$	$\dfrac{z(z-\cos\omega)}{(z-e^{j\omega})(z-e^{-j\omega})}$, $\lvert z\rvert > 1$
$\sin \omega n\, 1(n)$	$\dfrac{z\sin\omega}{(z-e^{j\omega})(z-e^{-j\omega})}$, $\lvert z\rvert > 1$
$-a^n 1(-n-1)$	$\dfrac{z}{z-a}$, $\lvert z\rvert < \lvert a\rvert$

4.4.2 Partial Fraction Expansion

Z-transforms of popular discrete time sequences are listed in Table 4.3. From such a table, we can directly read out the sequence, given its Z-transform. In case the given Z-transform is more complicated, it is decomposed into a sum of standard fractions, which appear in Table 4.3. One can then calculate the overall inverse Z-transform using the linearity property. We will restrict our attention to inversion of Z-transforms that are proper.

Because the inverse Z-transform of $z/(z-p)$ is given by p^n, we split $Y(z)/z$ into partial fractions, as given below:

$$\frac{Y(z)}{z} = \frac{A_1}{z-p_1} + \frac{A_2}{z-p_2} + \cdots + \frac{A_m}{z-p_m} \tag{4.43}$$

where we have assumed that $Y(z)/z$ has m simple poles at $p_1, p_2, \ldots, p_m$. The coefficients $A_1, \ldots, A_m$ are known as *residues* at the corresponding poles. The residues are calculated using the formula

$$A_i = (z-p_i)\frac{Y(z)}{z}\bigg|_{z=p_i}, \quad i = 1, \ldots, m \tag{4.44}$$

We will now present a few examples to illustrate this approach, before taking up the case of multiple poles.

Example 4.24 Find the inverse Z-transform of

$$Y(z) = \frac{2z^2+2z}{z^2+2z-3}, \quad |z| > 3$$

Because $Y(z)$ has z as a factor in the numerator, division by z is simplified:

$$\frac{Y(z)}{z} = \frac{2z+2}{(z+3)(z-1)} = \frac{A}{z+3} + \frac{B}{z-1} \tag{4.45}$$

with the region of convergence being $|z| > 3$. Multiply throughout by $z+3$ and let $z = -3$ to obtain

$$A = \left.\frac{2z+2}{z-1}\right|_{z=-3} = \frac{-4}{-4} = 1$$

Next we calculate B. Multiply Eq. 4.45 throughout by $z-1$ and let $z = 1$ to obtain $B = 4/4 = 1$. Thus, we have

$$\frac{Y(z)}{z} = \frac{1}{z+3} + \frac{1}{z-1}, \quad |z| > 3$$

Cross multiplying, we obtain

$$Y(z) = \frac{z}{z+3} + \frac{z}{z-1}, \quad |z| > 3 \tag{4.46}$$

It is straightforward to invert the fractions in this equation. We obtain the inverse of $Y(z)$ as

$$y(n) = (-3)^n 1(n) + 1(n)$$

M 4.6 sets up the problem discussed in this example and invokes M 4.5 to obtain the result. M 4.5 divides $Y(z)$ by z and carries out the residue computation. ∎

In the next example, we will see another approach to solve this problem.

Example 4.25 Determine the inverse Z-transform of the problem discussed in Example 4.24, namely

$$Y(z) = \frac{2z^2+2z}{z^2+2z-3}, \quad |z| > 3$$

after splitting it into a strictly proper transfer function.

Because the degrees of the numerator and the denominator polynomials are equal, we can begin by dividing, as follows:

$$Y(z) = 2 + \frac{-2z+6}{z^2+2z-3}$$

We define

$$Y_1(z) = \frac{-2z+6}{z^2+2z-3} = \frac{-2z+6}{(z+3)(z-1)} = \frac{A}{z+3} + \frac{B}{z-1} \tag{4.47}$$

We will now describe how to calculate the residues A and B. Multiply the above equation by $z-3$, cancel common terms and let $z=3$ to obtain

$$A = \left.\frac{-2z+6}{z-1}\right|_{z=-3} = \frac{12}{-4} = -3$$

We calculate B in a similar manner. Multiply Eq. 4.47 by $z+1$, cancel common terms and let $z=-1$ to obtain $B=4/4=1$. Putting it all together, we obtain,

$$Y(z) = 2 - \frac{3}{z+3} + \frac{1}{z-1}$$

Because the inverse Z-transform of $z/(z+3)$ is $(-3)^n 1(n)$, using the shifting property of Sec. 4.2.2 we see that $1/(z+3)$ has the inverse Z-transform $(-3)^{n-1}1(n-1)$. The same approach is to be used for the third term in the above equation as well. The inverse of $Y(z)$ is obtained as

$$\begin{aligned} y(n) &= 2\delta(n) - 3(-3)^{n-1}1(n-1) + 1^{n-1}1(n-1) \\ &= 2\delta(n) + (-3)^n 1(n-1) + 1(n-1) \end{aligned}$$

Note that we can write $2\delta(n) = (-3)^0\delta(n) + 1^0\delta(n)$. Substituting this in the above expression and using the fact that $1(n) = \delta(n) + 1(n-1)$, we obtain

$$y(n) = (-3)^n 1(n) + 1(n)$$

which is identical to the result obtained in Example 4.24. ∎

Now, we will take up the case of repeated poles. From Eq. 4.19 on page 80, we see that it is useful to have z in the numerator of fractions with multiple poles as well. In view of this, when we have multiple poles, we look for an expansion of the following form:

$$\frac{Y(z)}{z} = \frac{N(z)}{(z-\alpha)^p D_1(z)}$$

where α is not a root of $N(z)$ and $D_1(z)$ is a polynomial not containing α as its zero. On expanding this in partial fractions, we obtain

$$\frac{Y(z)}{z} = \frac{A_1}{z-\alpha} + \frac{A_2}{(z-\alpha)^2} + \cdots + \frac{A_p}{(z-\alpha)^p} + G_1(z)$$

where $G_1(z)$ is a rational that has poles corresponding to those of $D_1(z)$. On multiplying this by $(z-\alpha)^p$, we obtain

$$(z-\alpha)^p \frac{Y(z)}{z} = A_1(z-\alpha)^{p-1} + A_2(z-\alpha)^{p-2} + \cdots + A_p + G_1(z)(z-\alpha)^p \tag{4.48}$$

Substituting $z = \alpha$, we obtain

$$A_p = (z-\alpha)^p \frac{Y(z)}{z}\bigg|_{z=\alpha}$$

Now we differentiate Eq. 4.48 once and let $z = \alpha$:

$$A_{p-1} = \frac{d}{dz}\left((z-\alpha)^p \frac{Y(z)}{z}\right)\bigg|_{z=\alpha}$$

Then we differentiate Eq. 4.48 twice and let $z = \alpha$:

$$A_{p-2} = \frac{1}{2!}\frac{d^2}{dz^2}\left((z-\alpha)^p \frac{Y(z)}{z}\right)\bigg|_{z=\alpha}$$

In general, for an arbitrary integer m such that $0 \le m \le p-1$,

$$A_{p-m} = \frac{1}{m!}\frac{d^{p-1}}{dz^{p-1}}\left((z-\alpha)^p \frac{Y(z)}{z}\right)\bigg|_{z=\alpha}$$

Substituting $m = p-1$, we obtain

$$A_1 = \frac{1}{(p-1)!}\frac{d^{p-1}}{dz^{p-1}}\left((z-\alpha)^p \frac{Y(z)}{z}\right)\bigg|_{z=\alpha}$$

We now illustrate this with a few examples:

Example 4.26 Carry out a partial fraction expansion:

$$Y(z) = \frac{z^2+z}{(z-1)^3}$$

Expanding this in partial fractions, we obtain

$$\frac{Y(z)}{z} = \frac{z+1}{(z-1)^3} = \frac{A}{z-1} + \frac{B}{(z-1)^2} + \frac{C}{(z-1)^3} \tag{4.49}$$

On multiplying both sides of Eq. 4.49 by $(z-1)^3$, we obtain

$$z+1 = A(z-1)^2 + B(z-1) + C \tag{4.50}$$

On letting $z = 1$, we obtain $C = 2$. On differentiating Eq. 4.50 with respect to z and letting $z = 1$, we obtain $B = 1$. On differentiating Eq. 4.50 twice with respect to z and letting $z = 1$, we obtain $A = 0$. Substituting these in Eq. 4.49, we arrive at the partial fraction expansion:

$$Y(z) = \frac{z}{(z-1)^2} + \frac{2z}{(z-1)^3}$$

Using Eq. 4.19 on page 80, we obtain its inverse as

$$y(n) = n1(n) + n(n-1)1(n)$$

M 4.7 shows how to solve this problem through Matlab. ∎

Now we will consider a transfer function $Y(z)$ with two poles: one is a simple pole and the other has multiplicity two. If $Y(z)$ is not divisible by z, we can proceed with the methods developed earlier, but without first dividing by z. We illustrate this also in the next example.

Example 4.27 Obtain the inverse of $Y(z)$, defined by

$$Y(z) = \frac{11z^2 - 15z + 6}{(z-2)(z-1)^2}$$

We begin with partial fraction expansion:

$$Y(z) = \frac{B}{z-2} + \frac{A_1}{z-1} + \frac{A_2}{(z-1)^2}$$

Multiplying this equation by $z-2$ and letting $z = 2$, we obtain $B = 20$. Multiplying it by $(z-1)^2$, we obtain the following equation:

$$\frac{11z^2 - 15z + 6}{z-2} = A_1(z-1) + A_2 + B\frac{(z-1)^2}{z-2}$$

Substituting $z = 1$, we obtain $A_2 = -2$. On differentiating once with respect to z and substituting $z = 1$, we obtain

$$A_1 = \left. \frac{(z-2)(22z-15) - (11z^2 - 15z + 6)}{(z-2)^2} \right|_{z=1} = -9$$

Thus we obtain

$$Y(z) = \frac{20}{z-2} - \frac{9}{z-1} - \frac{2}{(z-1)^2}$$

Because we need z in the numerator for easy inversion, we multiply by z:

$$zY(z) = \frac{20z}{z-2} - \frac{9z}{z-1} - \frac{2z}{(z-1)^2}$$

To invert the last term, we make use of Eq. 4.19 on page 80. After inverting and making use of the shifting theorem of Sec. 4.2.2, we arrive at

$$y(n+1) = (20 \times 2^n - 9 - 2n)1(n)$$

Finally, we arrive at the solution we are looking for:

$$y(n) = (20 \times 2^{n-1} - 9 - 2(n-1))1(n-1)$$

M 4.8 shows how to solve this problem through Matlab. ∎

If $Y(z)$ is proper, but not divisible by z, we have to divide the numerator by the denominator and then carry out the partial fraction expansion, to be illustrated in the next example.

Example 4.28 Obtain the inverse of

$$Y(z) = \frac{(z^3 - z^2 + 3z - 1)}{(z-1)(z^2 - z + 1)}$$

Because it is proper, we first divide the numerator by the denominator and obtain

$$Y(z) = \left[1 + \frac{z(z+1)}{(z-1)(z^2-z+1)}\right] \triangleq (1 + Y'(z))$$

As $Y'(z)$ has a zero at the origin, we can divide by z:

$$\frac{Y'(z)}{z} = \frac{z+1}{(z-1)(z^2-z+1)} = \frac{z+1}{(z-1)(z-e^{j\pi/3})(z-e^{-j\pi/3})}$$

Note that complex poles or complex zeros, if any, would always occur in conjugate pairs for real sequences, see Problem 4.13. We obtain

$$Y'(z) = \frac{2}{z-1} - \frac{1}{z-e^{j\pi/3}} - \frac{1}{z-e^{-j\pi/3}}$$

We cross multiply by z and invert:

$$\begin{aligned} Y'(z) &= \frac{2z}{z-1} - \frac{z}{z-e^{j\pi/3}} - \frac{z}{z-e^{-j\pi/3}} \\ &\leftrightarrow \left(2 - e^{j\pi k/3} - e^{-j\pi k/3}\right) 1(k) = \left(2 - 2\cos\frac{\pi}{3}k\right) 1(k) \end{aligned}$$

Recall the fact that $Y(z) = 1 + Y'(z)$. Because the inverse Z-transform of 1 is $\delta(k)$, we obtain

$$y(k) = \delta(k) + \left(2 - 2\cos\frac{\pi}{3}k\right) 1(k)$$

∎

For the sake of completeness, we will now take an example where ROC lies in a ring.

Example 4.29 Determine the inverse Z-transform of

$$Y(z) = \frac{z^2 + 2z}{(z+1)^2(z-2)}, \quad 1 < |z| < 2$$

As the degree of the numerator polynomial is less than that of the denominator polynomial, and as it has a zero at the origin, first divide by z and do a partial fraction expansion:

$$\frac{Y(z)}{z} = \frac{z+2}{(z+1)^2(z-2)} = \frac{A}{z+1} + \frac{B}{(z+1)^2} + \frac{C}{z-2}$$

Multiplying by $z-2$ and letting $z=2$, we obtain $C=4/9$. Multiplying by $(z+1)^2$ and letting $z=-1$, we obtain $B=-1/3$. Multiplying by $(z+1)^2$, differentiating with respect to z and letting $z=-1$, we obtain

$$A = \left[\frac{d}{dz}\left\{\frac{z+2}{z-2}\right\}\right]\Bigg|_{z=-1} = \frac{(z-2)-(z+2)}{(z-2)^2}\Bigg|_{z=-1} = -\frac{4}{9}$$

Combining the above, we obtain

$$\frac{Y(z)}{z} = -\frac{4}{9}\frac{1}{z+1} - \frac{1}{3}\frac{1}{(z+1)^2} + \frac{4}{9}\frac{1}{z-2}$$

Because it is given to be the annulus $1<|z|<2$, and because it does not contain the poles, ROC should be as follows:

$$Y(z) = \underbrace{-\frac{4}{9}\frac{z}{z+1} - \frac{1}{3}\frac{z}{(z+1)^2}}_{|z|>1} + \underbrace{\frac{4}{9}\frac{z}{z-2}}_{|z|<2}$$

We make use of Eq. 4.19 on page 80 to invert the second term and the results of Sec. 4.1.2 to invert the third term. We obtain the inverse as

$$y(n) = -\frac{4}{9}(-1)^n 1(n) + \frac{1}{3}n(-1)^n 1(n) - \frac{4}{9}2^n 1(-n-1)$$

M 4.9 shows how to solve this problem through Matlab. ∎

Sometimes it helps to work directly in powers of z^{-1}. We illustrate this in the next example.

Example 4.30 Invert

$$Y(z) = \frac{3-\frac{5}{6}z^{-1}}{\left(1-\frac{1}{4}z^{-1}\right)\left(1-\frac{1}{3}z^{-1}\right)}, \quad |z| > \frac{1}{3}$$

A note on the notation used here is in order. Even though the right-hand side is a function of z^{-1}, we call it a function of z only.[5]

There are two poles, one at $z=\frac{1}{3}$ and one at $\frac{1}{4}$. As ROC lies outside the outermost pole, the inverse transform is a right handed sequence:

$$Y(z) = \frac{3-\frac{5}{6}z^{-1}}{\left(1-\frac{1}{4}z^{-1}\right)\left(1-\frac{1}{3}z^{-1}\right)} = \frac{A}{1-\frac{1}{4}z^{-1}} + \frac{B}{1-\frac{1}{3}z^{-1}} \tag{4.51}$$

Multiply both sides by $1-\frac{1}{4}z^{-1}$ and let $z=\frac{1}{4}$ to obtain

$$A = \frac{3-\frac{5}{6}z^{-1}}{1-\frac{1}{3}z^{-1}}\Bigg|_{z=\frac{1}{4}} = 1$$

[5] We use this notation throughout this book, unless stated otherwise.

Multiply both sides of Eq. 4.51 by $1 - \frac{1}{3}z^{-1}$ and let $z = \frac{1}{3}$ to obtain

$$B = \left. \frac{3 - \frac{5}{6}z^{-1}}{1 - \frac{1}{4}z^{-1}} \right|_{z=\frac{1}{3}} = 2$$

Substituting in Eq. 4.51, we obtain

$$Y(z) = \frac{1}{1 - \frac{1}{4}z^{-1}} + \frac{2}{1 - \frac{1}{3}z^{-1}} \leftrightarrow \left[\left(\frac{1}{4}\right)^n + 2\left(\frac{1}{3}\right)^n \right] 1(n)$$

M 4.10 shows how to solve this problem through Matlab. ∎

Now we will present another method of inversion. In this, both the numerator and the denominator are written in powers of z^{-1} and we divide the former by the latter through long division. Because we obtain the result in a power series, this method is known as the *power series method.* Wc illustrate it with an example.

Example 4.31 Invert the following transfer function by the power series method:

$$Y(z) = \frac{1}{1 - az^{-1}}, \quad |z| > |a|$$

Now the method of long division is applied:

$$\begin{array}{r|llll}
 & 1 + az^{-1} + a^2z^{-2} + \cdots & & \\
\hline
1 - az^{-1} & 1 & & \\
 & 1 & -az^{-1} & \\
\cline{2-3}
 & & az^{-1} & \\
 & & az^{-1} & -a^2z^{-2} \\
\cline{3-4}
 & & & a^2z^{-2}
\end{array}$$

To summarize,

$$\frac{1}{1 - az^{-1}} = 1 + az^{-1} + a^2z^{-2} + \cdots \tag{4.52}$$

from which it follows that

$$\begin{aligned}
y(n) &= 0, \quad n < 0 \\
y(0) &= 1 \\
y(1) &= a \\
y(2) &= a^2
\end{aligned}$$

Generalizing,

$$y(n) = a^n 1(n)$$

which is in agreement with the result obtained in Sec. 4.1.2. ∎

In the next example, we will explain how to carry out long division in Matlab.

Example 4.32 Obtain a long division solution of Example 4.27 through Matlab.

In Example 4.27, we have obtained the solution as $y(0) = 0$, and for $n \geq 1$, $y(n) = -9 - 2(n-1) + 20 \times 2^{n-1}$. Substituting for n we obtain $y(1) = 11$, $y(2) = 29$, $y(3) = 67$ and $y(4) = 145$.

M 4.11 implements this in Matlab. The command `filter` provides numerical values through long division. The command `impulse` provides the solution in a graphical form.

∎

How does one use the long division method for noncausal sequences? This is explained in the next example.

Example 4.33 In Example 4.31, we have solved for a causal sequence. If, instead, ROC is specified as $|z| < |a|$, we cannot use the expansion of Eq. 4.52. Instead, we would solve the problem as given below:

$$
\begin{array}{r|lll}
 & -a^{-1}z - a^{-2}z^2 - \cdots & & \\
\hline
-az^{-1} + 1 & 1 & & \\
 & 1 & -a^{-1}z & \\
\cline{2-3}
 & & a^{-1}z & \\
 & & a^{-1}z & +a^{-2}z^2 \\
\cline{3-4}
 & & & a^{-2}z^2
\end{array}
$$

From this, we obtain

$$\frac{1}{1 - az^{-1}} = -a^{-1}z - a^{-2}z^2 - \cdots \leftrightarrow -a^n 1(-n-1)$$

which is in agreement with the result obtained in Sec. 4.1.2.

∎

Although we have presented examples of both causal and noncausal systems, in view of the fact that control applications require mostly the former, we will assume causality from now on,[6] unless stated otherwise.

We will next present an example that shows how to convert a parametric model presented in Sec. 3.3.6 into a nonparametric model.

Example 4.34 Determine the impulse response of the system described by Eq. 3.42, reproduced here for convenience

$$y(n) + a_1 y(n-1) = b_1 u(n-1)$$

[6] In view of this observation, ROC will also not be explicitly stated.

Taking the Z-transform of this system and using the notation, as mentioned in Footnote 5 on page 100, we obtain

$$(1 + a_1 z^{-1})Y(z) = b_1 z^{-1} U(z)$$
$$Y(z) = \frac{b_1 z^{-1}}{1 + a_1 z^{-1}} U(z)$$

Using the power series method explained in Example 4.31 for causal systems, we obtain

$$Y(z) = b_1 z^{-1}(1 - a_1 z^{-1} + a_1^2 z^{-2} - \cdots)U(z)$$

Using the shifting theorem, we invert this and obtain

$$y(n) = b_1[u(n-1) - a_1 u(n-2) + a_1^2 u(n-3) - \cdots]$$

which can be written as

$$y(n) = b_1 \sum_{k=0}^{\infty} (-1)^k a_1^k u(n-k-1)$$

∎

In the latter part of this book, we will design controllers to improve the performance of plants of interest. We will often work with transfer functions of the plant. As a result, control design techniques will give rise to controller transfer functions. We cannot implement the controller if we know only its transfer function; we need to derive its time domain equivalent. Deriving the time domain equivalent of the transfer function is known as realization, as mentioned earlier. In the next section, we will look at two approaches to obtaining the time domain equivalent of transfer functions.

4.4.3 Realization

For discrete time systems, the time domain equivalent can be obtained by direct inversion. Suppose that the transfer function to be inverted is given by $G(z)$:

$$G(z) = \frac{b_0 + b_1 z^{-1} + b_2 z^{-2} + b_3 z^{-3} + \cdots}{1 + a_1 z^{-1} + a_2 z^{-2} + a_3 z^{-3} + \cdots} \triangleq \frac{B(z)}{A(z)} \tag{4.53}$$

If the input to this system is $U(z)$ and the corresponding output is $Y(z)$, we obtain

$$Y(z) = \frac{B(z)}{A(z)} U(z)$$

Cross multiplying,

$$A(z)Y(z) = B(z)U(z)$$

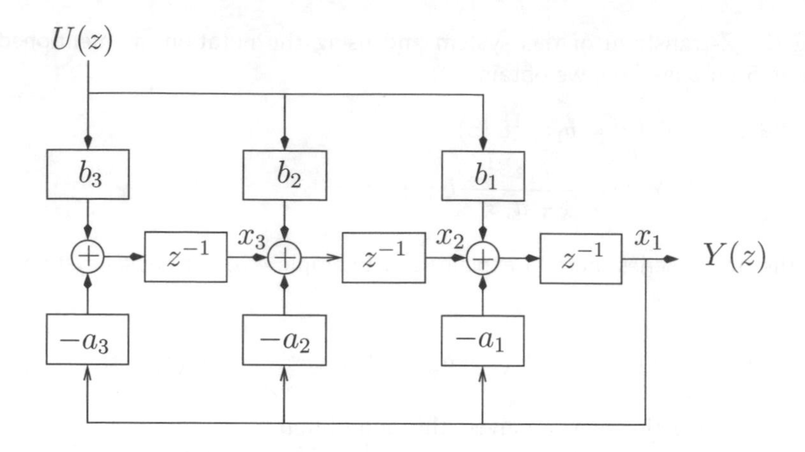

Figure 4.7: State space realization: observer canonical form

Using the expressions for A and B, the above equation becomes

$$\begin{aligned}Y(z) + a_1 z^{-1} Y(z) + a_2 z^{-2} Y(z) + a_3 z^{-3} Y(z) + \cdots \\ = b_0 U(z) + b_1 z^{-1} U(z) + b_2 z^{-2} U(z) + b_3 z^{-3} U(z) + \cdots\end{aligned}$$

From this, we obtain the following expression for $Y(z)$:

$$\begin{aligned}Y(z) &= -a_1 z^{-1} Y(z) - a_2 z^{-2} Y(z) - a_3 z^{-3} Y(z) - \cdots \\ &\quad + b_0 U(z) + b_1 z^{-1} U(z) + b_2 z^{-2} U(z) + b_3 z^{-3} U(z) + \cdots\end{aligned} \tag{4.54}$$

Using the shifting theorem, given by Eq. 4.9 on page 73, the above equation becomes

$$\begin{aligned}y(n) &= -a_1 y(n-1) - a_2 y(n-2) - a_3 y(n-3) + \cdots \\ &\quad + b_0 u(n) + b_1 u(n-1) + b_2 u(n-2) + b_3 u(n-3) + \cdots\end{aligned} \tag{4.55}$$

Suppose that u is the input to and y the output from the controller, whose transfer function is given by $G(z)$. Then, the current control action at time instant n is obtained as a function of previous control actions and current and previous inputs. This approach is applicable only to discrete time systems.

In Sec. 3.4, we have shown how to obtain a transfer function, given the state space description of a system. The next section is devoted to the inverse problem: how to obtain a state space equivalent, given the transfer function? This is another way to realize transfer functions.

Because most real life systems have at least one sample delay (see Sec. 3.4), we will take b_0 to be zero. The resulting expression can be realized as in Fig. 4.7, where we have taken the numerator and denominator degrees to be 3. If the states x_i are ignored for the time being, it is easy to check that this figure is just an implementation of Eq. 4.54.

Because the transfer function z^{-1} denotes a system with the input–output equation $y(n) = u(n-1)$, a block containing z^{-1} is known as the *delay block*. It is clear that if the output of a delay block is $x_1(n)$, its input should be $x_1(n+1)$.

Using this idea, we arrive at the following relations from Fig. 4.7, where the outputs of the delay blocks are denoted as x_1, x_2 and x_3:

$$\begin{aligned} y(k) &= x_1(k) \\ x_1(k+1) &= x_2(k) + b_1 u(k) - a_1 x_1(k) \\ x_2(k+1) &= x_3(k) + b_2 u(k) - a_2 x_1(k) \\ x_3(k+1) &= b_3 u(k) - a_3 x_1(k) \end{aligned}$$

which can be written in the form of state space equations:

$$\begin{aligned} \begin{bmatrix} x_1(k+1) \\ x_2(k+1) \\ x_3(k+1) \end{bmatrix} &= \begin{bmatrix} -a_1 & 1 & 0 \\ -a_2 & 0 & 1 \\ -a_3 & 0 & 0 \end{bmatrix} \begin{bmatrix} x_1(k) \\ x_2(k) \\ x_3(k) \end{bmatrix} + \begin{bmatrix} b_1 \\ b_2 \\ b_3 \end{bmatrix} u(k) \\ y(k) &= \begin{bmatrix} 1 & 0 & 0 \end{bmatrix} \begin{bmatrix} x_1(k) \\ x_2(k) \\ x_3(k) \end{bmatrix} \end{aligned} \tag{4.56}$$

This realization is known as the observer canonical form. It should be pointed out that this state space realization is not unique. A way to arrive at another realization, known as controller canonical form, is given in Problem 4.18. For other forms of realization, the reader is referred to [24].

4.5 Matlab Code

Matlab Code 4.1 To produce $a^n 1(n)$, discussed in Sec. 4.1.2. This code is available at `HOME/Z-trans/matlab/aconv1.m`[7]

```
1  a = 0.9;
2  n = -10:20;
3  y = zeros(size(n));
4  for i = 1:length(n)
5      if n(i)>=0,
6          y(i) = a^n(i);
7      end
8  end
9  axes('FontSize',18);
10 o = stem(n,y);
11 set(o(1),'Marker','.');
12 label('u1',18,'Time(n)','0.9^n1(n)',18)
```

Matlab Code 4.2 To produce $-(a)^n 1(-n-1)$, discussed in Sec. 4.1.2. This code is available at `HOME/Z-trans/matlab/aconv2.m`

```
1  a = 0.9;
2  n = -10:20;
3  y = zeros(size(n));
```

[7] `HOME` stands for `http://www.moudgalya.org/dc/` – first see the software installation directions, given in Appendix A.2.

```
4 for i = 1:length(n)
5     if n(i)<=-1,
6         y(i) = -(a^n(i));
7     end
8 end
9 axes('FontSize',18);
10 o = stem(n,y);
11 set(o(1),'Marker','.');
12 label('u2',18,'Time(n)','-(0.9)^n 1(-n-1)',18)
```

Matlab Code 4.3 To produce pole–zero plots of the form of Fig. 4.3 on page 67. This code is available at HOME/Z-trans/matlab/pz.m

```
1 % Pole zero plot
2 z = [0;5/12];
3 p = [1/2;1/3];
4 A = axes('FontSize',18);
5 [hz,hp,hl] = zplane(z,p);                        % Get handle
6 set(hz,'MarkerSize',12,'Color',[0 0 0]) % Set colour and size
7 set(hp,'MarkerSize',12,'color',[0 0 0])
8 label('Pole-Zero plot',18,'Real(z)','Imaginary(z)',18)
```

Matlab Code 4.4 Discrete transfer function of the continuous state space system, discussed in Example 4.18 on page 86. This code is available at HOME/Z-trans/matlab/disc1.m

```
1 F = [0 0;1 -0.1]; G = [0.1; 0];
2 C = [0 1]; D = 0; Ts = 0.2;
3 sys = ss(F,G,C,D);
4 sysd = c2d(sys,Ts,'zoh');
5 H = tf(sysd)
```

Matlab Code 4.5 Computation of residues, useful for calculations in Sec. 4.4.2. This code is available at HOME/Z-trans/matlab/respol.m

```
1 % This function computes residues for G(z)
2 % If G(z)=0 at z=0, residues are calculated for G(z)/z
3
4 function [res,pol,other] = respol(num,den)
5 len = length(num);
6 if num(len) == 0
7     num = num(1:len-1);
8 end
9 [res,pol,other] = residue(num,den);
```

Matlab Code 4.6 Partial fraction expansion for Example 4.24 on page 94. This code is available at HOME/Z-trans/matlab/respol1.m

```
% 2z^2 + 2z
% G(z) = ---------------
% z^2 + 2z - 3

num = [2 2 0];
den = [1 2 -3];
[res,pol] = respol(num,den) % respol is user defined
```

Matlab Code 4.7 Partial fraction expansion for Example 4.26 on page 97. This code is available at `HOME/Z-trans/matlab/respol2.m`

```
% z^2 + z A B C
% G(z) = ----------- = ------- + --------- + ---------
% (z - 1)^3 (z - 1) (z - 1)^2 (z - 1)^3
num = [1 1 0];
den = conv([1 -1],conv([1 -1],[1 -1])); % poly multiplication
[res,pol] = respol(num,den)

%% Output Interpretation:
%% res =
%% 0.0000 A = 0
%% 1.0000 B = 1
%% 2.0000 C = 2
%% pol =
%% 1.0000 (z - 1)
%% 1.0000 (z - 1)^2
%% 1.0000 (z - 1)^3
```

Matlab Code 4.8 Partial fraction expansion for Example 4.27 on page 98. This code is available at `HOME/Z-trans/matlab/respol3.m`

```
% 11z^2 - 15z + 6 A1 A2 B
% G(z) = ----------------- = ------- + --------- + -------
% (z - 2)(z - 1)^2 (z - 1) (z - 1)^2 (z - 2)

num = [11 -15 6];
den = conv([1 -2],conv([1 -1],[1 -1]));
[res,pol] = respol(num,den) %User defined function

%% res =
%% 20.0000 <------
%% -9.0000 <-----|--------
%% -2.0000 <-----|--------|--------
%% | | |
%% pol = | | |
%% 2.0000 <------ | |
%% 1.0000 <-------------- |
%% 1.0000 <----------------------
```

Matlab Code 4.9 Partial fraction expansion for Example 4.29 on page 99. This code is available at `HOME/Z-trans/matlab/respol5.m`

```
1 %                  z^2 + 2z
2 % G(z)  =  --------------------
3 %              (z + 1)^2 (z - 2)
4
5 num = [1 2 0];
6 den = conv(conv([1 1],[1 1]),[1 -2]);
7 [res,pol] = respol(num,den)
```

Matlab Code 4.10 Partial fraction expansion for Example 4.30 on page 100. This code is available at `HOME/Z-trans/matlab/respol6.m`

```
1 % Coefficients are in ascending power of z^-1
2 %               3 - (5/6)z^-1                 A                B
3 % G = ------------------------- = ----------- + -----------
4 %       (1-(1/4)z^-1)(1-(1/3)z^-1)   1-(1/4)z^-1    1-(1/3)z^-1
5
6 num = [3 -5/6];
7 den = conv([1 -1/4],[1 -1/3]); %Polynomial multiplication
8 [res,pol,other] = residuez(num,den)
```

Matlab Code 4.11 Long division of the problems discussed in Example 4.32 on page 102. This code is available at `HOME/Z-trans/matlab/division.m`

```
1 num = [11 -15 6];
2 den = conv([1 -2], conv([1 -1],[1 -1]));
3 u = [1 zeros(1,4)];
4 y = filter(num,den,u)
5 G = tf(num,den,-1)
6 impulse(G)
```

4.6 Problems

4.1. Find the Z-transform of the following and draw the pole–zero plot and ROC, if the transform exists:

$$u(n) = a^n 1(n) * b^n 1(-n)$$

4.2. Consider the sequence $u(n) = (n+1)1(n)$.

(a) Find its Z-transform using the fact that $u(n) = 1(n) * 1(n)$ (derive this).

(b) Find the Z-transform of u through the Z-transform of $n1(n)$ and $1(n)$. Compare the results.

4.3. Consider the equation

$$y(n) = \sum_{k=-\infty}^{n} u(k)$$

(a) Express the Z-transform of $y(n)$ in terms of $U(z)$. [Hint: Find the difference $y(n) - y(n-1)$.]

(b) Use the convolution property to determine the Z-transform of $y(n)$ in terms of $U(z)$. [Hint: If the right-hand side of the above equation can be written as $u(n) * g(n)$, what is $g(n)$?]

4.4. Find the causal sequence, whose Z-transform is given by

$$G(z) = \frac{1}{(1 - az^{-1})(1 - bz^{-1})}$$

where you can assume that $a \neq b$. Verify by computing $g(0)$ and $g(1)$ through some other means.

4.5. The inverse Z-transform of

$$G(z) = \frac{z^2 + 2z}{(z+1)^2(z-2)}$$

has already been computed for the case of $1 < |z| < 2$ in Example 4.29 on page 99. Now, find its inverse for two other cases: (a) $|z| > 2$; (b) $|z| < 1$.

4.6. Determine the inverse Z-transform of

$$Y(z) = \frac{4z^2 - 17z + 17}{(z-1)(z-2)(z-3)}, \quad |z| > 3$$

4.7. Recall that the ratio test [25] is one possible way of determining if a series is convergent. Use this test to show that the sequence $\{kp^k\}$ is absolutely summable if $|p| < 1$. Will this test also work for a sequence of the form $\{k^n p^k\}$, $|p| < 1$? Take $k \geq 0$.

4.8. Consider a right sided sequence $u(n)$ with the Z-transform

$$U(z) = \frac{1}{(1 - \frac{1}{2}z^{-1})(1 - z^{-1})} \tag{4.57}$$

(a) Carry out a partial fraction expansion expressed as a ratio of polynomials in z^{-1} and, from this expansion, determine $u(n)$.

(b) Rewrite Eq. 4.57 as a ratio of polynomials in z and carry out a partial fraction expansion of $U(z)$ expressed in terms of polynomials in z. From this expansion determine $u(n)$. How does this compare with the result obtained in part (a)?

4.9. Differentiating Eq. 4.17 on page 80 with respect to a, show that

$$\frac{z}{(z-a)^2} = \sum_{n=0}^{\infty} na^{n-1} z^{-n}$$

and hence that

$$na^{n-1}1(n) \leftrightarrow \frac{z}{(z-a)^2}$$

Successively differentiating with respect to a, show that

$$n(n-1)a^{n-2}1(n) \leftrightarrow \frac{2z}{(z-a)^3}$$

$$n(n-1)(n-2)a^{n-3}1(n) \leftrightarrow \frac{3!z}{(z-a)^4}$$

Generalize this to arrive at Eq. 4.18. Verify the condition on ROC.

4.10. If

$$z \in \mathrm{ROC}_G \cap \mathrm{ROC}_U$$

then show that

$$\sum |y(n)z^{-n}| < \infty, \quad i.e., \; z \in \mathrm{ROC}_Y$$

4.11. In Eq. 4.29 on page 85 we assumed the existence of $(zI - A)^{-1}$. Show that a sufficient condition is given by $|z| > \|A\|$. [Hint: Key theorem 5.8 on page 169 of [42] gives a sufficient condition for the existence of $(I - A/z)^{-1}$ to be $\|A/z\| < 1$.]

4.12. Show that with $z = e^{j\theta}$ in the stable function

$$G(z) = \frac{z - \frac{1}{a}}{z - a}, \quad 0 < a < 1 \tag{4.58}$$

$|G(e^{j\theta})|$ is independent of θ. Since its magnitude is not a function of θ, $G(z)$ is known as an all pass transfer function. [Hint: Show that $G(z)H\left(z^{-1}\right) = \frac{1}{a^2}$.]

4.13. This problem demonstrates a condition on complex poles of a real sequence [49]. Consider a real valued sequence $u(n)$ with a rational Z-transform $U(z)$.

(a) From the definition of the Z-transform, show that

$$U(z) = U^*(z^*)$$

where * indicates complex conjugation. The RHS means: take the transform with respect to z^* and then take the complex conjugate of the entire thing.

(b) From the result in part (a), show that if a pole of $U(z)$ occurs at $z = z$, then a pole must also occur at $z = z^*$. Show also that the same result is true for zeros.

4.14. Invert the following transfer functions:

(a) $\dfrac{z - be^{j\omega_1}}{z - ae^{j\omega_2}}, |z| > |a|$

(b) $\dfrac{z - be^{j\omega_1}}{z - ae^{j\omega_2}} + \dfrac{z - be^{-j\omega_1}}{z - ae^{-j\omega_2}}, |z| > |b|$

and determine whether the result is real. Explain.

4.15. Obtain the time domain equivalent of a system whose transfer function is

$$G(z) = \frac{z^2 + 3z + 1}{z^3 - 2z^2 + z + 1}$$

Use the following two approaches:

(a) Recursive formulation.

(b) State space realization.

4.16. This problem involves the solution of a famous difference equation.

(a) Consider

$$x(n+2) = x(n+1) + x(n) + \delta(n+2) \tag{4.59}$$

with $x(n) = 0$ for $n < 0$. Evaluate $x(n)$ recursively for $n = 0$ to 4. Do you know the name of this famous sequence?

(b) By Z-transforming Eq. 4.59 and simplifying, obtain an expression for $X(z)$.

(c) Invert the Z-transform obtained above and obtain an expression for $x(n)$.

(d) Find the ratio $x(n+1)/x(n)$ for large n. Do you know what the ancient Greeks called this famous ratio?

4.17. Suppose that in a sequence $\{u(k)\}$, $u(k) = 0$ for $k < 0$, *i.e.*, u is causal. Let

$$U(z) = \frac{N(z)}{D(z)}, \quad |z| > r_0$$

be its Z-transform with the degree of $N(z) = n$ and the degree of $D(z) = m$.

(a) Show that $m = n$ if and only if $u(0) \neq 0$. Notice that you have to show both the directions.

(b) Suppose that $u(0) = 0$, $u(1) \neq 0$. Then show that $n = m - 1$.

(c) Generalize the above result for $u(k) = 0$, $0 < k \leq k_0$.

If possible, give examples for each.

4.18. In this problem we obtain another realization of the transfer function given in Eq. 4.53. We do this as follows:

(a) Show that the following relation can be arrived at from the transfer function:

$$y(k+3) + a_1 y(k+2) + a_2 y(k+1) + a_3 y(k) = \\ b_3 u(k) + b_2 u(k+1) + b_1 u(k+2)$$

Show that

$$u(k) = \xi(k+3) + a_1 \xi(k+2) + a_2 \xi(k+1) + a_3 \xi(k)$$
$$y(k) = b_3 \xi(k) + b_2 \xi(k+1) + b_1 \xi(k+2)$$

satisfy the above relation.

(b) Fill in the boxes in the following block diagram so that it implements the above equation. Ignore x_i for now.

(c) The ξ at different time instants become different states as indicated by x in the diagram. Write expressions for states at $k+1$ in terms of the states at k and the input at k. Write this in matrix form. Similarly write an expression for $y(k)$ in terms of the states at k and the input at k. This is known as the controller canonical form.

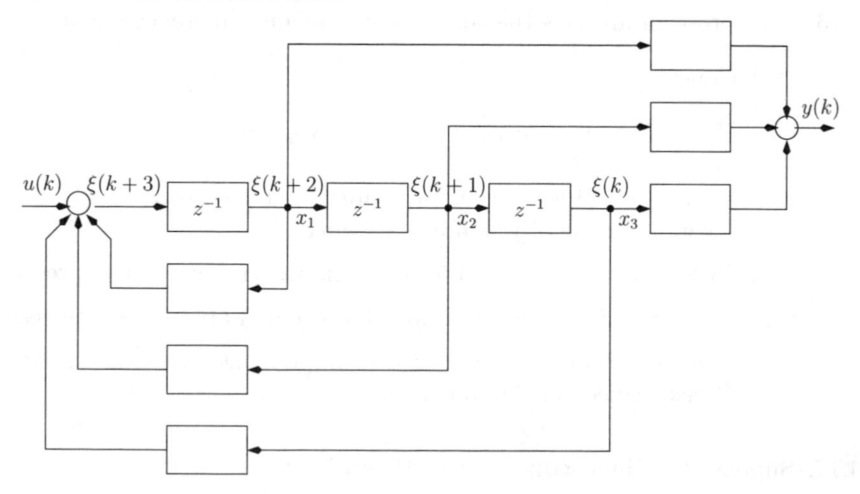

Chapter 5

Frequency Domain Analysis

In this chapter, we study frequency domain aspects of signals and systems. Such an insight is extremely important for the design of filters and controllers. We present the tools required for this study: Fourier series and Fourier transform. We also discuss Shannon's sampling theorem, which specifies the minimum speed of sampling. Finally, we present a brief introduction to filtering.

5.1 Basics

In this section, we discuss how oscillations naturally enter the system response. We also point out the differences between continuous time and discrete time sinusoids.

5.1.1 Oscillatory Nature of System Response

Let us first consider I/O LTI systems with a single real pole. These systems have a transfer function of the form

$$G(z) = \frac{z}{z - \rho} \tag{5.1}$$

see Footnote 6 on page 102. We will consider different cases of the location of the pole, ρ. If the impulse response of the system is denoted by $y(n)$, its Z-transform, $Y(z)$, is identical to $G(z)$. On inversion, we obtain $y(n) = \rho^n$. A plot of the response for different ρ values is presented in Fig. 5.1.

Notice that when $1 > \rho > 0$, y decays to zero monotonically. If, on the other hand, $\rho > 1$, the response grows monotonically. For negative values of ρ, we obtain an oscillatory response. For $0 \geq \rho > -1$, we obtain an oscillatory, but decaying exponential, while for $\rho < -1$, the output grows with oscillations.

We will next consider a system with a complex pole, at $\rho e^{j\omega}$. If complex poles are present in real systems, they should occur in conjugate pairs, see Problem 4.13. As a result, $\rho e^{-j\omega}$ will also be a pole, as shown in the diagram on the left-hand side of Fig. 5.2. Let the transfer function be given by

$$G(z) = \frac{z^2}{(z - \rho e^{j\omega})(z - \rho e^{-j\omega})} \tag{5.2}$$

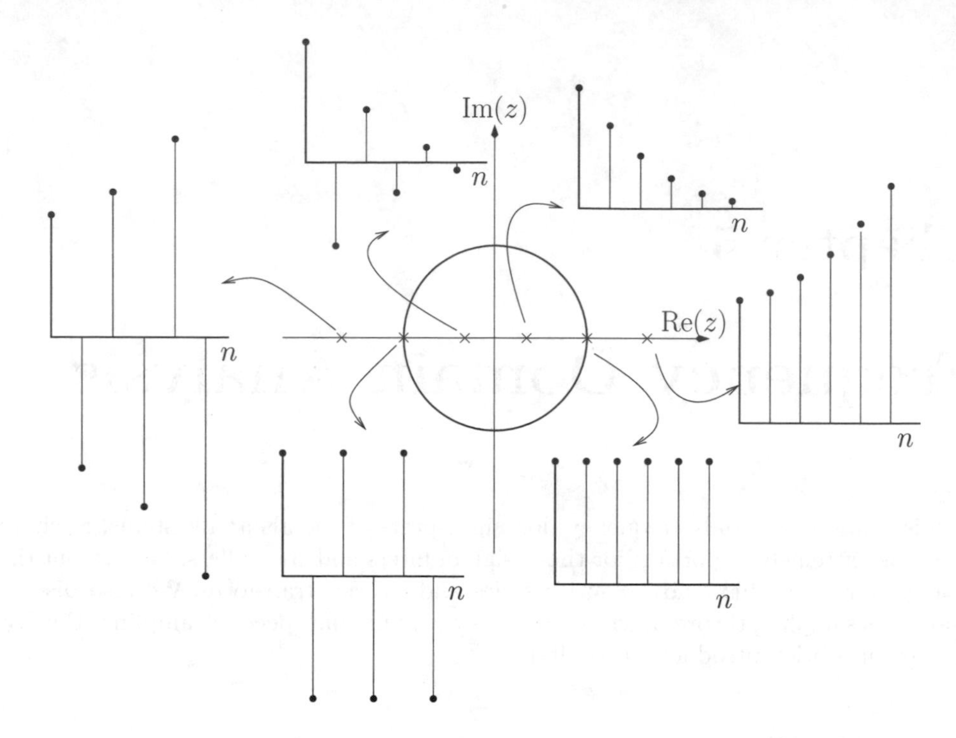

Figure 5.1: Impulse response for poles on the real axis

The response $Y(z)$ for an impulse input is

$$\frac{Y(z)}{z} = \frac{z}{(z - \rho e^{j\omega})(z - \rho e^{-j\omega})} = \frac{A_1}{z - \rho e^{j\omega}} + \frac{A_1^*}{z - \rho e^{-j\omega}}$$

where A_1 is some complex number, of the form $\alpha e^{j\theta}$, and A_1^* is the complex conjugate of A_1. On inverting, we obtain the impulse response $y(n)$ as

$$\begin{aligned} y(n) &= A_1 \rho^n e^{jn\omega} + A_1^* \rho^n e^{-jn\omega} = \alpha \rho^n \left[e^{j(n\omega+\theta)} + e^{-j(n\omega+\theta)} \right] \\ &= 2\alpha\rho^n \cos(n\omega + \theta), \quad n \geq 0 \end{aligned} \tag{5.3}$$

Notice that this response is sinusoidal. One can see ω to be the frequency of oscillation. For $\omega = 0$, there is no oscillation, which reaches the maximum for $\omega = 180°$. So long as the denominator is of the form assumed here, Eq. 5.3 holds true. For different numerator values, only the value of A_1 and hence those of α and θ will vary. The impulse responses for different pole locations are given in the diagram on the right-hand side of Fig. 5.2.

In Sec. 3.3.2, we have shown that the response to any signal can be written as a linear combination of impulse responses. As a result, one can expect the poles to have a say in the oscillatory nature of general responses as well. Problem 5.2 asks the reader to verify that the step response of Eq. 5.2 also is oscillatory.

Often, transfer functions having more than one or two poles can also be approximated by simpler transfer functions. As a result, the examples studied in this section are of general interest and the results are applicable to a larger class of systems.

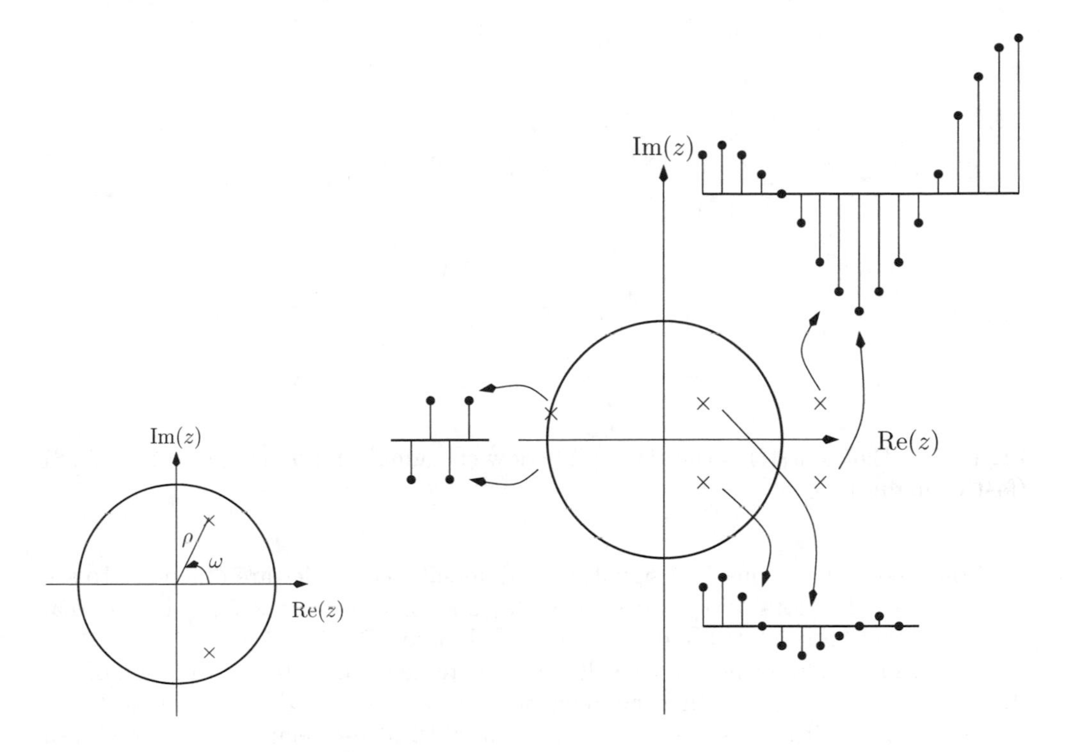

Figure 5.2: Pole locations of a typical stable second order discrete time LTI system, indicated within unit circle

These examples show that oscillations are inherent in a large class of systems. In order to understand this aspect, we undertake a frequency domain study of signals and systems.

5.1.2 Continuous and Discrete Time Sinusoidal Signals

We will first see some properties of continuous sinusoidal signals. Consider a continuous time sinusoidal signal of the form

$$u_a(t) = A\cos(\Omega t + \theta), \quad -\infty < t < \infty \tag{5.4}$$

where A is the amplitude, Ω is the angular frequency in rad/s and θ is the phase in rad. With $\Omega = 2\pi F$, where F is the frequency in cycles/s or hertz, we can write $u_a(t)$ also as

$$u_a(t) = A\cos(2\pi Ft + \theta) \tag{5.5}$$

Some properties of such signals are now listed. For every fixed value of F, $u_a(t)$ is periodic with a period $T_p = 1/F$:

$$\begin{aligned} u_a(t + T_p) &= A\cos(2\pi F(t + 1/F) + \theta) = A\cos(2\pi + 2\pi Ft + \theta) \\ &= A\cos(2\pi Ft + \theta) = u_a(t) \end{aligned}$$

Because $u_a(t + T_p) = u_a(t)$ for an arbitrary t, the signal is periodic with a period T_p.

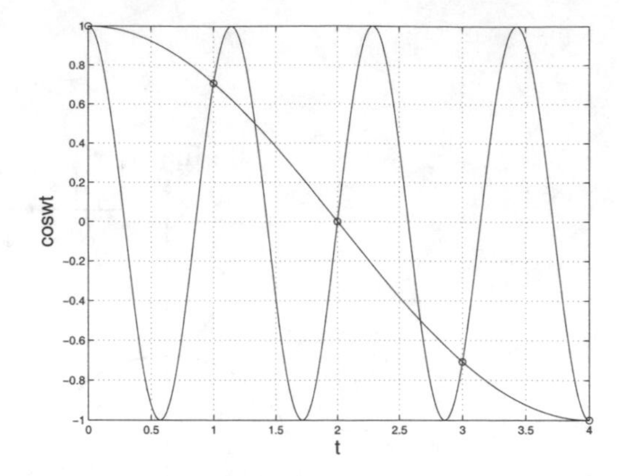

Figure 5.3: Plot of $u_1(t) = \cos(2\pi \times t/8)$ (slow changing) and $u_2(t) = \cos(2\pi \times 7t/8)$ (fast changing)

Continuous time sinusoidal signals with different frequencies are different. To see this, observe the plots of $u_1 = \cos(2\pi \times t/8)$ and $u_2 = \cos(2\pi \times 7t/8)$ in Fig. 5.3. It is clear that these plots are distinct from each other.

Increasing the frequency of u results in an increase in the rate of oscillation of the signal. We can go on decreasing the frequency, *i.e.*, $F \to 0$ and $T_p \to \infty$. Similarly, we can go on increasing F as t is a continuous variable. These relations hold good also for complex exponential signals of the following form:

$$u_a(t) = Ae^{j(\Omega t+\theta)} = A[\cos(\Omega t+\theta) + j\sin(\Omega t+\theta)]$$

For mathematical convenience, the concept of *negative frequency* is introduced. While positive frequency can be thought of as a counterclockwise rotation (see Fig. 5.2), negative frequency can be assumed to produce clockwise rotation. With this, sinusoidal signals can be expressed as a combination of complex exponential signals:

$$u_a(t) = A\cos(\Omega t+\theta) = \frac{A}{2}\left[e^{j(\Omega t+\theta)} + e^{-j(\Omega t+\theta)}\right]$$

If we refer to the real and the imaginary parts of u_a as u_R and u_I, we have

$$u_R(t) = A\cos(\Omega t+\theta) = \text{Re}\left[Ae^{j(\Omega t+\theta)}\right]$$

see Footnote 2 on page 65. We obtain,

$$u_I(t) = A\sin(\Omega t+\theta) = \text{Im}\left[Ae^{j(\Omega t+\theta)}\right]$$

We will now study discrete time sinusoidal signals briefly. Consider discrete time periodic signals of the form

$$u(n) = A\cos(\omega n+\theta), \quad -\infty < n < \infty \tag{5.6}$$

where the variables have the following meaning:

n integer variable, sample number
A amplitude of the sinusoid
ω frequency in radians per sample and
θ phase in radians.

If $\omega = 2\pi f$, where f is known as the *normalized frequency* with units of cycles/sample, we obtain

$$u(n) = A\cos(2\pi fn + \theta), \quad -\infty < n < \infty \tag{5.7}$$

The reason why f is called normalized frequency will be made clear shortly.

The signal $u(n)$ is said to be periodic with period N, $N > 0$, if and only if $u(n+N) = u(n)$ for all n. The smallest nonzero value for which this is true is known as the *fundamental period.* Some properties of a discrete time sinusoid are now listed.

Periodicity: A discrete time sinusoid is periodic only if its frequency f is a rational number. To see this, let

$$u(n) = \cos(2\pi f_0 n + \theta)$$

To check whether u has a period N, we need to calculate $u(n+N)$:

$$u(n+N) = \cos(2\pi f_0(n+N) + \theta)$$

The signals $u(n)$ and $u(n+N)$ will be equal if and only if there exists an integer k such that

$$2\pi f_0 N = 2k\pi \tag{5.8}$$

or equivalently,

$$f_0 = \frac{k}{N} \tag{5.9}$$

i.e., f_0 is rational. N obtained after cancelling the common factors in k/f_0, obtained from Eq. 5.8, is known as the fundamental period.

Identical signals: Discrete time sinusoids whose frequencies are separated by integer multiple of 2π are identical, as the following equation holds for all integer n:

$$\cos((\omega_0 + 2\pi)n + \theta) = \cos(\omega_0 n + \theta), \ \forall n$$

As a matter of fact, all sinusoidal sequences

$$u_k(n) = A\cos(\omega_k n + \theta)$$

where

$$\omega_k = \omega_0 + 2k\pi, \quad -\pi < \omega_0 < \pi \tag{5.10}$$

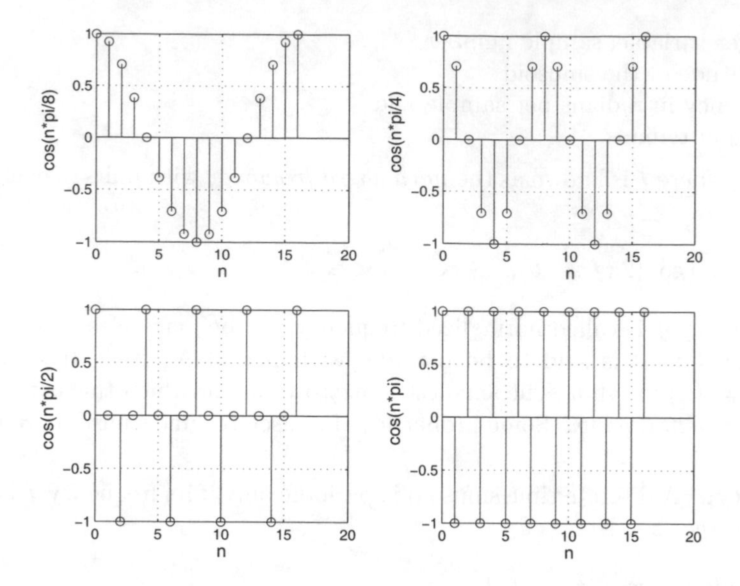

Figure 5.4: Plots of sinusoids of different frequencies

are indistinguishable or identical. In view of this observation, only the sinusoids in the frequency range $-\pi < \omega_0 < \pi$ are different, *i.e.*,

$$-\pi < \omega_0 < \pi \quad \text{or} \quad -\frac{1}{2} < f_0 < \frac{1}{2} \tag{5.11}$$

As a result, the above equation gives the unique frequency range. This property is different from the previous one: now we have a fixed n and a variable f while we had f fixed and n a variable earlier.

Alias: The highest rate of oscillation in a discrete time sinusoid is attained when $\omega = \pi$ or $\omega = -\pi$, or equivalently, $f = 1/2$ or $f = -1/2$. In Fig. 5.4, plots of $\cos \omega_0 n$ have been drawn for ω_0 values of $\pi/8$, $\pi/4$, $\pi/2$ and π with n taking integer values. Matlab code M 5.1 shows how these plots are produced.

As ω_0 increases, the frequency of oscillation also increases reaching a maximum at $\omega_0 = \pi$. What happens if ω_0 increases beyond π? Let $\omega_1 = \omega_0$ and $\omega_2 = 2\pi - \omega_0$. Then, $u_1 = A \cos \omega_1 n$ and $u_2 = A \cos \omega_2 n$ are identical, as shown below:

$$\begin{aligned} u_1(n) &= A \cos \omega_1 n = A \cos \omega_0 n \\ u_2(n) &= A \cos \omega_2 n = A \cos(2\pi - \omega_0) n = A \cos \omega_0 n \end{aligned}$$

Observe that $u_2(n) = u_1(n)$. We say that ω_2 is an *alias* of ω_1. We will study more about these properties shortly.

5.1.3 Sampling of Continuous Time Signals

The first decision to make in sampled data systems is the sampling period. Although we will study it in detail in Sec. 5.3, we introduce the topic of sampling now. Suppose that we sample the analog signal $u_a(t)$ at a uniform sampling rate of T_s. Then

$$u(n) = u_a(nT_s), \quad -\infty < n < \infty \tag{5.12}$$

where we have used

$$t = nT_s = \frac{n}{F_s} \tag{5.13}$$

where $F_s = 1/T_s$ is the sampling frequency, in samples/s or hertz.

Let the continuous time signal have a frequency of F cycles/s. We will derive a relation between the continuous time frequency F and the discrete time frequency f. Let us sample the continuous time signal, given by Eq. 5.4–5.5, at a rate of $F_s = 1/T_s$. We obtain

$$u(n) = A\cos(2\pi F T_s n + \theta) = A\cos\left(2\pi \frac{F}{F_s} n + \theta\right) \tag{5.14}$$

Comparing this with Eq. 5.7, we obtain

$$f = \frac{F}{F_s} \tag{5.15}$$

and, therefore, f is called normalized frequency or *relative frequency.* Multiplying both sides by 2π, we obtain the following relationship in angular frequency:

$$\omega = \frac{\Omega}{F_s} = \Omega T_s \tag{5.16}$$

We know from the earlier discussion that while the continuous frequencies can vary all the way up to infinity, *i.e.*, $-\infty < F < \infty$ or $-\infty < \Omega < \infty$, the discrete frequencies are bounded, as given by Eq. 5.11. In view of this, the maximum continuous time frequency that can be accommodated is given by

$$F_{\max} = \frac{F_s}{2}, \quad \Omega_{\max} = 2\pi F_{\max} = \pi F_s = \frac{\pi}{T_s} \tag{5.17}$$

We see that by sampling, frequency values in an infinite range are mapped into a finite range. We illustrate this with an example.

Example 5.1 Consider sampling at the rate of $T_s = 1$ s the two functions plotted in Fig. 5.3, namely $u_1(t) = \cos(2\pi \times t/8)$ and $u_2(t) = \cos(2\pi \times 7t/8)$. On sampling, we obtain the discrete time signals as

$$\begin{aligned} u_1(n) &= \cos 2\pi \frac{1}{8} n \\ u_2(n) &= \cos 2\pi \frac{7}{8} n = \cos 2\pi \left(1 - \frac{1}{8}\right) n \\ &= \cos\left(2\pi n - 2\pi \frac{1}{8} n\right) = \cos\left(2\pi \frac{1}{8} n\right) \end{aligned}$$

Thus, we see that $u_2(n) = u_1(n)$ at all the sampling instants, which can be seen from Fig. 5.3 as well. In this figure, the two waveforms intersect exactly at the sampling instants. Thus if we have a sequence generated by sampling $\cos 2\pi \times n/8$ at the rate of 1 s, we will not know whether it comes from $u_1(t)$ or from $u_2(t)$. We see that $f_2 = 7/8$ Hz is an aliasof $f_1 = 1/8$ Hz. In fact, $f_1 + nF_s$ are aliases of f_1 for all integer values of n.

∎

The above example shows that we lose information when u_2 is sampled at a slow rate. We are interested in knowing the rate that we should maintain while sampling a continuous signal. The Fourier transform is a tool that will help us study this problem. We discuss the topic of Fourier transforms in the next section.

5.2 Fourier Series and Fourier Transforms

Fourier series is a powerful tool to study periodic signals. Many useful signals are not periodic, though. Fourier transforms can be used to study these. We begin the discussion with continuous time signals. We motivate the discrete time Fourier transform with frequency response and point out how it can be obtained from the Z-transform. We also discuss some properties of Fourier transforms.

5.2.1 Fourier Series for Continuous Time Periodic Signals

Let $u(t)$ be a periodic signal with a fundamental period, T_p, given by

$$T_p = \frac{1}{F_0} \tag{5.18}$$

where F_0 is the frequency of oscillations. We can write it in a *Fourier series* as follows:

$$u(t) = \sum_{k=-\infty}^{\infty} C_k e^{j2\pi kF_0 t} \tag{5.19}$$

To find C_k, we multiply both sides by $e^{-j2\pi lF_0 t}$ and integrate from t_0 to $t_0 + T_p$:

$$\int_{t_0}^{t_0+T_p} u(t)e^{-j2\pi lF_0 t}dt = \int_{t_0}^{t_0+T_p} e^{-j2\pi lF_0 t}\left(\sum_{k=-\infty}^{\infty} C_k e^{j2\pi kF_0 t}\right)dt$$

Exchanging integration and summation,

$$\int_{t_0}^{t_0+T_p} u(t)e^{-j2\pi lF_0 t}dt = \sum_{k=-\infty}^{\infty} C_k \int_{t_0}^{t_0+T_p} e^{j2\pi(k-l)F_0 t}dt$$

Separating the integral corresponding to $k = l$, we obtain

$$\int_{t_0}^{t_0+T_p} u(t)e^{-j2\pi lF_0 t}dt = C_l \int_{t_0}^{t_0+T_p} dt + \sum_{k=-\infty, k\neq l}^{\infty} C_k \int_{t_0}^{t_0+T_p} e^{j2\pi(k-l)F_0 t}dt$$

Carrying out the integration, we obtain

$$\int_{t_0}^{t_0+T_p} u(t)e^{-j2\pi lF_0 t}dt = C_l T_p + \sum_{k=-\infty, k\neq l}^{\infty} C_k \left.\frac{e^{j2\pi(k-l)F_0 t}}{j2\pi(k-l)F_0}\right|_{t_0}^{t_0+T_p} \tag{5.20}$$

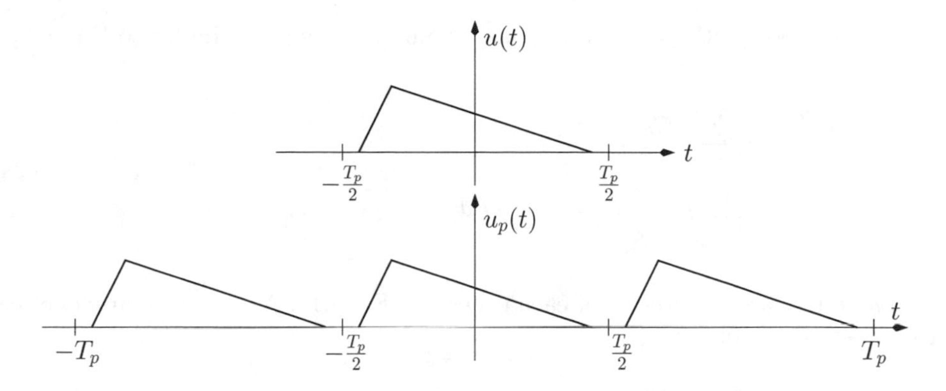

Figure 5.5: Conversion of an aperiodic signal into a periodic signal

Note that if we had not separated the $k = l$ term, we would have had a zero in the denominator of the second term. On taking limits, the second term becomes zero, as we show now:

$$e^{j2\pi nF_0t}\Big|_{t_0}^{t_0+T_p} = e^{j2\pi nF_0(t_0+T_p)} - e^{j2\pi nF_0t_0} = e^{j2\pi nF_0t_0}(e^{j2\pi n} - 1) = 0$$

because, from Eq. 5.18, $F_0T_p = 1$. In view of this result, Eq. 5.20 becomes

$$\int_{t_0}^{t_0+T_p} u(t)e^{-j2\pi lF_0t}dt = C_lT_p$$

from which it follows that

$$C_l = \frac{1}{T_p}\int_{t_0}^{t_0+T_p} u(t)e^{-j2\pi lF_0t}dt = \frac{1}{T_p}\int_{T_p} u(t)e^{-j2\pi lF_0t}dt \tag{5.21}$$

where the last integral is over any one period T_p, as $u(t)$ is periodic. Eq. 5.19 and Eq. 5.21 make the Fourier series pair, which we reproduce below in one place:

$$\begin{aligned} u(t) &= \sum_{k=-\infty}^{\infty} C_k e^{j2\pi kF_0t} \\ C_l &= \frac{1}{T_p}\int_{T_p} u(t)e^{-j2\pi lF_0t}dt \end{aligned} \tag{5.22}$$

5.2.2 Fourier Transform of Continuous Time Aperiodic Signals

To extend the above analysis to aperiodic signals, which are common in control applications, we produce a periodic extension of $u(t)$, as shown in Fig. 5.5, and call it $u_p(t)$. Here we have assumed that $u(t)$ vanishes outside the interval $(-T_p/2, T_p/2)$, where T_p can be arbitrarily large. We can recover $u(t)$ from $u_p(t)$, as follows:

$$\lim_{T_p\to\infty} u_p(t) = u(t) \tag{5.23}$$

As $u_p(t)$ is periodic with period T_p, it has a Fourier series pair, similar to Eq. 5.22:

$$\begin{aligned} u_p(t) &= \sum_{k=-\infty}^{\infty} C_k e^{j2\pi k F_0 t} \\ C_k &= \frac{1}{T_p} \int_{-T_p/2}^{T_p/2} u_p(t) e^{-j2\pi k F_0 t} dt \end{aligned} \tag{5.24}$$

where F_0 and T_p are reciprocals of each other, see Eq. 5.18. As u_p and u are identical over one period, we obtain

$$C_k = \frac{1}{T_p} \int_{-T_p/2}^{T_p/2} u(t) e^{-j2\pi k F_0 t} dt$$

As u vanishes outside one period, this becomes

$$C_k = \frac{1}{T_p} \int_{-\infty}^{\infty} u(t) e^{-j2\pi k F_0 t} dt \tag{5.25}$$

We define a function $U(F)$, called the Fourier transform of $u(t)$, as

$$U(F) = \int_{-\infty}^{\infty} u(t) e^{-j2\pi F t} dt \tag{5.26}$$

where $U(F)$ is a function of the continuous variable F. It doesn't depend on T_p or F_0. But if we compare Eq. 5.25 and Eq. 5.26, we obtain

$$C_k = \frac{1}{T_p} U\left(k \frac{1}{T_p}\right) \tag{5.27}$$

where we have used Eq. 5.18. Thus, Fourier coefficients are samples of $U(F)$ taken at multiples of F_0 and scaled by F_0. Substituting Eq. 5.27 in Eq. 5.24, we obtain

$$u_p(t) = \frac{1}{T_p} \sum_{k=-\infty}^{\infty} U\left(\frac{k}{T_p}\right) e^{j2\pi k F_0 t}$$

Defining $\Delta F = 1/T_p = F_0$, the above equation becomes

$$u_p(t) = \sum_{k=-\infty}^{\infty} U(k\Delta F) e^{j2\pi k \Delta F t} \Delta F$$

Invoking Eq. 5.23, we obtain

$$u(t) = \lim_{T_p \to \infty} u_p(t) = \lim_{\Delta F \to 0} \sum_{k=-\infty}^{\infty} U(k\Delta F) e^{j2\pi k \Delta F t} \Delta F$$

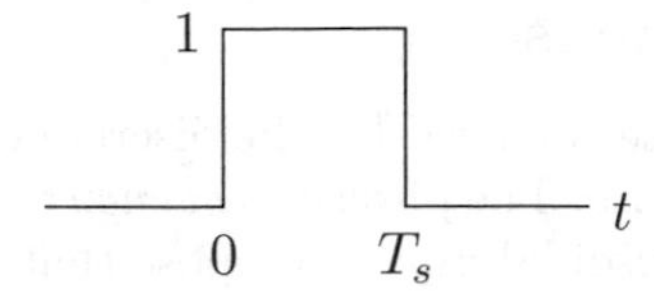

Figure 5.6: Impulse response of zero order hold

This is nothing but rectangular approximation of an integral. We obtain, on taking limits,

$$u(t) = \int_{-\infty}^{\infty} U(F)e^{j2\pi Ft}dF$$

Thus, along with Eq. 5.26, we obtain the following Fourier transform pair:

$$\begin{aligned} u(t) &= \int_{-\infty}^{\infty} U(F)e^{j2\pi Ft}dF \\ U(F) &= \int_{-\infty}^{\infty} u(t)e^{-j2\pi Ft}dt \end{aligned} \tag{5.28}$$

With radian frequency Ω given by $\Omega = 2\pi F$, we obtain the following Fourier transform pair:

$$\begin{aligned} u(t) &= \frac{1}{2\pi}\int_{-\infty}^{\infty} U(\Omega)e^{j\Omega t}d\Omega \\ U(\Omega) &= \int_{-\infty}^{\infty} u(t)e^{-j\Omega t}dt \end{aligned} \tag{5.29}$$

We will now illustrate these ideas with an example.

Example 5.2 Find the Fourier transform of a pulse of height 1 and width T_s, as given in Fig. 5.6.

We can think of this function as the response of zero order hold to an impulse sequence. In other words, this can be thought of as the impulse response of ZOH. We obtain its Fourier transform as

$$\mathsf{ZOH}(j\Omega) = \int_0^{T_s} e^{-j\Omega t}dt = \left.\frac{e^{-j\Omega t}}{-j\Omega}\right|_0^{T_s} = \frac{1-e^{-j\Omega T_s}}{j\Omega}$$

We can write this in polar form by factoring out $e^{-j\Omega T_s/2}$ and multiplying and dividing by 2:

$$\mathsf{ZOH}(j\Omega) = e^{-j\Omega T_s/2}\left(\frac{e^{j\Omega T_s/2}-e^{-j\Omega T_s/2}}{2j}\right)\frac{2}{\Omega} = e^{-j\Omega T_s/2}\frac{\sin(\Omega T_s/2)}{\Omega/2}$$

Thus, we arrive at the Fourier transform of ZOH:

$$\mathsf{ZOH}(j\Omega) = T_s e^{-j\Omega T_s/2}\frac{\sin(\Omega T_s/2)}{\Omega T_s/2} \tag{5.30}$$

∎

5.2.3 Frequency Response

In this section, we will motivate the need for the discrete time Fourier transform with the concept of frequency response. The phrase *frequency response* denotes the response of an I/O LTI system to a sinusoidal input. Suppose that a complex frequency signal

$$u(n) = e^{j\omega n} \tag{5.31}$$

is applied to an LTI system with impulse response $g(n)$. The output y is given by

$$y(n) = g(n) * u(n) = \sum_{k=-\infty}^{\infty} g(k)u(n-k) = \sum_{k=-\infty}^{\infty} g(k)e^{j\omega(n-k)}$$

Pulling the constant part out of the summation sign, we obtain

$$y(n) = e^{j\omega n} \sum_{k=-\infty}^{\infty} g(k)e^{-j\omega k} \tag{5.32}$$

Notice that the term inside the summation is a function of ω alone. It is of the form $\sum_{k=-\infty}^{\infty} g(k)z^{-k}$, which is the Z-transform of $\{g(n)\}$. Motivated by this fact, we define the discrete time Fourier transform of $\{g(n)\}$ as

$$G(e^{j\omega}) \triangleq \sum_{k=-\infty}^{\infty} g(k)e^{-j\omega k} \tag{5.33}$$

We will discuss the convergence condition for this series in Sec. 5.2.5. Combining Eq. 5.32 and Eq. 5.33, we obtain

$$y(n) = e^{j\omega n} G(e^{j\omega}) \tag{5.34}$$

As $G(e^{j\omega})$ is a complex number, it can be written as $|G(e^{j\omega})|e^{j\varphi}$, where φ is the phase angle of $G(e^{j\omega})$ at the given ω. With this, the above equation becomes

$$y(n) = e^{j\omega n}|G(e^{j\omega})|e^{j\varphi} = |G(e^{j\omega})|e^{j(wn+\varphi)} \tag{5.35}$$

We see that when the input is a sinusoid with frequency ω (see Eq. 5.31), the output of the LTI system also is a sinusoid with the same frequency ω. In addition, we observe the following:

1. The amplitude of the output is multiplied by the magnitude of $G(e^{j\omega})$.
2. The output sinusoid shifts by φ, the phase angle of $G(e^{j\omega})$, with respect to the input.

The frequency response behaviour is fundamental to the design of filters, which are ubiquitous. This topic is discussed in detail in Sec. 5.4. Central to frequency response is the Fourier transform of discrete time signals, defined in Eq. 5.33. Problem 5.4 is concerned with the Fourier transform of discrete time periodic signals. The next section is devoted to the topic of Fourier transform of discrete time aperiodic signals.

5.2.4 Fourier Transform of Discrete Time Aperiodic Signals

In the previous section, we motivated the need for the Fourier transform of discrete time signals. Such a transform is known as the discrete time Fourier transform. Unless otherwise stated, we will refer to this simply as the Fourier transform. The Fourier transform of a finite energy discrete time signal $u(n)$ is defined as

$$U(e^{j\omega}) \triangleq \sum_{n=-\infty}^{\infty} u(n)e^{-j\omega n} \tag{5.36}$$

Recall the definition given earlier in Eq. 5.35. We would like to find an expression for $u(n)$ in terms of its Fourier transform, U. First observe that U is periodic in ω with a period of 2π:

$$U\left(e^{j(\omega+2\pi k)}\right) = \sum_{n=-\infty}^{\infty} u(n)e^{-j(\omega+2\pi k)n} = U(e^{j\omega}) \tag{5.37}$$

This periodicity is just a consequence of the frequency of any discrete time signal being unique only in the range $(-\pi, \pi)$ or $(0, 2\pi)$. The Fourier transform in this case is a summation instead of an integral. Because U is periodic, it has a Fourier series expansion, the same as in Eq. 5.36. The Fourier series coefficients, namely $u(n)$, can be calculated by integrating both sides of Eq. 5.36 as follows:

$$\int_{-\pi}^{\pi} U(e^{j\omega})e^{j\omega m}d\omega = \int_{-\pi}^{\pi} \left[\sum_{n=-\infty}^{\infty} u(n)e^{-j\omega n}\right] e^{j\omega m}d\omega \tag{5.38}$$

If the infinite sum in this equation converges, we can change the order of integration and summation to obtain

$$\int_{-\pi}^{\pi} U(e^{j\omega})e^{j\omega m}d\omega = \sum_{n=-\infty}^{\infty} u(n) \int_{-\pi}^{\pi} e^{j\omega(m-n)}d\omega$$

Using the procedure used to arrive at Eq. 5.20, we can split the right-hand side into a term consisting of only m and the rest and then show that the entire right-hand side is equal to $2\pi u(m)$. We arrive at the expression we are looking for:

$$u(m) = \frac{1}{2\pi}\int_{-\pi}^{\pi} U(e^{j\omega})e^{j\omega m}d\omega = \int_{-1/2}^{1/2} U\left(e^{j2\pi f}\right)e^{j2\pi fm}df \tag{5.39}$$

because $\omega = 2\pi f$. To summarize, we obtain the Fourier transform pair

$$\begin{aligned} U(e^{j\omega}) &= \sum_{n=-\infty}^{\infty} u(n)e^{-j\omega n} \\ u(m) &= \int_{-1/2}^{1/2} U\left(e^{j2\pi f}\right)e^{j2\pi fm}df \end{aligned} \tag{5.40}$$

5.2.5 Convergence Conditions for Fourier Transform[1]

We now briefly address the conditions under which the Fourier transform of discrete time sequences converges. Recall that while defining the Z-transform, we have enforced the condition of absolute convergence. The objectives of this requirement are that the Z-transform should exist and that we can uniquely invert it.

In case of the Z-transform, we have had the luxury of choosing z so as to achieve absolute convergence. As a result, we can obtain Z-transforms of unstable sequences as well, see Footnote 4 on page 68. In the case of Fourier transforms, we substitute $z = e^{j\omega}$ and hence get constrained to the unit circle. As a result, absolute convergence becomes restrictive for Fourier transforms. A weaker condition is that of *mean square convergence*. This results in the sequence converging to the mean at the points of discontinuity. Signals with finite energy (see Eq. 3.2 on page 35) can be accommodated through this relaxation, even if they do not satisfy the condition of absolute convergence. Unfortunately, even this condition is not sufficient to accommodate useful signals, such as step and sinusoidal sequences. These sequences may be accommodated by requiring the convergence in a distributional sense, through special functions such as Dirac delta functions. This is the least restrictive criterion of all.

This topic is not pursued any further here. The interested reader is referred to texts such as [38].

5.2.6 Fourier Transform of Real Discrete Time Signals

As the Fourier transform is the Z-transform evaluated at $z = e^{j\omega}$, all properties of the Z-transform hold, with appropriate substitutions. In this section we will discuss some additional properties, in the case of real signals. Starting from the definition of Fourier transform, we obtain

$$U(e^{j\omega}) = \sum_{n=-\infty}^{\infty} u(n)e^{-j\omega n} = \sum_{n=-\infty}^{\infty} u(n)\cos\omega n - j\sum_{n=-\infty}^{\infty} u(n)\sin\omega n \tag{5.41}$$

Similarly, we obtain for negative frequency:

$$U(e^{-j\omega}) = \sum_{n=-\infty}^{\infty} u(n)e^{j\omega n} = \sum_{n=-\infty}^{\infty} u(n)\cos\omega n + j\sum_{n=-\infty}^{\infty} u(n)\sin\omega n \tag{5.42}$$

Next, we will derive the inverse Fourier transform for real valued signals. From Eq. 5.39, we obtain for a real valued signal u,

$$u(m) = \frac{1}{2\pi}\int_{-\pi}^{\pi} U(e^{j\omega})e^{j\omega m}d\omega$$

Denoting the real and imaginary parts of U as U_R and U_I, respectively, we obtain

$$u(m) = \frac{1}{2\pi}\int_{-\pi}^{\pi} (U_R(e^{j\omega}) + jU_I(e^{j\omega}))(\cos\omega m + j\sin\omega m)d\omega$$

[1]This section may be skipped in a first reading.

Let us compare the real and imaginary parts. Because u is real, we obtain

$$u(m) = \frac{1}{2\pi} \int_{-\pi}^{\pi} (U_R(e^{j\omega}) \cos \omega m - U_I(e^{j\omega}) \sin \omega m) d\omega \tag{5.43}$$

Symmetry of real and imaginary parts for real valued sequences: Notice that the real parts of Eq. 5.41 and Eq. 5.42 are identical. Also, the imaginary parts of these two expressions are negatives of each other:

$$\begin{aligned} U_R(e^{j\omega}) &= U_R(e^{-j\omega}) \\ U_I(e^{j\omega}) &= -U_I(e^{-j\omega}) \end{aligned} \tag{5.44}$$

From the above equations, we see that the real part is an even function of ω, while the imaginary part is an odd function of ω. We can summarize these properties as

$$U(e^{j\omega}) = U^*(e^{-j\omega}) \tag{5.45}$$

where the asterisk denotes complex conjugation.

Symmetry of magnitude and phase angle for real valued sequences: By the definition of magnitude and phase of complex signals, we have

$$\begin{aligned} |U(e^{j\omega})| &= \sqrt{U_R^2(e^{j\omega}) + U_I^2(e^{j\omega})} \\ \text{Arg}(U(e^{j\omega})) &= \tan^{-1} \frac{U_I(e^{j\omega})}{U_R(e^{j\omega})} \end{aligned}$$

In view of Eq. 5.44, we obtain the following symmetry properties for both magnitude and phase:

$$\begin{aligned} |U(e^{j\omega})| &= |U(e^{-j\omega})| \\ \text{Arg}(U(e^{j\omega})) &= -\text{Arg}(U(e^{-j\omega})) \end{aligned} \tag{5.46}$$

In other words, the magnitude and the phase are even and odd functions, respectively.

Example 5.3 Find the Fourier transform of the unit impulse sequence, $\{\delta(n)\}$.

The Z-transform of $\{\delta(n)\}$ is one, see Eq. 4.7 on page 70. Because the Fourier transform is obtained by substituting $z = e^{j\omega}$, the Fourier transform of $\{\delta(n)\}$ also is one.

We see that the Fourier transform of the impulse sequence is unity at all frequencies. Because the time sequence is real, as expected, the symmetry properties implied by Eq. 5.44 and Eq. 5.46 are trivially satisfied: the imaginary part of the Fourier transform is zero.

∎

Example 5.4 Find the Fourier transform of

$$u(n) = \begin{cases} A & 0 \le n \le N-1 \\ 0 & \text{otherwise} \end{cases}$$

where A can be taken to be real. As $u(n)$ is a finite sequence, its Fourier transform exists. We obtain

$$\begin{aligned} U(e^{j\omega}) &= \sum_{n=0}^{N-1} Ae^{-j\omega n} = A\frac{1-e^{-j\omega N}}{1-e^{-j\omega}} \\ &= A\frac{e^{-j\omega N/2}\left(e^{j\omega N/2} - e^{-j\omega N/2}\right)}{e^{-j\omega/2}\left(e^{j\omega/2} - e^{-j\omega/2}\right)} \\ &= Ae^{-j\omega(N-1)/2}\frac{\sin \omega N/2}{\sin \omega/2} \end{aligned}$$

Using

$$\lim_{\omega \to 0} \frac{\sin \omega N/2}{\sin \omega/2} = \lim_{\omega \to 0} \frac{N/2 \cos \omega N/2}{1/2 \cos \omega/2} = N,$$

we obtain

$$|U(e^{j\omega})| = \begin{cases} AN & \omega = 0 \\ A\left|\dfrac{\sin \omega N/2}{\sin \omega/2}\right| & \text{otherwise} \end{cases}$$

$$\mathsf{Arg}(U) = -\frac{\omega}{2}(N-1) + \mathsf{Arg}\left(\frac{\sin \omega N/2}{\sin \omega/2}\right)$$

The symmetry properties of Eq. 5.44 and Eq. 5.46 are once again satisfied. ∎

Properties of real and even signals: If u is real, we obtain from Eq. 5.41 the real and imaginary parts of U, respectively, as

$$\begin{aligned} U_R(e^{j\omega}) &= \sum_{n=-\infty}^{\infty} u(n) \cos \omega n \\ U_I(e^{j\omega}) &= -\sum_{n=-\infty}^{\infty} u(n) \sin \omega n \end{aligned} \tag{5.47}$$

If $u(n)$ is real and even, *i.e.*, $u(-n) = u(n)$, then $u(n) \cos \omega n$ is even and $u(n) \sin \omega n$ is odd. Hence, from Eq. 5.47, we obtain

$$\begin{aligned} U_R(e^{j\omega}) &= u(0) + 2\sum_{n=1}^{\infty} u(n) \cos \omega n \\ U_I(e^{j\omega}) &= 0 \end{aligned} \tag{5.48}$$

This shows that for real and even signals, the Fourier transform is real. In view of the fact that the imaginary part is zero, using Eq. 5.43, we obtain a simplified expression for the inverse Fourier transform as

$$u(m) = \frac{1}{\pi} \int_0^{\pi} U_R(e^{j\omega}) \cos \omega m \, d\omega \tag{5.49}$$

The fact that the Fourier transform of real and even signals is real has been demonstrated already in Example 5.3. We present one more example to illustrate this fact.

Example 5.5 A moving average filter of order n averages n consecutive inputs. Find the Fourier transform of the moving average filter of order three,

$$y(n) = \frac{1}{3}[u(n+1) + u(n) + u(n-1)]$$

We would like to determine the impulse response coefficients first. Let us write $y(n)$ as a convolution of impulse response and input:

$$y(n) = \sum_{k=-\infty}^{\infty} g(k)u(n-k) = g(-1)u(n+1) + g(0)u(n) + g(1)u(n-1)$$

Comparing the above two expressions for $y(n)$, we obtain

$$g(-1) = g(0) = g(1) = \frac{1}{3}$$

and g is zero at all other time instants. We can now calculate the Fourier transform:

$$\begin{aligned} G\left(e^{j\omega}\right) &= G(z)|_{z=e^{j\omega}} = \sum_{n=-\infty}^{\infty} g(n)z^{-n}|_{z=e^{j\omega}} \\ &= \frac{1}{3}\left(e^{j\omega} + 1 + e^{-j\omega}\right) = \frac{1}{3}(1 + 2\cos\omega) \end{aligned}$$

Observe that G is real for ω, which is not surprising as the time sequence is real and even. For ω in the range $[0, 2\pi/3)$, $\cos\omega > -1/2$. As a result, $G(e^{j\omega}) > 0$ in this range. For ω in the range $(2\pi/3, \pi]$, $\cos\omega < -1/2$ and, as a result, $G(e^{j\omega}) < 0$. In view of this, we obtain the phase of G as

$$Arg(G) = \begin{cases} 0 & 0 \le \omega < \dfrac{2\pi}{3} \\ \pi & \dfrac{2\pi}{3} < \omega \le \pi \end{cases}$$

As G is real, but with sign changes, we obtain

$$|G(e^{j\omega})| = \frac{1}{3}|(1 + 2\cos\omega)|$$

Plots of magnitude of G *vs.* ω using a log–log scale and the phase of G *vs.* ω using a semi-log scale are together known as the *Bode plot*. A Bode plot of this example, evaluated using M 5.3, is given in Fig. 5.7. ∎

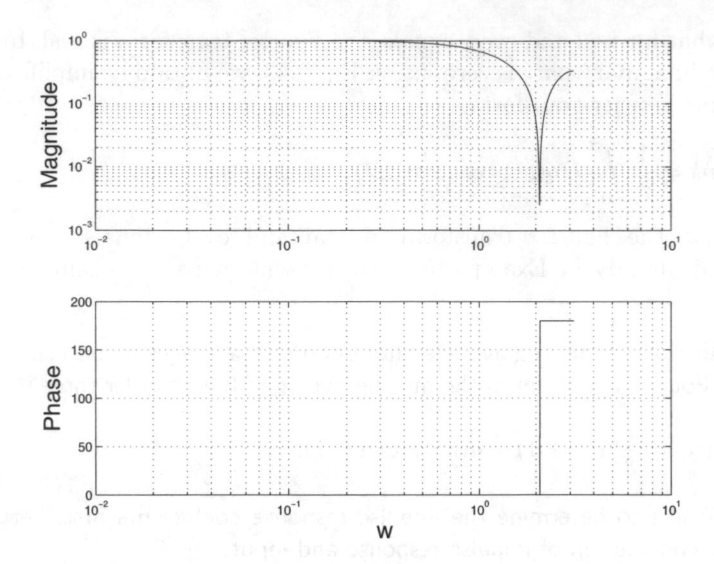

Figure 5.7: Bode plot of moving average filter

Example 5.6 Find the Fourier transform of the differencing filter,

$$y(n) = u(n) - u(n-1)$$

Using the procedure of Example 5.5, it is easy to verify that $g(0) = 1$, $g(1) = -1$ and all other terms are zero. We obtain

$$\begin{aligned} G\left(e^{j\omega}\right) &= G(z)|_{z=e^{j\omega}} = \sum_{n=-\infty}^{\infty} g(n)z^{-n}|_{z=e^{j\omega}} \\ &= 1 - e^{-j\omega} = e^{-j\omega/2}\left(e^{j\omega/2} - e^{-j\omega/2}\right) = 2je^{-j\omega/2}\sin\frac{\omega}{2} \\ |G| &= 2\left|\sin\frac{\omega}{2}\right| \end{aligned}$$

A Bode plot of this example is given in Fig. 5.8. M 5.4 is used to draw this. From this figure, one can see that the differencing filter is high pass. ∎

5.2.7 Parseval's Theorem

In this section, we will derive a relation for finite energy signals. The energy of a discrete time signal $\{u(n)\}$ has been defined in Eq. 3.2 on page 35; we reproduce it for convenience:

$$E_u = \sum_{n=-\infty}^{\infty} |u(n)|^2 \tag{5.50}$$

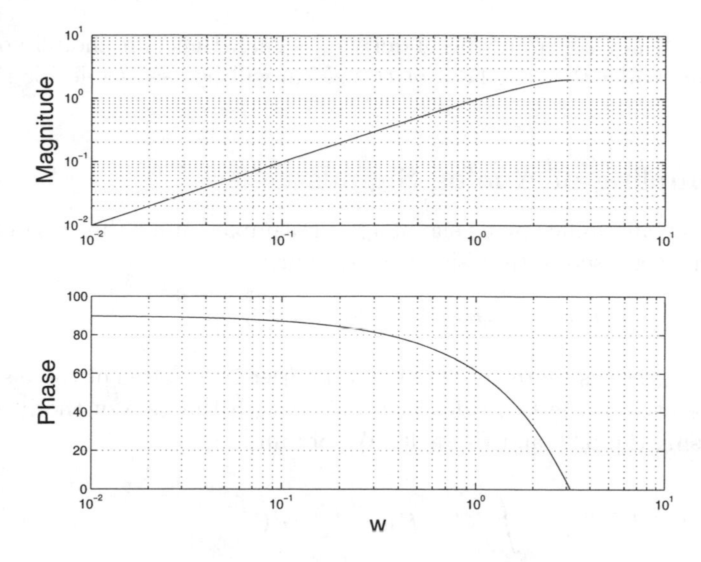

Figure 5.8: Bode plot of differencing filter

We write $|u(n)|^2$ as $u(n)u^*(n)$ and express $u^*(n)$ with the inverse Fourier transform, given by Eq. 5.39:

$$E_u = \sum_{n=-\infty}^{\infty} u(n)u^*(n) = \sum_{n=-\infty}^{\infty} u(n) \left[\frac{1}{2\pi} \int_{-\pi}^{\pi} U^*(e^{j\omega}) d\omega \right]$$

As u is a finite energy signal, it is possible to exchange the order of integration and summation. We obtain

$$E_u = \frac{1}{2\pi} \int_{-\pi}^{\pi} U^*(e^{j\omega}) \left[\sum_{n=-\infty}^{\infty} u(n)e^{-j\omega n} \right] d\omega = \frac{1}{2\pi} \int_{-\pi}^{\pi} |U(e^{j\omega})|^2 d\omega$$

We arrive at the relation

$$E_u = \sum_{n=-\infty}^{\infty} |u(n)|^2 = \frac{1}{2\pi} \int_{-\pi}^{\pi} |U(e^{j\omega})|^2 d\omega \tag{5.51}$$

5.3 Sampling and Reconstruction

We saw in Example 5.1 on page 119 the pitfalls that result from not sampling a signal at a fast enough rate. Using Fourier transforms, we will study this problem in more detail. This will also help determine the minimum rate at which a continuous signal has to be sampled and how to reconstruct the continuous signal, given the discrete time signal.

It is not difficult to understand that a frequency domain perspective is useful in the sampling rate selection problem. For example, we expect that large sampling periods T_s will be sufficient for slowly varying signals. Similarly, small T_s will be required for fast changing signals. It is obvious that the phrases *slowly varying* and *fast changing* denote the frequency aspect of the underlying signal. Hence, it should not come as a surprise to the reader that the Fourier transform will be used to decide the sampling rate.

5.3.1 Sampling of Analog Signals

Suppose that an aperiodic analog signal $u_a(t)$ is sampled at a uniform sampling period of T_s to produce a discrete time signal $u(n)$. Then,

$$u(n) = u_a(nT_s), \quad -\infty < n < \infty \tag{5.52}$$

Suppose that $u_a(t)$ has finite energy. Then, Fourier transforms of these signals exist. We will write each of the two sides in Eq. 5.52 with the help of the inverse Fourier transform, using Eq. 5.29 and Eq. 5.40. We obtain

$$\int_{-1/2}^{1/2} U(f)e^{j2\pi fn}df = \int_{-\infty}^{\infty} U_a(F)e^{j2\pi FnT_s}dF$$

where we have replaced $U(e^{j2\pi f})$ with the simplified notation of $U(f)$. Because $T_s = 1/F_s$, we obtain

$$\int_{-1/2}^{1/2} U(f)e^{j2\pi fn}df = \int_{-\infty}^{\infty} U_a(F)e^{j2\pi nF/F_s}dF \tag{5.53}$$

The left-hand side of the above equation becomes

$$\int_{-1/2}^{1/2} U(f)e^{j2\pi fn}df = \frac{1}{F_s}\int_{-F_s/2}^{F_s/2} U\left(\frac{F}{F_s}\right) e^{j2\pi nF/F_s}dF \tag{5.54}$$

while the right-hand side becomes

$$\int_{-\infty}^{\infty} U_a(F)e^{j2\pi nF/F_s}dF = \sum_{k=-\infty}^{\infty} \int_{(k-1/2)F_s}^{(k+1/2)F_s} U_a(F)e^{j2\pi nF/F_s}dF$$

Substituting $Q = F - kF_s$,

$$= \sum_{k=-\infty}^{\infty} \int_{-F_s/2}^{F_s/2} U_a(Q + kF_s)e^{j2\pi n(Q+kF_s)/F_s}dQ$$

Because $e^{j2\pi n} = 1$,

$$= \sum_{k=-\infty}^{\infty} \int_{-F_s/2}^{F_s/2} U_a(Q + kF_s)e^{j2\pi nQ/F_s}dQ$$

Because Q is dummy,

$$= \sum_{k=-\infty}^{\infty} \int_{-F_s/2}^{F_s/2} U_a(F + kF_s)e^{j2\pi nF/F_s}dF$$

Finally, exchanging the integral and the sum, we obtain

$$\int_{-\infty}^{\infty} U_a(F) e^{j2\pi nF/F_s} dF = \int_{-F_s/2}^{F_s/2} \left(\sum_{k=-\infty}^{\infty} U_a(F + kF_s) \right) e^{j2\pi nF/F_s} dF \tag{5.55}$$

Because of Eq. 5.53, Eq. 5.54 and Eq. 5.55 are equal. So we obtain

$$\frac{1}{F_s} \int_{-F_s/2}^{F_s/2} U\left(\frac{F}{F_s}\right) e^{j2\pi nF/F_s} dF = \int_{-F_s/2}^{F_s/2} \left(\sum_{k=-\infty}^{\infty} U_a(F + kF_s) \right) e^{j2\pi nF/F_s} dF$$

From this it follows that

$$\frac{1}{F_s} U\left(\frac{F}{F_s}\right) = \sum_{k=-\infty}^{\infty} U_a(F + kF_s)$$

or equivalently,

$$U\left(\frac{F}{F_s}\right) = F_s \sum_{k=-\infty}^{\infty} U_a(F + kF_s) \tag{5.56}$$

A study of this equation will help us determine the minimum required speed of sampling. The above equation can be written for a specific F_0, $F_o/2 > F_0 > -F_o/2$, as follows:

$$U\left(\frac{F_0}{F_s}\right) = F_s[\cdots + U_a(F_0 - F_s) + U_a(F_0) + U_a(F_0 + F_s) + \cdots] \tag{5.57}$$

In Fig. 5.9, a function $u_a(t)$ and its Fourier transform $U_a(F)$ are plotted. It is assumed that $U_a(F) = 0$ for $F > B$, *i.e.*, B is its bandwidth. It is also assumed that $\frac{F_s}{2} > B$, *i.e.*, a fast sampling rate is used and hence $U_a(F) = 0$ for $F > \frac{F_s}{2}$. In other words, $U_a(F_0 + kF_s) = 0,\ k = \pm 1, \pm 2, \ldots$. Using this fact in Eq. 5.57, we obtain

$$U\left(\frac{F_0}{F_s}\right) = F_s U_a(F_0)$$

i.e., U is a scaled version of U_a. In other words, the shape of the Fourier transform of the analog signal is not affected by sampling. Thus for $F_s/2 > F > -F_s/2$, the plot of $U(F/F_s)$ looks similar to $U_a(F)$. From Eq. 5.56, however, notice that $U(F/F_s)$ is periodic in F with a period F_s, *i.e.*,

$$U\left(\frac{F}{F_s}\right) = U\left(\frac{F+F_s}{F_s}\right) = U\left(\frac{F-F_s}{F_s}\right) = \cdots = U\left(\frac{F+kF_s}{F_s}\right),\ k = \pm 1,\ \pm 2$$

A plot of $U(F/F_s)$ reflects this fact in Fig. 5.9. The right-hand side consists of a periodic repetition of the scaled spectrum of $F_s U_a(F)$ with period F_s.

If sampled fast enough, we will not lose any information, as shown in Fig. 5.9. We can recover $U_a(F)$ from $U(F/F_s)$. If not sampled fast enough, aliasing takes place as shown in Fig. 5.10. We cannot recover $U_a(F)$ from $U(F/F_s)$, because the high frequency components have changed.

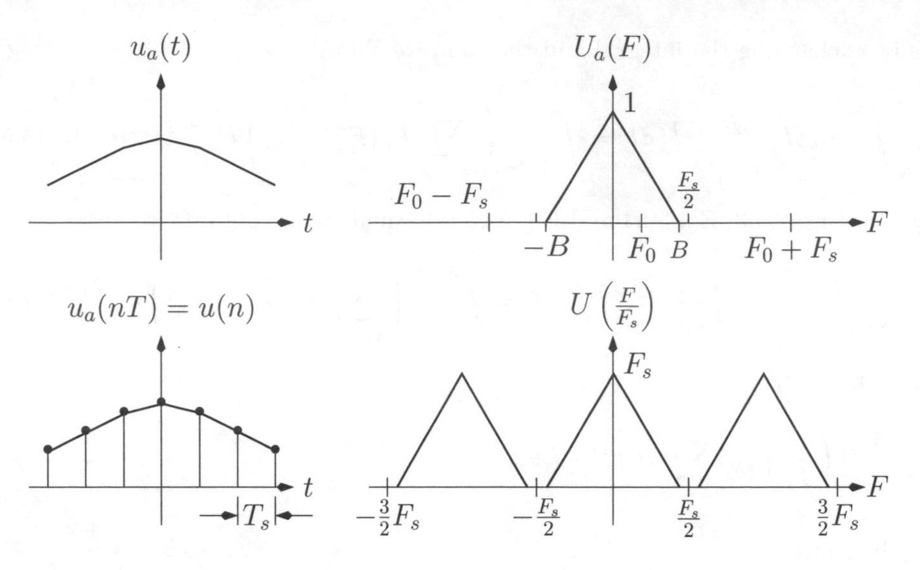

Figure 5.9: Fast sampling preserves all required information. The central part of U is identical in shape to that of U_a.

Sometimes aliasing cannot be avoided. This will happen when the Fourier transform of the continuous signal never vanishes. For example, consider Fig. 5.11. We can see that the Fourier transform of the sampled signal is a superposition of frequency shifted signals. In a situation like this, there will always be aliasing irrespective of the sampling rate. The only way to handle this problem is to filter the high frequency components before sampling, as shown in Fig. 5.12.

5.3.2 Reconstruction of Analog Signal from Samples

In the previous section, we have presented the condition to be satisfied so that the Fourier transform of the discrete time signal will have no distortions. Suppose that we sample at the required rate and obtain discrete time signals. Observe that this is the only measurement we have of the continuous time signal, *i.e.*, we do not measure the values in between the sampling instants. Given that we have not lost any information while sampling, can we recover the continuous time signal completely? Because this method of reconstructing the continuous time signal is likely to be more accurate than ZOH, presented in Sec. 2.4.2 and Sec. 2.5, can we use it for control purposes?

To answer these questions, let us first see if it is possible to reconstruct the continuous time signal from the sampled data. Let us assume that we have sampled the analog signals fast enough and hence that there is no aliasing. As a result, let

$$U_a(F) = \begin{cases} \dfrac{1}{F_s} U\left(\dfrac{F}{F_s}\right) & |F| \le F_s/2 \\ 0 & |F| > F_s/2 \end{cases} \tag{5.58}$$

Let us proceed to construct the analog signal from the samples. Recall the inverse Fourier transform relation given in Eq. 5.28 on page 123:

$$u_a(t) = \int_{-\infty}^{\infty} U_a(F) e^{j2\pi F t} dF$$

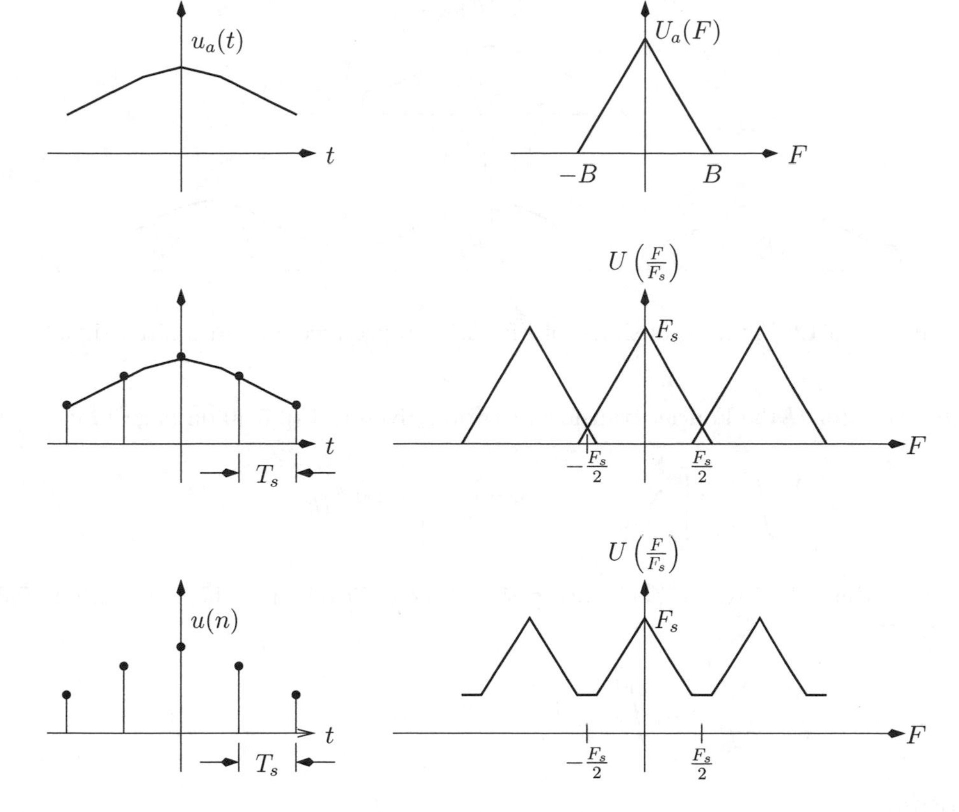

Figure 5.10: Slow sampling results in aliasing. The central part of U is no longer identical to U_a and, as a result, U_a cannot be recovered from U.

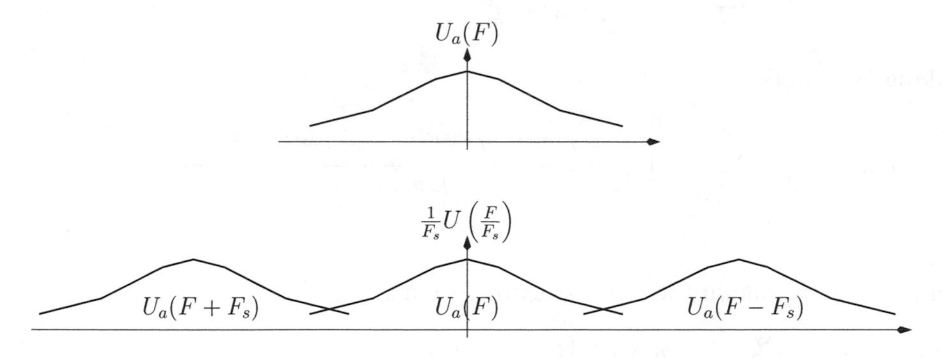

Figure 5.11: Signal that is not band limited results in aliasing, whatever the sampling rate

Next, substitute for U_a from Eq. 5.58:

$$u_a(t) = \frac{1}{F_s}\int_{-F_s/2}^{F_s/2} U\left(\frac{F}{F_s}\right) e^{j2\pi Ft} dF$$

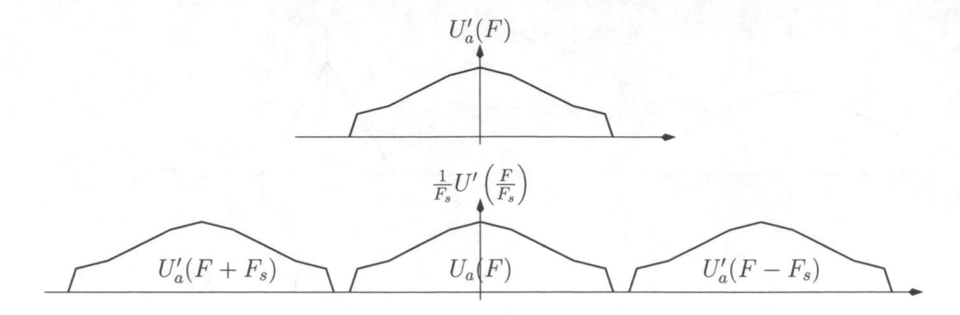

Figure 5.12: Filtering of signal of Fig. 5.11 to produce a band limited signal

Substitute for U the Fourier transform relation given in Eq. 5.40 on page 125:

$$u_a(t) = \frac{1}{F_s}\int_{-F_s/2}^{F_s/2}\left[\sum_{n=-\infty}^{\infty} u(n)e^{-j2\pi nF/F_s}\right]e^{j2\pi Ft}dF$$

Assume that this expression converges and hence exchange the summation and integration:

$$u_a(t) = \frac{1}{F_s}\sum_{n=-\infty}^{\infty} u(n)\int_{-F_s/2}^{F_s/2} e^{j2\pi F(t-n/F_s)}dF$$

Integrating,

$$u_a(t) = \frac{1}{F_s}\sum_{n=-\infty}^{\infty} u_a(nT)\frac{1}{j2\pi\left(t-\frac{n}{F_s}\right)}\left.e^{j2\pi F(t-n/F_s)}\right|_{-F_s/2}^{F_s/2}$$

Taking the limits,

$$u_a(t) = \frac{1}{F_s}\sum_{n=-\infty}^{\infty} u_a\left(\frac{n}{F_s}\right)\frac{e^{j2\pi(t-n/F_s)F_s/2} - e^{-j2\pi(t-n/F_s)F_s/2}}{j2\pi(t-n/F_s)}$$

Simplifying, we obtain the reconstruction formula,

$$u_a(t) = \sum_{n=-\infty}^{\infty} u_a\left(\frac{n}{F_s}\right)\frac{\sin\left\{\pi\left(t-\frac{n}{F_s}\right)F_s\right\}}{\pi\left(t-\frac{n}{F_s}\right)F_s} \tag{5.59}$$

This is also known as the *ideal interpolation formula*. Let us summarize our findings.

If the highest frequency contained in any analog signal $u_a(t)$ is $F_{\max} = B$ and the signal is sampled at a rate $F_s > 2F_{\max} = 2B$, then $u_a(t)$ can be exactly recovered from its sample values using the interpolation function

$$g(t) = \frac{\sin(2\pi Bt)}{2\pi Bt} \tag{5.60a}$$

Thus

$$u_a(t) = \sum_{n=-\infty}^{\infty} u_a\left(\frac{n}{F_s}\right) g(t - n/F_s) \tag{5.60b}$$

where

$$u_a\left(\frac{n}{F_s}\right) = u_a(nT) = u(n) \tag{5.60c}$$

are the samples of $u_a(t)$. Thus, we have answered the first question of whether we can uniquely reconstruct the analog signal in the affirmative.

Now, let us look at the second question of whether we can use this reconstruction procedure to replace ZOH. The reconstruction procedure indicated in Eq. 5.59–5.60 is not causal: the sum requires *all* samples, including the ones yet to be obtained. For example, the reconstruction formula requires samples from the infinite future as well. As a result, this reconstruction is not useful for real time applications, such as control.

What is the utility of this result, if it is not causal and hence not useful for real time applications? This important result provides the absolute minimum limit of sampling rate. If the sampling rate is lower than this minimum, then no filter (whether causal or not) can achieve exact reproduction of the continuous function from the sampled signals. Moreover, in certain applications, such as image processing, the reconstruction takes place offline, *i.e.*, reconstruction is attempted on all sampled data already obtained. Here, the variable t can be taken to denote the space and the results derived earlier can be used.

We conclude this section by stating that if $F_s = 2F_{\max}$, F_s is denoted by F_N, the *Nyquist rate.*

5.3.3 Frequency Domain Perspective of Zero Order Hold

Although the Shannon sampling theorem gives a way of reconstructing the original signal from samples, ZOH is more practical as it is a causal function. Suppose $u(k)$ is the sample sequence and we want to produce a continuous function out of this. We define $u_h(t)$ as

$$u_h(t) = u(kT_s), \quad kT_s \le t < (kT_s + T_s)$$

As u_h is composed of a zero order polynomial passing through the samples $u(kT_s)$, this hold operation is called the zero order hold or ZOH. The result of applying ZOH on sampled signals is that we obtain a staircase waveform. It is clear that if we sample fast enough, the staircase becomes a better approximation of the original, smooth, waveform. It is also clear that if the original signal sample is one of high frequency, we need to sample it fast so as to make the staircase waveform a good approximation. It is clear from this that the goodness of ZOH is dependent on the frequency content of the original waveform. In order to get a better feel for this, we would like to study the effect of ZOH in the frequency domain.

The process of sampling and ZOH, which will be abbreviated as SH, is not a time invariant function. This will be clear if one studies Fig. 5.13, where a continuous time

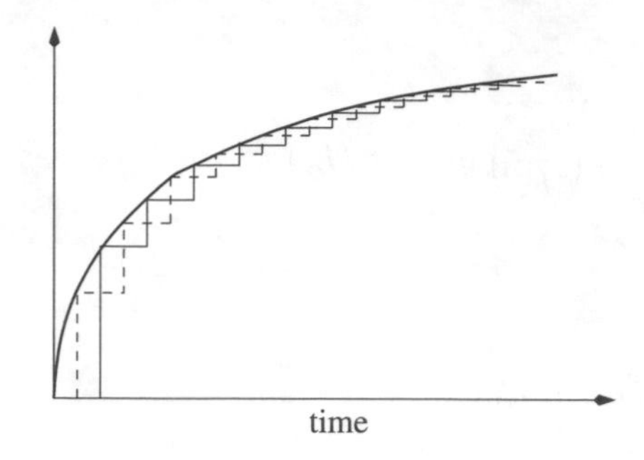

Figure 5.13: Staircase waveforms obtained for different starting sampling times. The sampled waveforms are not time shifted versions of each other.

function is sampled, followed by ZOH, for two different starting points, arriving at two staircase forms. If the operation of SH, considered as a single operation, is time invariant, the two staircases will be time shifted versions of each other. It is clear from this figure, though, that this is not the case. As a result, SH is not time invariant. Nevertheless, we will pretend that SH is LTI and carry out the following approximate analysis, as it provides a frequency domain perspective of SH.

Note that when a unit impulse $\{\delta(n)\}$ is sent through a ZOH, the resulting output is as shown in Fig. 5.6 on page 123. In other words, Fig. 5.6 has the impulse response of a ZOH. Its Fourier transform has been calculated in Eq. 5.30. Using $\Omega = 2\pi F = 2\pi f/T_s$, we obtain $\Omega T_s/2 = \pi f$. Thus Eq. 5.30 becomes

$$\mathrm{ZOH} = T_s e^{-j\pi f}\frac{\sin \pi f}{\pi f} \tag{5.61}$$

Now we will see the effect of sampling and ZOH together. If the sampling is carried out fast enough, the transfer function of sampling is F_s, see Fig. 5.9. Thus the transfer function of sampling and ZOH, denoted as SH, is given as

$$\mathrm{SH} = F_s T_s e^{-j\pi f}\frac{\sin \pi f}{\pi f} = e^{-j\pi f}\frac{\sin \pi f}{\pi f} \tag{5.62}$$

since $F_s = 1/T_s$. If SH were equal to one, there would be zero distortion while going from a continuous to a discrete time signal. Let us now evaluate the actual distortion in the SH operation. From the above equation, we obtain the magnitude of the SH operation as

$$|\mathrm{SH}| = \left|\frac{\sin \pi f}{\pi f}\right| \tag{5.63}$$

A plot of |SH|, as a function of f, is shown in Fig. 5.14. For f values close to 0, $|SH|$ is approximately one and hence there is not much distortion. For values of f close to the maximum value of 0.5, however, the gain is far from one. If only we can operate the system at low values of f, the distortion due to sampling will be small. We now show that this can be achieved by increasing the sampling frequency.

Consider a band limited continuous signal as shown in Fig. 5.9. The frequency over which the sampled function is nonzero is $(0, B)$. Recall from Eq. 5.15 on page 119

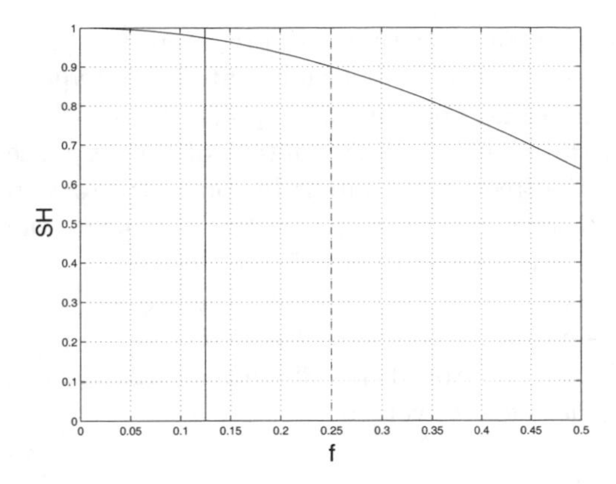

Figure 5.14: Magnitude Bode plot of SH and frequency limits when sampled at $F_s = 4B$ or twice the minimum rate (dashed vertical line) and at $F_s = 8B$ or four times the minimum rate (solid vertical line)

that the normalized frequency is $f = F/F_s$. As a result, the maximum normalized frequency becomes $f_{\max} = B/F_s$. Hence, if the sampling rate F_s is twice the band, *i.e.*, $F_s = 2B$, we obtain $f_{\max} = 1/2$. As a result, the normalized frequency is nonzero only in the interval (0,0.5). Note that this is the absolute minimum sampling frequency, as dictated by the sampling theorem. Let this minimum sampling frequency be denoted by F_{s_0}. In other words, we obtain

$$\frac{B}{F_{s_0}} = 0.5 \tag{5.64}$$

Suppose that we now explore the use of a sampling frequency F_{s_1} at twice the minimum frequency, *i.e.*, $F_{s_1} = 2F_{s_0}$. Then, using Eq. 5.64, we obtain $B/F_{s_1} = 0.25$. Because B is the largest frequency present in the continuous time signal, the largest f that we have to deal with is precisely 0.25. In other words, for all other frequencies present in the input signal, the f value will be even smaller. This frequency limit is indicated by the dashed vertical line in Fig. 5.14. The maximum distortion is produced at $f = 0.25$ and it is about 10%. If, instead, we use four times the minimum frequency F_{s_2}, *i.e.*, $F_{s_2} = 4F_{s_0}$, using the same argument as above, we obtain $B/F_{s_2} = 0.125$. We once again indicate the working frequency range by drawing a solid vertical line at $f = 0.125$ in Fig. 5.14. The maximum distortion is now produced at $f = 0.125$ and it is about 3%. It is easy to see that as the sampling frequency increases, the distortion in the reconstructed signal decreases.

Thus, even though the above analysis is only approximate, it agrees with our intuition that as the sampling frequency increases, the discrete time signal, obtained through SH, gets closer to the continuous time signal.

5.4 Filtering

Measurements are often corrupted by high frequency noise, and have to be filtered before further processing. Systems that transmit the low frequency information while removing the effect of high frequency noise are known as *low pass filters* and this action is known as *low pass filtering*. Sometimes we are interested in monitoring a transient response so as to take early corrective action. Often this requires a derivative action that works on the basis of the slope of the response curve. We will see later that this requires the usage of the high frequency content of the response. Indeed we may be interested in filtering the frequency content in some arbitrary frequency range while passing the others. We now show how to achieve this using the frequency response property of LTI systems, discussed in Sec. 5.2.3.

We have seen in Sec. 5.2.3 that when the input to an LTI system is a sinusoid, the output also is a sinusoid of the same frequency, but with a phase shift and an amplitude change. Recall also from Eq. 5.35 on page 124 the expression for the filter output y, reproduced here for convenience:

$$y(n) = |G(e^{j\omega n})|e^{j(wn+\varphi)} \tag{5.65}$$

where the input is e^{jwn} and G is the transfer function of the LTI system. Observe from this equation that at frequencies where $|G(e^{j\omega})|$ is large, the sinusoid is amplified and at those frequencies where it is small, the sinusoid is attenuated:

1. The systems with large gains at low frequencies and small gains at high frequencies are called *low pass filters* as these transmit low frequency signals without much loss in information.
2. On the other hand, the systems that have high gain at high frequencies and small gain at low frequencies are called *high pass filters*.
3. A filter that filters out a specific frequency is called a *notch filter*, often used to filter the harmonics associated with the power source.
4. A filter that amplifies frequencies in a specified frequency range $[\omega_1, \omega_2]$ is known as a *band pass filter*. This frequency range is known as the pass band of this filter. The range $[\omega_3, \omega_4]$ is known as a stop band when signals with frequency in this range are stopped or filtered out.

Other filtering characteristics lead to other kinds of filters. We will look at some more definitions associated with filtering. The cutoff frequency of ω_c of a filter is defined as the value of ω such that $|G(e^{j\omega_c})| = \hat{G}/2$, where $\hat{G}$ is $|G(e^{j0})|$ for low pass, $|G(e^{j\pi})|$ for high pass and the maximum value of $|G(e^{j\omega})|$ in the pass band for the band pass filter.

The bandwidth of a low pass filter is the frequency range $[0, \omega_c]$, where ω_c is the cutoff frequency. Pass band filters have two cutoff frequencies, $\omega_{c2} > \omega_{c1} > 0$, and for these, $[\omega_{c1}, \omega_{c2}]$ is the bandwidth.

5.4.1 Pole–Zero Location Based Filter Design

There are many different ways to design filters. Indeed, there are entire books devoted exclusively to this topic. In this section, we will look at the method of filter design through appropriate choice of pole and zero locations. This quick and simple method helps arrive at first-cut filters, capable of meeting all needs, except the most demanding ones.

Consider the application of the input $u(k) = a^k 1(k)$ to an LTI plant with transfer function $G(z)$. As we are interested in finding out what happens to the frequency content of u by the action of $G(z)$, we will take a to be of the form $e^{j\omega}$. Let the output from the LTI system be $y(k)$. Then

$$Y(z) = G(z)\frac{z}{z-a} \tag{5.66}$$

Suppose that $G(z)$ does not have a pole at a. On expanding by partial fractions, we obtain

$$Y(z) = e_0 + e_1\frac{z}{z-a} + \{\text{terms due to the poles of } G(z)\} \tag{5.67}$$

Notice that the effect of the input u on the output y depends on e_1. If e_1 is large, this input component is present in the output y, while if e_1 is small, the effect of u is removed in the output. Substituting for Y in Eq. 5.67 from Eq. 5.66, we obtain

$$G(z)\frac{z}{z-a} = e_0 + e_1\frac{z}{z-a} + \{\text{terms due to the poles of } G(z)\}$$

From this, we arrive at the value of e_1 as

$$e_1 = \frac{z-a}{z}G(z)\frac{z}{z-a}\bigg|_{z=a} = G(a) \tag{5.68}$$

From this equation, it is clear that if we want to pass the input signal a^k in the output, $G(a)$ should be large, while a small $G(a)$ would result in the attenuation of this input signal in the output. Large $G(a)$ can be achieved if $G(z)$ has a pole close to a while a zero of $G(z)$ near a will ensure the reduction of the effect of u on the output.

In summary, a designer should place the poles of the transfer function near the frequencies of the input signal that they want to transmit. Similarly, the zeros should be placed at the frequencies of the input that they want to reject.

We have already seen that the frequency response characteristics of a system with impulse response $g(n)$ are given by its Fourier transform $G(e^{j\omega})$. We also know that the range of unique values of G are obtained over $-\pi < \omega \leq \pi$ only. We have seen earlier that values of ω close to 0 correspond to low frequencies while those close to $\pm\pi$ correspond to high frequencies as far as the sampled signal $g(n)$ is concerned. Notice that $e^{j\omega}$ with $\omega \in (-\pi, \pi]$ defines the unit circle. As a result, we can mark the low and high frequency regions as in Fig. 5.15. From the earlier discussion, to pass signals of frequency ω_1, we should place poles around that frequency and, to reject ω_2, we should place zeros around that. In a low pass filter, we pass low frequency signals and reject high frequency signals. Thus a low pass filter is of the form given on the left-hand side of Fig. 5.16. Notice that the poles are generally placed inside the unit circle so that the system is stable. Although we have shown three poles and three zeros, there can be more or less number of poles. The only thing we have to ensure is that the complex poles or zeros should occur in conjugate pairs. An example of a high pass filter is given on the right-hand side of the same figure.

Example 5.7 Draw the Bode plots of two filters with transfer function $G_1(z) = 0.5/(z-0.5)$ and $G_2(z) = 0.25(z+1)/(z-0.5)$ and comment on their filtering characteristics.

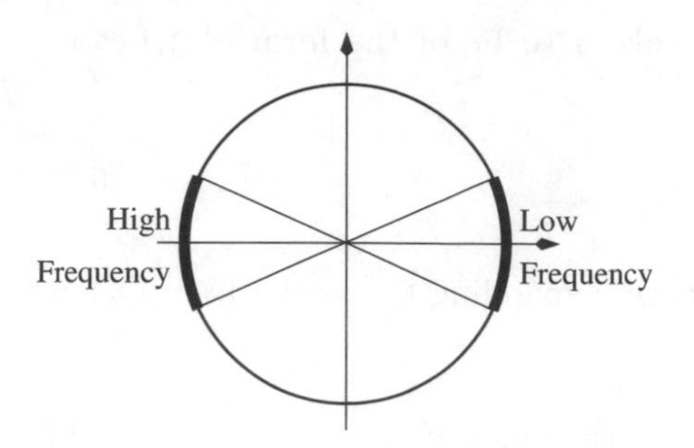

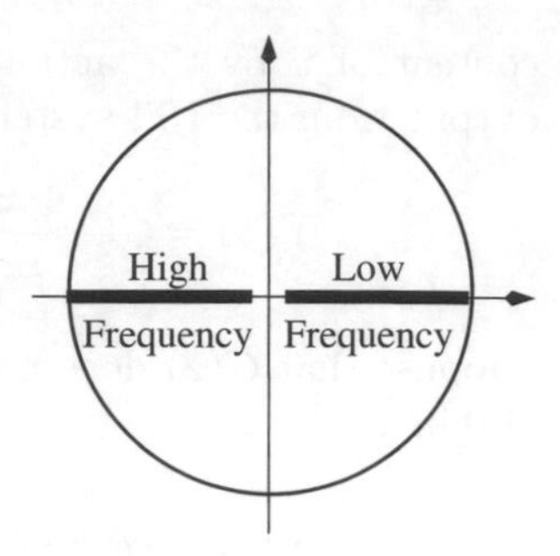

Figure 5.15: High and low frequency regions

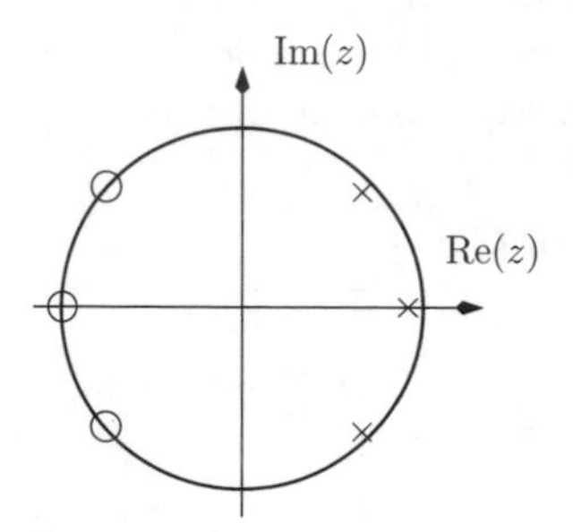

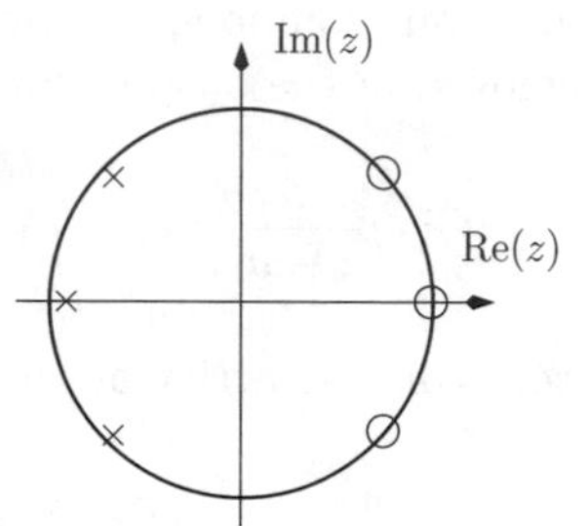

Figure 5.16: Pole and zero locations of a low pass filter (left) and of a high pass filter (right)

Note that $G_1(z)|_{z=1} = 1$, so that its steady state gain is 1, see Sec. 4.3.1. Substituting $z = e^{j\omega}$, we obtain

$$\begin{aligned} G_1(e^{j\omega}) &= \frac{0.5}{e^{j\omega} - 0.5} = \frac{0.5}{(\cos\omega - 0.5) + j\sin\omega} \\ &= 0.5\frac{(\cos\omega - 0.5) - j\sin\omega}{(\cos\omega - 0.5)^2 + \sin^2\omega} \end{aligned}$$

As we want to draw the Bode plot, let us evaluate the magnitude and the phase:

$$\begin{aligned} |G_1(e^{j\omega})| &= \frac{0.5}{\sqrt{1.25 - \cos\omega}} \\ \angle G_1(e^{j\omega}) &= -\tan^{-1}\left(\frac{\sin\omega}{\cos\omega - 0.5}\right) \end{aligned}$$

The Bode plot for this system, obtained using M 5.1, is drawn with solid lines in Fig. 5.17. From this plot it is clear that this filter only magnifies the signal frequencies near $\omega = 0$ in relation to other frequencies. As a result, it is a low pass filter.

The filter G_2, on the other hand, actively rejects high frequencies with the help of a zero. Note that the steady state gain of G_2 also is one, *i.e.*, $|G_2(z)|_{z=1} = 1$. Let the filter now have the transfer function. Let us now calculate the magnitude

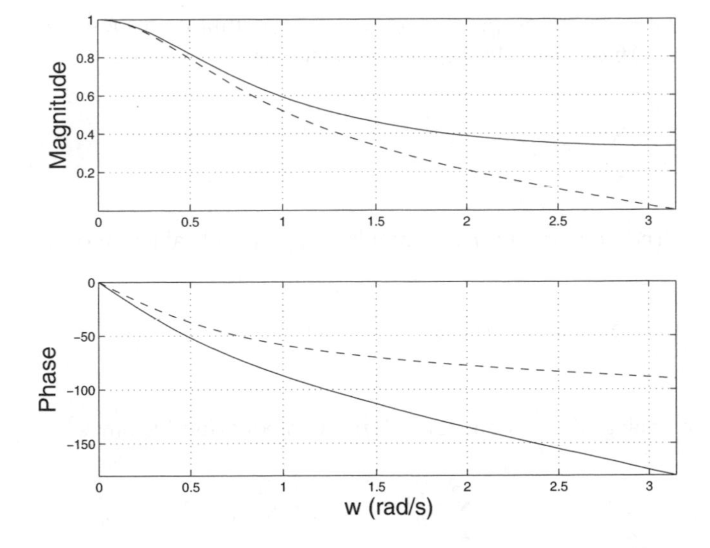

Figure 5.17: Bode plot of $G_1(z) = 0.5/(z-0.5)$ (solid line) and $G_2(z) = 0.25/(z+1)(z-0.5)$ (broken line)

and phase of G_2:

$$\frac{G_2(e^{j\omega})}{K} = \frac{e^{j\omega}+1}{e^{j\omega}-0.5} = \frac{\cos\omega + j\sin\omega + 1}{\cos\omega + j\sin\omega - 0.5}$$

$$= \frac{[(\cos\omega+1) + j\sin\omega][(\cos\omega-0.5) - j\sin\omega]}{(\cos\omega-0.5)^2 + \sin^2\omega}$$

$$\frac{G_2(e^{j\omega})}{0.25} = \frac{(\cos^2\omega + 0.5\cos\omega - 0.5) + \sin^2\omega}{\sin^2\omega + \cos^2\omega + 0.25 - \cos\omega} + \frac{j\sin\omega(\cos\omega - 0.5 - \cos\omega - 1)}{\sin^2\omega + \cos^2\omega + 0.25 - \cos\omega}$$

$$G_2 = \frac{(0.5 + 0.5\cos\omega) - 1.5j\sin\omega}{1.25 - \cos\omega} 0.25$$

The Bode plot of G_2 is shown with dashed lines in M 5.1. Notice that $|G_2(e^{j\omega})| < |G_1(e^{j\omega})|$, $\forall\omega > 0$. Thus, G_2, which actively rejects high frequency input with the help of a zero, is a better low pass filter. ∎

Example 5.8 Study how the filter $G_3(z) = (z+1)/(z-1)$ handles the input signal $u(n) = (-1)^n 1(n)$.

First, we split $G_3(z)$ as

$$G_3(z) = \frac{z}{z-1} + \frac{1}{z-1}$$

Inverting this, we obtain the following impulse response:

$$g_3(n) = 1(n) + 1(n-1)$$

Recall the expression for output as a convolution of impulse response and the input, given by Eq. 3.26 on page 48, reproduced here for convenience:

$$y(n) = \sum_{i=-\infty}^{\infty} g(i)u(n-i)$$

Substituting the expression for the impulse response obtained above, we obtain

$$y(n) = \sum_{i=-\infty}^{\infty} [1(i) + 1(i-1)]u(n-i)$$

Using the meaning of the step signal $1(n)$, this equation becomes

$$y(n) = \sum_{i=0}^{\infty} u(n-i) + \sum_{i=1}^{\infty} u(n-i) = 2\sum_{i=0}^{n} u(n-i) - u(n)$$

We will now use the statement of the problem, namely $u(n) = (-1)^n 1(n)$:

$$y(n) = \left[2\sum_{i=0}^{n}(-1)^{n-i}\right] 1(n) - (-1)^n 1(n)$$

With the substitution of $k = n - i$, this becomes

$$\begin{aligned} y(n) &= \left[2\sum_{k=n}^{0}(-1)^{k}\right] 1(n) - (-1)^n 1(n) \\ &= 2\frac{1-(-1)^{n+1}}{1-(-1)} 1(n) - (-1)^n 1(n) \\ &= 1(n)\left[1 - (-1)^{n+1} - (-1)^n\right] = 1(n) \end{aligned}$$

This shows that $(-1)^n$ has been filtered in the output. This is only expected, of course, because the filter has a zero at $z = -1$.

∎

5.4.2 Classification of Filters by Phase

We will now classify the FIR or all zero systems, defined in Sec. 4.3, according to their phase characteristic. An FIR filter that has the smallest phase change as ω goes from 0 to π is known as the *minimum phase* filter. Minimum phase filters have all their zeros inside the unit circle. Filters that have some of their zeros outside the unit circle are called *nonminimum phase* filters. The phase Bode plot of these systems shows a net phase change as ω goes from 0 to π. An example of minimum and nonminimum phase filters is given in the next example.

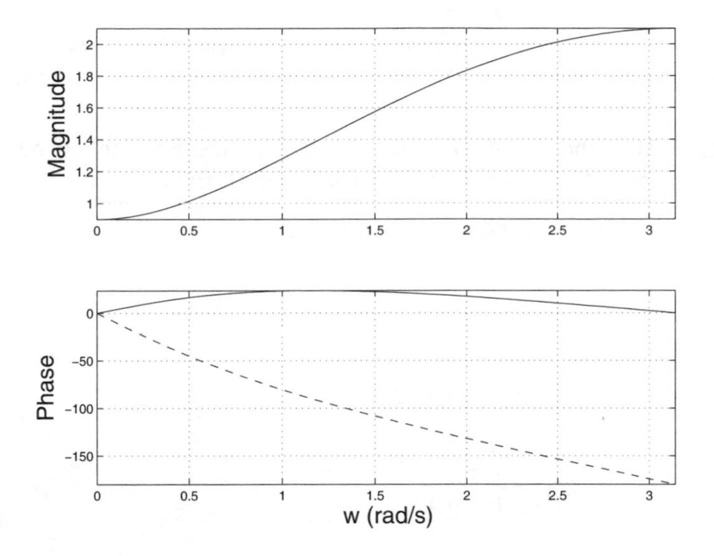

Figure 5.18: Bode plot of $G_1(z) = 1.5(1 - 0.4z^{-1})$ (solid line) and $G_2(z) = -0.6(1 - 2.5z^{-1})$ (broken line), as discussed in Example 5.9

Example 5.9 Draw the Bode plots of the following transfer functions

$$G_1(z) = 1.5(1 - 0.4z^{-1})$$
$$G_2(z) = -0.6(1 - 2.5z^{-1})$$

and verify the above mentioned phase characteristic.

In Fig. 5.18, the Bode plots of G_1 and G_2 are drawn with solid and broken lines, respectively. Note that the magnitude plots of these two functions coincide. This is because the zeros of G_1 and G_2 are reciprocals of each other and the formula for the magnitude of $|G(e^{j\omega})|^2 = |G(z)G(z^{-1})|_{z=e^{-j\omega}}$. Note that in this expression, $G(z)$ and $G(z^{-1})$ are polynomials in z and z^{-1}, respectively. As expected, the phase angle of G_1 does not show a net change, which is not the case with G_2. M 5.5 shows how these plots are created using Matlab. ∎

It is more difficult to design controllers for nonminimum phase systems, compared to the minimum phase systems. For example, see Sec. 9.2 and Sec. 11.2.3. Because the magnitude plots of a nonminimum phase and of the corresponding minimum phase filters are identical, there is a loss of uniqueness while identifying systems from input–output data. The standard practice is to choose minimum phase systems. The details are in Sec. 6.4.4. A continuous time minimum phase system could lose this property on sampling. Problem 8.4 shows how this happens. A standard reference that discusses the fundamentals of nonminimum phase systems is [46].

If every zero of an FIR system is outside the unit circle, it is known as a *maximum phase* filter. The phase of this filter changes the maximum as ω is varied from 0 to π. When some zeros are inside and the rest outside the unit circle, it is known as a *mixed phase* or nonminimum phase system.

A filter that has the same magnitude at all frequencies is known as an *all pass* filter. Because the magnitude is constant at all frequencies, the zeros and the poles are reciprocals of each other. As a result of this, it is possible to write any filter as a product of minimum phase and all pass filters, as we see in the next example.

Example 5.10 Show that G_1 and G_2 of Example 5.9 can be related by a transfer function, whose magnitude is constant at all frequencies.

G_1 and G_2 are related by the following equation:

$$G_2 = G_1(-0.4)\frac{1-2.5z^{-1}}{1-0.4z^{-1}}$$

We need to show that $G_3(z)$, defined as

$$G_3 = \frac{1-2.5z^{-1}}{1-0.4z^{-1}}$$

has a constant magnitude at all frequencies. We have been using the argument z to indicate polynomials in powers of z^{-1} as well, see Footnote 5 on page 100. In this example, we have to use polynomials in z as well. In view of this, we will explicitly use the dependence on z^{-1} and z. In other words, we let

$$G_3(z^{-1}) = \frac{1-2.5z^{-1}}{1-0.4z^{-1}}$$

and

$$G_3(z) = \frac{1-2.5z}{1-0.4z}$$

We rewrite this relation as follows:

$$G_3(z) = \frac{2.5z(0.4z^{-1}-1)}{0.4z(2.5z^{-1}-1)} = 6.25\frac{1-0.4z^{-1}}{1-2.5z^{-1}}$$

It is easy to see that

$$G_3(z^{-1})G_3(z) = 6.25$$

By letting $z = e^{j\omega}$, we obtain

$$G_3(e^{-j\omega})G_3(e^{j\omega}) = |G_3(e^{j\omega})|^2 = 6.25$$

which shows that G_3 has a constant magnitude at all frequencies. ∎

Problem 5.12 generalizes this result.

5.5 Discrete Fourier Transform

The Fourier transform of the discrete time signal $\{g(n)\}$ is a continuous signal

$$G(e^{j\omega}) = \sum_{n=-\infty}^{\infty} g(n)e^{-j\omega n}$$

We know that this is a periodic signal with period 2π. The important thing to note is that $G(e^{j\omega})$ is a *continuous* signal. Indeed, the inverse Fourier transform of this function is given by

$$g(n) = \frac{1}{2\pi}\int_{-\pi}^{\pi} G(e^{j\omega})e^{j\omega n}d\omega$$

As we cannot store all of a continuous signal but only samples of it, we would like to sample $G(e^{j\omega})$. As we have to store $G(e^{j\omega})$ only over a *finite interval* of length 2π, we need to store only a finite number of its values. The discrete Fourier transform, DFT, is the finite duration discrete frequency sequence that is obtained by sampling one period of the Fourier transform. This sampling is carried out at N equally spaced points over a period of $0 \le \omega < 2\pi$ or at $\omega_k = 2\pi k/N$ for $0 \le k \le N-1$. In other words, if $\{g(n)\}$ is a discrete time sequence with a Fourier transform $G(e^{j\omega})$, then the DFT denoted by $\{G(k)\}$ is defined as

$$G(k) = G(e^{j\omega})|_{\omega=\omega_k=2\pi k/N}, \quad 0 \le k \le N-1$$

The DFT starts at $\omega = 0$ but does not include 2π. It was shown in Sec. 5.3.2 that when a continuous function of time is sampled with a sampling period T_s, then the spectrum of the resulting discrete time sequence becomes a periodic function of frequency $F_s = 1/T_s$. In a similar way, it can be shown that if the Fourier transform $G(e^{j\omega})$ is sampled with a sampling period of $1/N$, *i.e.*, N sampling points, the inverse of the sampled function, denoted by $\{\tilde{g}(n)\}$, becomes a periodic function of frequency N. This periodic discrete time sequence can be expressed in terms of $\{g(n)\}$ as

$$\tilde{g}(n) = \sum_{m=-\infty}^{\infty} g(n+mN)$$

The sequence $\{\tilde{g}(n)\}$ is called the periodic extension of $\{g(n)\}$. It has a period N. As we are free to choose N, we should select it to be reasonably large. Ideally N should be larger than the time during which $g(n)$ is nonzero. Suppose $g(n)$ is a causal finite duration sequence containing M samples, choose $N \ge M$ and let

$$\tilde{g}(n) = \begin{cases} g(n) & 0 \le n \le M-1 \\ 0 & M \le n \le N-1 \end{cases}$$

If $N > M$, we say that $\{\tilde{g}(n)\}$ is obtained by padding $\{g(n)\}$ with zeros. If $N \ge M$, $g(n)$ can be recovered uniquely from $\tilde{g}$: choose the values corresponding to the first period. If $M > N$, time aliasing will take place. Since $\tilde{g}$ is periodic, it has a Fourier series:

$$\tilde{g} = \frac{1}{N}\sum_{k=-\infty}^{\infty} \tilde{G}(k)e^{j2\pi kn/N}$$

Since $e^{j2\pi kn/N}$ is periodic in k with a period of N, this can be written as

$$\tilde{g} = \frac{1}{N}\sum_{k=0}^{N-1} \tilde{G}(k)e^{j2\pi kn/N}$$

$\tilde{G}(k)$ is calculated by multiplying both sides by $e^{-j2\pi nr/N}$ and summing over $0 \le n \le N-1$. This is similar to the procedure we have used to arrive at Eq. 5.21 on page 121. We obtain

$$\tilde{G}(r) = \sum_{n=0}^{N-1} \tilde{g}e^{-j2\pi nr/N}$$

It is instructive to compare this with the Fourier transform of the original sequence. Consider the Fourier transform of a causal, finite time sequence over 0 to $M-1$, $M \le N$:

$$G(e^{j\omega}) = \sum_{n=0}^{M-1} g(n)e^{-j\omega n}$$

Padding with zeros,

$$G(e^{j\omega}) = \sum_{n=0}^{N-1} g(n)e^{-j\omega n}$$

Sampling this at N equally spaced points over $0 \le w < 2\pi$ produces the sequence

$$G(k) = G(e^{j\omega})|_{\omega=\omega\pi k/N} = \sum_{n=0}^{N-1} g(n)e^{-j\omega\pi kn/N}, \quad 0 \le k < N-1$$

Notice that the first N samples of $\tilde{g}$ are equal to $g(n)$, *i.e.*,

$$g(n) = \tilde{g}(n), \quad 0 \le n \le N-1$$

But only the first N values are required in the sum. Therefore $G(k) = \tilde{G}(k)$, or

$$g(n) = \frac{1}{N}\sum_{k=0}^{N-1} G(k)e^{j2\pi kn/N}$$

It is a common practice to call this the discrete Fourier transform and abbreviate it as DFT. We collect the DFT and inverse DFT pair as

$$G(k) = \sum_{n=0}^{N-1} g(n)e^{-j2\pi kn/N}$$

$$g(n) = \frac{1}{N}\sum_{k=0}^{N-1} G(k)e^{j2\pi kn/N}$$

Example 5.11 Find the DFT of the causal three point averager:

$$g(n) = \begin{cases} \frac{1}{3} & 0 \le n \le 2 \\ 0 & \text{otherwise} \end{cases}$$

$$G(k) = \sum_{n=0}^{N-1} g(n) e^{-j2\pi nk/N} = \sum_{n=0}^{2} \frac{1}{3} e^{-j2\pi nk/N}$$
$$= \frac{1}{3}\left(1 + e^{-j2\pi k/N} + e^{-j4\pi k/N}\right)$$

∎

5.6 Matlab Code

Matlab Code 5.1 Sinusoidal plots for increasing frequency, shown in Fig. 5.4 on page 118. This code is available at HOME/freq/matlab/incr_freq.m[2]

```
1 n=0:16;
2 subplot(2,2,1), stem(n,cos(n*pi/8))
3 grid,xlabel('n'),ylabel('cos(n*pi/8)')
4 subplot(2,2,2), stem(n,cos(n*pi/4))
5 grid,xlabel('n'),ylabel('cos(n*pi/4)')
6 subplot(2,2,3), stem(n,cos(n*pi/2))
7 grid,xlabel('n'),ylabel('cos(n*pi/2)')
8 subplot(2,2,4), stem(n,cos(n*pi))
9 grid,xlabel('n'),ylabel('cos(n*pi)')
```

Matlab Code 5.2 Bode plots for Example 5.7 on page 141. This code is available at HOME/freq/matlab/filter1.m

```
1 omega = linspace(0,pi);
2 g1 = 0.5 ./ (cos(omega)-0.5+j*sin(omega));
3 mag1 = abs(g1);
4 angle1 = angle(g1) * 180/pi;
5 g2 = (0.5+0.5*cos(omega)-1.5*j*sin(omega)) ...
6     * 0.25 ./ (1.25-cos(omega));
7 mag2 = abs(g2);
8 angle2 = angle(g2) * 180/pi;
9 subplot(2,1,1)
10 plot(omega,mag1,omega,mag2,'--')
11 axis tight, label('',18,' ','Magnitude',18)
12 subplot(2,1,2)
13 plot(omega,angle1,omega,angle2,'--')
14 axis tight, label('',18,'w (rad/s)','Phase',18)
```

[2]HOME stands for http://www.moudgalya.org/dc/ – first see the software installation directions, given in Appendix A.2.

Matlab Code 5.3 Bode plot of the moving average filter, discussed in Example 5.5 on page 129. This code is available at HOME/freq/matlab/ma_bode.m

```
w = 0:0.01:pi;
subplot(2,1,1)
loglog(w,abs(1+2*cos(w))/3)
label('',18,' ','Magnitude',18)
subplot(2,1,2)
semilogx(w,angle(1+2*cos(w))*180/pi)
label('',18,'w','Phase',18)
```

Matlab Code 5.4 Bode plot of the differencing filter, discussed in Example 5.6 on page 130. This code is available at HOME/freq/matlab/derv_bode.m

```
w = 0:0.01:pi;
G = 1-exp(-j*w);
subplot(2,1,1)
loglog(w,abs(G))
label('',18,' ','Magnitude',18)
subplot(2,1,2)
semilogx(w,180*angle(G)/pi);
label('',18,'w','Phase',18)
```

Matlab Code 5.5 Bode plots of minimum and nonminimum phase filters, discussed in Example 5.9 on page 145. This code is available at HOME/freq/matlab/nmp.m

```
omega = linspace(0,pi);
ejw = exp(-j*omega);

G1 = 1.5*(1-0.4*ejw);
mag1 = abs(G1); angle1 = (angle(G1))*180/pi;
G2 = -0.6*(1-2.5*ejw);
mag2 = abs(G2); angle2 = (angle(G2))*180/pi;

subplot(2,1,1)
plot(omega,mag1,omega,mag2,'--')
axis tight, label('',18,' ','Magnitude',18)
subplot(2,1,2)
plot(omega,angle1,omega,angle2,'--')
axis tight, label('',18,'w (rad/s)','Phase',18)
```

5.7 Problems

5.1. Consider an LTI system with the following transfer function:

$$G(z) = \frac{z^2}{z^2 + r^2}$$

Show that the output $Y(z)$ for an impulse input is given by

$$\frac{Y(z)}{z} = \frac{z}{(z - jr)(z + jr)} = \frac{1}{2}\left[\frac{1}{z - jr} + \frac{1}{z + jr}\right]$$

Invert it and arrive at

$$\begin{aligned} y(n) &= \frac{1}{2}\left[(jr)^n + (-jr)^n\right] = \frac{1}{2}r^n\left[j^n + (-j)^n\right] \\ y(4m) &= r^{4m} \\ y(4m+1) &= 0 \\ y(4m+2) &= -r^{4m+2} \\ y(4m+3) &= 0 \end{aligned}$$

Check that this is an oscillatory output.

5.2. Show that the step response of a system with transfer function given by Eq. 5.2 on page 113 is given by

$$\begin{aligned} \frac{y(z)}{z} &= \frac{A_2}{z - \rho e^{j\omega}} + \frac{A_2^*}{z - \rho e^{-j\omega}} + \frac{C}{z - 1} \\ y(n) &= A_2 \rho^n e^{jn\omega} + A_2^* \rho^n e^{-jn\omega} + C \end{aligned}$$

Observe that this response is also oscillatory in general, just like the impulse response, given by Eq. 5.3.

5.3. Plot the following waveforms: $\cos(2\pi 5n + \theta)$, $\cos(2\pi\frac{1}{4}n + \theta)$, $\cos(2\pi\sqrt{2}n + \theta)$ and $\cos(2\pi\sqrt{2}t + \theta)$, and verify whether the periodicity property of discrete time signals, explained in Sec. 5.1.2, is satisfied [49].

5.4. This problem is concerned with the Fourier transform of discrete time periodic signals, with period N, of the form $u(n + N) = u(n)$.

(a) Start with the Fourier series for the discrete time periodic signal $u(n)$ as follows:

$$u(n) = \sum_{k'=-\infty}^{\infty} C'_k e^{j2\pi k' n/N}$$

Compare this with Eq. 5.19 on page 120. Notice that in the place of t and $T_p = 1/F_0$, we have n and N, respectively. For a fixed k', show that $e^{j2\pi k' n/N}$ is periodic in n with period N. That is, $e^{j2\pi k'(n+N)/N} = e^{j2\pi k' n/N}$ for all n.

(b) Because $e^{j2\pi k'n/N}$ is periodic with period N, by summing all its coefficients for every $k' \in [0, N-1]$, obtain the following simplified Fourier series:

$$u(n) = \sum_{k=0}^{N-1} C_k e^{j2\pi kn/N}$$

where

$$C_k = \sum_{r=-\infty}^{\infty} C_{k'+rN}, \quad C_k, C_k' \in [0, N-1]$$

(c) Multiply both sides of the simplified Fourier series by $e^{-j2\pi ln/N}$ and sum over 0 to $N-1$ to arrive at

$$\sum_{n=0}^{N-1} u(n) e^{-j2\pi ln/N} = \sum_{n=0}^{N-1}\sum_{k=0}^{N-1} C_k e^{j2\pi(k-l)n/N}$$

Show that the right-hand side of the above equation is equal to NC_l and thus arrive at an expression for C_l.

(d) Show that C_i is periodic with period N, *i.e.*, $C_k = C_{k+N}$, for all k.

In view of C_k being periodic, we see that N consecutive signals provide complete information in the time and frequency domains. Some of the N intervals of interest are

$$0 \le k \le N-1 \Leftrightarrow 0 \le \omega_k = \frac{2\pi k}{N} < 2\pi$$

when N is odd and

$$-\frac{N}{2} < k \le \frac{N}{2} \Leftrightarrow -\pi \le \omega_k = \frac{2\pi k}{N} < \pi$$

when N is even.

5.5. Find the minimum degree transfer function $G(z)$ of an LTI system whose magnitude function $|G(e^{j\omega})|$ satisfies the following relation:

$$|G(e^{j\omega})| = |\cos\omega|$$

In case the solution is not unique, state all the solutions.

5.6. Another approach to frequency response is given in this problem. Consider a system with a transfer function $G(z)$ whose poles are inside the unit circle. Suppose that it is subjected to the input $u(k) = M\cos kw$ where M is the amplitude, ω the frequency in rad/sample and k the sample number.

(a) Show that the Z-transform of the output is given by

$$Y(z) = \frac{M}{2}\left(\frac{z}{z-e^{j\omega}} + \frac{z}{z-e^{-j\omega}}\right) G(z)$$

(b) Show that this can be written as

$$Y(z) = \frac{\alpha z}{z - e^{j\omega}} + \frac{\alpha^* z}{z - e^{-j\omega}} + \sum_{i=1}^{n} \frac{D_i z}{z - p_i}$$

where α is given by $\alpha = MG(e^{j\omega})/2$.

(c) Show that the Z-transform of the steady state portion of the output is

$$\begin{aligned} Y(z) &= \frac{M}{2}\left[\frac{G(e^{j\omega})z}{z - e^{j\omega}} + \frac{G(e^{-j\omega})z}{z - e^{-j\omega}}\right] \\ &= \frac{M}{2}|G(e^{j\omega})|\left[\frac{e^{j\phi}z}{z - e^{j\omega}} + \frac{e^{-j\phi}z}{z - e^{-j\omega}}\right] \end{aligned}$$

(d) Invert this and show that

$$y(k) = M|G(e^{j\omega})| \cos(kw + \phi)$$

This shows that the output also is a sinusoid and shifted in phase by ϕ, the phase angle of $G(e^{j\omega})$, and amplified by $|G(e^{j\omega})|$.

(e) Where did you use the fact that all the poles of $G(z)$ are inside the unit circle?

5.7. In Example 5.5 on page 129, we have seen the moving average filter to be low pass. Is this in agreement with the findings of Sec. 5.4.1?

5.8. Repeat the above approach to verify the high pass property of the differencing filter, discussed in Example 5.6 on page 130.

5.9. An LTI system, initially at rest and with impulse response $g(n)$,

$$g(n) = \delta(n) - \sqrt{2}\delta(n-1) + \delta(n-2)$$

is subjected to an input

$$u(n) = \left[\cos\frac{\pi}{4}n + \cos\frac{\pi}{2}n\right] 1(n)$$

(a) Calculate the output by convolution techniques.

(b) Find the zeros of the transfer function $G(z)$ and using this explain the results of part (a).

(c) Draw a Bode plot of $G(z)$ and using this explain the results of part (a).

5.10. This problem is concerned with the design of an *all-zero* filter $G(z)$.

(a) Using the pole–zero placement method developed in this chapter, design a filter $G(z)$ that filters out the frequency

$$\omega_0 = \frac{2\pi}{3} \tag{5.69}$$

and at $\omega = 0$, $G(e^{j\omega}) = 1$. [Hint: Make the filter realizable, if required, by including the necessary number of poles at $z = 0$.]

(b) Let $u(n)$ be the input to and $y(n)$ be the output from the above filter $G(z)$. This filter can be expressed in the input–output notation as

$$y(n) = b_0u(n) + b_1u(n-1) + b_2u(n-2) \tag{5.70}$$

Find the coefficients b_0, b_1 and b_2.

(c) By directly substituting $u(n) = \cos\omega_0 n$ into Eq. 5.70 and simplifying it, show that $y(n) = 0$. Note that ω_0 is given by Eq. 5.69. Explain if this result is expected. [Hint: You may want to use the relation $\cos A + \cos B = 2\cos\frac{A+B}{2}\cos\frac{A-B}{2}$).]

(d) Show that the phase angle of this filter is a linear function of frequency. [Hint: You may find the form in Eq. 5.70 to be easier to manipulate compared to the product form derived in part (a).]

$G(z)$ discussed above is known as an all zero filter because it has only zeros – the poles at zero don't count. It is also a linear phase filter, see the next problem.

5.11. This problem is concerned with linear phase filters.

(a) If the impulse response $\{g(n)\}$ of an LTI system is given by

$$\begin{aligned} g(-1) &= a \\ g(0) &= b \\ g(1) &= a \end{aligned}$$

and all other terms are zero, find the system transfer function in the frequency domain, *i.e.*, $G(e^{j\omega})$ using real terms only.

(b) Suppose that we make

$$g(-2) = g(2) = c \neq 0$$

in the above. Repeat the calculations of the above part.

(c) Repeat the above calculations for the following general case:

$$\begin{aligned} g(0) &= g_0 \\ g(1) &= g_1 \\ &\vdots \\ g(M) &= g_M \end{aligned}$$

with $g(k) = 0$, $\forall k > M$, and $g(-i) = g(i)$, $\forall i$.

(d) Is it possible to use a filter with the impulse response as in the above part for real time applications?

(e) Suppose that we shift the sequence $\{g(n)\}$ to the right so that it becomes causal and that we call the shifted sequence $\{g_1(n)\}$, *i.e.*,

$$g_1(k) = g(k - M),\ \forall k$$

What is the relation between the Fourier transforms of $\{g_1(n)\}$ and $\{g(n)\}$?

(f) Let the impulse response of an arbitrary linear phase filter be denoted by $\{g^+(n)\}$. Suppose that its Fourier transform is given by

$$G^+(e^{j\omega}) = K(\omega)e^{-jD\omega}$$

where D is a positive integer and $K(\omega)$ is real and positive, *i.e.*, $K(\omega) > 0$ for all ω. Using the result obtained in the above parts, argue that this filter produces no phase distortion.

This is known as the *linear phase* filter, as the phase is a linear function of frequency. Such filters have the property that the delay through them is the same at all frequencies. These filters are useful because they preserve edges and bright spots in image processing applications.

5.12. This question is concerned with filters that have constant magnitude at all frequencies. Suppose that

$$G(z) = z^{-N}\frac{A(z^{-1})}{A(z)}$$

where

$$A(z) = \sum_{k=0}^{N} a_k z^{-k}, \quad a_0 = 1$$

Show that $G(z)G(z^{-1}) = 1$ and hence that $|G(e^{j\omega})| = 1$. Recall from Sec. 5.4.2 that a filter, whose magnitude function is a constant for all frequencies, is known as an all pass filter. Thus G is an all pass filter.

5.13. This problem is concerned with the phase angle property of all-pass transfer functions. You will consider the following transfer function for this purpose:

$$G_{\text{ap}}(z) = \frac{z^{-1} - a}{1 - az^{-1}}, \quad 0 < |a| < 1$$

(a) Show that the magnitude Bode plot of $G_{\text{ap}}(z)$ is a constant. [Hint: Evaluate $G_{\text{ap}}(z)G_{\text{ap}}(z^{-1})$.]

(b) You have to determine the sign of the phase angle Bode plot of $G_{\text{ap}}(z)$. You can do this directly or follow the steps given below:

i. Show that the phase angle of the transfer function $1 - az^{-1}$, evaluated at $z = e^{j\omega}$, is given by $\tan^{-1}\dfrac{a\sin\omega}{1 - a\cos\omega}$.

ii. Similarly, determine the angle contribution of the numerator term of $G_{\text{ap}}(z)$. It may be useful to write $G_{\text{ap}}(z)$ as

$$G_{\text{ap}}(z) = -a\frac{1 - bz^{-1}}{1 - az^{-1}}, \quad b = \frac{1}{a}$$

iii. Using the results of (i) and (ii) obtained above, determine an expression for the phase angle of $G_{\text{ap}}(z)$, with $z = e^{j\omega}$.

iv. Does the phase angle Bode plot of $G_{\mathrm{ap}}(z)$ remain negative or positive or does it change sign over the range $\omega \in (0, \pi)$? Examine for $0 < a < 1$ and $-1 < a < 0$ separately.

5.14. Consider the infinite duration impulse response of an LTI system given by

$$g(n) = a^n 1(n), \quad 1 > a > 0$$

(a) Find its Fourier transform $G(e^{j\omega})$.

(b) Sample this at N equal points and produce the DFT sequence given by

$$G(k) = G(e^{j\omega})\big|_{\omega=2\pi k/N}, \quad 0 \le k \le N-1$$

(c) Apply the inverse DFT to this sequence and call the result $g_1(n)$. We would want $g_1(n)$ to be as close to $g(n)$ as possible. Show that $g_1(n)$ can be written as

$$g_1(n) = \frac{1}{N} \sum_{k=0}^{N-1} e^{j2\pi kn/N} \left[\sum_{m=0}^{\infty} a^m e^{-j2\pi km/N} \right], \quad 0 \le n \le N-1$$

(d) Simplify the above expression. Is it the same as $g(n)$? If so, explain why. If not, explain under what conditions they will become equal.

Part II

Identification

Chapter 6

Identification

In order to design a good controller for a plant, a model is usually required. A controller designed on the basis of a model often works better than the one designed without a model. Unfortunately, model development from first principles is a difficult and an expensive task, requiring participation of experts in fields such as engineering, thermodynamics, materials, simulation and numerical analysis. This is also a time consuming process, and often, as a result, model development from first principles is attempted only if a company is interested in a major investment. The above discussion suggests that although model based controllers are desirable, detailed first principles models are generally not available. Another factor that complicates the modelling process is the presence of noise. Typically, deterministic models cannot explain noise processes – probabilistic models are required.

Statistical model development techniques have been developed to address both of the above issues, namely, the quick determination of approximate models and modelling of the noise. Although at best approximate, the models developed in this fashion are often sufficient for control purposes. In this computerized age, a lot of plant data is generally available, meeting the basic need of statistical modelling techniques. It is no wonder that the data driven model development techniques appeal to those people who want to quickly model their plants so as to improve the performance. The process of construction of models from the input (u) and the output (y) data is known as model *identification*, see Fig. 6.1.

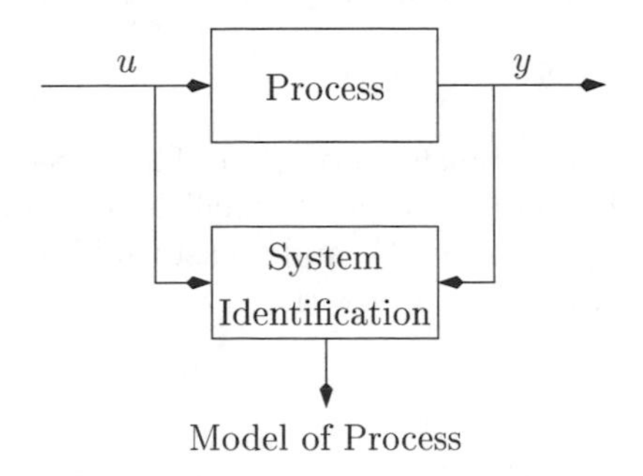

Figure 6.1: System identification, a schematic

Digital Control Kannan M. Moudgalya

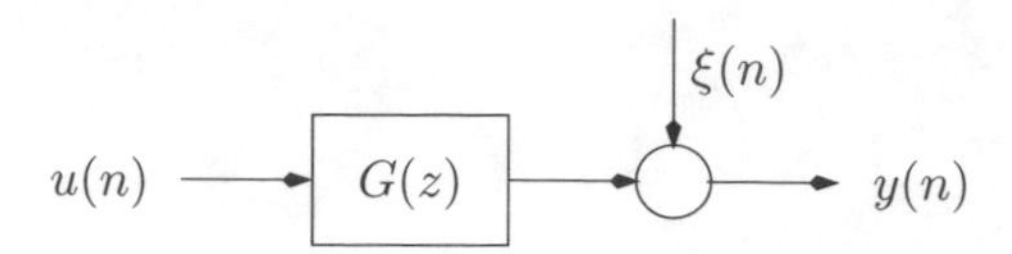

Figure 6.2: Transfer functions of plant with noise affecting the output

Once appropriate measurements are made, we proceed to obtain the plant model. There are two steps in this process: 1. Identification of an appropriate model structure. 2. Estimation of values of parameters that pertain to the above model structure.

This chapter is organized as follows: The topic of identification is motivated with the correlation approach applied to finite impulse response (FIR) models. This naturally leads to a least squares estimation approach. We next present a summary of time series terminology. We study LTI systems driven by white noise. In the subsequent section, we present parametric models and demonstrate how identification involving these can be posed in the least squares framework. A procedure to select the order of the noise model $H(z)$ that relates the white noise $\xi(n)$ and the residual $v(n)$ is presented next. This is followed by a procedure for simultaneous determination of $G(z)$ and $H(z)$. Interpreting identification from the frequency domain point of view, we arrive at filtering functions required for identification. We present two case studies to illustrate the identification techniques.

We assume that the reader has access to Matlab and the System Identification Toolbox [32] and make extensive use of these to illustrate the principles of identification. The area of identification is vast and there are entire books that are devoted to it [32, 43, 53, 57]. We present just a summary of this topic in this chapter.

6.1 Introduction

In Sec. 2.2, we have discussed a general procedure to derive first principles models. In real systems, there is usually some noise that acts on the output. An example of this is the measurement noise. This can be depicted diagrammatically as in Fig. 6.2, where $G(z)$ is the transfer function of the plant. We have assumed the noise process to be *white*, the concept of which will be explained in detail in Sec. 6.3.2. Roughly speaking, a white noise process can be thought of as something that cannot be decomposed further into simpler processes.

The mathematical model corresponding to Fig. 6.2 is given by

$$y(n) = g(n) * u(n) + \xi(n) \tag{6.1}$$

where, as usual, $G(z)$ is the Z-transform of $g(n)$. The objective of the identification problem is to determine the impulse response coefficients, namely $\{g(n)\}$, by exciting the process by an input sequence $\{u(n)\}$, and measuring the output sequence $\{y(n)\}$. When the plant is stable, Eq. 6.1 can be written as

$$y(n) = \sum_{l=0}^{N} g(l)u(n-l) + \xi(n) \tag{6.2}$$

We will now discuss a possible method to determine $\{g(n)\}$.

Example 6.1 Writing the equations for $y(n)$, $y(n-1)$, $y(n-2)$, ... and stacking them one below another, we obtain

$$\begin{bmatrix} y(n) \\ y(n-1) \\ y(n-2) \\ \vdots \end{bmatrix} = \begin{bmatrix} u(n) & \cdots & u(n-N) \\ u(n-1) & \cdots & u(n-N-1) \\ u(n-2) & \cdots & u(n-N-2) \\ \vdots & & \end{bmatrix} \begin{bmatrix} g(0) \\ g(1) \\ \vdots \\ g(N) \end{bmatrix} + \begin{bmatrix} \xi(n) \\ \xi(n-1) \\ \xi(n-2) \\ \vdots \end{bmatrix} \tag{6.3}$$

We have not mentioned how many rows we should include in the above matrix equation, although it is easy to see that it should be greater than or equal to $N+1$. We will defer the answer to this question. The above equation can be seen to be in the following form:

$$Y(n) = \Phi(n)\theta + \Xi(n) \tag{6.4}$$

where

$$Y(n) = \begin{bmatrix} y(n) \\ y(n-1) \\ y(n-2) \\ \vdots \end{bmatrix}, \quad \Phi(n) = \begin{bmatrix} u(n) & \cdots & u(n-N) \\ u(n-1) & \cdots & u(n-N-1) \\ u(n-2) & \cdots & u(n-N-2) \\ \vdots & & \end{bmatrix}$$

$$\theta = \begin{bmatrix} g(0) \\ g(1) \\ \vdots \\ g(N) \end{bmatrix}, \quad \Xi(n) = \begin{bmatrix} \xi(n) \\ \xi(n-1) \\ \xi(n-2) \\ \vdots \end{bmatrix}$$

Note that θ consists of impulse response coefficients $g(0)$, ..., $g(N)$. By convention, the argument n of $Y(n)$, $\Phi(n)$ and $\Xi(n)$ indicates that the data obtained at and before the time instant n have been used to construct Eq. 6.4 [36].

∎

The most popular way to solve the above equation, so as to determine the impulse response coefficients, is the least squares estimation[1] approach. Although this approach may appear ad hoc at present, a rigorous justification for the use of LSE, from the viewpoint of identification, will be provided in Sec. 6.6. In fact, it turns out that LSE is the most popular tool used in identification problems. In view of this, we will begin discussing LSE.

6.2 Least Squares Estimation

LSE is probably the most popular estimation technique used in parameter estimation. We assemble the measurement information and minimize the residual between the actual data and that predicted by the model in a least squares sense to arrive at the solution.

[1] We will abbreviate least squares estimation as *LSE* .

Table 6.1: Data for Example 6.2

i	1	2	3	4	5	6	7
ϕ_i	2	1	2.5	−0.8	1.25	−2.3	−1.75
y_i	4.01	1.98	5.03	−1.61	2.52	−4.63	−3.49

6.2.1 Linear Model for Least Squares Estimation

LSE is a convenient method to determine model parameters from experimental data. Let the model that relates the parameters and the experimental data at time instant j be given by

$$y(j) = \phi(j)\theta + \xi(j) \tag{6.5}$$

where y and ϕ consist of measurements and θ is a vector of parameters to be estimated. $\xi(j)$ can be thought of as a mismatch between the best that the underlying model, characterized by θ, can predict and the actual measurement $y(j)$. $\xi(j)$ can also be thought of as a random measurement noise. By stacking these equations one below another for $j = n, n-1, \ldots, n-N$, where N is a large integer, we obtain

$$Y(n) = \Phi(n)\theta + \Xi(n) \tag{6.6}$$

In the statistics literature, this is known as the *linear model*, even though it should strictly be called an *affine model* [36]. We will refer to it as the linear model in this book. An example of this is Eq. 6.3. By solving this equation, we can determine θ. We will present more examples of this equation shortly.

Example 6.2 Find the best value of a scalar θ that satisfies Eq. 6.5 given the data of Table 6.1.

It is clear from the data that we will get an unreliable answer if we use just one data set $(\phi(j), y(j))$ at some j. Although θ seems to vary from measurement to measurement, an approximate value of θ is seen to be 2, since y is nearly twice ϕ for every entry. If the measurement errors are large, it is not clear how θ will be affected. This idea is explored further in Example 6.4.

∎

We will now present the solution to Eq. 6.6.

6.2.2 Least Squares Problem: Formulation and Solution

It is clear that $\Xi(n)$ is an unknown in Eq. 6.6. Moreover, because it is often random, it is not clear how to model it. Because of these reasons, we neglect it while forming an estimate:

$$\hat{Y}(n) = \Phi(n)\hat{\theta}(n) \tag{6.7}$$

where $\hat{\theta}(n)$ is said to be an *estimate* of θ. Recall our convention that the argument n indicates that this is an estimate of Eq. 6.6 constructed with measurement data up to and including the time instant n. Because we would like the predicted model $\hat{Y}(n)$ to be close to the actual $Y(n)$, we try to obtain $\hat{\theta}$ by minimizing a function of

$$\tilde{Y}(n) \triangleq Y(n) - \hat{Y}(n) \tag{6.8}$$

where $\tilde{Y}(n)$ is the mismatch between the data and the model prediction. Let $\tilde{Y}(n) = (\tilde{y}(n), \tilde{y}(n-1), \ldots, \tilde{y}(n-N))$. A popular approach is to minimize the objective function

$$J[\hat{\theta}(n)] = w(n)\tilde{y}^2(n) + \cdots + w(n-N)\tilde{y}^2(n-N) \tag{6.9}$$

where $w(j)$ is the positive constant used to weight the mismatch $\tilde{y}(j)$ at time instant j. Using this, we can weight the data obtained at different times differently. If $w(j)$ are constant, all the errors are weighted equally. Small variations in the variance of θ can be accommodated by varying $w(j)$. Letting $W(n) = \text{diag}(w(n), w(n-1), \ldots, w(n-N))$, where diag stands for diagonal, we obtain

$$J[\hat{\theta}(n)] = \tilde{Y}(n)^T W(n)\tilde{Y}(n) \tag{6.10}$$

For notational simplicity, we will drop the argument n. The above equation becomes, after substituting Eq. 6.8,

$$J[\hat{\theta}] = [Y - \hat{Y}]^T W[Y - \hat{Y}] \tag{6.11}$$

In the least squares approach, we minimize the objective function J so as to determine the model parameter θ. This is formally stated in the following equation:

$$\hat{\theta}_{\text{WLS}} = \arg\min_{\theta} J[\hat{\theta}] \tag{6.12}$$

The subscript WLS indicates that the parameter θ is obtained by minimizing a sum of weighted squares. Substituting for $\hat{Y}(n)$ from Eq. 6.7 into Eq. 6.11, we obtain

$$J[\hat{\theta}] = [Y - \Phi\hat{\theta}]^T W[Y - \Phi\hat{\theta}] \tag{6.13}$$

Multiplying out the right-hand side, we obtain

$$J[\hat{\theta}] = Y^T WY - 2Y^T W\Phi\hat{\theta} + \hat{\theta}^T\Phi^T W\Phi\hat{\theta} \tag{6.14}$$

We would like to find $\hat{\theta}$ at which J is minimum. This requires that the derivative of J with respect to $\hat{\theta}$ is zero. Following the approach of Sec. A.1.1, we obtain

$$\frac{\partial J}{\partial \hat{\theta}} = -2\Phi^T WY + 2\Phi^T W\Phi\hat{\theta}_{\text{WLS}} = 0 \tag{6.15}$$

From this, we arrive at the *normal equation*,

$$\Phi^T(n)W(n)\Phi(n)\hat{\theta}_{\text{WLS}}(n) = \Phi^T(n)W(n)Y(n) \tag{6.16}$$

We will next assume that $\Phi^T W \Phi$ is nonsingular. If it is singular, the usual procedure is to include more sets of data. This is equivalent to adding more rows in Eq. 6.3, for example. If that also does not solve the problem, one has to check whether all the components of θ are really required and whether some of them can be removed.

In identification problems, $\Phi^T W \Phi$ could turn out to be singular if the input signal, used to excite the plant, is not designed properly. If this matrix product is nonsingular instead, we call the input *persistently exciting.*

With the assumption that $\Phi^T W \Phi$ is nonsingular, we solve for $\hat{\theta}$ to arrive at

$$\hat{\theta}_{\mathrm{WLS}}(n) = [\Phi^T(n)W(n)\Phi(n)]^{-1}\Phi^T(n)W(n)Y(n) \tag{6.17}$$

where we have once again explicitly indicated the dependence on n. We will now demonstrate what normal equation is arrived at in the identification example discussed earlier.

Example 6.3 Determine the equations to solve for impulse response coefficients $g(0)$, $g(1)$ and $g(2)$ when $N = 2$ in Eq. 6.2.

Recall that we developed the linear model for a general FIR model earlier. In particular, Eq. 6.3 becomes

$$\begin{bmatrix} y(n) \\ y(n-1) \\ y(n-2) \end{bmatrix} = \begin{bmatrix} u(n) & u(n-1) & u(n-2) \\ u(n-1) & u(n-2) & u(n-3) \\ u(n-2) & u(n-3) & u(n-4) \end{bmatrix} \begin{bmatrix} g(0) \\ g(1) \\ g(2) \end{bmatrix} + \begin{bmatrix} \xi(n) \\ \xi(n-1) \\ \xi(n-2) \end{bmatrix} \tag{6.18}$$

Premultiplying by the transpose of the coefficient matrix and ignoring the noise term, we arrive at

$$\begin{aligned} &\begin{bmatrix} u(n) & u(n-1) & u(n-2) \\ u(n-1) & u(n-2) & u(n-3) \\ u(n-2) & u(n-3) & u(n-4) \end{bmatrix} \begin{bmatrix} u(n) & u(n-1) & u(n-2) \\ u(n-1) & u(n-2) & u(n-3) \\ u(n-2) & u(n-3) & u(n-4) \end{bmatrix} \begin{bmatrix} g(0) \\ g(1) \\ g(2) \end{bmatrix} \\ &\qquad = \begin{bmatrix} u(n) & u(n-1) & u(n-2) \\ u(n-1) & u(n-2) & u(n-3) \\ u(n-2) & u(n-3) & u(n-4) \end{bmatrix} \begin{bmatrix} y(n) \\ y(n-1) \\ y(n-2) \end{bmatrix} \end{aligned} \tag{6.19}$$

If the 3×3 matrix product, which is the coefficient of g on the left-hand side, is nonsingular, the input is said to be persistently exciting. ∎

In the next example, we continue with the discussion presented in Example 6.2 using Matlab.

Example 6.4 Randomly generate a set of 100 ϕ and 100 e values and, using Eq. 6.5, generate the corresponding set of y for $\theta = 2$. Thus, $N = 100$ in this equation. Evaluate through Matlab the least squares solution of θ as well as the maximum and minimum of y_i/ϕ_i for different magnitudes of e.

As this is a batch experiment problem, as opposed to continuous identification, the argument n in Eq. 6.17 can be dropped. Let us also take all the weights in Eq. 6.11 to be equal to one. In view of these observations, the normal equation and the solution to the least squares problem, given by Eq. 6.16–6.17, become

$$\Phi^T \Phi \hat{\theta}_{\mathrm{WLS}} = \Phi^T Y, \quad \hat{\theta}_{\mathrm{WLS}} = [\Phi^T \Phi]^{-1} \Phi^T Y$$

Table 6.2: Solution to problem presented in Example 6.4. These values have been obtained by executing M 6.1.

V	$\hat{\theta}$	$\max(y_i/\phi_i)$	$\min(y_i/\phi_i)$
1	2.0049	3.1846	−8.4116
2	1.9649	3.8134	0.3274
5	1.8816	150.7621	−1.7748
10	2.1445	10.6843	−46.6836

In view of the fact that θ is a scalar, the above two equations, respectively, become

$$\hat{\theta}_{\mathrm{WLS}} \sum_{j=1}^{N} \phi(j)^2 = \sum_{j-1}^{N} \phi(j)y(j), \quad \hat{\theta}_{\mathrm{WLS}} = \frac{\sum_{j=1}^{N} \phi(j)y(j)}{\sum_{j=1}^{N} \phi(j)^2}$$

M 6.1 implements this problem. The least squares solution of θ as well as the maximum and minimum of y_i/ϕ_i, for different magnitudes, are shown in Table 6.2. It is easy to see that while the LSE is more or less equal to the actual value, the maximum and minimum of individual solutions are arbitrary, especially when the noise magnitude becomes large.

It should be observed that because of the presence of random numbers, it may not be possible to reproduce these numbers exactly. Nevertheless, the above observation that the least squares solution is close to the correct solution and the calculation through individual division will give wrong answers should still hold. ∎

Recall that we have been exploring the use of LSE to determine model parameters. In Example 6.1, we showed how to establish the linear model for FIR for an assumed noise process of the type depicted in Fig. 6.2. Often, this simple characterization of the noise process is insufficient – we need more elaborate models. We need some concepts from the area of time series to study these models. The next section is devoted to this topic.

6.3 Covariance

A stochastic process is a statistical phenomenon that evolves in time according to probabilistic laws. A realization is a sample of the many possibilities that a process can take (population). A time series is a set of values sampled from the process sequentially. A time series is a particular realization of the process. We can classify time series in the following way. A discrete time series is a set of observations made at discrete times. While a deterministic time series is one whose future values can be generated by a known (mathematical) function, a stochastic time series is one whose future values can be described only by some probabilistic distribution.

6.3.1 Covariance in Stationary, Ergodic Processes

Consider a discrete time series $\{u(t_1), u(t_2), \ldots, u(t_N)\}$ comprising N observations of a stochastic process. The joint probability distribution function (*pdf*) of $\{u(t_k)\}_{k=1}^{N}$ describes the probability that the random variable U takes on the values $U = u(t_k)$ jointly in a sequence of samples.

A stochastic process is said to be *stationary*, in the strict sense, if the joint pdf associated with the N observations taken at $t_1, t_2, \ldots, t_N$ is identical to the joint pdf associated with another set of N observations taken at times $t_1 + k, t_2 + k, \ldots, t_N + k$ for integers N and k. We will assume in this book that we deal with second order stationary processes, which means that the mean and covariance, to be explained below, do not depend on time. We will refer to such processes as simply *stationary.*

We will now present some important properties of a stationary time series. The first property that is of interest in a stationary stochastic series is the mean. The *mean* of a stationary process is defined as

$$\mu_u = \mathscr{E}(u) = \int_{-\infty}^{\infty} up(u)\, du \tag{6.20}$$

where, p stands for probability distribution function. We will refer to the above sum as the *statistical average.* To calculate using the above formula, a lot of information, such as the pdf $p(u)$, is required. This, in turn, assumes several realizations of the random signal. But in reality, we usually have only one realization of the random signal. In view of this, one works with an estimate of the mean, given by the following equation:

$$m_u = \frac{1}{2N+1} \sum_{n=-N}^{N} u(n) \tag{6.21}$$

Thus, the estimate of the mean is just the average. In other words, we estimate a statistical average with a *time average.* The estimate becomes more accurate when N is chosen large.

The next property of interest is *variance*, which gives a measure of variation of data from the mean. The variance of a stationary process is defined as

$$\sigma_u^2 = \mathscr{E}((u(k) - \mu_u)^2) = \int_{-\infty}^{\infty} (u(k) - \mu_u)^2 p(u)\, du \tag{6.22}$$

As mentioned above, when knowledge of the pdf is not available, one looks for a simpler way to calculate it. The estimate of the variance, given by

$$\hat{\sigma}_u^2 = \frac{1}{2N} \sum_{k=-N}^{N} (u(k) - m_u)^2 \tag{6.23}$$

comes to the rescue. Note that $2N$ is one less than the number of terms being summed.

We will next discuss the *auto covariance function* (ACF) that helps us understand the interdependence of samples of time series. The ACF for a general stochastic time series is defined as $\gamma(k, j) = \mathscr{E}((u(k) - \mu_{u(k)})(u(j) - \mu_{u(j)}))$. For a stationary time series, the mean is constant and the dependence is only a function of the lag $l = k - j$. In view of this, for a stationary stochastic process, we obtain

$$\gamma_{uu}(k,j) = \gamma_{uu}(l) = \mathscr{E}((u(k) - \mu_u)(u(k-l) - \mu_u)) \tag{6.24}$$

As in the case of the mean, the estimate of ACF is given by

$$r_{uu}(l) = \frac{1}{2N} \sum_{k=-N}^{N} (u(k) - m_u)(u(k-l) - m_u) \tag{6.25}$$

Note that we need only one realization to calculate the above sum. The ACF is used in detecting the underlying process, *i.e.*, whether it is periodic, integrating, independent, etc. We present a detailed study of a periodic process in Sec. 6.3.3. In Sec. 6.3.4, we show that the ACF takes the largest value at lag $l = 0$.

In practice, a normalized function, known as the *auto correlation function*,

$$\rho_{uu}(l) = \frac{\gamma_{uu}(l)}{\gamma_{uu}(0)} \tag{6.26}$$

is used. We abbreviate the auto correlation function also as ACF – the context will explain what is intended.

It is easy to verify the following symmetry properties:

$$\begin{aligned}
\gamma_{uu}(l) &= \gamma_{uu}(-l) \\
r_{uu}(l) &= r_{uu}(-l) \\
\rho_{uu}(l) &= \rho_{uu}(-l)
\end{aligned} \tag{6.27}$$

Now we illustrate these ideas with a finite length sequence.

Example 6.5 Find the ACF of the sequence $\{u(n)\} = \{1, 2\}$.

Because there are only two nonzero elements, we obtain $N = 2$. First, we determine an estimate of the mean, given by Eq. 6.21:

$$m_u = \frac{1}{2}\sum_{k=0}^{1} u(k) = \frac{1}{2}(u(0) + u(1)) = 1.5$$

Next, using Eq. 6.25, we calculate the ACF for every lag:

$$\begin{aligned}
r_{uu}(0) &= \sum_{k=0}^{1} (u(k) - 1.5)^2 = (-0.5)^2 + 0.5^2 = 0.5 \\
r_{uu}(1) &= \sum_{k=0}^{1} (u(k) - 1.5)(u(k-1) - 1.5) \\
&= (u(1) - 1.5)(u(0) - 1.5) = 0.5 \times (-0.5) = -0.25 \\
r_{uu}(-1) &= \sum_{k=0}^{1} (u(k) - 1.5)(u(k+1) - 1.5) \\
&= (u(0) - 1.5)(u(1) - 1.5) = (-0.5) \times 0.5 = -0.25
\end{aligned}$$

Thus, we see that $r_{uu}(n) = \{-0.25, 0.5, -0.25\}$, where the starting value of -0.25 corresponds to $n = -1$. The Matlab command `xcov` carries out these calculations. We can scale this sequence by dividing by $r_{uu}(0)$ to arrive at $\rho_{uu}(n) = \{-0.5, 1, -0.5\}$.

We can also obtain this result using the Matlab command `xcov` with the optional parameter `coeff` enabled. M 6.2 carries out this calculation. It can be seen that the symmetry properties of both r and ρ are satisfied. ∎

The calculations are easier if we work with sequences of zero mean. Zero mean equivalent of a sequence is obtained by subtracting the mean from every entry of the sequence. For example, by subtracting the mean of u discussed in the above example from every entry, we obtain $\{-0.5, 0.5\}$. The ACF of this sequence is identical to that obtained in the above example. Problem 6.5 shows that working with zero mean sequences could help prevent some mistakes.[2]

Now we present the concept of *cross covariance function* (CCF). It is a measure of dependence between samples of two time series. The CCF of two stationary time series u and y is given by

$$\gamma_{uy}(l) = \mathscr{E}((u(k) - \mu_u)(y(k-l) - \mu_y)) \tag{6.28}$$

while its estimate is given by

$$r_{uy}(l) = \frac{1}{2N} \sum_{k=-N}^{N} (u(k) - m_u)(y(k-l) - m_y) \tag{6.29}$$

Thus, to calculate $r_{uy}(l)$, $l > 0$, we need to shift y by l time points to the right, or equivalently, introduce l zeros in front, multiply the corresponding elements and add. For $l < 0$, we need to shift y to the left.

When the CCF between two sequences is large, we say that they are *correlated*, when it is small, we say that they are *less correlated* and when it is small, we say that they are *uncorrelated*. When two signals u and y are completely uncorrelated, we obtain

$$\begin{aligned} \gamma_{uy}(l) &= 0, \ \forall l \\ r_{uy}(l) &= 0, \ \forall l \end{aligned} \tag{6.30}$$

We would also say that u and y are independent, although usually this word refers to a stricter condition. The CCF assumes the largest value when two time series have the strongest correlation. This has wide application in determining the time delay of a system, an important step in identification, see Sec. 6.3.4.

For negative arguments, the CCF result is somewhat different from that of the ACF, given by Eq. 6.27. It is easy to verify the following property of the CCF:

$$\begin{aligned} \gamma_{uy}(l) &= \gamma_{yu}(-l) \\ r_{uy}(l) &= r_{yu}(-l) \end{aligned} \tag{6.31}$$

[2] In view of this observation, we will assume all sequences to be of zero mean in the rest of this chapter, unless explicitly stated otherwise .

The largest value occurs at the lag where the dependency is strongest. Suppose that u and y refer to the input to and the output from an I/O LTI system. If the system is causal, the current output cannot be correlated with a future input. As a result, we obtain the following relationship:

$$\begin{aligned} \gamma_{uy}(l) &= \gamma_{yu}(-l) = 0, \ \forall l > 0 \\ r_{uy}(l) &= r_{yu}(-l) = 0, \ \forall l > 0 \end{aligned} \tag{6.32}$$

In practice, a normalized function, known as the *cross correlation function*, which is also be abbreviated as CCF,

$$\rho_{uy}(l) = \frac{\gamma_{uy}(l)}{\sqrt{\gamma_{uu}(0)}\sqrt{\gamma_{yy}(0)}} \tag{6.33}$$

is used.

Recall that we have suggested the use of time averages to estimate statistical averages. We will now state when such an estimation is valid. A random signal $u(n)$ is said to be *ergodic* if all the statistical averages can be determined from a single realization with probability 1. That is, for an ergodic signal, time averages obtained from a single realization are equal to the statistical averages. For ergodic processes, the estimates of the statistical properties approach the actual values when a sufficiently large number of samples are taken while evaluating the summation. In the rest of this book, we will assume that we have a sufficient number of samples.

Unless otherwise stated, we will also assume that the noise process being studied is of zero mean. This helps simplify the calculations. In case the process under study does not obey this condition, we will subtract the mean, so that it becomes a zero mean process. The first zero mean process that we will study is white noise.

6.3.2 White Noise

The discrete time *white noise* sequence $\{\xi(k)\}$ is a set of independent identically distributed (*iid*) values belonging to a stationary stochastic process, with the following properties. The mean of white noise is zero. That is,

$$\mu_\xi = 0 \tag{6.34}$$

Because $\xi(k)$ is independent, its ACF is an impulse function:

$$\gamma_{\xi\xi}(k) = \sigma_\xi^2 \delta(k) = \begin{cases} \sigma_\xi^2 & k = 0 \\ 0 & \text{otherwise} \end{cases} \tag{6.35}$$

where σ_ξ^2 is the variance of $\{\xi(k)\}$. On taking the Z-transform, we obtain

$$\Gamma_{\xi\xi}(z) = \sigma_\xi^2 \tag{6.36}$$

We obtain the Fourier transform of $\{\gamma_{\xi\xi}(k)\}$, known as the *power density spectrum*

$$\Gamma_{\xi\xi}(e^{j\omega}) = \sigma_\xi^2, \ \forall\omega \tag{6.37}$$

The reason why $\Gamma(e^{j\omega})$ and ξ are called the power density spectrum and white noise, respectively, will be explained in Sec. 6.5.2.

Because white noise is uncorrelated with itself, it is easy to see that it is uncorrelated with any other sequence. For example, for an arbitrary sequence u different from ξ, we obtain

$$\gamma_{u\xi}(k) = 0, \ \forall k \tag{6.38}$$

Note that the above equation is true for $k = 0$ as well.

White noise, as defined in this section, is an idealization and it is difficult to create it. We approximate it with random number sequences. Even though not iid, these sequences satisfy Eq. 6.34–6.35. In Matlab, white noise is approximated using the command `randn`.

The concept of white noise is indispensable in system identification. Most noise sequences of interest can be modelled as filtered white noise. Moreover, because white noise is uncorrelated with every other sequence, we can simplify calculations quite a bit, as we will see in the rest of this chapter. We present a small example that demonstrates how white noise, combined with the concepts of ACF and CCF, can be used in estimating the model parameters of a simple system.

Example 6.6 Determine a procedure to find the model parameter a in the LTI system given by

$$y(n) - ay(n-1) = \xi(n) \tag{6.39}$$

where ξ is white and y is stationary.

Because y is stationary, by applying the expectation operation to Eq. 6.39, we see that y is of zero mean. Multiplying Eq. 6.39 by $\xi(n-k)$ and taking the expectation, we obtain

$$\gamma_{y\xi}(k) - a\gamma_{y\xi}(k-1) = \gamma_{\xi\xi}(k) \tag{6.40}$$

where we have made use of the definition of CCF, given in Eq. 6.29. In the rest of this chapter, we will carry out such calculations mechanically, without any further explanation. By evaluating Eq. 6.40 for $k = 0$, we obtain

$$\gamma_{y\xi}(0) = \gamma_{\xi\xi}(0) = \sigma_\xi^2 \tag{6.41}$$

where we have used the fact that

$$\gamma_{y\xi}(-k) = 0, \ \forall k > 0 \tag{6.42}$$

which is nothing but the causality condition of Eq. 6.32. Next, we substitute $k = 1$ in Eq. 6.40 to obtain $\gamma_{y\xi}(1) - a\gamma_{y\xi}(0) = 0$, where we have used Eq. 6.36. Using Eq. 6.41, we obtain

$$\gamma_{y\xi}(1) = a\gamma_{y\xi}(0) = a\sigma_\xi^2$$

One may be tempted to think that the above equation can be used to determine the model parameter a. Unfortunately, however, the white noise ξ is usually not measured and hence it will not be possible to calculate $\gamma_{y\xi}(k)$, or even its estimate, $r_{y\xi}(k)$. Nevertheless, we now explain how to make use of the above relation. Multiplying Eq. 6.39 by $y(n-k)$ and, as before, taking the expectation,

$$\gamma_{yy}(k) - a\gamma_{yy}(k-1) = \gamma_{y\xi}(-k) \tag{6.43}$$

By evaluating this for $k = 1$, we obtain

$$\gamma_{yy}(1) - a\gamma_{yy}(0) = 0 \tag{6.44}$$

where we have used Eq. 6.42. We are now in a position to calculate the model parameter a. From the above equation, we obtain

$$a = \frac{\gamma_{yy}(1)}{\gamma_{yy}(0)} \tag{6.45}$$

Using the above calculations, we can also get an idea of the behaviour of ACF for AR(1) processes. By evaluating Eq. 6.43 for $k = 0$, we obtain

$$\gamma_{yy}(0) - a\gamma_{yy}(1) = \sigma_\xi^2 \tag{6.46}$$

Solving Eq. 6.44 and Eq. 6.46 simultaneously, we obtain

$$\gamma_{yy}(0) = \frac{\sigma_\xi^2}{1-a^2} \tag{6.47}$$

Using Eq. 6.42, we obtain also from Eq. 6.43 the following recursive relations:

$$\gamma_{yy}(k) = a\gamma_{yy}(k-1) = a^2\gamma_{yy}(k-2) = \cdots = a^k r_{yy}(0) \tag{6.48}$$

Substituting the expression for $\gamma_{yy}(0)$ from Eq. 6.47, we obtain

$$\gamma_{yy}(k) = \frac{\sigma_\xi^2}{1-a^2}a^k \tag{6.49}$$

This equation shows that if $|a| < 1$, the ACF decays monotonically when $a > 0$ and with oscillations when $a < 0$.

Calculating model parameters in this manner is known as the theoretical prediction. In reality, however, we work with actual data and hence replace all the quantities of γ obtained above with their estimates, r. Thus, a may be determined from experimental data, through ACF calculations.

∎

The above example shows how the ACF naturally gets to be used in parameter estimation problems and how the idealization of white noise plays a role in it.

We will next show that the ACF can be used to detect the periodicity of the underlying process.

6.3.3 Detection of Periodicity Through ACF

The ACF of a periodic function is also periodic, as we will show now. Consider a periodic function u. We will assume that u is of zero mean. If it is not, we can subtract the mean at every time and make it obey this property. Thus, Eq. 6.24 becomes

$$\gamma_{uu}(l) = \mathscr{E}\left[u(k)u(k-l)\right] \tag{6.50}$$

Let the period of u be M, $M > 0$. That is,

$$u(l) = u(l-M), \ \forall l \tag{6.51}$$

Replacing l with $l+M$ in Eq. 6.50, we obtain

$$\gamma_{uu}(l+M) = \mathscr{E}\left[u(k)u(k-l-M)\right] = \mathscr{E}\left[u(k)u(k-l)\right] \tag{6.52}$$

using the periodicity property of u. In view of the above two relations, we see that

$$\gamma_{uu}(l) = \gamma_{uu}(l+M) \tag{6.53}$$

and hence conclude that the ACF of u is also periodic, with the same period as u.

There is another property that is equally interesting: The ACF of a periodic function exhibits periodicity even in the presence of noise. Let the noise affected signal be $\{u'(k)\}$; that is,

$$u'(k) = u(k) + \xi(k) \tag{6.54}$$

where $\xi(k)$ is white noise of zero mean and variance $\sigma_{\xi\xi}^2$. We make use of this relation in the expression for $\gamma_{u'u'}(l)$ and obtain

$$\gamma_{u'u'}(l) = \mathscr{E}\left[[u(k) + \xi(k)][u(k-l) + \xi(k-l)]\right]$$

Expanding the terms, we obtain

$$\gamma_{u'u'}(l) = \mathscr{E}\left[u(k)u(k-l) + u(k)\xi(k-l) + \xi(k)u(k-l) + \xi(k)\xi(k-l)\right] \tag{6.55}$$

Because ξ is white, it is uncorrelated with u, see Eq. 6.38. As a result, the cross terms involving u and ξ vanish. We are left with the ACF of u and ξ only. We make use of Eq. 6.35 and obtain

$$\gamma_{u'u'}(l) = \gamma_{uu}(l) + \sigma_{\xi\xi}^2\delta(l) \tag{6.56}$$

Only when $l = 0$ is the ACF of u' different from that of u. For all other values of l, the ACFs of u and u' are identical. In view of this result, we observe that $\gamma_{u'u'}(l)$ also is periodic, except for a spike at $l = 0$. This result is independent of the magnitude of $\sigma_{\xi\xi}^2$. This shows that even if noise corrupts a periodic signal, we can decipher the periodicity through ACF calculations.

The above result is due to the averaging nature of the ACF. Because of this averaging nature, ACF and CCF are used extensively in the identification problems with experimental data, which are usually noisy.

In reality, however, we will only use an estimate of γ. In addition, we will choose a large but finite N in the summations that define the estimates. In view of this, all of the above relations will hold only approximately. Now we illustrate this idea with a simple example.

Example 6.7 Plot the function

$$y(n) = \sin 0.1n + m\xi(n)$$

and its ACF for m values of 0.1, 0.5 and 1. Determine the periodicity property for all of these m values.

M 6.3 carries out the required calculations. Fig. 6.3 shows the plots of y and the ACF for m values of 0.1, 0.5 and 1. It is easy to see that as m increases, y gets noisier. In fact, if one were to see only Fig. 6.3(e), and not all the other figures in the sequence that we have developed, one would not see the periodicity in y.

We can also see that the period of the ACF for all three m values is the same and is equal to the period of the noise-free y.

As m is the variance of ξ, $r_{yy}(0)$ should increase with m. Because all the ACF plots have been scaled by $r_{yy}(0)$, the amplitude of the sinusoids in the ACF gets smaller as m increases.

We conclude this example by observing that the period of a periodic function can be determined from its ACF. ∎

Although the periodicity property has been illustrated with a single harmonic in the above example, one can extend it to the general case as well. Typically, one takes a Fourier transform of the ACF to identify the frequency content.

The averaging property of the ACF results in $\{u(l)\}$ being greatest at zero lag, *i.e.*, at $l = 0$, compared to all other lags. This property allows the ACF and CCF to detect pure delays in transmission, even in the presence of measurement noise. The next section explains this idea in detail.

6.3.4 Detection of Transmission Delays Using ACF

In this section, we will show how to detect transmission delays in the passage of a signal. We will first demonstrate that $r_{uu}(l)$ takes the largest value at zero lag, *i.e.*, at $l = 0$, compared to all other lags. Let $\{u(n)\}$ and $\{y(n)\}$ be real, zero mean signals, with finite energy. We define a signal $\{w(n)\}$

$$w(n) = au(n) + by(n-k)$$

where a and b are real constants, not equal to zero. Let us calculate the energy of w:

$$\begin{aligned} E_w &= \sum_{n=-N}^{N} [au(n) + by(n-k)]^2 \\ &= a^2 \sum_{n=-N}^{N} u^2(n) + b^2 \sum_{n=-N}^{N} y^2(n-k) + 2ab \sum_{n=-N}^{N} u(n)y(n-k) \\ &= (a^2 r_{uu}(0) + b^2 r_{yy}(0) + 2ab r_{uy}(k))(2N) \end{aligned}$$

From the definition of energy, $E_w > 0$. Because $N > 0$, we obtain the relation

$$a^2 r_{uu}(0) + b^2 r_{yy}(0) + 2ab r_{uy}(k) > 0 \tag{6.57}$$

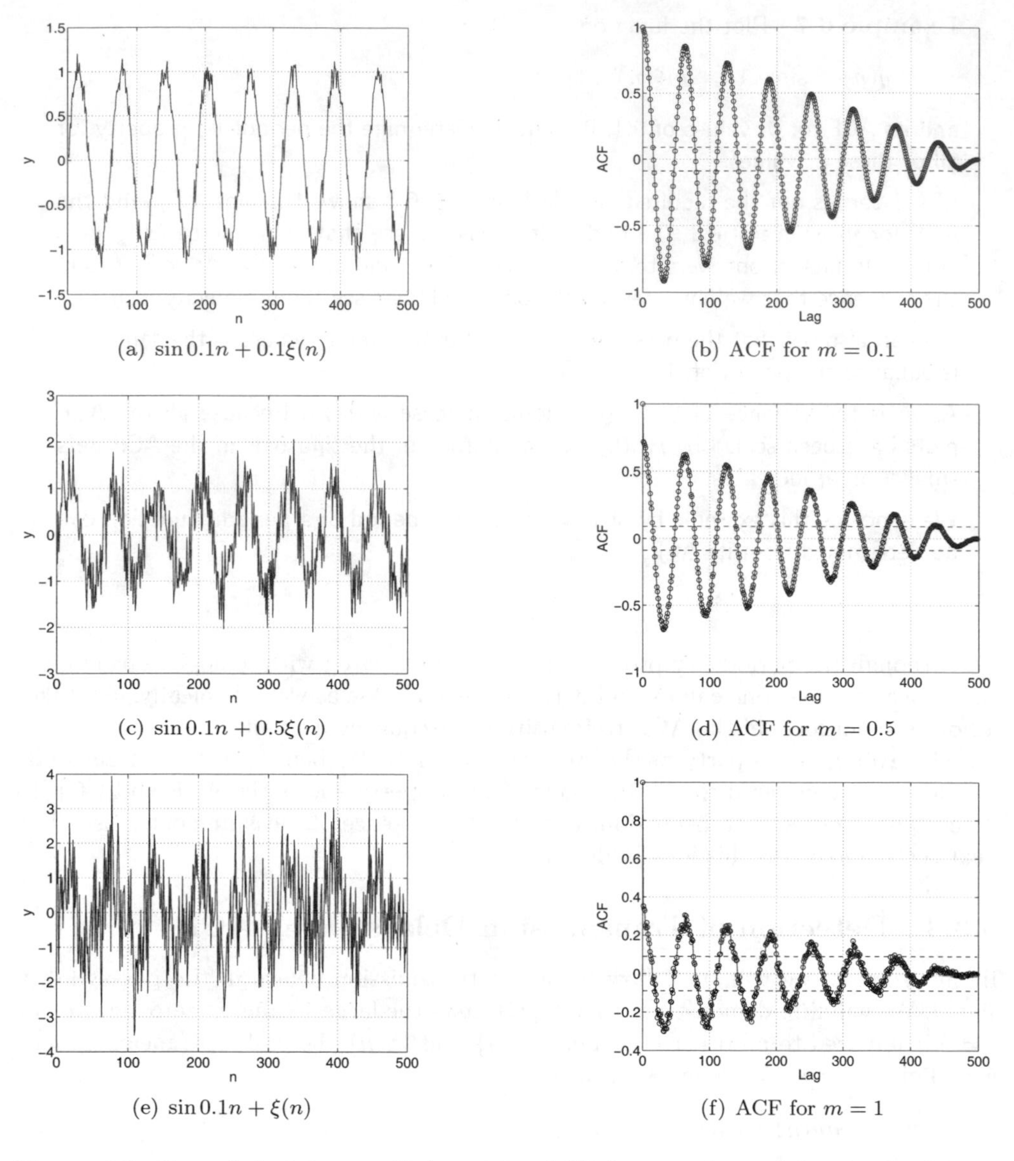

(a) $\sin 0.1n + 0.1\xi(n)$ (b) ACF for $m = 0.1$

(c) $\sin 0.1n + 0.5\xi(n)$ (d) ACF for $m = 0.5$

(e) $\sin 0.1n + \xi(n)$ (f) ACF for $m = 1$

Figure 6.3: Plot of $\sin 0.1n + m\xi(n)$ and its ACF for m values of 0.1, 0.5 and 1, as discussed in Example 6.7

Assuming that $b \neq 0$, we can divide by b^2 (otherwise, we can divide by a^2 to arrive at the same result):

$$\left(\frac{a}{b}\right)^2 r_{uu}(0) + 2\left(\frac{a}{b}\right) r_{uy}(k) + r_{yy}(0) \geq 0$$

This is a quadratic in (a/b), with coefficients $r_{uu}(0)$, $2r_{uy}(k)$ and $r_{yy}(0)$. As the quadratic is nonnegative, it follows that its discriminant is nonpositive. That is,

$$4[r_{uy}(k)^2 - r_{uu}(0)r_{yy}(0)] \leq 0 \tag{6.58}$$

That is,

$$|r_{uy}(k)| \le \sqrt{r_{uu}(0) r_{yy}(0)} \tag{6.59}$$

When $u = y$, the above relation reduces to

$$|r_{uu}(k)| \le r_{uu}(0) \tag{6.60}$$

This shows that $r_{uu}(k)$ takes the largest value at $k = 0$. In other words, the largest value of $r_{uu}(k)$ is reached when $k = 0$. An intuitive explanation is that at zero lag, the ACF is equal to the energy of the signal, while there could be cancellations of terms for other l values. Another approach to arrive at Eq. 6.60 is presented in Problem 6.4.

As in the case of periodicity, the property described above generally holds true even in the presence of noise. Also note that we do not need N to be infinite for this property to hold. We now illustrate these ideas with a simple example.

Example 6.8 Study the ACF of the following sequences

$$y_1 = \{1, 2, 3, 4\} + m\{\xi\}$$
$$y_2 = \{1, -2, 3, -4\} + m\{\xi\}$$
$$y_3 = \{-1, -2, 3, 4\} + m\{\xi\}$$

for m values of 0.1 and 1 and interpret.

M 6.4 implements this problem. The results are plotted in Fig. 6.4. It is easy to see that the ACF at zero lag is the largest compared to all other lags. This property holds true for all three sequences. The presence of noise does not change these observations.

∎

Now it is easy to see how Eq. 6.60 can be used to detect pure delays. Suppose that the input to the delay system is given by $\{u(n)\} = \{u_1, u_2, \ldots\}$ and the corresponding output is $\{y(n)\} = \{0, \ldots, 0, u_1, u_2, \ldots\}$, with d zeros in front. Note that d is an unknown number at this point. We have to calculate $r_{uy}(l)$ for $l = 0, -1, \ldots$. The lag at which this sequence attains the maximum will correspond to d, the delay.

It is important to note that a zero mean signal should be used for the detection of time delays. Alternatively, the mean of input and output signals should be calculated in a consistent fashion, as explained in Problem 6.5.

The objective of this section is to bring out the important property that the ACF is largest at zero lag. The above described method cannot be used to detect delays in dynamical systems, however. The causality property of dynamical systems may be used to estimate the delays. For example, one can make a step change in the input of the plant at steady state and see the time it takes for the output to start responding. The time that the output takes to respond to a change in input is the delay present in the plant.

6.3.5 Covariance of Zero Mean Processes Through Convolution

We conclude the discussion on covariance with a useful property that will help simplify derivations. We will show in this section that the ACF can be obtained through convolution.

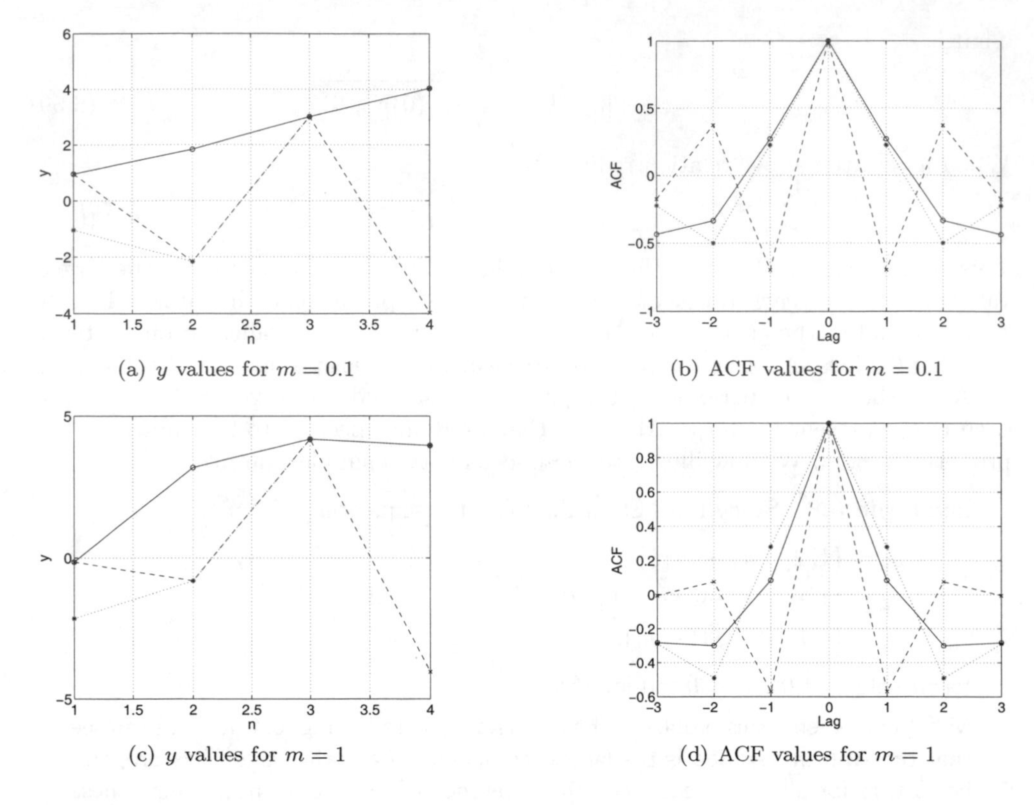

(a) y values for $m = 0.1$

(b) ACF values for $m = 0.1$

(c) y values for $m = 1$

(d) ACF values for $m = 1$

Figure 6.4: Plot of sequences (y_1 solid line, y_2 dashed, y_3 dotted) and their ACFs for different noise variances, as discussed in Example 6.8

As mentioned earlier, we can assume the signals to be of zero mean, without loss of generality. As a result, Eq. 6.29 becomes

$$r_{uy}(l) = \frac{1}{2N} \sum_{k=-N}^{N} u(k)y(k-l) \tag{6.61}$$

Comparing Eq. 6.61 with Eq. 3.27 on page 48, we see that

$$r_{uy}(l) = \frac{1}{2N} u(l) * y(-l) \tag{6.62}$$

Let us now focus on a single series. When the mean m_u is zero, Eq. 6.25 becomes

$$r_{uu}(l) = \frac{1}{2N} \sum_{k=-N}^{N} u(k)u(k-l) \tag{6.63}$$

Substituting u for y, we obtain from Eq. 6.62 the following result:

$$r_{uu}(n) = \frac{1}{2N} u(n) * u(-n) \tag{6.64}$$

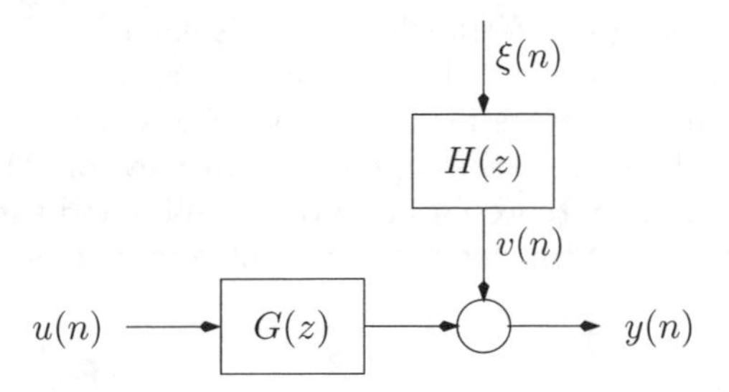

Figure 6.5: Transfer functions of system and noise process

Finally, from the definition of auto covariance, see Eq. 6.63, and energy defined by Eq. 3.2 on page 35, we arrive at

$$r_{uu}(0) = \frac{1}{2N} E_u \tag{6.65}$$

where N is taken as a large value, as mentioned earlier. Because E_u denotes energy of u and N denotes time, $r_{uu}(0)$ can be thought of as power.

Recall that the main objective in introducing the discussion on time series is to help identify plant models from measured data. In order to facilitate this, we examine the effect of exciting a linear system with white noise.

6.4 ARMA Processes

One does not measure noise directly. For example, even though in Fig. 6.2 on page 160 we have shown that noise ξ affects the measurement, only the plant output y and the input u are measured. If we know the transfer function G, the noise is obtained as the difference between the measurement and what is expected.

Although the noise in Fig. 6.2 is modelled as white, it is usually inadequate in practice. Fortunately, a large amount of noise behaviour can be modelled as auto regressive moving average or ARMA processes. These noise processes, denoted by the symbol v in Fig. 6.5, can be modelled by white noise filtered by a linear system, with a transfer function, say, $H(z)$.

The identification problem is to determine $G(z)$ and $H(z)$, given the measurements $u(n)$ and $y(n)$. In this section, we restrict our attention to the estimation of $H(z)$, given $v(n)$. We present the conditions under which we can represent $H(z)$ as an all zero and all pole model, respectively. We conclude this section with a combination of these two models. We begin this section with a brief discussion on the notation we will use in the rest of this book.

6.4.1 Mixed Notation

Recall that the system presented in Fig. 6.2 on page 160 has been modelled with the help of Eq. 6.1, reproduced here for convenience:

$$y(n) = g(n) * u(n) + \xi(n) \tag{6.66}$$

We have difficulty in taking the Z-transform of this equation. The reason is that the Z-transform of $\xi(n)$ may not exist, because it is a random number. For example, the Z-transform may not converge for any value of z. On the other hand, it is also inconvenient to carry the convolution operation. In view of this, a mixed notation, consisting of the time domain representation of variables and the transfer function of the linear systems, is used. With this notation, the above equation becomes

$$y(n) = G(z)u(n) + \xi(n) \tag{6.67}$$

where $G(z)$ denotes the Z-transform of $g(n)$. Some control and identification books follow the convention of using an *operator notation* $G(q)$, in the place of $G(z)$. In this book, however, we have avoided using this notation. On seeing an equation such as the one above, given in mixed notation, the reader should recognize that it denotes Eq. 6.66. In a similar way, when the noise goes through a linear system, such as the one shown in Fig. 6.5, we arrive at the following model for the output of the process:

$$y(n) = g(n) * u(n) + h(n) * \xi(n) \tag{6.68}$$

Once again, we will write the above equation in the following mixed form:

$$y(n) = G(z)u(n) + H(z)\xi(n) \tag{6.69}$$

where Y, U, G, H are, respectively, the Z-transforms of the output, input, process impulse response and noise process impulse response. Once again, it is understood that when we write the above equation, we mean Eq. 6.68.

6.4.2 What is an ARMA Process?

The noise process v of Fig. 6.5 is modelled in the most general form as an auto regressive moving average process, abbreviated as an ARMA process:

$$v(n) + a_1 v(n-1) + \cdots + a_p v(n-p) = \xi(n) + c_1 \xi(n-1) + \cdots + c_q \xi(n-q) \tag{6.70}$$

The object of this model is to predict the latest value taken by the noise process, namely $v(n)$. As this prediction depends on the past measurements of v, we call it *auto regressive* (AR). The phrase auto regressive refers to the fact that the current value of v depends on its previous values. The phrase *moving average* refers to the fact that a weighted average of ξ, over a moving window, is taken. For example, we need the current value, $\xi(n)$, as well as q previous values of it.

We refer to this as an ARMA(p, q) process. The symbol p refers to the number of previous v values used or the length of the auto regressive part. The symbol q refers to the number of previous ξ used in these calculations.

We will also study simpler forms of ARMA processes. Suppose that in Eq. 6.70, $p = 0$, *i.e.*, the model is

$$v(n) = \xi(n) + c_1 \xi(n-1) + \cdots + c_q \xi(n-q) \tag{6.71}$$

which can be written in the mixed notation of the previous section as

$$v(n) = \left(1 + \sum_{n=1}^{q} c_n z^{-n}\right) \xi(n) = C(z)\xi(n) \tag{6.72}$$

This is known as an MA(q) process. In other words, a random sequence whose value $v(n)$ can be represented as a finite combination of the past q entries of a white noise sequence plus a random error $\xi(n)$ is said to be an MA process of order q. An MA process is also known as an all zero process, because the transfer function of the process is a polynomial, $C(z)$.

With $p = 0$ in Eq. 6.70, we have obtained an MA process. If instead $q = 0$, we obtain

$$v(n) + a_1 v(n-1) + \cdots + a_p v(n-p) = \xi(n) \tag{6.73}$$

This is known as an AR(p) process. In other words, a random sequence whose value $v(n)$ can be represented as a weighted finite aggregate of the p previous values plus a white noise sequence $\xi(n)$ is said to be an AR process of order p. Eq. 6.73 can be written as

$$\left(1 + \sum_{n=1}^{p} a_n z^{-n}\right) v(n) = \xi(n)$$

or, equivalently, using the mixed notation of Sec. 6.4.1,

$$v(n) = \frac{1}{\left(1 + \sum_{n=1}^{p} a_n z^{-n}\right)} \xi(n) = \frac{1}{A(z)} \xi(n) \tag{6.74}$$

A variety of practical processes can be represented using this structure. The process is stationary if and only if the weights of the infinite polynomial $1/A(z)$ form a convergent series. If at least one root of $A(z)$ lies on the unit circle, the process is said to contain an integrator. An AR process can also be represented as an infinite summation of $\xi(n)$, as discussed in Sec. 3.3.6. An AR process is also known as an all pole process, because the transfer function has a polynomial only in the denominator.

Eq. 6.70 is known as the *ARMA* process as it contains both AR and MA components. Using the terminology defined above, it can be written as

$$v(n) = \frac{C(z)}{A(z)} \xi(n) = \frac{1 + \sum_{n=1}^{q} c_n z^{-n}}{1 + \sum_{n=1}^{p} a_n z^{-n}} \xi(n) \tag{6.75}$$

6.4.3 Moving Average Processes

In this section, we will present a technique to determine the order q of MA processes, *i.e.*, how *many* parameters are required to define the model. We will illustrate this idea with a simple example first and then generalize it.

Example 6.9 Determine a procedure to find the order of the MA(1) process

$$v(n) = \xi(n) + c_1 \xi(n-1)$$

making use only of the output data generated, namely $\{v(n)\}$.

We follow a procedure similar to the one outlined in Example 6.6 on page 170. We begin with the calculation of the ACF at zero lag:

$$\gamma_{vv}(0) = \mathscr{E}(v(n), v(n)) = \mathscr{E}\left[(\xi(n) + c_1\xi(n-1))(\xi(n) + c_1\xi(n-1))\right]$$

Because $\xi(n)$ is white, the expectation of cross products are zero. We obtain

$$\gamma_{vv}(0) = (1 + c_1^2)\sigma_\xi^2$$

Next, we determine the ACF at lag 1:

$$\begin{aligned}\gamma_{vv}(1) &= \mathscr{E}(v(n), v(n-1)) \\ &= \mathscr{E}\left[(\xi(n) + c_1\xi(n-1))(\xi(n-1) + c_1\xi(n-2))\right]\end{aligned}$$

Once again invoking the fact that ξ is white and cancelling the cross terms, we obtain

$$\gamma_{vv}(1) = \mathscr{E}(c_1\xi^2(n-1)) = c_1\sigma_\xi^2$$

These are the only nonzero terms. For all other lags, the ACF is zero. That is,

$$\gamma_{vv}(l) = 0, \quad l > 1$$

The ACF is simply obtained as

$$\rho_{vv}(l) = \frac{\gamma_{vv}(1)}{\gamma_{vv}(0)}$$

We observe from this example that for an MA(1) process, the ACF becomes zero for lags greater than 1, *i.e.*, $|l| > 1$. ∎

We will now generalize the above result for MA(q) processes. Multiplying Eq. 6.71 by $v(n)$ and taking the expectation, we obtain

$$\gamma_{vv}(0) = \gamma_{v\xi}(0) + c_1\gamma_{v\xi}(1) + \cdots + c_q\gamma_{v\xi}(q) \tag{6.76}$$

Multiplying Eq. 6.71 by $v(n-1)$ and taking the expectation,

$$\gamma_{vv}(1) = c_1\gamma_{v\xi}(0) + c_2\gamma_{v\xi}(1) + \cdots + c_q\gamma_{v\xi}(q-1) \tag{6.77}$$

where $\mathscr{E}[v(n-1)\xi(n)] = 0$, using the causality principle of Eq. 6.32 on page 169: for causal systems, the output cannot depend on future input $\xi(n)$. Continuing the above process and stacking the resulting equations, we arrive at

$$\begin{bmatrix}\gamma_{vv}(0) \\ \gamma_{vv}(1) \\ \vdots \\ \gamma_{vv}(q)\end{bmatrix} = \begin{bmatrix}1 & c_1 & \cdots & c_{q-1} & c_q \\ c_1 & c_2 & \cdots & c_q & 0 \\ \vdots & & & & \vdots \\ c_q & 0 & \cdots & & 0\end{bmatrix} \begin{bmatrix}\gamma_{v\xi}(0) \\ \gamma_{v\xi}(1) \\ \vdots \\ \gamma_{v\xi}(q)\end{bmatrix} \tag{6.78}$$

All the terms below the secondary diagonal are zero. It is clear that

$$\gamma_{vv}(n) = 0, \ \forall n > q \tag{6.79}$$

Thus we obtain the rule that for MA(q) processes, a plot of $\{\gamma_{vv}(n)\}$ *vs.* n becomes zero for all $n > q$. In other words, the index corresponding to the last nonzero value is q. We will now illustrate this idea with an example.

Example 6.10 Calculate $\{\gamma_{vv}(k)\}$ for

$$v(n) = \xi(n) + \xi(n-1) - 0.5\xi(n-2) \tag{6.80}$$

and verify that Eq. 6.79 is satisfied.

Multiplying Eq. 6.80 by $v(n-k)$, $k \geq 0$, and taking the expectation, we arrive at

$$\begin{aligned}
\gamma_{vv}(0) &= \gamma_{v\xi}(0) + \gamma_{v\xi}(1) - 0.5\gamma_{v\xi}(2) \\
\gamma_{vv}(1) &= \gamma_{v\xi}(0) - 0.5\gamma_{v\xi}(1) \\
\gamma_{vv}(2) &= \quad 0.5\gamma_{v\xi}(0) \\
\gamma_{vv}(k) &= 0, \ k \geq 3
\end{aligned} \tag{6.81}$$

As explained in Example 6.6 on page 170, measurement of the white noise ξ is not available and hence the CCFterms in the above equation cannot be calculated. Nevertheless, we can come up with equivalent expressions for the CCF terms using only the available quantities. We multiply Eq. 6.80 by $\xi(n)$ and take the expectation to arrive at

$$\gamma_{v\xi}(0) = \gamma_{\xi\xi}(0) = \sigma_\xi^2 \tag{6.82a}$$

using Eq. 6.35 on page 169. Multiplying Eq. 6.80 by $\xi(n-1)$ and $\xi(n-2)$, one at a time, and taking expectations, we obtain

$$\begin{aligned}
\gamma_{v\xi}(1) &= \gamma_{\xi\xi}(0) = \sigma_\xi^2 \\
\gamma_{v\xi}(2) &= -0.5\gamma_{\xi\xi}(0) = -0.5\sigma_\xi^2
\end{aligned} \tag{6.82b}$$

Substituting Eq. 6.82 in Eq. 6.81, we obtain

$$\begin{aligned}
\gamma_{vv}(0) &= (1 + 1 + 0.25)\sigma_\xi^2 = 2.25\sigma_\xi^2 \\
\gamma_{vv}(1) &= (1 - 0.5)\sigma_\xi^2 = 0.5\sigma_\xi^2 \\
\gamma_{vv}(2) &= -0.5\sigma_\xi^2 \\
\gamma_{vv}(k) &= 0, \ k \geq 3.
\end{aligned}$$

Recall that this procedure is known as theoretical prediction. Using the definition of ACF, given in Eq. 6.26,

$$\begin{aligned}
\rho_{yy}(0) &= 2.25/2.25 = 1 \\
\rho_{yy}(1) &= 0.5/2.25 = 0.22 \\
\rho_{yy}(2) &= -0.5/2.25 = -0.22
\end{aligned} \tag{6.83}$$

For all other k values, $\rho_{yy}(k) = 0$. We conclude this example with the observation that the output of the noise process v is either directly measured, or estimated using a procedure to be outlined in Sec. 6.6. ∎

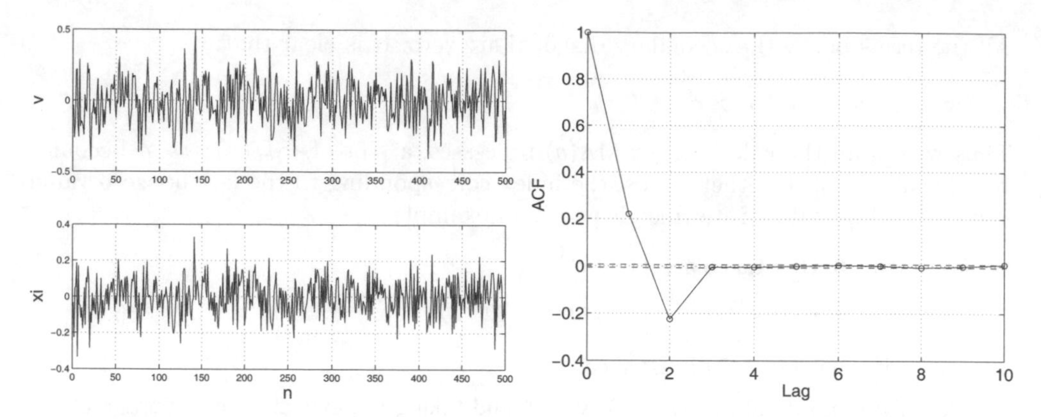

Figure 6.6: Input–output profiles (left) and the ACF of the system described by Eq. 6.80. M 6.5 is used to generate it.

In the next example, we solve the above problem using Matlab.

Example 6.11 Simulate the process given in Eq. 6.80, calculate an estimate of $\{\gamma_{vv}(k)\}$ and compare with the results obtained in Example 6.10.

M 6.5 carries out this task. Pseudo random numbers are taken as a good approximation to white noise process. The white noise process, thus generated, should not be directly used to determine the parameters. For example, one should avoid the temptation to apply least squares techniques to Eq. 6.80 in order to directly estimate the parameters. The reason is that in reality, white noise is not measured. Indeed, white noise is an idealization. The only available measurement is v.

The Matlab identification toolbox is required to carry out the simulation. The system is simulated for 100,000 time instants. First 500 instances of $\{\xi(n)\}$ and $\{v(n)\}$ are shown in the left hand plot of Fig. 6.6. A plot of the ACF, *i.e.*, $\{\rho_{vv}(k)\}$, generated by M 6.6, is shown in the right hand plot of Fig. 6.6. It is easy to see that the estimates reported in this figure are in agreement with the theoretically predicted values, given by Eq. 6.83.

∎

Once we know the order of an MA process, it is straightforward to determine the coefficients. Let us start with Eq. 6.78. If we can express $\gamma_{v\xi}(n)$ as a function of $\gamma_{vv}(n)$, we are done. Multiplying Eq. 6.71 with $\xi(n-k)$, $k \geq 0$, and taking the expectation, we obtain $\gamma_{vv}(\xi)(k) = c_k\sigma_\xi^2$. For $k = 0$, c_k is taken as 1. Substituting this result in Eq. 6.78, we obtain

$$\begin{bmatrix} \gamma_{vv}(0) \\ \gamma_{vv}(1) \\ \vdots \\ \gamma_{vv}(q) \end{bmatrix} = \begin{bmatrix} 1 & c_1 & \cdots & c_{q-1} & c_q \\ c_1 & c_2 & \cdots & c_q & 0 \\ \vdots & & & & \vdots \\ c_q & 0 & \cdots & & 0 \end{bmatrix} \begin{bmatrix} 1 \\ c_1 \\ \vdots \\ c_q \end{bmatrix} \sigma_\xi^2 \tag{6.84}$$

What we have above is a system of $q+1$ equations in $q+1$ unknowns, c_1 to c_q and σ_ξ^2. The left-hand side of the above equation can be determined from the experimental data, using a procedure similar to the one given in Example 6.71.

It should be pointed out that we do not calculate the coefficients using the above method. Computationally efficient procedures are normally employed to estimate the parameters of AR, MA and ARMA processes [5]. This topic is beyond the scope of this book and hence will not be pursued here.

We will illustrate the above approach with an example.

Example 6.12 Determine the model parameters c_1 and c_2 in the following MA(2) process:

$$v(n) = \xi(n) + c_1\xi(n-1) + c_2\xi(n-2) \tag{6.85}$$

Using Eq. 6.84, we arrive at the following equations:

$$\begin{aligned}\gamma_{vv}(0) &= \sigma_\xi^2 + c_1^2\sigma_\xi^2 + c_2^2\sigma_\xi^2 = (1 + c_1^2 + c_2^2)\sigma_\xi^2\\ \gamma_{vv}(1) &= c_1\gamma_{v\xi}(0) + c_2\gamma_{v\xi}(1) = c_1(1 + c_2)\sigma_\xi^2\\ \gamma_{vv}(2) &= c_2\gamma_{v\xi}(0) = c_2\sigma_\xi^2\end{aligned}$$

We have three equations in three unknowns, c_1, c_2 and σ_ξ^2. ∎

6.4.4 Is Unique Estimation Possible?

We would like to address the question of whether we can always determine the model parameters of an MA process from the experimental data, as explained in the previous section. The MA process, given by Eq. 6.72, implies

$$v(n) = c(n) * \xi(n) \tag{6.86}$$

where $c(n)$ is the inverse Z-transform of $C(z)$ and $\xi(n)$ is white noise of variance 1. We also obtain

$$v(-n) = c(-n) * \xi(-n)$$

Convolving the expressions for $v(n)$ and $v(-n)$ and using the commutativity property of convolution, we obtain

$$v(n) * v(-n) = c(n) * c(-n) * \xi(n) * \xi(-n)$$

Using the definition of auto covariance, as given by Eq. 6.64, we obtain

$$\gamma_{vv}(n) = c(n) * c(-n) * \gamma_{\xi\xi}(n)$$

Taking the Z-transform of both sides, we obtain

$$\Gamma_{vv}(z) = C(z)C(z^{-1}) \tag{6.87}$$

where we have used the fact that $\gamma_{\xi\xi}(n) = \delta(n)$ from Eq. 6.35 on page 169, for white noise of variance 1, and that the Z-transform of $\delta(n)$ is 1, obtained in Example 4.2 on page 70. We have also made use of the result of Sec. 4.2.8 to arrive at $C(z^{-1})$. The power of z^{-1} as an argument of C indicates that we have to replace the occurrences of z in $C(z)$ with z^{-1}; compare this with Footnote 5 on page 100.

Because the zeros of $C(z)$ are reciprocals of the corresponding zeros of $C(z^{-1})$, we should expect a loss in uniqueness. We illustrate this idea with a simple example.

Example 6.13 Study the ACFs of two processes $v_1(n)$ and $v_2(n)$, modelled as

$$v_1(n) = \xi(n) + c_1\xi(n-1) = (1 + c_1 z^{-1})\xi(n)$$
$$v_2(n) = \xi(n) + c_1^{-1}\xi(n-1) = (1 + c_1^{-1} z^{-1})\xi(n)$$

where we have used the mixed notation of Sec. 6.4.1. Using Eq. 6.87, we obtain

$$\Gamma_{v_2 v_2}(z) = (1 + c_1^{-1} z^{-1})(1 + c_1^{-1} z)$$

In a similar way, we obtain

$$\Gamma_{v_1 v_1}(z) = (1 + c_1 z^{-1})(1 + c_1 z)$$

Pulling out $c_1 z^{-1}$ and $c_1 z$, respectively, from the first and second terms on the right-hand side, we obtain

$$\Gamma_{v_1 v_1}(z) = c_1 z^{-1}(c_1^{-1} z + 1) c_1 z (c_1^{-1} z^{-1} + 1)$$

Comparing this with the expression for $\Gamma_{v_2 v_2}$, we obtain

$$\Gamma_{v_1 v_1}(z) = c_1^2 \Gamma_{v_2 v_2}(z)$$

It is clear that the ACFs of v_1 and v_2 are identical, *i.e.*,

$$\rho_{v_1 v_1}(l) = \rho_{v_2 v_2}(l), \ \forall l$$

because scaling results in the removal of constant factors – see the definition of the ACF in Eq. 6.26 on page 167. As a result, given the ACF, it is not possible to say whether the underlying noise process is v_1 or v_2.

∎

In the above example, if c_1 lies outside the unit circle, c_1^{-1} will lie inside it. Because we cannot say which one has given rise to the ACF, by convention, we choose the zeros that are inside the unit circle. Although this discussion used a first degree polynomial C, it holds good even if the degree is higher. We illustrate these ideas with a Matlab based example.

Example 6.14 The MA(2) process described by

$$v_1(n) = \xi(n) - 3\xi(n-1) + 1.25\xi(n-2) = (1 - 3z^{-1} + 1.25z^{-2})\xi(n)$$
$$= (1 - 0.5z^{-1})(1 - 2.5z^{-1})\xi(n)$$

is used to generate data as in M 6.7. The same code determines the model parameters. We present the resulting model using the command `present`:

```
Discrete-time IDPOLY model: v(t) = C(q)e(t)
C(q) = 1 - 0.8923 (+-0.009942) q^-1 + 0.1926 (+-0.009935) q^-2
```

The values within the brackets are the standard deviation for each parameter. The identified model parameters are different from the ones used to generate the data. Observe also that one of the zeros lies outside the unit circle. We repeat this exercise with the process described by

$$\begin{aligned} v_2(n) &= \xi(n) - 0.9\xi(n-1) + 0.2\xi(n-2) = (1 - 0.9z^{-1} + 0.2z^{-2})\xi(n) \\ &= (1 - 0.5z^{-1})(1 - 0.4z^{-1})\xi(n) \end{aligned}$$

Note that this process is identical to v_1, but with the zero outside the unit circle (2.5) being replaced by its reciprocal (0.4). M 6.7 generates data for this model as well, and estimates the parameters. We obtain the following result:

```
Discrete-time IDPOLY model: v(t) = C(q)e(t)
C(q) = 1 - 0.8912 (+-0.009939) q^-1 + 0.1927 (+-0.009935) q^-2
```

Observe that Matlab estimates the parameters correctly this time. Although not shown here, using M 6.7, one can see that the ACF plots of these two models are identical. Similarly, the PACF plots, to be introduced in Sec. 6.4.5, are also identical. ∎

It is important to point out the sequence of calculations in the above discussed model identification. From the data, one calculates the ACF, with which the parameters are calculated. Recall from Sec. 5.4.2 that systems with zeros inside the unit circle are known as minimum phase systems and those with all their zeros outside the unit circle are known as maximum phase systems. Splitting the ACF into minimum and maximum phase systems is known as *spectral factorization*, to be studied in more detail in Sec. 13.1. As explained in this section, using spectral factorization, one identifies minimum phase systems. We will make use of this fact while deriving prediction error models, to be discussed in Sec. 6.6.1.

We take this opportunity to point out that every parameter estimated in the above example comes with an uncertainty band. This can be helpful at times to decide on the model order. We illustrate this with an example.

Example 6.15 Simulate the following MA(2) process,

$$v_2(n) = (1 - 0.9z^{-1} + 0.2z^{-2})\xi(n)$$

and estimate the model parameters assuming the data to have come first from an MA(2) process and then from an MA(3) process.

This problem is solved in M 6.8. This system is simulated and the data thus generated are used to estimate the model parameters. When a second order model is fitted to these data, using the command `present`, we obtain the following model parameters:

```
Discrete-time IDPOLY model: y(t) = C(q)e(t)
C(q) = 1 - 0.9045 (+-0.00313) q^-1 + 0.2063 (+-0.003131) q^-2

Estimated using ARMAX from data set v
```

```
Loss function 0.00995515 and FPE 0.00995554
Sampling interval: 1
Created:       23-Nov-2006 22:38:31
Last modified: 23-Nov-2006 22:38:37
```

In this, the *loss function* refers to the sum of squares of residuals and *FPE* denotes Akaike's final prediction error criterion [32]. We note that the estimated parameters are close to the actual values. Next, we fit the same data with an MA(3) model and obtain the model parameters:

```
Discrete-time IDPOLY model: y(t) = C(q)e(t)
C(q) = 1 - 0.9046 (+-0.0032) q^-1 + 0.2068 (+-0.004267) q^-2
                             - 0.0003695 (+-0.003201) q^-3
Estimated using ARMAX from data set v
Loss function 0.00995514 and FPE 0.00995574
Sampling interval: 1
```

We make two observations:

1. The third parameter is estimated to be small, -0.0003695.
2. The uncertainty in the third parameter, namely 0.003201, is larger than the estimate itself, suggesting that the third parameter is not trustworthy.

In this example, the uncertainty has turned out to be larger than the parameter itself. Generally, if the uncertainty turns out to be comparable to the value of the parameter being estimated, the validity of the coefficient is questionable. When the estimate is good, however, the uncertainty is generally much smaller than the parameter value.

∎

From this section, we conclude that the ACF can be used as an effective tool to determine the order of MA processes. We will now devote our attention to AR processes.

6.4.5 Auto Regressive Processes

In this section, we will present a method to determine the order of AR processes. Let us first explore whether it is possible to do this through the ACF. We will begin with a simple example.

Example 6.16 Calculate the ACF of the AR(1) process

$$v(n) + a_1 v(n-1) = \xi(n) \tag{6.88}$$

Multiplying both sides of this equation successively by $v(n-1)$, $v(n-2), \ldots,$ $v(n-l)$ and taking the expectation, we obtain

$$\begin{aligned}
\gamma_{vv}(1) + a_1\gamma_{vv}(0) &= 0 \\
\gamma_{vv}(2) + a_1\gamma_{vv}(1) &= 0 \\
&\vdots \\
\gamma_{vv}(l) + a_1\gamma_{vv}(l-1) &= 0
\end{aligned}$$

where the right-hand side of every equation is zero, because of the causality condition, given by Eq. 6.32 on page 169. Starting from the last equation and recursively working upwards, we obtain

$$\begin{aligned}\gamma_{vv}(l) &= -a_1\gamma_{vv}(l-1) = -a_1(-a_1\gamma_{vv}(l-2)) = a_1^2\gamma_{vv}(l-2)\\ &= \cdots = (-1)^l a_1^l \gamma_{vv}(0)\end{aligned}$$

Dividing both sides by $\gamma_{vv}(0)$, we obtain

$$\rho_{vv}(l) = (-1)^l a_1^l$$

Thus, the ACF never dies out and hence cannot be used for detecting the order of an AR process.

Although there is no direct correlation between $v(n)$ and $v(n-l)$ for $l > 1$, it appears to exist due to auto regression. ∎

As the ACF is not useful to determine the order of AR processes, we take an alternative approach. Given the AR process

$$v(n) + a_1 v(n-1) + \cdots + a_p v(n-p) = \xi(n)$$

we multiply both sides with $v(n-l)$ for $l = 0, \ldots, M$ and take the expectation to obtain

$$\begin{aligned}\gamma_{vv}(0) + a_1\gamma_{vv}(1) + \cdots + a_p\gamma_{vv}(p) &= \sigma_\xi^2\\ \gamma_{vv}(1) + a_1\gamma_{vv}(0) + \cdots + a_p\gamma_{vv}(p-1) &= 0\\ &\vdots\\ \gamma_{vv}(M) + a_1\gamma_{vv}(M-1) + \cdots + a_p\gamma_{vv}(p-M) &= 0\end{aligned}$$

At any lag $l > p$, we replace $p - l$ with $l - p$ since the ACF is symmetric, see Eq. 6.27 on page 167. We may solve the above set of equations simultaneously to obtain the ACF.

We see that although the ACF does not work for an AR process, the process we have described above involves an ACF-like approach. We explore this further and compute the covariance between $v(n)$ and $v(n-l)$ by taking the simultaneous effects of the intermediate lags. The lag after which the correlation between $v(n)$ and $v(n-l)$ dies out is the order of the AR process. The correlation computed this way is known as the *partial autocorrelation function* (*PACF*) and is denoted by $\phi_{vv}(l)$.

We now summarize this approach. The procedure to determine the order p in Eq. 6.73 is as follows:

1. Let $j = 1$.
2. Assume that the system is an AR(j) model:

$$v(n) + a_{1j} v(n-1) + \cdots + a_{jj} v(n-j) = \xi(n) \tag{6.89}$$

3. Multiplying this equation by $v(n-k)$ and taking the expectation, we obtain

$$\gamma_{vv}(k) + a_{1j}\gamma_{vv}(k-1) + \cdots + a_{jj}\gamma_{vv}(k-j) = 0, \ \forall k \geq 1 \tag{6.90}$$

   where the right-hand side is zero because of the causality relationship, see Eq. 6.32 on page 169.

4. Writing down the above equation for $k = 1$ to j, we arrive at j equations. We solve them and determine a_{jj}.

5. If $j < j_{max}$, increment j by 1 and go to step 2 above.

Note that j_{max} should be chosen to be greater than the expected p. A plot of a_{jj} *vs.* j will have a cutoff from $j = p+1$ onwards. We now illustrate this procedure with an example.

Example 6.17 Demonstrate the procedure discussed above for the system

$$v(n) - v(n-1) + 0.5v(n-2) = \xi(n) \tag{6.91}$$

For $j = 1$, Eq. 6.90 becomes

$$\gamma_{vv}(k) + a_{11}\gamma_{vv}(k-1) = 0, \ \forall k \geq 1$$

For $k = 1$, the above equation becomes

$$\begin{aligned} \gamma_{vv}(1) + a_{11}\gamma_{vv}(0) &= 0 \\ a_{11} &= -\frac{\gamma_{vv}(1)}{\gamma_{vv}(0)} \end{aligned} \tag{6.92}$$

For $j = 2$, Eq. 6.90 becomes

$$\gamma_{vv}(k) + a_{12}\gamma_{vv}(k-1) + a_{22}\gamma_{vv}(k-2) = 0$$

for $k \geq 1$. For $k = 1, \ 2$, this equation becomes

$$\begin{bmatrix} \gamma_{vv}(0) & \gamma_{vv}(1) \\ \gamma_{vv}(1) & \gamma_{vv}(0) \end{bmatrix} \begin{bmatrix} a_{12} \\ a_{22} \end{bmatrix} = - \begin{bmatrix} \gamma_{vv}(1) \\ \gamma_{vv}(2) \end{bmatrix} \tag{6.93}$$

For $j = 3$, Eq. 6.90 becomes

$$\gamma_{vv}(k) + a_{13}\gamma_{vv}(k-1) + a_{23}\gamma_{vv}(k-2) + a_{33}\gamma_{vv}(k-3) = 0$$

for all $k \geq 1$. For $k = 1, \ 2, \ 3$, it becomes

$$\begin{bmatrix} \gamma_{vv}(0) & \gamma_{vv}(1) & \gamma_{vv}(2) \\ \gamma_{vv}(1) & \gamma_{vv}(0) & \gamma_{vv}(1) \\ \gamma_{vv}(2) & \gamma_{vv}(1) & \gamma_{vv}(0) \end{bmatrix} \begin{bmatrix} a_{13} \\ a_{23} \\ a_{33} \end{bmatrix} = - \begin{bmatrix} \gamma_{vv}(1) \\ \gamma_{vv}(2) \\ \gamma_{vv}(3) \end{bmatrix} \tag{6.94}$$

If we have experimental data, we can calculate $\gamma_{vv}(k)$, $k = 0$ to 3, using which, a_{jj}, $j = 1, 2, 3$ can be calculated. In this example, let us calculate these using the method of theoretical prediction. Multiplying Eq. 6.91 by $\xi(n)$ and taking the expectation, we arrive at

$$\gamma_{v\xi}(0) = \gamma_{\xi\xi}(0) = \sigma_\xi^2$$

We also use Eq. 6.90 from $k = 0$ onwards. For example, multiplying Eq. 6.91 by $v(n)$, $v(n-1)$ and $v(n-2)$, one at a time, and taking the expectation, we obtain

$$\begin{bmatrix} 1 & -1 & 0.5 \\ -1 & 1.5 & 0 \\ 0.5 & -1 & 1 \end{bmatrix} \begin{bmatrix} \gamma_{vv}(0) \\ \gamma_{vv}(1) \\ \gamma_{vv}(2) \end{bmatrix} = \sigma_\xi^2 \begin{bmatrix} 1 \\ 0 \\ 0 \end{bmatrix}$$

Solving this matrix equation, we obtain

$$\begin{bmatrix} \gamma_{vv}(0) \\ \gamma_{vv}(1) \\ \gamma_{vv}(2) \end{bmatrix} = \begin{bmatrix} 2.4 \\ 1.6 \\ 0.4 \end{bmatrix} \sigma_\xi^2 \tag{6.95}$$

Multiplying Eq. 6.91 by $v(n-3)$ and taking the expectation, we obtain

$$\gamma_{vv}(3) - \gamma_{vv}(2) + 0.5\gamma_{vv}(1) = 0$$

the solution of which is

$$\gamma_{vv}(3) = -0.4\sigma_\xi^2 \tag{6.96}$$

We substitute Eq. 6.95–6.96 in Eq. 6.92–6.94 and determine a_{ij}, $i \le j$, $j = 1, 2, 3$. We obtain

$$\begin{aligned} a_{11} &= -0.67 \\ a_{22} &= 0.5 \\ a_{33} &= 0 \end{aligned}$$

As expected, for the AR(2) process, $a_{jj} = 0$ for $j > 2$. ∎

For large j values, we will have to solve large systems of equations, a tedious procedure if we adopt manual methods. In the next example, we implement the calculations through Matlab where we let j go up to 10.

Example 6.18 Simulate the process given in Eq. 6.91, calculate a_{jj} defined in Eq. 6.90 and compare with the results obtained in Example 6.17.

M 6.9 carries out this task. The identification toolbox is required to carry out these calculations. The system is simulated for 100,000 time instants. First 500 instances of $\{\xi(n)\}$ and $\{v(n)\}$ are shown in the left hand plot of Fig. 6.7. M 6.11 builds the square matrices given in Eq. 6.93–6.94. A plot of a_{jj} *vs.* j is shown in the right hand plot of Fig. 6.7, for j values up to 10. Note that the PACF is defined from lag 1 only. It is easy to see that the values reported in this figure are in agreement with those calculated in Example 6.17. ∎

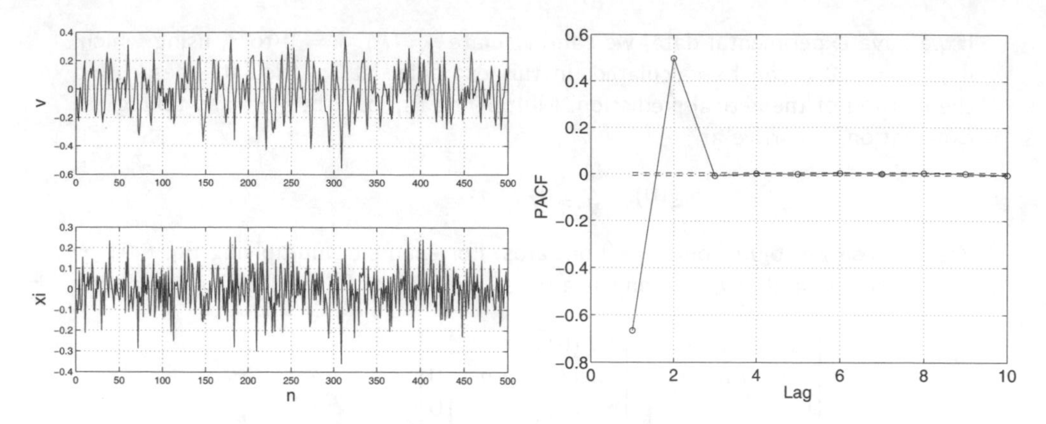

Figure 6.7: Input–output profiles (left) and the PACF (right) of the system described by Eq. 6.91. M 6.9 is used to generate it.

Before concluding this section, we would like to discuss what the PACF plots of MA processes look like. We can get a hint of this from the discussion in Sec. 3.3.6, where we have used a Taylor series expansion to convert an infinite order MA process into a finite order AR process. We can also do the reverse: convert a finite order MA process into an infinite order AR process. In view of this, the PACF plots of MA processes decay slowly, either monotonically or with oscillations. This is illustrated in the next example.

Example 6.19 Calculate the ACF and the PACF of the process

$$v_2(n) = (1 - 0.9z^{-1} + 0.2z^{-2})\xi(n)$$

discussed in Example 6.15 on page 185.

M 6.12 implements these calculations. We obtain the ACF and the PACF plots, as in Fig. 6.8. The ACF plot has only two nonzero coefficients, as expected. The PACF, on the other hand, decays slowly.

∎

The above trend is true in all MA systems: while the ACF will show a definite cutoff, the PACF plot will decay slowly, either monotonically or with oscillations. In the next section, we take up the case of ARMA processes.

6.4.6 Auto Regressive Moving Average Processes

In the previous sections, we have developed procedures to determine the order for either AR or MA processes. Now we will consider the situation of both AR and MA occurring simultaneously. In other words, we will now develop methods to determine the orders p and q of an ARMA(p,q) process. We begin with a simple trial and error procedure, which can be summarized as follows:

1. Plot the ACF and PACF to check if it is a pure AR, MA or a mixed process.

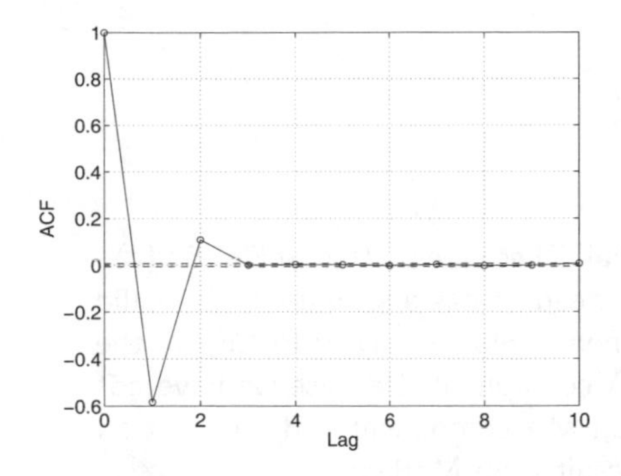

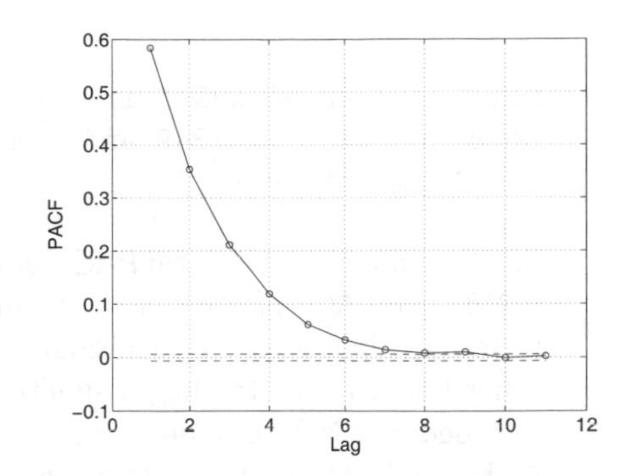

Figure 6.8: ACF (left) and PACF (right) plots of MA(2) process discussed in Example 6.19

2. For a mixed process, start with an ARMA(1,1) model (use the `arma` function in Matlab).

3. Compute the residuals of this model (use the `pe` function in Matlab).

4. If the ACF of the residual shows a slow decay, increase the AR component. If, on the other hand, the PACF has a slow decay, increase the MA component. If both show a slow decay, increase the orders of both AR and MA components by 1. Go to step 3.

Because most realistic processes can be modelled with ARMA processes of small AR and MA orders, the above indicated method should converge. We now illustrate this procedure with an example, taken from [56].

Example 6.20 Determine the order of the following ARMA(1,1) process, whose transfer function is given by

$$H(z) = \frac{1 - 0.3z^{-1}}{1 - 0.8z^{-1}}$$

This can, equivalently, be written as

$$v(n) - 0.8v(n-1) = \xi(n) - 0.3\xi(n-1)$$

We explore the trial and error procedure described above, through M 6.13. First we define the transfer function through `idpoly`. We set up the noise input ξ and simulate its effect through `idsim`. The ACF and PACF are calculated and plotted in Fig. 6.9(a)–6.9(b).

Because both the ACF and PACF decay slowly, we could start with an ARMA(1,1) process. But suppose that we start with an AR(1) process, because the ACF has a much slower decay. We obtain the difference between this model prediction and the actual data. We obtain the following Matlab report:

```
Discrete-time IDPOLY model: A(q)v(t) = e(t)
A(q) = 1 - 0.6567 (+-0.01685) q^-1
```

```
Estimated using ARMAX from data set y
Loss function 1.01349 and FPE 1.01448
Sampling interval: 1
```

We determine the ACF and PACF of the residual. These are plotted in Fig. 6.9(c)–6.9(d). The ACF plot has a small cutoff. If, instead, it has a slow decay, it would have been indicative of additional AR components. Slow decay of PACF, on the other hand, points to the possibility of an MA component. Because we have not included any MA component so far, we add an MA component and arrive at an ARMA(1,1) model. We obtain the following result from Matlab:

```
Discrete-time IDPOLY model: A(q)v(t) = C(q)e(t)
A(q) = 1 - 0.7999 (+-0.01976) q^-1
C(q) = 1 - 0.2625 (+-0.03186) q^-1

Estimated using ARMAX from data set y
Loss function 0.985375 and FPE 0.987304
Sampling interval: 1
```

We find the loss function to be smaller now. We calculate and plot the ACF and the PACF of the residuals in Fig. 6.9(e)–6.9(f). We conclude that the residuals obtained with the ARMA(1,1) are white and hence we have arrived at the following model

$$v(n) - 0.7999v(n-1) = \xi(n) - 0.2625\xi(n-1)$$

which is close to the actual process used to generate the input–output data. ∎

In practice, commonly occurring stochastic processes can be adequately represented using ARMA(2,2), or lower order, processes.

6.5 Nonparametric Models

Recall that we are developing tools to determine the transfer functions $G(z)$ and $H(z)$ in Fig. 6.5 on page 177. In Sec. 6.4, we have explained in detail the tools available for the determination of the noise model $H(z)$. In this section, we will devote our attention to the plant model $G(z)$, restricting ourselves to nonparametric models (see Sec. 3.3.6).

6.5.1 Covariance Between Signals of LTI Systems

Let us suppose that an LTI system with an impulse response $\{g(n)\}$ is excited by an input $\{u(n)\}$ to produce the output $\{y(n)\}$. We obtain the relation for output as a convolution of the impulse response and the input:

$$y(n) = g(n) * u(n) + h(n) * \xi(n) \tag{6.97}$$

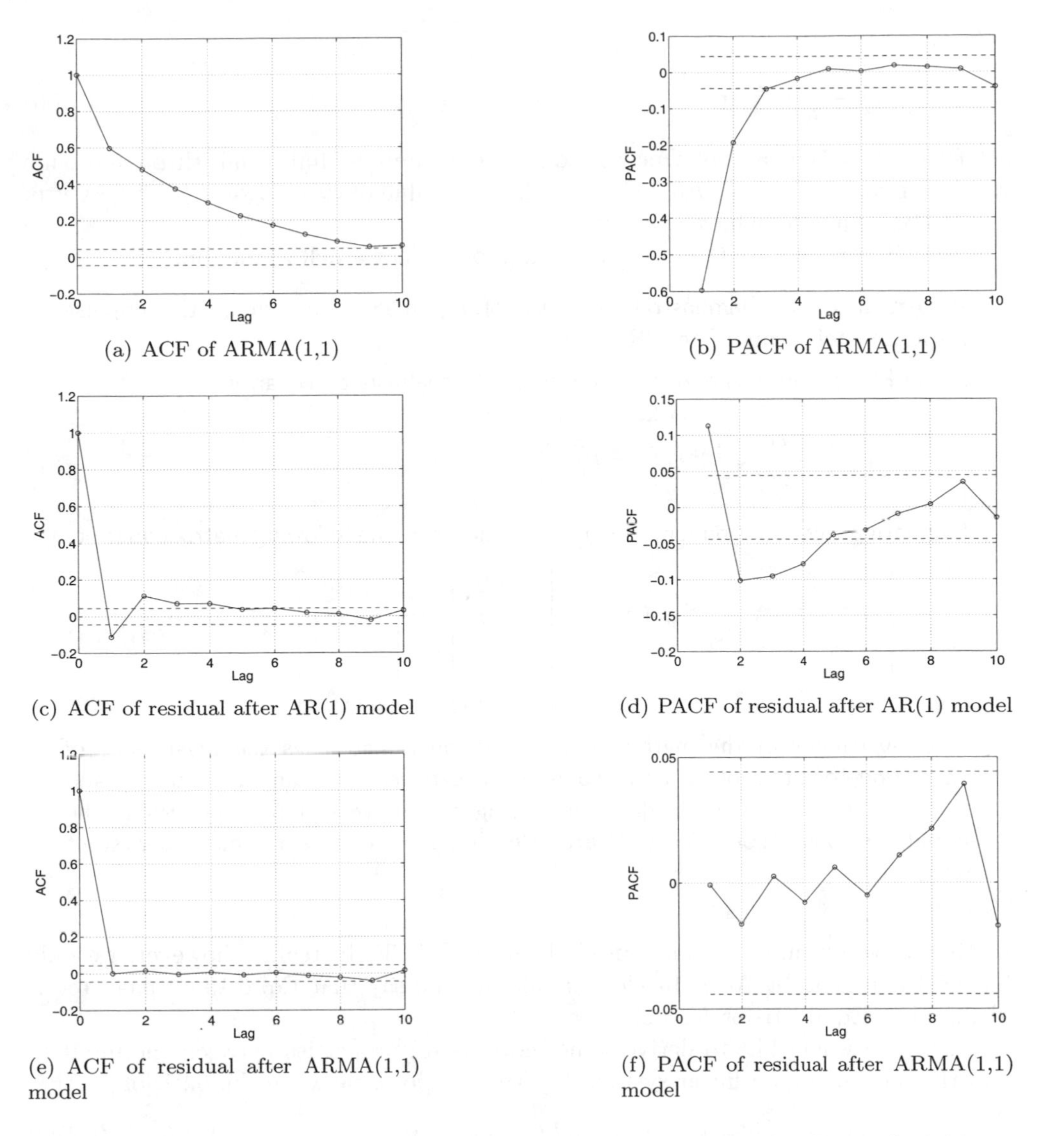

(a) ACF of ARMA(1,1)

(b) PACF of ARMA(1,1)

(c) ACF of residual after AR(1) model

(d) PACF of residual after AR(1) model

(e) ACF of residual after ARMA(1,1) model

(f) PACF of residual after ARMA(1,1) model

Figure 6.9: ACF and PACF for different processes, explored in Example 6.20

Convolving both sides with $u(-n)$ and making use of Eq. 6.62 and 6.63 on page 176, we obtain

$$\gamma_{yu}(n) = g(n) * \gamma_{uu}(n) \tag{6.98}$$

where we have made use of the fact that u and ξ are uncorrelated, see Eq. 6.38 on page 170. As it can be used to determine the impulse response, this is an important relation. For example, if the input is chosen such that its auto correlation is a delta function, *i.e.*,

$$\gamma_{uu}(n) = K\delta(n) \tag{6.99}$$

then

$$g(n) = \frac{1}{K}\gamma_{yu}(n) \tag{6.100}$$

see Example 3.7 on page 49. One way to realize the impulse function is through pseudo random binary sequence (*PRBS*) that takes the value of either 1 or -1. The ACF of the PRBS sequence behaves like an impulse [53].

We will now discuss the case when the input is not a delta function.

Example 6.21 Demonstrate the use of Eq. 6.98 to determine the impulse response coefficients of an FIR system.

As an FIR system has only a finite number of coefficients, we arrive at

$$\gamma_{yu}(n) = \sum_{k=0}^{N} g(k)\gamma_{uu}(n-k)$$

Evaluating this equation for different n, we arrive at the following matrix equation:

$$\begin{bmatrix} \gamma_{uu}(0) & \cdots & \gamma_{uu}(N) \\ \gamma_{uu}(-1) & \cdots & \gamma_{uu}(N-1) \\ \vdots & & \\ \gamma_{uu}(-N) & \cdots & \gamma_{uu}(0) \end{bmatrix} \begin{bmatrix} g(0) \\ g(1) \\ \vdots \\ g(N) \end{bmatrix} = \begin{bmatrix} \gamma_{yu}(0) \\ \gamma_{yu}(1) \\ \vdots \\ \gamma_{yu}(N) \end{bmatrix} \tag{6.101}$$

The invertibility of the matrix in this equation is known as the *persistence of excitation* condition of u. If this condition is satisfied, one can calculate r_{uu} and r_{yu} from the experimental data and, using the above equation, determine the impulse response coefficients. Use of PRBS signals can help achieve persistent excitation.

∎

In the above example, we have used theoretical ACF. In reality, however, we only have estimates of the ACF. In view of this, we can say that Eq. 6.19 on page 164 is identical to Eq. 6.101 for $N = 2$.

Next, we would like to derive some more useful properties between the input to and the output from a linear system. For this purpose, we write Eq. 6.97 for $-n$:

$$y(-n) = g(-n) * u(-n) + h(-n) * \xi(-n) \tag{6.102}$$

Convolving this with $u(n)$, and making use of Eq. 6.62 on page 176, we arrive at

$$\gamma_{yu}(-n) = g(-n) * \gamma_{uu}(n)$$

where we have once again used the fact that u and ξ are uncorrelated. As $\gamma_{yu}(-n) = \gamma_{uy}(n)$, we obtain

$$\gamma_{uy}(n) = g(-n) * \gamma_{uu}(n) \tag{6.103}$$

We can also get a useful relation for the ACF of the output signal. For this purpose, we first convolve Eq. 6.97 with Eq. 6.102 to obtain

$$y(n) * y(-n) = g(n) * u(n) * g(-n) * u(-n) + h(n) * \xi(n) * h(-n) * \xi(-n)$$

where once again, the cross terms are zero as u and ξ are uncorrelated. Now we make use of Eq. 6.62 on page 176 to obtain

$$\gamma_{yy}(n) = g(n) * g(-n) * \gamma_{uu}(n) + h(n) * h(-n) * \gamma_{\xi\xi}(n) \tag{6.104}$$

where we have made use of the fact that convolution is commutative.

6.5.2 Frequency Response of LTI Systems Excited by White Noise

Consider an LTI I/O system with impulse response $\{g(n)\}$, input $\{u(n)\}$ and output $\{y(n)\}$, with corresponding Z-transforms $G(z)$, $U(z)$ and $Y(z)$, respectively. Let $\{\gamma_{yu}(n)\}$ be the cross covariance between y and u and $\Gamma_{yu}(z)$ be its Z-transform. Let also $\{\gamma_{uu}(n)\}$ be the auto covariance of u and $\Gamma_{uu}(z)$ be its Z-transform. We would like to take the Z-transform of Eq. 6.104. For this purpose, we take the Z-transform of Eq. 6.64 on page 176, use the results of Sec. 4.2.8, and obtain

$$\Gamma_{uu}(z) = \frac{1}{2N} U(z)U(z^{-1}) \tag{6.105}$$

In a similar way, the Z-transform of Eq. 6.98 gives rise to

$$\Gamma_{yu}(z) = G(z)\Gamma_{uu}(z) \tag{6.106}$$

Taking the Z-transform of Eq. 6.104 and cancelling the common divisor $2N$, we obtain

$$\Gamma_{yy}(z) = G(z)G(z^{-1})\Gamma_{uu}(z) + H(z)H(z^{-1})\Gamma_{\xi\xi}(z) \tag{6.107}$$

We would like to visualize the above parameters as a function of frequency. Invoking the definition of Fourier transform as evaluation at $z = e^{j\omega}$, Eq. 6.105 becomes

$$\Gamma_{uu}(e^{j\omega}) = \frac{1}{2N} U(e^{j\omega})U^*(e^{j\omega}) = \frac{1}{2N} |U(e^{j\omega})|^2 \tag{6.108}$$

Substituting this in Eq. 5.51 on page 131, we obtain an expression for the energy of u as

$$E_u = \frac{1}{2\pi}\int_{-\pi}^{\pi} |U(e^{j\omega})|^2 d\omega = \frac{1}{2\pi}\int_{-\pi}^{\pi} 2N\,\Gamma_{uu}(e^{j\omega}) d\omega \tag{6.109}$$

We see that $\Gamma_{uu}(e^{j\omega}) = |U(e^{j\omega})|^2/(2N)$ represents the distribution of power as a function of frequency and hence is called the *power density spectrum* of $u(n)$. When $\Gamma_{uu}(e^{j\omega})$ is a constant, it is easy to see that

$$E_u = |U(e^{j\omega})|^2 = 2N\,\Gamma_{uu}(e^{j\omega}) \tag{6.110}$$

Now it is easy to see why white noise is so called. Recall from Sec. 6.3.2 that $\Gamma_{\xi\xi}\left(e^{j\omega}\right) = \sigma_\xi^2$, which says that the power is constant at all frequencies, the property of white light.

It is possible to make use of the relations derived above for identification. When Eq. 6.107 is evaluated at $z = e^{j\omega}$, we obtain

$$\Gamma_{yy}(e^{j\omega}) = \left|G(e^{j\omega})\right|^2 \Gamma_{uu}(e^{j\omega}) + \left|H(e^{j\omega})\right|^2 \Gamma_{\xi\xi}(e^{j\omega}) \tag{6.111}$$

Similarly, Eq. 6.106 becomes

$$\Gamma_{yu}(e^{j\omega}) = G(e^{j\omega})\Gamma_{uu}(e^{j\omega}) \tag{6.112}$$

Recall that this expression does not change even if ξ is present as an input, so long as it is uncorrelated with u. If $|\Gamma_{uu}(e^{j\omega})| = K$, a constant, the above equation becomes

$$G(e^{j\omega}) = \frac{1}{K}\Gamma_{yu}(e^{j\omega}) \tag{6.113}$$

This method of estimating the impulse response reduces the noise, because covariance calculation is a smoothing operation, see Sec. 6.3.3.

In order that the estimated transfer function fits the plant well at the select frequency range, we have to use an input that does not have constant energy at all frequencies, to be discussed in Example 6.34 on page 223. The ACF of such an input u is not an impulse function. As a result, we cannot determine the model parameters with ease. To recover this desirable property, we use a *pre-whitening filter*. This involves use of the ideas of Sec. 6.4, expressing u as a filtered version of the white noise ξ and inverting it to obtain a W, such that $Wu = \xi$. We call W the pre-whitening filter. How is this useful? Let us apply W to the input u and the plant output y. That is, let $u_F = Wu$ and $y_F = Wy$. Applying the same filtering operation to $y = Gu$ we obtain $Wy = WGu = GWu$, which is the same as $u_F = Gy_F$. In other words, the relation between u and y remains unchanged if we use filtered data u_F and y_F in their places, respectively. Now, the ACF of u_F is an impulse function and hence using Eq. 6.100, we obtain

$$g(n) = \frac{1}{K}\gamma_{y_F u_F}(n)$$

which makes the calculation of $g(n)$ an easy affair. The identification toolbox uses the concept of pre-whitening factor extensively.

6.6 Prediction Error Models

In Sec. 6.5, we have presented the topic of identification of nonparametric models. In this section, we will discuss parametric models.

In Example 6.1 on page 161, we have constructed a linear model for LSE of a general FIR model. Nevertheless, we have not presented any rigorous argument why we should follow that method and why we could not use something else. This question becomes important in more difficult models, such as ARMAX, to be discussed in Sec. 6.6.4. Indeed, we would like to extend the approach that we have developed for FIR system to general systems. We will demonstrate in this section that by minimizing the error that results from a one step ahead prediction error, we can arrive at the required linear model.

6.6.1 One Step Ahead Prediction Error Model

In this section, we build the one step ahead prediction error model for a general system, with a schematic as in Fig. 6.5 on page 177. Recall the mathematical representation of this system, given in Eq. 6.69, reproduced here for convenience:

$$y(n) = G(z)u(n) + H(z)\xi(n) \tag{6.114}$$

The variables u, y and ξ refer to input, output and white noise, respectively. The input $u(n)$ can be taken to be known, because it is under the user's control. On the other hand, $\xi(n)$ is unknown, as it is a random variable. Because of the presence of $\xi(n)$, the output $y(n)$ is also nondeterministic. We make the following assumptions about G and H:

1. $G(z)$ is strictly proper, which means that $g(n) = 0$ for $n \le 0$.

2. $H(z)$ is stable and minimum phase, which follows from the way we identify the noise models, see Sec. 6.4.4. Equivalently, the poles and the zeros of H can be taken to be inside the unit circle.

3. $H(z)$ is *monic*, which means that $h(0) = 1$. This not a restriction, because if the gain is different from 1, it can be incorporated in the variance of ξ.

Through minimization of the error in estimating $y(n)$, we can construct a one step ahead *prediction error model*. Consistent with Fig. 6.5 on page 177, we define a variable $v(n)$,

$$v(n) = H(z)\xi(n) = h(n) * \xi(n) = \sum_{l=0}^{\infty} h(l)\xi(n-l) \tag{6.115}$$

so that Eq. 6.114 becomes

$$y(n) = G(z)u(n) + v(n) \tag{6.116}$$

Suppose that the current time is denoted by n. In the above equation, we know $y(j)$ and $u(j)$ for $j = n-1, n-2, \ldots$. We can calculate v as follows:

$$v(l) = y(l) - G(z)u(l) \tag{6.117}$$

From this, it is clear that $v(l)$ can be calculated if $y(l)$ and $u(l)$ are known. Because for $l = n-1$, $n-2, \ldots,$ $y(l)$ and $u(l)$ would have been measured, $v(l)$ can be calculated for past values. What we will not know, however, is the noise for the current time instant, so we estimate it; call it $\hat{v}(n|n-1)$. With this, the output y can also be estimated; call it $\hat{y}(n|n-1)$. Because $G(z)$ is strictly proper, $y(n)$ does not depend on the current value of input, namely $u(n)$, but only on the previous values, namely $u(n-1)$, $u(n-2)$, etc. Substituting these estimates in Eq. 6.116, we obtain

$$\hat{y}(n|n-1) = G(z)u(n) + \hat{v}(n|n-1) \tag{6.118}$$

This is known as the one step ahead prediction model of y. Using the condition that $h(0) = 1$, the noise model given by Eq. 6.115 can be written in the time domain as

$$v(n) = \xi(n) + \sum_{l=1}^{\infty} h(l)\xi(n-l) \tag{6.119}$$

The best prediction of $v(n)$ is given by its expectation and hence we obtain

$$\hat{v}(n|n-1) = \mathscr{E}\left[v(n)\right] = \mathscr{E}\left[\xi(n)\right] + \mathscr{E}\left[\sum_{l=1}^{\infty} h(l)\xi(n-l)\right] \tag{6.120}$$

As ξ is a zero mean white noise, its expectation is zero. The second term is deterministic as it involves only the measurements in the previous time. Thus we obtain

$$\hat{v}(n|n-1) = \sum_{l=1}^{\infty} h(l)\xi(n-l) \tag{6.121}$$

Using Eq. 6.119, we see that $\hat{v}(n|n-1) = v(n) - \xi(n)$ and from Eq. 6.115 we obtain

$$\hat{v}(n|n-1) = h(n) * \xi(n) - \xi(n) \tag{6.122}$$

In the mixed notation of Sec. 6.4.1, this can be written as

$$\hat{v}(n|n-1) = H(z)\xi(n) - \xi(n) = (H(z) - 1)\xi(n) \tag{6.123}$$

Substituting for $\xi(n)$ from Eq. 6.115, we obtain

$$\hat{v}(n|n-1) = (H(z) - 1)H^{-1}(z)v(n) = (1 - H^{-1}(z))v(n) \tag{6.124}$$

where $H^{-1}(z)$ is stable, owing to the assumption that $H(z)$ is minimum phase. Substituting this expression for $\hat{v}(n|n-1)$ in Eq. 6.118, we obtain

$$\hat{y}(n|n-1) = G(z)u(n) + (1 - H^{-1}(z))v(n) \tag{6.125}$$

Substituting for $v(n)$ from Eq. 6.117, we obtain

$$\hat{y}(n|n-1) = G(z)u(n) + [1 - H^{-1}(z)][y(n) - G(z)u(n)] \tag{6.126}$$

which can be rewritten as

$$\hat{y}(n|n-1) = H^{-1}(z)G(z)u(n) + [1 - H^{-1}(z)]y(n) \tag{6.127}$$

This is the one step ahead predictor for the general model given by Eq. 6.114. From this, the prediction error can be written as

$$\hat{\varepsilon}(n|n-1) = y(n) - \hat{y}(n|n-1) = H^{-1}(z)\left[y(n) - G(z)u(n)\right] \tag{6.128}$$

We will apply this approach to a general model of the following form:

$$A(z)y(n) = \frac{B(z)}{F(z)}u(n) + \frac{C(z)}{D(z)}\xi(n) \tag{6.129}$$

where $A(z)$, $B(z)$, $C(z)$, $D(z)$ and $F(z)$ are polynomials in z^{-1} defined as

$$A(z) = 1 + a_1 z^{-1} + a_2 z^{-2} + \cdots + a_{\mathrm{d}A} z^{-\mathrm{d}A} \tag{6.130}$$

$$B(z) = b_1 z^{-1} + b_2 z^{-2} + \cdots + b_{\mathrm{d}B} z^{-\mathrm{d}B} \tag{6.131}$$

$$C(z) = 1 + c_1 z^{-1} + c_2 z^{-2} + \cdots + c_{\mathrm{d}C} z^{-\mathrm{d}C} \tag{6.132}$$

$$D(z) = 1 + d_1 z^{-1} + d_2 z^{-2} + \cdots + d_{\mathrm{d}D} z^{-\mathrm{d}D} \tag{6.133}$$

$$F(z) = 1 + f_1 z^{-1} + f_2 z^{-2} + \cdots + f_{\mathrm{d}F} z^{-\mathrm{d}F} \tag{6.134}$$

where d denotes the degree, see Footnote 3 on page 68. Also, recall the notation on using z as the argument even for polynomials in powers of z^{-1}, as mentioned in Footnote 5 on page 100. Notice the structure of B in the above: the constant term is zero. This means that the transfer function is strictly rational or that the input u does not affect the output immediately, *i.e.*, there is at least one sample delay before the input affects the output, see Problem 4.17. Note also that the constant term of all other variables is 1. From Eq. 6.114 and 6.129, we obtain

$$G(z) = \frac{B(z)}{A(z)F(z)} \tag{6.135}$$

$$H(z) = \frac{C(z)}{A(z)D(z)} \tag{6.136}$$

By specializing this, we obtain different models, which we will study in detail in subsequent sections. We begin with the popular FIR model.

6.6.2 Finite Impulse Response Model

By letting $A = C = D = F = 1$ and $B(z)$ be an arbitrary polynomial, we obtain from Eq. 6.129,

$$y(n) = B(z)u(n) + \xi(n) \tag{6.137}$$

Comparing this equation with Eq. 6.114, we obtain

$$G(z) = B(z) \tag{6.138}$$

$$H(z) = 1 \tag{6.139}$$

Eq. 6.137 is known as the FIR model, as it can be written in the form

$$y(n) = b(n) * u(n) + \xi(n) \tag{6.140}$$

and the convolution involves a finite number of terms. Because the constant term is zero in B, see Eq. 6.131, only the previous inputs affect the current output. We will now derive the one step ahead predictor for FIR processes: using Eq. 6.138 and 6.139 in Eq. 6.127, we obtain

$$\hat{y}(n|n-1) = B(z)u(n) \tag{6.141}$$

Eq. 6.140–6.141 are, respectively, in the form of Eq. 6.6 and Eq. 6.7 on page 162, the equations we derived while discussing the least squares approach. The details of the linear prediction equation for this FIR model have already been presented in Example 6.1 on page 161. We now illustrate a procedure to estimate the FIR model using Matlab. For the time being, we will assume that the model order is known and determine the parameters. We will take up the issue of order determination in a later section.

Example 6.22 Generate input $u(n)$, white noise $\xi(n)$ and the corresponding output $y(n)$ of the following model:

$$y(n) = \frac{0.6 - 0.2z^{-1}}{1 - 0.5z^{-1}} u(n-1) + \xi(n)$$

From the input–output data, determine the underlying FIR model.

Expanding the model in power series,

$$\begin{aligned} y(n) &= (0.6 - 0.2z^{-1})(1 + 0.5z^{-1} + 0.25z^{-2} + \cdots)u(n-1) + \xi(n) \\ &= (0.6 + 0.1z^{-1} + 0.05z^{-2} + 0.025z^{-3} + \cdots)u(n-1) + \xi(n) \end{aligned}$$

As this is a fast decaying sequence, it can be approximated well by a FIR model. M 6.14 shows a way to solve this problem. The routine `idpoly` is used to generate the model under study. As in the previous examples, a random number sequence is used to approximate the white noise. We choose the deterministic input u to be PRBS. We have seen the utility of this in Sec. 6.5. We have used the function `idinput` to generate the PRBS sequence. Plots of the inputs and the corresponding output, generated using `sim`, are presented in the plot on the left-hand side of Fig. 6.10.

The FIR parameters are calculated using the command `cra`, which essentially establishes the linear model as in Example 6.1 and solves it using the least squares procedure. The result of `cra` is compared with the actual plot on the right-hand side of Fig. 6.10. It is easy to see that the calculated values are in close agreement with the actual values.

∎

FIR models turn out to be unbiased, to be discussed in Sec. 6.7. Because of this reason, FIR models are useful, even though they may have a large number of parameters, as compared to the parametric models. The parametric models are also useful, because they require only a few parameters. In the next section, we present several parametric models.

6.6.3 Auto Regressive, Exogeneous (ARX) Input, Model

In this section, we will study the *ARX* model, in which the current output y is a function of previous outputs, previous inputs u and a random noise ξ. The following is the general form of an ARX model:

$$A(z)y(n) = B(z)u(n) + \xi(n) \tag{6.142}$$

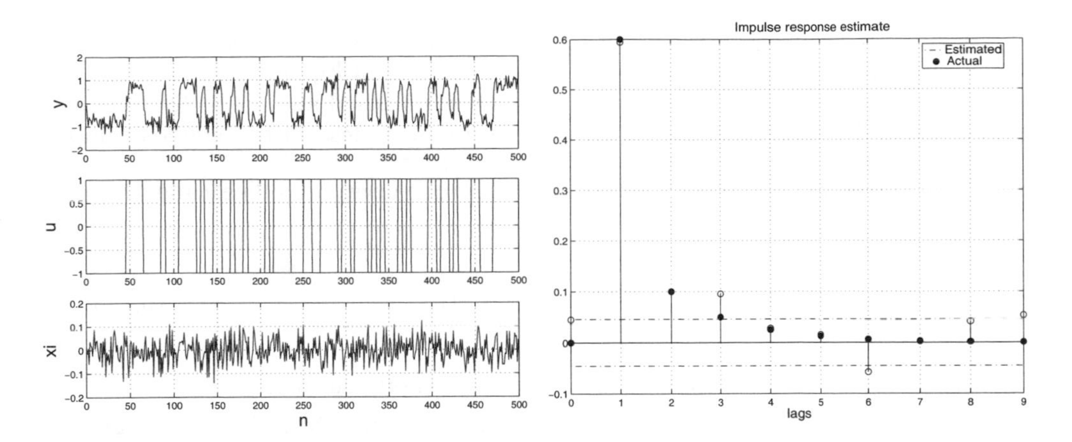

Figure 6.10: Simulated output of the FIR model, discussed in Example 6.22, with PRBS and noise inputs (left) and calculated and estimated FIR parameters (right). M 6.14 is used to generate these plots.

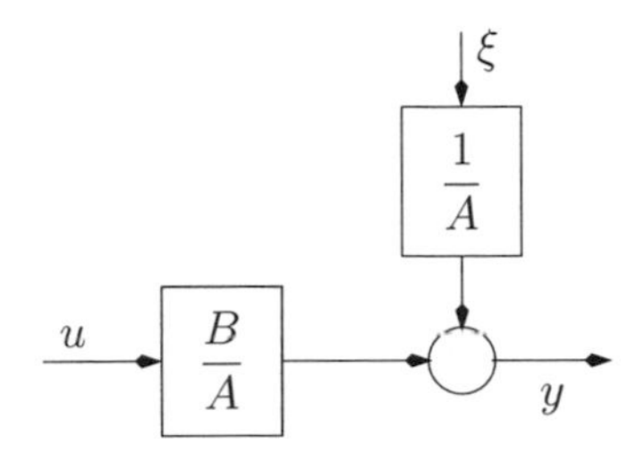

Figure 6.11: ARX structure. The output y is a sum of past inputs (Bu) and white noise (ξ), filtered through $1/A$.

Fig. 6.11 shows a schematic of this model structure. Because it can be written as

$$y(n) = \frac{1}{A(z)}(B(z)u(n) + \xi(n))$$

we see that the noise enters the model through an equation. In view of this, this is known as an equation error model. Using the standard notation, as in Eq. 6.129, we see that Eq. 6.142 satisfies $C = D = F = 1$. Comparing this equation with Eq. 6.114, we obtain

$$G(z) = \frac{B(z)}{A(z)} \tag{6.143}$$

$$H(z) = \frac{1}{A(z)} \tag{6.144}$$

Substituting these in Eq. 6.127, we obtain the prediction model as

$$\hat{y}(n|n-1) = B(z)u(n) + [1 - A(z)]y(n) \tag{6.145}$$

Recall from Eq. 6.130 that the leading coefficient of A is 1. As a result, the right-hand side of the above equation has only past values of $y(n)$. Because B has at least one delay, $u(n)$ does not affect $\hat{y}(n|n-1)$. Substituting the values of A and B using Eq. 6.130–6.131, we obtain the linear model in the form of Eq. 6.7 on page 162. Thus we see that the use of the one step prediction error method automatically results in the linear model for estimation. Now we illustrate these ideas with a simple example.

Example 6.23 From the prediction error model, develop the linear model for the ARX system

$$y(n) = -a_1 y(n-1) + \sum_{l=1}^{N} b_l u(n-l) + \xi(n) \tag{6.146}$$

where the output is assumed to depend also on the previous output.

As in Example 6.1 on page 161, writing the equations for $y(n)$, $y(n-1), \ldots$ and stacking them one below the other, we obtain

$$\begin{bmatrix} y(n) \\ y(n-1) \\ \vdots \end{bmatrix} = \begin{bmatrix} -y(n-1) & u(n-1) & \cdots & u(n-N) \\ -y(n-2) & u(n-2) & \cdots & u(n-N-1) \\ \vdots & & & \end{bmatrix} \begin{bmatrix} a_1 \\ b_1 \\ \vdots \\ b_N \end{bmatrix} + \begin{bmatrix} \xi(n) \\ \xi(n-1) \\ \vdots \end{bmatrix}$$

This is in the form of Eq. 6.6 with

$$Z(n) = \begin{bmatrix} y(n) \\ y(n-1) \\ \vdots \end{bmatrix}, \quad \theta = \begin{bmatrix} a_1 \\ b_1 \\ \vdots \\ b_N \end{bmatrix}, \quad \Xi(n) = \begin{bmatrix} \xi(n) \\ \xi(n-1) \\ \vdots \end{bmatrix},$$

$$\Phi(n) = \begin{bmatrix} -y(n-1) & u(n-1) & \cdots & u(n-N) \\ -y(n-2) & u(n-2) & \cdots & u(n-N-1) \\ \vdots & & & \end{bmatrix}$$

∎

The general ARX case is considered in Problem 6.9. We now illustrate with an example the equivalence between the correlation and the LSE methods.

Example 6.24 Establish the linear equations required to determine the parameters in the following model:

$$y(i) = -a_1 y(i-1) - a_2 y(i-2) + b_1 u(i-1) + b_2 u(i-2), \quad 1 \le i \le n \tag{6.147}$$

Using the least squares approach, we arrive at

$$\begin{bmatrix} -y(n-1) \\ -y(n-2) \\ u(n-1) \\ u(n-2) \end{bmatrix} \begin{bmatrix} -y(n-1) & -y(n-2) & u(n-1) & u(n-2) \end{bmatrix} \begin{bmatrix} a_1 \\ a_2 \\ b_1 \\ b_2 \end{bmatrix} = \begin{bmatrix} -y(n-1) \\ -y(n-2) \\ u(n-1) \\ u(n-2) \end{bmatrix} y(n) \tag{6.148}$$

By solving this equation, we can determine the impulse response coefficients, provided the coefficient matrix is nonsingular. It is easy to verify that the above system is equivalent to (see Problem 6.10)

$$\begin{bmatrix} r_{yy}(0) & r_{yy}(1) & -r_{yu}(0) & -r_{yu}(1) \\ r_{yy}(1) & r_{yy}(0) & -r_{yu}(-1) & -r_{yu}(0) \\ -r_{yu}(0) & -r_{yu}(-1) & r_{uu}(0) & r_{uu}(1) \\ -r_{yu}(1) & -r_{yu}(0) & r_{uu}(1) & r_{uu}(0) \end{bmatrix} \begin{bmatrix} a_1 \\ a_2 \\ b_1 \\ b_2 \end{bmatrix} = \begin{bmatrix} -r_{yy}(1) \\ -r_{yy}(2) \\ r_{yu}(1) \\ r_{yu}(2) \end{bmatrix}, \tag{6.149}$$

establishing the equivalence between LSE and correlation methods for ARX systems as well. ∎

We will now present an example that illustrates how to determine the ARX parameters using Matlab.

Example 6.25 Generate input $u(n)$, white noise $\xi(n)$ and the corresponding output $y(n)$ of the ARX model

$$y(n) - 0.5y(n-1) = 0.6u(n-2) - 0.2u(n-3) + \xi(n)$$

From the input–output data, determine the underlying ARX model.

M 6.15 shows a way to solve this problem. As in Example 6.22, input u, noise ξ and the output y are generated. Using the `cra` function, the impulse response coefficients are calculated and displayed in the right-hand side plot of Fig. 6.12. It is easy to see from this plot that a delay of two sampling intervals is present.

Next, we assume the order of this system and determine the model parameters using the function call `arx`. Using the function call `present`, we see that the model parameters are

```
Discrete-time IDPOLY model: A(q)y(t) = B(q)u(t) + e(t)
A(q) = 1 - 0.4878 (+-0.01567) q^-1
B(q) = 0.604 (+-0.00763) q^-2 - 0.1887 (+-0.01416) q^-3

Estimated using ARX from data set zd
Loss function 0.0529887 and FPE 0.0531133
Sampling interval: 1
```

We see that the estimated values are close to the actual values. A check on the validity is made with the command `resid`. The result is a plot of the ACF of the residual, and the cross covariance of the residual with the input, shown on the right-hand side of Fig. 6.12. If the residuals are white, the ACF should be an impulse function and the CCF should be zero. From this figure, we see that both conditions are satisfied. Thus the identified model can be taken to be correct. ∎

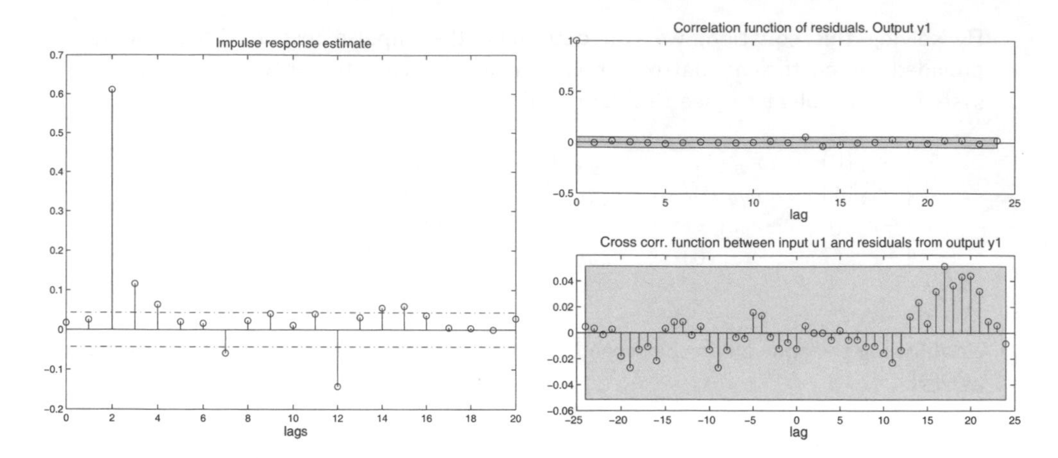

Figure 6.12: Impulse response coefficients (left) and ACF of residuals and CCF of residuals with input (right) for the ARX model discussed in Example 6.25. M 6.15 is used to generate these plots.

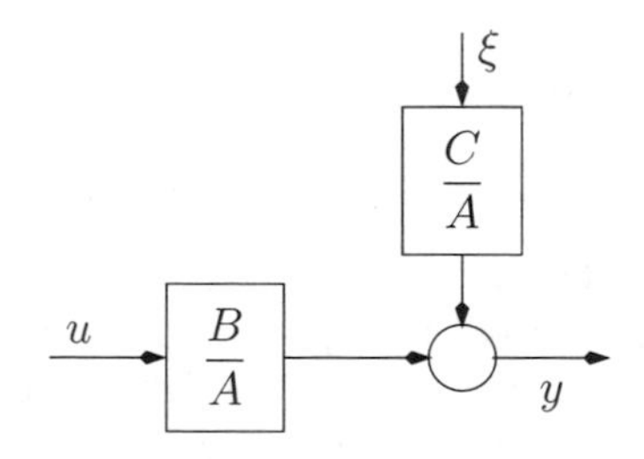

Figure 6.13: ARMAX Structure. The output y is a sum of past inputs (Bu) and current and past white noise ($C\xi$), filtered through $1/A$

6.6.4 Auto Regressive Moving Average, Exogeneous (ARMAX) Input, Model

In the ARMAX model, the current output is a function of previous outputs (auto regressive part), past inputs (exogeneous part) and current and previous noise terms (moving average part). A general ARMAX model can be written as

$$A(z)y(n) = B(z)u(n) + C(z)\xi(n) \tag{6.150}$$

A schematic of the model structure is given in Fig. 6.13. As in the case of the ARX model, the noise enters the model through an equation. In view of this, the ARMAX model is said to belong to the family of equation error models.

While for FIR and ARX models the LSE scheme can be written down simply by observation, this is not the case for the ARMAX model. The reason is that now we have a weighted sum of previous noise signals. The method of one step prediction error comes to our rescue.

Comparing Eq. 6.150 with the general expression given by Eq. 6.129, we obtain $F = D = 1$. First we rewrite Eq. 6.150 as

$$y(n) = \frac{B(z)}{A(z)}u(n) + \frac{C(z)}{A(z)}\xi(n) \tag{6.151}$$

Comparing this equation with Eq. 6.114, we obtain

$$G(z) = \frac{B(z)}{A(z)} \tag{6.152}$$

$$H(z) = \frac{C(z)}{A(z)} \tag{6.153}$$

for the ARMAX model. Substituting these in Eq. 6.127, we obtain the prediction model to be

$$\hat{y}(n|n-1) = \frac{B(z)}{C(z)}u(n) + \left[1 - \frac{A(z)}{C(z)}\right]y(n) \tag{6.154}$$

Multiplying it by $C(z)$ and adding $[1 - C(z)]\hat{y}(n|n-1)$ to both sides, we obtain

$$\hat{y}(n|n-1) = B(z)u(n) + [1 - A(z)]y(n) + [C(z) - 1][y(n) - \hat{y}(n|n-1)] \tag{6.155}$$

Defining the prediction error to be

$$\varepsilon(n|\theta) = y(n) - \hat{y}(n|n-1) \tag{6.156}$$

where we have included θ to indicate that ε is a function of model parameters, Eq. 6.155 becomes

$$\hat{y}(n|n-1) = B(z)u(n) + [1 - A(z)]y(n) + [C(z) - 1]\varepsilon(n|\theta) \tag{6.157}$$

Recall from Eq. 6.130–6.132 that the leading coefficients of A and C are unity. As a result, the right-hand side of the above equation does not involve the current terms of y and the residual ε. Nevertheless, as ε is a function of θ, which can be calculated only when ε is known, we see that we will not obtain a linear model of the form given in Eq. 6.6 on page 162. In other words, the regression matrix Φ is itself a function of the unknown parameters θ. As a result, $\hat{y}$ is not a linear function of θ. In view of this, the ARMAX prediction model is called a pseudo linear model.

With $A(z)$ and $B(z)$ as given in Eq. 6.130– 6.131 on page 199, we can arrive at the linear model of the form given in Problem 6.11. The solution begins by guessing $C(z)$. A good initial guess for C is 1. That is, we start with an ARX model. We then calculate θ, and then, using Eq. 6.156, determine ε. Now, using Eq. 6.157, we calculate θ once again and repeat this procedure. Because we solve a least squares problem at every step, this approach is known as pseudo linear regression, which is implemented in the Matlab System Identification Toolbox, using the command `armax`. We now illustrate its use with a simple example.

Example 6.26 Find the prediction error model for the following system:

$$y(n) + ay(n-1) = bu(n-1) + \xi(n) + c\xi(n-1) \tag{6.158}$$

We write this as

$$(1 + az^{-1})y(n) = bz^{-1}u(n) + (1 + cz^{-1})\xi(n)$$

We have

$$\begin{aligned} A(z) &= 1 + az^{-1} \\ B(z) &= bz^{-1} \\ C(z) &= 1 + cz^{-1} \end{aligned}$$

Using Eq. 6.157, we obtain the prediction model as

$$\hat{y}(n|n-1) = bz^{-1}u(n) - az^{-1}y(n) + cz^{-1}\varepsilon(n)$$

Note that the right-hand side involves only the past terms. The prediction error ε requires model output $\hat{y}$, which requires knowledge of θ, not available until the above equation is solved. We have to resort to a trial and error procedure to solve this equation. ∎

We will now present an example that illustrates how to use Matlab to determine the ARMAX parameters.

Example 6.27 Generate input $u(n)$, white noise $\xi(n)$ and the corresponding output $y(n)$ of the ARMAX model

$$y(n) - 0.5y(n-1) = 0.6u(n-2) - 0.2u(n-3) + \xi(n) - 0.3\xi(n-1)$$

From the input–output data, determine the model.

We generate input u, noise ξ and output y, as in Example 6.25, to solve this problem. The details are in M 6.16.

As far as calculation of impulse response by the function `cra` is concerned, it does not matter whether the system is ARX or ARMAX, so long as the input u is uncorrelated with the noise ξ. The impulse response coefficients are calculated and displayed in the left-hand side plot of Fig. 6.14. It is easy to see from this plot that a delay of two sampling intervals is present in this system.

We assume the order of this system and determine the model parameters using the function call `armax`. Using the function call `present`, we see that the model parameters are

```
Estimated using ARMAX from data set zd
Loss function 0.0504576 and FPE 0.0506161
Sampling interval: 1

Discrete-time IDPOLY model: A(q)y(t) = B(q)u(t) + C(q)e(t)
A(q) = 1 - 0.4636 (+-0.0286) q^-1
B(q) = 0.5779 (+-0.007621) q^-2 - 0.1517 (+-0.02408) q^-3
C(q) = 1 - 0.2377 (+-0.03594) q^-1
```

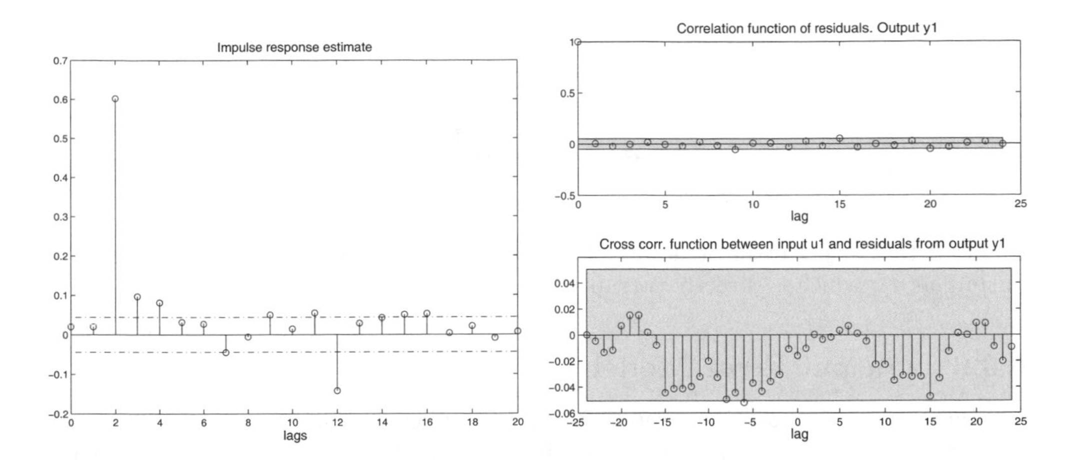

Figure 6.14: Impulse response coefficients of ARMAX model (left) and ACF of residuals and CCF of residuals with input for the ARMAX model (right), as discussed in Example 6.27. M 6.16 is used to generate these plots.

We see that the estimated values are close to the actual values. A check on the validity is made with the command `resid`. The result is a plot of the ACF of the residual, and the cross covariance of the residual with the input, shown on the right-hand side of Fig. 6.14. If the residuals are white, the ACF should be an impulse function and the CCF should be zero. From this figure, we see that both conditions are satisfied. Thus the identified model can be taken to be correct. ∎

6.6.5 Auto Regressive Integrated Moving Average, Exogeneous (ARIMAX) Input, Model

Often the noise enters the ARMAX model in an integrated form. In view of this, we will refer to these as ARIMAX models. Such an integrated noise manifests itself in the form of drifts. A general form of the ARIMAX model is

$$A(z)y(n) = B(z)u(n) + \frac{C(z)}{\Delta(z)}\xi(n) \tag{6.159}$$

where $\Delta = 1 - z^{-1}$. The only difference between Eq. 6.150 and Eq. 6.159 is that, in the latter, there is an extra Δ term. One of the common ways of handling this model is to multiply this equation throughout by Δ:

$$A(z)\Delta y(n) = B(z)\Delta u(n) + C(z)\xi(n) \tag{6.160}$$

Comparing with Eq. 6.150, we see that the methods of ARMAX can be applied to Eq. 6.160. As $\Delta y(n)$ is equal to $y(n) - y(n-1)$, and similarly for input u, Eq. 6.160 suggests applying ARMAX techniques to differenced input and output data sets. It should be clear, however, that the above approach is applicable only when the input and the output do not have much high frequency information. We will see an application of this model in Sec. 6.6.8.

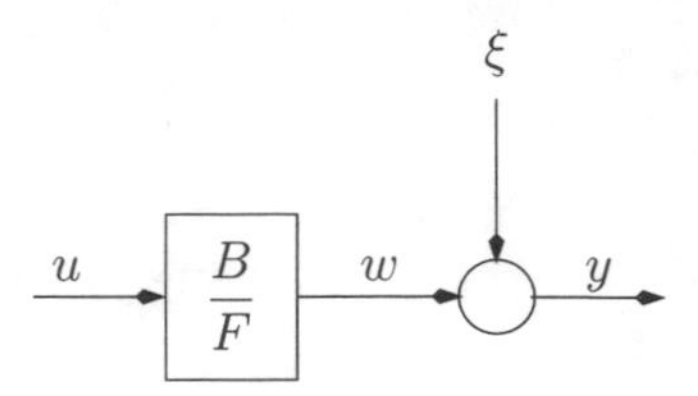

Figure 6.15: Output error Structure. The output y is a combination of an undisturbed output $w(n)$, which is directly corrupted by white noise

6.6.6 Output Error Model

The next popular model of interest is the output error (OE) model, which is also known as the transfer function model. In this, the noise affects the output of the transfer function directly. It can be modelled as

$$y(n) = \frac{B(z)}{F(z)} u(n) + \xi(n) \tag{6.161}$$

and can be represented as in Fig. 6.15. This can be written as

$$F(z)w(n) = B(z)u(n) \tag{6.162}$$
$$y(n) = w(n) + \xi(n) \tag{6.163}$$

Comparing this with Eq. 6.114 on page 197, we obtain

$$\begin{aligned} G(z) &= \frac{B(z)}{F(z)} \\ H(z) &= 1 \end{aligned} \tag{6.164}$$

Using Eq. 6.127, the prediction model is obtained as

$$\hat{y}(n) = \frac{B(z)}{F(z)} u(n) \tag{6.165}$$

which, in view of Eq. 6.162, becomes

$$\hat{y}(n) = w(n|\theta) \tag{6.166}$$

Although we do not measure w, we can calculate it using Eq. 6.162. We arrive at the following linear model:

$$\hat{y}(n) = w(n|\theta) = \phi^T(n|\theta)\theta \tag{6.167}$$

where

$$\begin{aligned} \phi(n|\theta) &= [u(n-1) \quad \cdots \quad u(n-\mathrm{d}B) \\ &\quad -w(n-1|\theta) \quad \cdots \quad -w(n-\mathrm{d}F|\theta)]^T \\ \theta &= \begin{bmatrix} b_1 & \cdots & b_{\mathrm{d}B} & f_1 & \cdots & f_{\mathrm{d}F} \end{bmatrix} \end{aligned} \tag{6.168}$$

Because w needs to be calculated using Eq. 6.162, it implies knowledge of model parameters θ. As a result, we see that this is also a pseudo linear model. We will now explain how to use Matlab to determine the OE model parameters.

Example 6.28 Generate input $u(n)$, white noise $\xi(n)$ and the corresponding output $y(n)$ of the OE model

$$y(n) = \frac{0.6 - 0.2z^{-1}}{1 - 0.5z^{-1}} u(n-2) + \xi(n)$$

From the input–output data, determine the underlying OE model.

The procedure to be followed is similar to the one given in M 6.16. It has been implemented in M 6.17. As in Example 6.27, input u, noise ξ and the output y are generated.

As in Example 6.27, using the `cra` function, the impulse response coefficients are calculated and plotted in the left-hand side of Fig. 6.16. It is easy to see from this plot that a delay of two sampling intervals is present in this system.

We assume that we know the order of this system and determine the model parameters using the function call `oe`. Using the function call `present`, we see that the model parameters are

```
Discrete-time IDPOLY model: y(t) = [B(q)/F(q)]u(t) + e(t)
B(q) = 0.5908 (+-0.006944) q^-2 - 0.1841 (+-0.0238) q^-3
F(q) = 1 - 0.4897 (+-0.02685) q^-1

Estimated using OE from data set zd
Loss function 0.0497585 and FPE 0.0498757
Sampling interval: 1
```

We see that the estimated values are close to the actual values. A check on the validity is made with the command `resid`. The result is a plot of the ACF of the residual, and the cross covariance of the residual with the input, shown on the right-hand side of Fig. 6.16. If the residuals are white, the ACF should be an impulse function and the CCF should be zero. From this figure, we see that both conditions are satisfied. Thus the identified model can be taken to be correct. ∎

6.6.7 Box–Jenkins Model

Recall that in Eq. 6.129 on page 199, reproduced here for convenience, we introduced a general model structure:

$$A(z)y(n) = \frac{B(z)}{F(z)} u(n) + \frac{C(z)}{D(z)} \xi(n)$$

When $A = 1$, we obtain the Box–Jenkins (BJ) model, a schematic of which is given in Fig. 6.17. The prediction model for this general structure is a little involved. The interested reader is referred to [32].

We restrict our attention to illustrating how to use Matlab to determine the BJ parameters through a simple example.

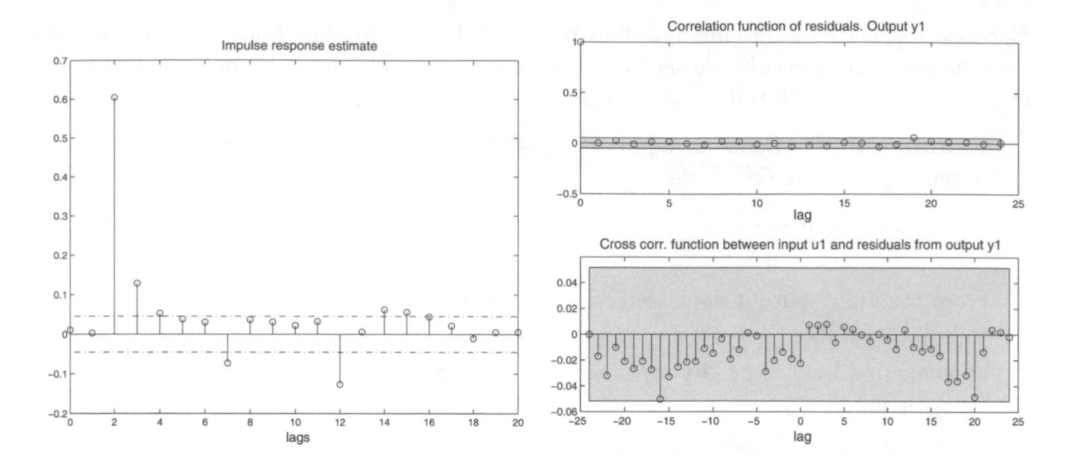

Figure 6.16: Impulse response coefficients of OE model (left), and ACF of residuals and CCF of residuals with input (right), for the OE model discussed in Example 6.28. M 6.17 is used to generate these plots.

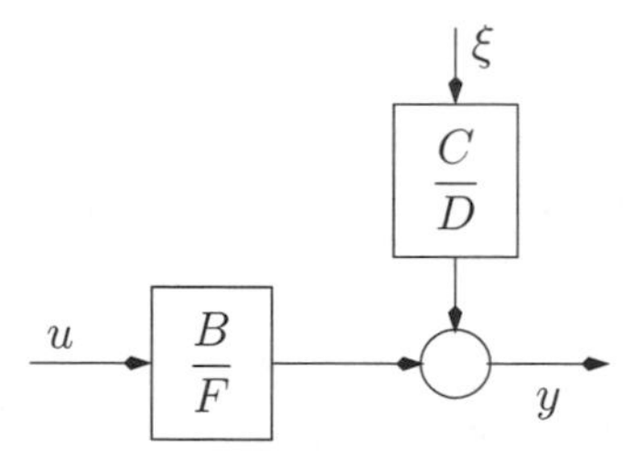

Figure 6.17: Box–Jenkins structure, which generalizes the OE structure

Example 6.29 Generate input $u(n)$, white noise $\xi(n)$ and the corresponding output $y(n)$ of the BJ model

$$y(n) = \frac{0.6 - 0.2z^{-1}}{1 - 0.7z^{-1}}u(n-2) + \frac{1 - 0.3z^{-1}}{1 - 0.5z^{-1}}\xi(n)$$

From the input–output data, determine the underlying BJ model.

The procedure to be followed is similar to the one given in Example 6.28. M 6.18 shows the details of the calculations, starting from the generation of u, y and ξ.

Using the `cra` function, the impulse response coefficients are calculated and displayed in the left-hand side plot of Fig. 6.18. It is easy to see from this plot that a delay of two sampling intervals is present in this system.

We assume that we know the order of this system and determine the model parameters using the function call `armax`. Using the function call `present`, we see that the model parameters are

```
Discrete-time IDPOLY model:
y(t) = [B(q)/F(q)]u(t) + [C(q)/D(q)]e(t)
B(q) = 0.6006 (+-0.007073) q^-2 - 0.1993 (+-0.01177) q^-3
```

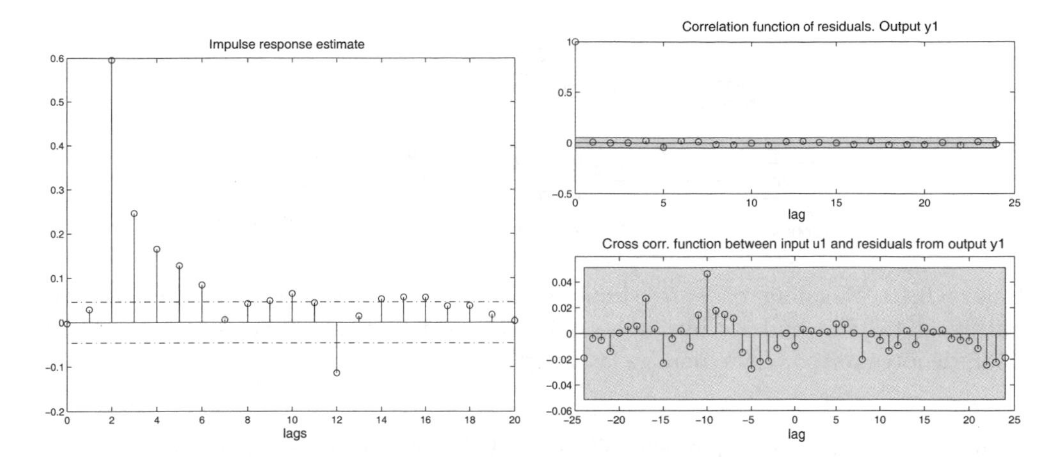

Figure 6.18: Impulse response coefficients of BJ model (left) and the ACF of residuals and CCF of residuals with input for the BJ model (right), discussed in Example 6.29. M 6.18 is used to generate these plots.

```
C(q) = 1 - 0.3108 (+-0.0695) q^-1
D(q) = 1 - 0.5083 (+-0.0605) q^-1
F(q) = 1 - 0.6953 (+-0.006909) q^-1

Estimated using BJ from data set zd
Loss function 0.048982 and FPE 0.0491745
Sampling interval: 1
```

We see that the estimated values are close to the actual values. A check on the validity is made with the command `resid`. The result is a plot of the ACF of the residual, and the cross covariance of the residual with the input, shown on the right-hand side of Fig. 6.18. If the residuals are white, the ACF should be an impulse function and the CCF should be identically zero. From this figure, we see that both conditions are satisfied. Thus the identified model can be taken to be correct.

∎

We will conclude the section on prediction error models with a summary. By assigning the values as in Table 6.3, we can obtain the different models we presented earlier.

6.6.8 Case Study: Drifting Noise Model

In this section, we will illustrate how to use the theory presented so far to identify unknown plants. We will use a simulated case study for this purpose.

Let us explore the problem of determining the model from the data generated using

$$y(n) = \frac{1.2z^{-1} + 0.1z^{-2}}{1 - z^{-1} + 0.2275z^{-2}}u(n) + \frac{1}{1 - 0.94z^{-1}}\xi(n) \tag{6.169}$$

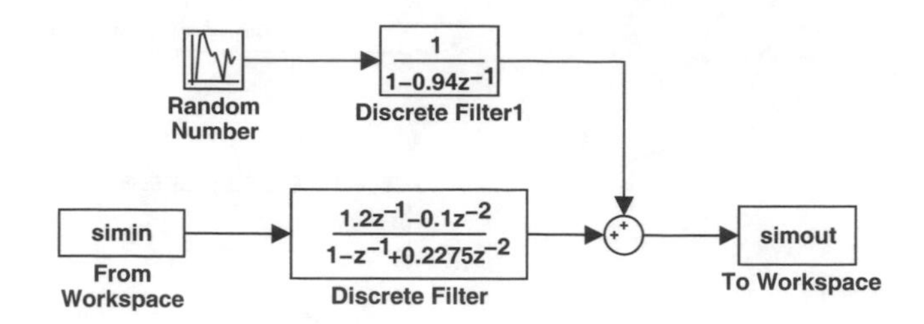

Figure 6.19: Simulink code to simulate the plant in Sec. 6.6.8 and to generate output data. The code is available at `HOME/ident/matlab/drift_ex1.mdl`, where `HOME` denotes `http://www.moudgalya.org/dc/`

Unlike the previous example, we will pretend that we do not know the order of the model. It has been implemented through M 6.20. Note that after creating the input data in M 6.20 (line 3), the Simulink code, given in Fig. 6.19, has to be executed to generate the output data. After this, the rest of M 6.20 has to be executed.

We now describe the effect of executing the above mentioned routines. Using the function `cra`, the impulse response coefficients are found and plotted in Fig. 6.20(a). We see that there is one sample delay in the process. The step response has been independently found using the function `step` in the identification toolbox and plotted in Fig. 6.20(b). We observe that this system is stable, indicated by a steady state. Since there is a drift at steady state, we guess the presence of, possibly, an integrated noise.

As a first step, we assume that the noise model is white and attempt to fit the data with an OE model. We first explore the possibility of using the following model:

$$y(n) = \frac{b_1}{1 + f_1 z^{-1}} u(n-1) + \xi(n) \tag{6.170}$$

Note that this model has one unknown in each of the B and F polynomials. We have selected these as they are amongst the smallest number of coefficients required to model an OE system. We have included one sample delay as suggested by the impulse response plot. The command `oe` gives rise to the following model parameters:

```
Discrete-time IDPOLY model: y(t) = [B(q)/F(q)]u(t) + e(t)
B(q) = 1.441 (+-0.02501) q^-1
F(q) = 1 - 0.7136 (+-0.006516) q^-1
```

Table 6.3: Values to be assigned in Eq. 6.129 on page 199 to arrive at different models

Model	Polynomial values
FIR	$A(z) = F(z) = C(z) = D(z) = 1$
ARX	$F(z) = C(z) = D(z) = 1$
ARMAX	$F(z) = D(z) = 1$
ARIMAX	$F(z) = 1$, $D(z) = \Delta$
OE	$A(z) = C(z) = D(z) = 1$
BJ	$A(z) = 1$

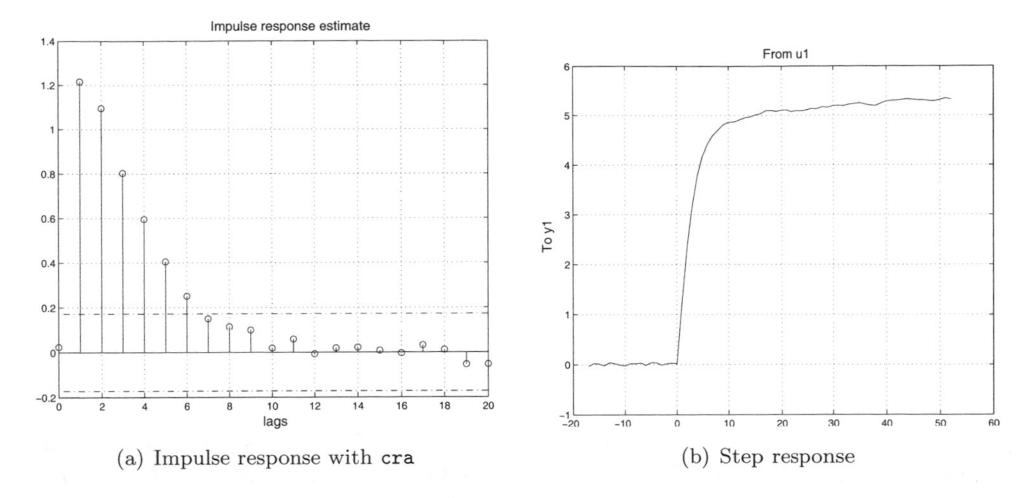

(a) Impulse response with `cra` (b) Step response

Figure 6.20: Impulse response coefficients in the drifting noise case study, obtained using `cra` and step response, obtained using `step`, both from the identification toolbox. Impulse response indicates one sample delay and step response indicates the presence of drifting noise.

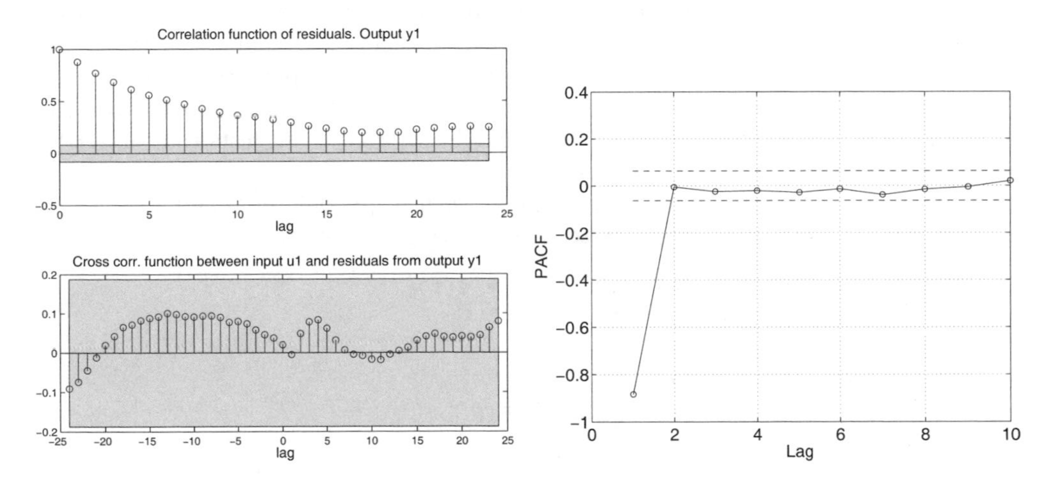

Figure 6.21: Analysis of the residuals between the plant data and that predicted by Eq. 6.170: ACF of the residuals (left top), CCF of the residuals and input u (left bottom), PACF of the residuals (right)

```
Estimated using OE from data set datatrain
Loss function 0.523117 and FPE 0.52522
Sampling interval: 1
```

We now try to validate the assumption of whiteness of noise. For this purpose, we compute the difference between the data and the value predicted by this model and refer to it as the residual. We next determine the ACF of this residual and plot it on the left-hand side of Fig. 6.21. Since it is not an impulse function, we conclude that the residuals are not white. Because of the exponential decay, we see the need

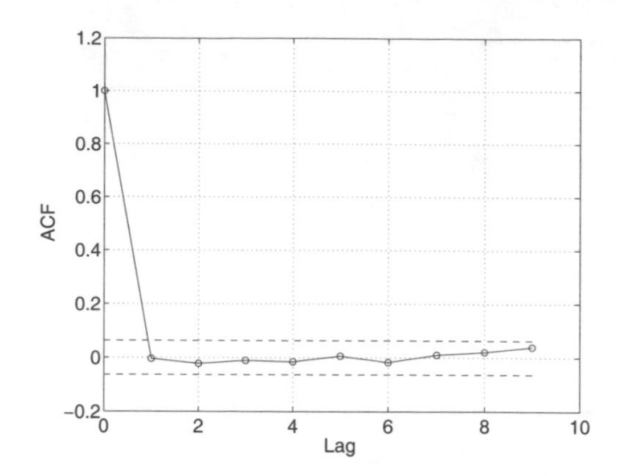

Figure 6.22: ACF of the difference between the AR(1) process and the residuals obtained by modelling the original data with the OE process, modelled with Eq. 6.170. ACF is an impulse function, indicating the sufficiency of modelling the residuals as an AR(1) process.

to model the residual, possibly, as an AR process. The CCF of the residual and the input suggests that the input u and the residuals are uncorrelated, an easier condition to meet. These computations have been carried out using the command `resid`.

To conform to the need to use an AR process, we compute the PACF of the residuals and plot in on the right-hand side of Fig. 6.21. We see a nonzero value at a lag of 1 and zero values for most other lags. This strongly suggests that an AR(1) process may be used to model the residuals.

As suggested by the above observations, we next model the residuals with an AR(1) process, using the command `ar`. The values are computed as follows:

```
Discrete-time IDPOLY model: A(q)y(t) = e(t)
A(q) = 1 - 0.8771 (+-0.01519) q^-1

Estimated using AR ('fb'/'now')
Loss function 0.120275 and FPE 0.120516
Sampling interval: 1
```

The ACF of the difference between the residuals under consideration and the AR(1) model of it is plotted in Fig. 6.22. Note that the ACF is an impulse function. This suggests that the above calculated difference is white. In other words, the residuals obtained with the OE model of Eq. 6.170 and the original data can be modelled as an AR(1) process. As the pole of this AR(1) process is near 1, the effect is close to integration, as guessed from the step response, earlier.

Combining the OE model and the AR(1) model of the resulting residuals, we obtain the following estimate of the process

$$y(n) = \frac{1.441}{1 - 0.7136z^{-1}} u(n-1) + \frac{1}{1 - 0.8771z^{-1}} \xi(n) \tag{6.171}$$

whose actual model is given in Eq. 6.169.

One problem with the above estimate is that the plant model $G(z)$ and the noise model $H(z)$ have been computed in two steps, possibly resulting in some inconsistency.

In view of this, we now try to re-estimate the model parameters simultaneously, assuming the above structure to be correct. We do this through the command `bj`, as the above structure is a BJ model. We obtain the following model parameters:

```
Discrete-time IDPOLY model: y(t) = [B(q)/F(q)]u(t) + [1/D(q)]e(t)
B(q) = 1.341 (+-0.01788) q^-1
D(q) = 1 - 0.8906 (+-0.01486) q^-1
F(q) = 1 - 0.7395 (+-0.005) q^-1

Estimated using BJ from data set datatrain
Loss function 0.115904 and FPE 0.116604
Sampling interval: 1
```

Let us refer to this as the BJ1 model. Now we calculate the residuals between the original data and that predicted by BJ1 model. We plot the ACF of the residuals and the CCF between these residuals and the input, using the `resid` command, as on the left-hand side of Fig. 6.23. Although the ACF is an impulse, because the CCF is not zero we conclude that the residuals are not white. We next explore whether a larger order BJ model can overcome this shortcoming.

Because the effect of the input is not fully captured, we increase the order of F by one and re-estimate the model parameters by BJ. We obtain the following model parameters:

```
Discrete-time IDPOLY model: y(t) = [B(q)/F(q)]u(t) + [1/D(q)]e(t)
B(q) = 1.206 (+-0.021) q^-1
D(q) = 1 - 0.8927 (+-0.01442) q^-1
F(q) = 1 - 0.9156 (+-0.01734) q^-1 + 0.1624 (+-0.01533) q^-2

Estimated using BJ from data set datatrain
Loss function 0.105265 and FPE 0.106115
Sampling interval: 1
```

Let us refer to this as the BJ2 model. We notice that the BJ2 model has smaller residuals than those of the BJ1 model. As a result, we expect the BJ2 model to be somewhat better than the BJ1 model. To confirm the adequacy of this new model, we first calculate residuals between the original data and that predicted by the BJ2 model. We calculate the ACF of the residuals and the CCF between the residuals and the input using `resid` as on the right-hand side of Fig. 6.23. Unlike the BJ1 model, the BJ2 model shows that it is sufficient: the ACF is an impulse and the CCF is zero. We have arrived at the following model:

$$y(n) = \frac{1.206z^{-1}}{1 - 0.9156z^{-1} + 0.1624z^{-2}}u(n) + \frac{1}{1 - 0.8927z^{-1}}\xi(n) \tag{6.172}$$

We can stop the identification procedure at this point.

As an alternative approach, we would like to see if we can use the fact that there is an integrated noise process right from the beginning. We proceed once again, trying to model the system with an ARIMAX structure, as explained in Sec. 6.6.5.

We determine the successive difference between the input values and we do the same with the output values. We will refer to these as the differenced data. We first attempt to model the differenced data with a simple OE model. We obtain the following parameters:

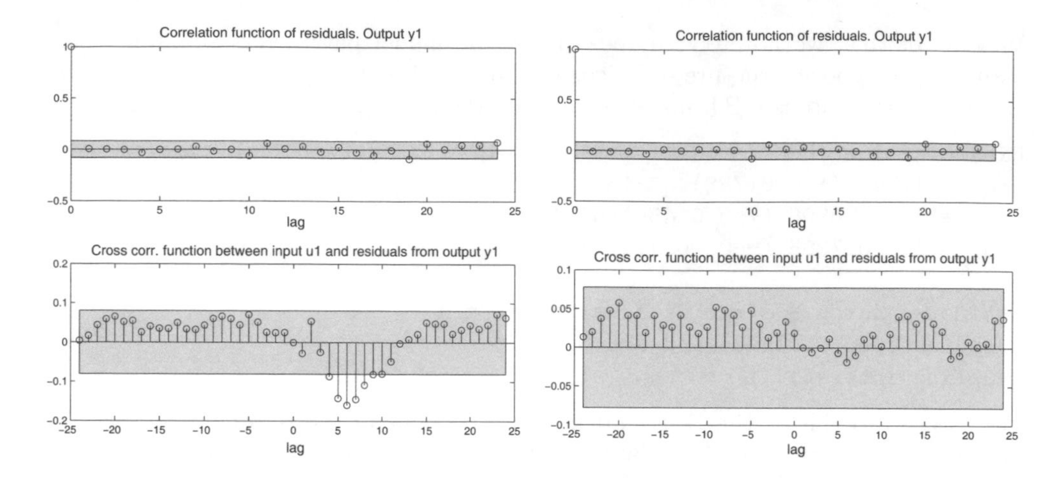

Figure 6.23: Left: ACF of the residuals between original data and the BJ1 model and CCF of input and the residuals. Although ACF is an impulse, because CCF is not zero we conclude that the residuals are not white. Right: ACF of the residuals between original data and BJ2 model and CCF of input and the residuals. Residuals are white, confirming the sufficiency of the BJ2 model.

```
Discrete-time IDPOLY model: y(t) = [B(q)/F(q)]u(t) + e(t)
B(q) = 1.333 (+-0.0177) q^-1
F(q) = 1 - 0.7422 (+-0.005223) q^-1

Estimated using OE from data set datadifftrain
Loss function 0.127381 and FPE 0.127893
Sampling interval: 1
```

We will refer to this as the OE-D1 model. To check the sufficiency of this model, we calculate the residual between its prediction and the differenced data and plot the ACF and CCF on the left-hand side of Fig. 6.24. From these plots, we see that the residuals are not white, confirming the inadequacy of the model.

We would like to see if the shortcomings of the previous model can be overcome through a higher order OE model. We obtain the following model:

```
Discrete-time IDPOLY model: y(t) = [B(q)/F(q)]u(t) + e(t)
B(q) = 1.2 (+-0.02035) q^-1
F(q) = 1 - 0.9157 (+-0.0169) q^-1 + 0.163 (+-0.01496) q^-2

Estimated using OE from data set datadifftrain
Loss function 0.115816 and FPE 0.116517
Sampling interval: 1
```

We will refer to this as the OE-D2 model. We see that the resulting residuals are smaller than that obtained with OE-D1 model. To confirm the adequacy, using `resid`, we calculate and plot ACF and CCF on the right-hand side of Fig. 6.24. The OE-D2 model that we have obtained above can be written as

$$\Delta y(n) = \frac{1.2z^{-1}}{1 - 0.9157z^{-1} + 0.163z^{-2}}\Delta u(n) + \xi(n) \tag{6.173}$$

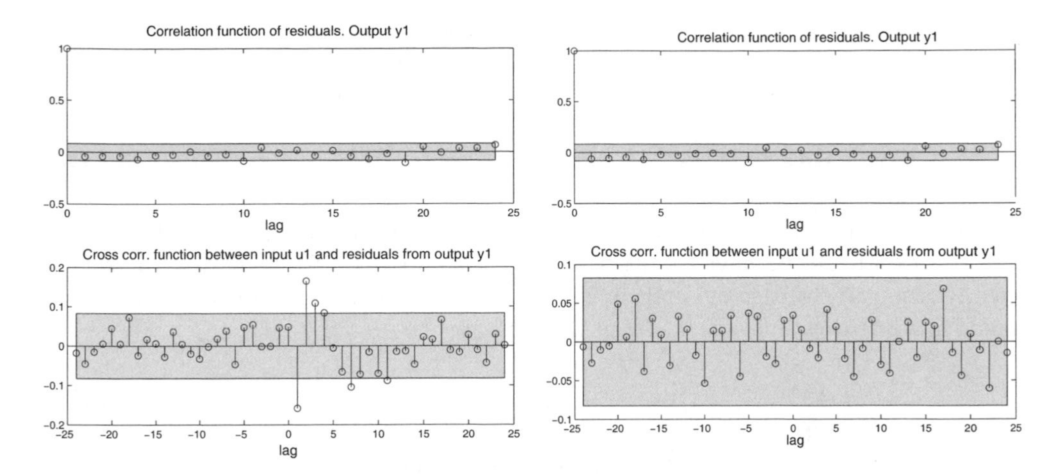

Figure 6.24: Left: ACF of the residuals between differenced data and OE-D1 model and the CCF of input and the residuals. While the ACF is an impulse, the CCF is not zero, confirming the residuals to be coloured. Right: ACF of the residuals between differenced data and OE-D2 model and the CCF of input and the residuals. The residuals are white, confirming the adequacy of the model.

or, equivalently,

$$y(n) = \frac{1.2z^{-1}}{1 - 0.9157z^{-1} + 0.163z^{-2}}u(n) + \frac{1}{\Delta}\xi(n) \tag{6.174}$$

where $\Delta = 1 - z^{-1}$. Compare this model with the BJ2 model given in Eq. 6.172 and the actual model, given in Eq. 6.169.

We conclude this section with the remark that the identification toolbox of Matlab has a program `selstruc` that helps determine the model order. This toolbox uses subspace identification methods to determine the initial model order. These topics are beyond the scope of this book. The interested reader is referred to [32].

6.7 Revisiting Least Squares Estimation

We have discussed in great detail different models and the LSE approach to estimate the model parameters. Next, we would like to find out how close the estimated parameters are to the real parameters. Even if the model parameters are deterministic, because of the possible presence of noise, the estimate will be random. In view of this, the closeness is judged by a statistical test to be presented in this section.

6.7.1 Statistical Properties of Least Squares Estimate

We would like to begin this section by addressing the question whether the estimate of a statistical quantity is a *fair estimate* or whether it is biased. The estimate $\hat{\theta}(n)$ is an *unbiased estimate* of θ if

$$\mathscr{E}(\hat{\theta}(n)) = \theta, \ \forall n \tag{6.175}$$

where we have taken θ to be deterministic. Consider the linear model that we have been using for parameter determination, namely that given by Eq. 6.6, reproduced here for convenience. We will drop the dependence on n for notational simplicity.

$$Y = \Phi\theta + \Xi \tag{6.176}$$

We would like to arrive at the conditions that the estimation model of the form given by Eq. 6.7 on page 162 have to satisfy so that it is unbiased. To start with, we restrict our attention to the case of deterministic Φ. In order to arrive at a general result, we consider a linear estimation model of the form

$$\hat{\theta} = \Psi Y \tag{6.177}$$

where Ψ also is deterministic. Substituting for Y from Eq. 6.176 we arrive at

$$\hat{\theta} = \Psi\Phi\theta + \Psi\Xi \tag{6.178}$$

Taking expectation on both sides, we obtain

$$\mathscr{E}(\hat{\theta}) = \Psi\Phi\theta \tag{6.179}$$

where we have used the fact that Ψ and Φ are deterministic and that the mean of the noise term is zero. The above equation is reduced to the unbiasedness condition, namely Eq. 6.175, when

$$\Psi\Phi = I \tag{6.180}$$

Example 6.30 Verify whether $\hat{\theta}_{\text{WLS}}$ given by Eq. 6.17 on page 164 is an unbiased estimate of θ characterized by Eq. 6.6 when Φ is deterministic and Ξ is of zero mean.

Eq. 6.177 and Eq. 6.17 on page 164 give expressions for $\hat{\theta}_{\text{WLS}}$. Comparing the right-hand sides of these equations, we obtain

$$\Psi = [\Phi^T W \Phi]^{-1} \Phi^T W \tag{6.181}$$

Postmultiplying both sides by Φ, we see that Eq. 6.180 is satisfied. As Ψ is also deterministic, we see that $\hat{\theta}_{\text{WLS}}$ is unbiased.

∎

Typically, a large number of parameters are required to describe a system if the FIR model is used. FIR is an all zero model. If, on the other hand, poles are also used, the required number of parameters will be much smaller. This is the same as representing an infinite number of impulse response coefficients with a finite number of poles and zeros, see Sec. 3.3.6. In view of the parsimony in the number of model parameters, we would typically like to explore the possibility of incorporating a pole. This is nothing but the use of the ARX model. We would like to know whether the ARX model is unbiased. The next example explores this possibility.

Example 6.31 Determine if the least square technique can give unbiased estimates of the model parameters a_1, $g(0), \ldots, g(N)$, discussed in Example 6.23 on page 202.

The output y is nondeterministic because of the presence of the noise terms in Eq. 6.146 on page 202. As a result, the matrix $\Phi(n)$ is no longer deterministic. In view of this, the condition for unbiasedness, namely Eq. 6.180, is no longer applicable. As a result, we are not in a position to say whether the estimate is unbiased.

∎

As seen in the above example, it is too restrictive to demand that Φ be deterministic for the estimate to be unbiased. We will now explore the condition required for an unbiased estimate when Φ is not deterministic. Substituting for Y from Eq. 6.6 on page 162 into Eq. 6.17, we obtain

$$\hat{\theta} = [\Phi^T W \Phi]^{-1} \Phi^T W [\Phi\theta + \Xi] = \theta + [\Phi^T W \Phi]^{-1} \Phi^T W \Xi$$

If we assume that Ξ and Φ are independent, on taking the expectation, we obtain

$$\mathscr{E}[\hat{\theta}] = \theta + \mathscr{E}\left\{[\Phi^T W \Phi]^{-1} \Phi^T W\right\} \mathscr{E}(\Xi) \tag{6.182}$$

If we further assume that Ξ is of zero mean, we obtain

$$\mathscr{E}[\hat{\theta}] = \theta \tag{6.183}$$

This is nothing but Eq. 6.175.

Example 6.32 Determine whether the least squares technique can give unbiased estimates of the model parameters a_1, $g(0), \ldots, g(N-1)$, discussed in Example 6.23.

The matrix Φ has $y(n-1)$ as an entry. By specializing Eq. 6.146 on page 202 for $n-1$, we arrive at

$$y(n-1) = -a_1 y(n-2) + \sum_{l=0}^{N-1} g(l) u(n-1-l) + \xi(n-1)$$

We see that $y(n-1)$ depends on $\xi(n-1)$. Multiplying both sides by $\xi(n-1)$ and taking the expectation, we obtain

$$\gamma_{y\xi}(0) = \sigma_\xi^2$$

because of the assumption of causality and the fact that u and ξ are uncorrelated. Unfortunately, however, the above equation says that y, an element of Φ, and ξ, an element of Ξ, are correlated. Thus, Eq. 6.183 is no longer valid. In view of this, we cannot guarantee the unbiasedness of the estimate.

∎

Example 6.33 Examine whether the LSE of the FIR model, given in Eq. 6.2 on page 160 and discussed in Example 6.1, is unbiased.

Φ consists of only the inputs. If the inputs are deterministic, the result is unbiased. Even if the input consists of some random components, in an open loop, it is reasonable to assume that it is uncorrelated with the noise vector Ξ. In view of this, use of the least squares procedure gives an unbiased estimate of the impulse response coefficients g.

▮

We have discussed the conditions that ensured the mean of the estimated parameters would be close to the actual parameter. It is clear that it is not enough if the mean is the same; we would like the variance also to be small. In view of this, we define the concept of *efficiency* .

An unbiased estimator $\hat{\theta}$ is said to be more efficient than any other estimator $\breve{\theta}$ of θ if

$$\mathscr{E}\left\{[\theta-\hat{\theta}][\theta-\hat{\theta}]^T\right\} \leq \mathscr{E}\left\{[\theta-\breve{\theta}][\theta-\breve{\theta}]^T\right\} \tag{6.184}$$

Thus the most efficient estimator has the smallest error covariance amongst all unbiased estimators.

Suppose that Φ is deterministic and V is of zero mean with positive definite covariance matrix R. It is easy to show that if we choose

$$W = R^{-1} \tag{6.185}$$

the estimate given by Eq. 6.17 is both unbiased and of smallest error covariance [36]. For this reason, the least squares method with Eq. 6.185 is known as the best linear unbiased estimator (BLUE).

Solving the normal equation repeatedly could involve a lot of effort, especially for online applications. In such a situation, we use the recursive least squares approach to reduce the calculations. The next section is devoted to this topic.

6.7.2 Recursive Least Squares

Suppose that we have found $\hat{\theta}_{\text{WLS}}(n)$, an estimate of parameters, using the data available until the time instant n. We would like to know whether it is possible to update it with the data set obtained at time instant $n+1$, so as to get $\hat{\theta}_{\text{WLS}}(n+1)$. Let

$$\Phi(n+1) = \begin{bmatrix}\phi^T \\ \Phi\end{bmatrix}, \quad Y(n+1) = \begin{bmatrix}y \\ Y\end{bmatrix}, \quad W(n+1) = \begin{bmatrix}w & 0 \\ 0 & W\end{bmatrix} \tag{6.186}$$

where we have used the convention that Φ, Y and W, which have been written without any arguments, correspond to matrix/vector values obtained at time instant n. The small letters correspond to the values obtained at time instant $n+1$. At time instant $n+1$, the solution given by Eq. 6.17 on page 164 becomes

$$\begin{aligned}\hat{\theta}_{\text{WLS}}(n+1) = {} & [\Phi^T(n+1)W(n+1)\Phi(n+1)]^{-1} \\ & \times \Phi^T(n+1)W(n+1)Y(n+1)\end{aligned} \tag{6.187}$$

Using the identity

$$\begin{aligned}\Phi^T(n+1)W(n+1)Y(n+1) &= \begin{bmatrix}\phi & \Phi^T\end{bmatrix}\begin{bmatrix}w & 0\\ 0 & W\end{bmatrix}\begin{bmatrix}y\\ Y\end{bmatrix}\\ &= \phi w y + \Phi^T W Y\end{aligned} \tag{6.188}$$

$\hat{\theta}_{\text{WLS}}(n+1)$ becomes

$$\hat{\theta}_{\text{WLS}}(n+1) = [\Phi^T(n+1)W(n+1)\Phi(n+1)]^{-1}[\phi w y + \Phi^T W Y] \tag{6.189}$$

With the definition

$$P(n+1) \triangleq [\Phi^T(n+1)W(n+1)\Phi(n+1)]^{-1} \tag{6.190}$$

Eq. 6.16 on page 163 is reduced to

$$\Phi^T W Y = \Phi^T W \Phi \hat{\theta}_{\text{WLS}} = P^{-1}\hat{\theta}_{\text{WLS}} \tag{6.191}$$

Substituting this in Eq. 6.189 and making use of Eq. 6.190, we obtain

$$\hat{\theta}_{\text{WLS}}(n+1) = P(n+1)[P^{-1}\hat{\theta}_{\text{WLS}} + \phi w y] \tag{6.192}$$

It is easy to verify that

$$P^{-1} = P^{-1}(n+1) - \phi w \phi^T \tag{6.193}$$

Substituting in Eq. 6.192, we obtain

$$\hat{\theta}_{\text{WLS}}(n+1) = P(n+1)[(P^{-1}(n+1) - \phi w \phi^T)\hat{\theta}_{\text{WLS}} + \phi w y]$$

which can be simplified to

$$\hat{\theta}_{\text{WLS}}(n+1) = \hat{\theta}_{\text{WLS}} + P(n+1)\phi w[y - \phi^T\hat{\theta}_{\text{WLS}}] \tag{6.194}$$

This shows how to get an update to the parameter vector at the $(n+1)$st instant, given its value at the nth. The second term in Eq. 6.194 can be thought of as a correction factor that helps achieve this.

It is easy to see that the recursive form of $\hat{\theta}_{\text{WLS}}$ also is unbiased, using the result in Problem 6.14. Substituting for $A(n+1)$ from Eq. 6.216 into Eq. 6.212, we obtain

$$\begin{aligned}\hat{\theta}_{\text{WLS}}(n+1) &= [I - b\phi^T]\hat{\theta}_{\text{WLS}}(n) + b(n+1)y(n+1)\\ &= \hat{\theta}_{\text{WLS}}(n) + b(n+1)[y(n+1) - \phi^T(n+1)\hat{\theta}_{\text{WLS}}(n)]\end{aligned}$$

which is in the same form as Eq. 6.194.

6.8 Weight Selection for Iterative Calculations

Frequency considerations are extremely important in identification. For example, we often characterize our plants in a certain, often low, frequency range. In contrast, the influence of noise is felt generally at high frequencies. In view of this, it is not surprising that we get useful tips on estimation procedures in the frequency domain. In this section, we arrive at a weighting strategy for estimation through frequency domain analysis.

Let us consider the problem of choosing the model parameters by minimizing the prediction error given by Eq. 6.128 on page 198, reproduced here for convenience:

$$\hat{\varepsilon}(n|n-1) = y(n) - \hat{y}(n|n-1) = H^{-1}(z)\left[y(n) - G(z)u(n)\right]$$

Suppose that the real process is given by $y(n) = G_0(z)u(n) + v(n)$. The above equation becomes

$$\varepsilon(n,\theta) = \hat{H}^{-1}(z,\theta)\left[(G_0(z) - \hat{G}(z,\theta))u(n) + v(n)\right] \tag{6.195}$$

where we have not shown the dependence on $n-1$, but indicated the dependence on model parameters with θ. Finally, in view of the fact that G and H are estimated, we have put a hat over them. Suppose that we determine θ by minimizing the sum of squares of ε, *i.e.*, $\sum_n \varepsilon^2(n,\theta)$. For zero mean signals, this is achieved by minimizing $\gamma_{\varepsilon\varepsilon}$ at zero lag, namely $\gamma_{\varepsilon\varepsilon}(0,\theta)$, see Eq. 6.24–6.25 on page 167. Thus the objective function to minimize becomes

$$J(\theta) = \gamma_{\varepsilon\varepsilon}(0,\theta) \tag{6.196}$$

We express $\gamma_{\varepsilon\varepsilon}(0,\theta)$ in terms of its inverse Fourier transform, $\Gamma_{\varepsilon\varepsilon}(e^{j\omega},\theta)$ using Eq. 5.39 on page 125. We obtain

$$J(\theta) = \frac{1}{2\pi}\int_{-\pi}^{\pi} \Gamma_{\varepsilon\varepsilon}(e^{j\omega},\theta)d\omega \tag{6.197}$$

Using a procedure similar to the one used to arrive at Eq. 6.111 on page 196, we obtain $\Gamma_{\varepsilon\varepsilon}$ as

$$\Gamma_{\varepsilon\varepsilon} = \frac{|G_0(e^{j\omega}) - \hat{G}(e^{j\omega},\theta)|^2\Gamma_{uu}(e^{j\omega}) + \Gamma_{vv}(e^{j\omega})}{|\hat{H}(e^{j\omega},\theta)|^2} \tag{6.198}$$

with the assumption that u and v are independent. Substituting this in Eq. 6.197, we obtain

$$J(\theta) = \frac{1}{2\pi}\int_{-\pi}^{\pi} \frac{|G_0(e^{j\omega}) - \hat{G}(e^{j\omega},\theta)|^2\Gamma_{uu}(e^{j\omega}) + \Gamma_{vv}(e^{j\omega})}{|\hat{H}(e^{j\omega},\theta)|^2}d\omega \tag{6.199}$$

Suppose that the noise model v is not a function of the model parameters θ. The objective function to minimize then becomes

$$J(\theta) = \frac{1}{2\pi}\int_{-\pi}^{\pi} |G_0(e^{j\omega}) - \hat{G}(e^{j\omega},\theta)|^2\frac{\Gamma_{uu}(e^{j\omega})}{|\hat{H}(e^{j\omega},\theta)|^2}d\omega \tag{6.200}$$

The above integral has the error between the actual transfer function and its estimate, weighted by the signal to noise ratio, *SNR*,

$$\mathrm{SNR} = \frac{\Gamma_{uu}(e^{j\omega})}{|\hat{H}(e^{j\omega},\theta)|^2} \tag{6.201}$$

We make SNR large in the frequency range where we want the error between the transfer function and its estimate to be small. This is achieved by making the input have frequency components in the required frequency range.

We can specialize the objective function defined in Eq. 6.200 for different parametric models. For FIR models, the objective function becomes

$$J_\infty(\theta) = \frac{1}{2\pi}\int_{-\pi}^{\pi} |G_0(e^{j\omega}) - \hat{G}(e^{j\omega},\theta)|^2 \Gamma_{uu}(e^{j\omega}) d\omega \tag{6.202}$$

see Eq. 6.139 on page 199. For ARX models, the objective function will include the noise term as well, because the noise model is no longer independent of θ. For the current discussion, however, we need only the first term that involves the plant–model mismatch. For ARX models, $H = 1/A$ (see Eq. 6.144 on page 201) and, as a result, Eq. 6.200 becomes

$$J_\infty(\theta) = \frac{1}{2\pi}\int_{-\pi}^{\pi} |G_0(e^{j\omega}) - \hat{G}(e^{j\omega},\theta)|^2 \Gamma_{uu}(e^{j\omega}) |A(e^{j\omega},\theta)|^2 d\omega \tag{6.203}$$

Normally, $1/A$ is low pass and hence A is high pass. As a result, the prediction error model with ARX structure is expected to produce a model that is valid at high frequencies. The fit may not be good at the low frequencies, however. Unfortunately, however, we generally need a good fit at low frequencies. In view of this, it is made to consist of only low frequency components by sending the input through a low pass filter. An ideal choice is $1/A$. But unfortunately, A is not known *a priori*. Nevertheless, it is possible to guess A and then iterate. The procedure to carry out this filtering is similar to the pre-whitening operation, explained in Sec. 6.5.2.

We will conclude this section with an example [51] to illustrate the fact that the predicted model will approximate well the actual plant in the frequency range of the input signal.

Example 6.34 Suppose that the plant model is given by the following equation:

$$y(n) - 0.9y(n-1) + 0.8y(n-2) = 0.1u(n-1) + 0.1u(n-2) + \xi(n) \tag{6.204}$$

Perturb this system with the input u at a certain frequency, identify using the input–output data, and compare the prediction with the actual model through the Nyquist plot. Repeat this exercise for input that has strength at different frequencies. M 6.19 implements this problem.

First Eq. 6.204 is excited by u_2 that has frequency components at all frequencies. As π is the maximum possible frequency for discrete time signals, the frequency range is 0 to π, which is the same as Nyquist frequency, ω_N. In other words, u_2 has components in the range of $[0, 1\,\omega_N]$. This input and the corresponding plant output y are shown in Fig. 6.25(a). The input (u_3), output profiles for the

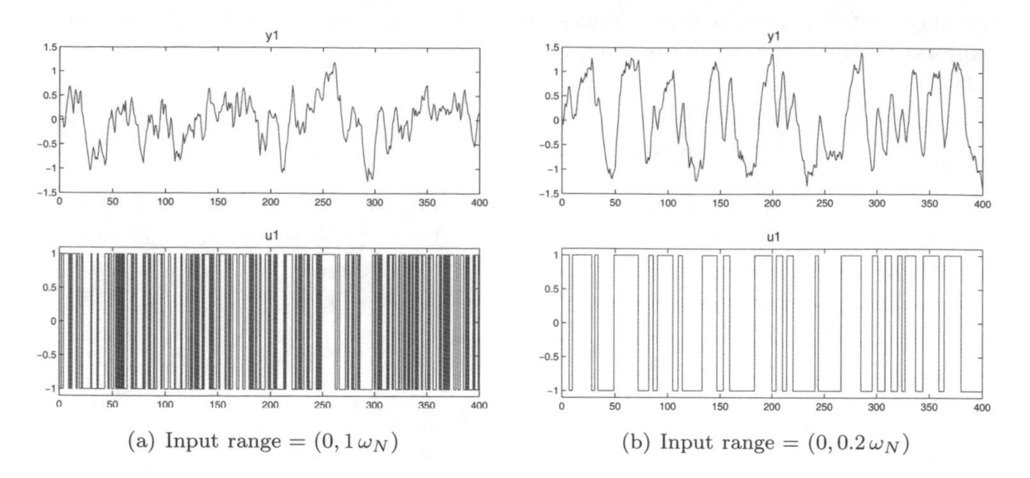

(a) Input range $= (0, 1\,\omega_N)$

(b) Input range $= (0, 0.2\,\omega_N)$

Figure 6.25: Output of the system for input in two different ranges in Example 6.34. M 6.19 is used to generate it.

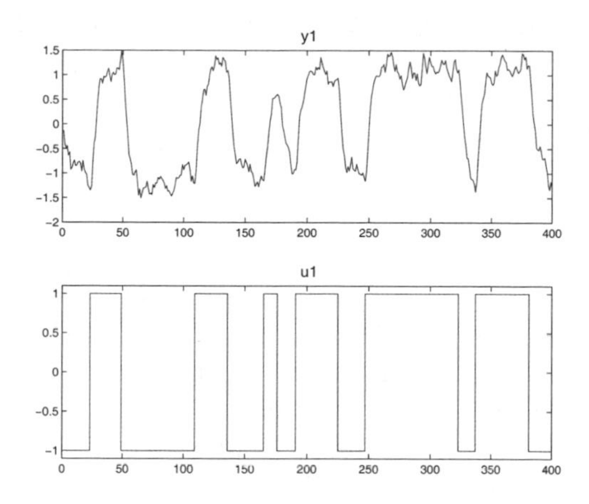

Figure 6.26: Output of the system for input in the range $[0, 0.05\omega_N]$ in Example 6.34. M 6.19 is used to generate it.

frequency range $[0, 0.2\omega_N]$ are shown in Fig. 6.25(b). Similarly, input (u_4), output curves for $[0, 0.05\omega_N]$ are shown in Fig. 6.26.

For each of these data sets, a model of the following form is identified:

$$y(n) + a_1 y(n-1) = u(n-1) + b_1 u(n-2) + \xi(n)$$

The Nyquist plots of the three models corresponding to the above presented three input–output data sets (dashed lines) are compared with the original model (solid line) in Fig. 6.27.

As an ARX model is used in this example, the discussion corresponding to Eq. 6.203 is applicable. In particular, the match is best at high frequencies when the input also has components at high frequencies. As it has all components, u_1 naturally

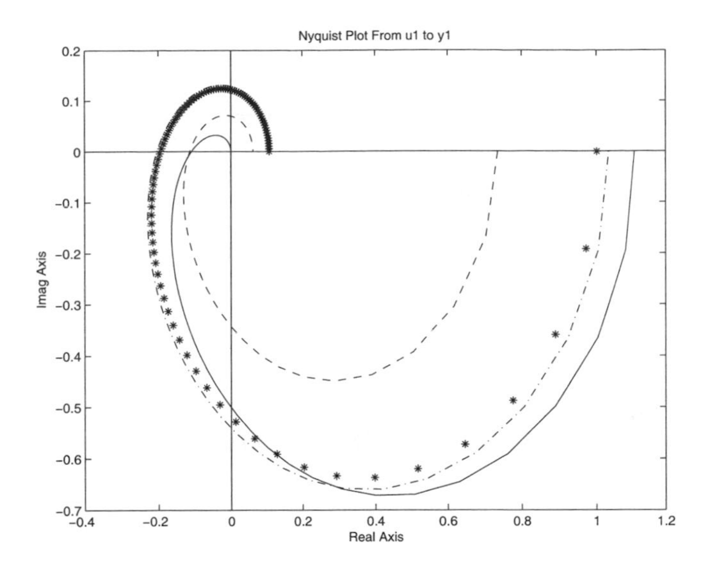

Figure 6.27: Comparison of Nyquist plots of actual system (solid line) with models obtained with input (u_2) in frequency range $[0, 1\,\omega_N]$ (dashed line), input (u_3) in range $[0, 0.2\omega_N]$ (star), input (u_4) in range $[0, 0.05\omega_N]$ (dash dot). These, respectively, fit the original model best at high, intermediate and low frequencies.

has components at the highest frequency as well. Note that in Fig. 6.27, the model identified with u_2, shown by dashes, fits best at the highest frequency. As the input u_3 has intermediate frequency components, the model matches the actual system at this range (star). Finally, u_4 has only low frequency components. Naturally, as predicted by Eq. 6.203, the fit is best (dash dot) at the lowest frequency.

We summarize the findings by saying that the selection of an appropriate frequency range is important in identification. ∎

The topic of identification is vast and hence only a glimpse of it can be given in this book. Interested readers are encouraged to read the standard books [32, 53] in this area to further their understanding.

6.9 Matlab Code

Matlab Code 6.1 Least squares solution of the simple problem discussed in Example 6.4 on page 164. Available at `HOME/ident/matlab/LS_ex.m`[3]

```
1 Mag = 10; V = 10; No_pts = 100; theta = 2;
2 Phi = Mag * (1-2*rand([No_pts,1]));
3 E = V * (1-2*rand([No_pts,1]));
4 Z = Phi*theta + E;
```

[3] `HOME` stands for `http://www.moudgalya.org/dc/` – first see the software installation directions, given in Appendix A.2.

```
LS = Phi \ Z
Max = max(Z ./ Phi), Min = min(Z ./ Phi)
```

Matlab Code 6.2 ACF calculation for the problem discussed in Example 6.5 on page 167. Available at `HOME/ident/matlab/ACF_def.m`

```
u = [1 2];
r = xcov(u);
rho = xcov(u,'coeff');
```

Matlab Code 6.3 To demonstrate the periodicity property of ACF as discussed in Example 6.7 on page 173. Available at `HOME/ident/matlab/acf_ex.m`

```
L = 500;
n = 1:L;
w = 0.1;
S = sin(w*n);
m = 1;
xi = m*randn(L,1);
Spxi = S+xi';
axes('FontSize',18);
plot(Spxi);
label('',18,'n','y',18)
figure
plotacf(Spxi,1,L,1);
```

Matlab Code 6.4 To demonstrate the maximum property of ACF at zero lag, as discussed in Example 6.8 on page 175. Available at `HOME/ident/matlab/max_ex.m`

```
S1 = [1 2 3 4];
S2 = [1,-2,3,-4];
S3 = [-1,-2,3,4];
len = length(S1)-1;
xv = -len:len;
m = 1;
xi = randn(4,1);
Spxi1 = S1 + m*xi';
Spxi2 = S2 + m*xi';
Spxi3 = S3 + m*xi';
axes('FontSize',18);
n = 1:length(S1);
plot(n,Spxi1,'o-',n,Spxi2,'x--',n,Spxi3,'*:')
label('',18,'n','y',18)

ACF1 = xcov(Spxi1,'coeff');
ACF2 = xcov(Spxi2,'coeff');
ACF3 = xcov(Spxi3,'coeff');
figure
```

```
20 axis([-len len -1 1])
21 axis off
22 axes('FontSize',18);
23 plot(xv,ACF1,'o-',xv,ACF2,'x—',xv,ACF3,'*:')
24 label('',18,'Lag','ACF',18)
```

Matlab Code 6.5 Determination of order of MA(q) as discussed in Example 6.11 on page 182. This requires the identification toolbox and the plotting routine in M 6.6. This code is available at `HOME/ident/matlab/ma.m`

```
1 % Define the model
2 m = idpoly(1,[],[1,1,-0.5]);
3
4 % Generate noise and the response
5 xi = 0.1*randn(100000,1);
6 v = sim(m, xi); z = [v xi];
7
8 % Plot noise, plant output and ACF
9
10 subplot(2,1,1), plot(v(1:500))
11 label('',18,'','v',18)
12 subplot(2,1,2), plot(xi(1:500))
13 label('',18,'n','xi',18)
14 figure, plotacf(v,1,11,1);
```

Matlab Code 6.6 Procedure to plot the ACF, as discussed in Sec. 6.4.3. An example usage is given in M 6.5. This code is available at `HOME/ident/matlab/plotacf.m`

```
1 % PLOTACF.M Plots normalized autocorrelation function
2 %
3 % [acf]=plotacf(x,errlim,len,print_code)
4 %
5 %  acf = autocorrelation values
6 %  x = time series data
7 %  errlim > 0; error limit = 2/sqrt(data_len)
8 %  len = length of acf that need to be plotted
9 %  NOTE: if len=0 then len=data_length/2;
10 %  print_code = 0 ==> does not plot OR ELSE plots
11
12 function [x]=plotacf(y,errlim,len,code)
13
14 x=xcov(y); l=length(y); x=x/x(1);
15 r=l:2*(l-1); lim=2/sqrt(l); rl=1:length(r) ;
16 N=length(rl); x=x(r);
17 if len>0 & len<N, rl=1:len; x=x(rl); N=len; end;
18 axis([0 length(rl) min(min(x),-lim-0.1) 1.1])
19 axis off, axes('FontSize',18);
20 if(code > 0 )
```

```
21    if(errlim > 0 )
22      rl=rl-1;
23      plot(rl,x,rl,x,'o' , rl,lim*ones(N,1),'--', ...
24           rl,-lim*ones(N,1),'--')
25      grid
26    else
27      plot(rl,x )
28    end
29  end;
30  label(' ',18,'Lag','ACF',18)
```

Matlab Code 6.7 Illustration of nonuniqueness in estimation of MA model parameters using ACF, discussed in Example 6.14 on page 184. This code is available at HOME/ident/matlab/unique_ma.m

```
1  xi = 0.1*randn(10000,1);
2
3  % Simulation and estimation of first model
4  m1 = idpoly(1,[],[1,-3,1.25]);
5  v1 = sim(m1,xi);
6  M1 = armax(v1,[0 2]);
7  present(M1)
8
9  % Simulation and estimation of second model
10 m2 = idpoly(1,[],[1,-0.9,0.2]);
11 v2 = sim(m2,xi);
12 M2 = armax(v2,[0 2]);
13 present(M2)
14
15 % ACF and PACF of both models
16 figure, plotacf(v1,1,11,1);
17 figure, plotacf(v2,1,11,1);
18 figure, pacf(v1,11);
19 figure, pacf(v2,11);
```

Matlab Code 6.8 Estimation with a larger order model results in large uncertainty, as discussed in Example 6.15 on page 185. This code is available at HOME/ident/matlab/ma_larger.m

```
1  m = idpoly(1,[],[1 -0.9 0.2]);
2  xi = 0.1*randn(100000,1);
3  v = sim(m,xi);
4  M1 = armax(v,[0 2]);
5  present(M1)
6  M2 = armax(v,[0 3]);
7  present(M2)
```

Matlab Code 6.9 Determination of order of AR(p) process, as discussed in Example 6.18 on page 189, using identification toolbox and the function given in M 6.10. Available at HOME/ident/matlab/pacf_ex.m

```
% Define model and generate data
m = idpoly([1,-1,0.5],[],1);
xi = 0.1*randn(100000,1);
v = sim(m,xi);

% Plot noise, plant output and PACF
subplot(2,1,1), plot(v(1:500))
label('',18,'','v',18);
subplot(2,1,2), plot(xi(1:500))
label('',18,'n','xi',18);
figure, pacf(v,10);
```

Matlab Code 6.10 Determination of the PACF of AR(p) process, as explained in Sec. 6.4.5. M 6.11 shows one usage of it. Available at HOME/ident/matlab/pacf.m

```
function [ajj] = pacf(v,M)
rvv = xcorr(v,'coeff');
len = length(rvv);
zero = (len+1)/2;
rvv0 = rvv(zero);
rvv_one_side = rvv(zero+1:len);
ajj = [];
for j = 1:M,
  ajj = [ajj pacf_mat(rvv0,rvv_one_side,j,1)];
end
p = 1:length(ajj);
N = length(p);
lim = 2/sqrt(length(v));

% Plot the figure

axes('FontSize',18);
plot(p,ajj,p,ajj,'o',p,lim*ones(N,1),'--',...
              p,-lim*ones(N,1),'--')
label('',18,'Lag','PACF',18)
```

Matlab Code 6.11 Construction of square matrix required to compute PACF a_{jj}, useful for the calculations in Sec. 6.4.5. It is used in M 6.10. This code is available at HOME/ident/matlab/pacf_mat.m

```
function ajj = pacf_mat(rvv0,rvv_rest,p,k)
if nargin == 3,
  k = 1;
end
for i = 1:p
```

```
6     for j = 1:p
7       index = (k+i-1)-j;
8       if index == 0,
9         A(i,j) = rvv0;
10      elseif index < 0,
11        A(i,j) = rvv_rest(-index);
12      else
13        A(i,j) = rvv_rest(index);
14      end
15    end
16    b(i) = -rvv_rest(k+i-1);
17  end
18  a = A\b';
19  ajj = a(p);
```

Matlab Code 6.12 PACF plot of an MA process decays slowly, as discussed in Example 6.19 on page 190. This code is available at `HOME/ident/matlab/ma_pacf.m`

```
1  m = idpoly(1,[],[1,-0.9,0.2]);
2  xi = 0.1*randn(100000,1);
3  v = sim(m,xi);
4  plotacf(v,1,11,1);
5  figure
6  pacf(v,11);
```

Matlab Code 6.13 Implementation of trial and error procedure to determine ARMA(1,1) process, presented in Example 6.20 on page 191. This requires the System Identification Toolbox, M 6.6 and M 6.10. This code is available at `HOME/ident/matlab/arma_ex.m`

```
1  % Set up the model for simulation
2  arma_mod = idpoly(1,0,[1 -0.3],[1 -0.8],1,1);
3
4  % Generate the inputs for simulation
5  % Deterministic Input can be anything
6  u = zeros(2048,1);
7  e = randn(2048,1);
8
9  % Simulate the model
10 v = sim([u e],arma_mod);
11
12 % Plot ACF and PACF for 10 lags
13 figure, plotacf(v,1e-03,11,1);
14 figure, pacf(v,10);
15
16 % Estimate AR(1) model and present it
17 mod_est1 = armax(v,[1 0]); present(mod_est1)
18
```

```
19 % compute the residuals
20 err_mod1 = pe(mod_est1,v);
21
22 % Plot ACF and PACF for 10 lags
23 figure, plotacf(err_mod1,1e-03,11,1);
24 figure, pacf(err_mod1,10);
25
26 % Check ACF and PACF of residuals
27 mod_est2 = armax(v,[1 1]); present(mod_est2)
28 err_mod2 = pe(mod_est2,v);
29
30 % Plot ACF and PACF for 10 lags
31 figure, plotacf(err_mod2,1e-03,11,1);
32 figure, pacf(err_mod2,10);
```

Matlab Code 6.14 Determination of FIR parameters as described in Example 6.22 on page 200. This code is available at `HOME/ident/matlab/fir_cra_ex1.m`. This requires the identification toolbox.

```
1 % Create the plant and noise model objects
2 var = 0.05;
3 process_mod = idpoly(1,[0 0.6 -0.2], 1, 1,...
4       [1 -0.5],'Noisevariance',var, 'Ts',1);
5
6 % Create input sequence
7 u = idinput(2555,'prbs',[0 0.2],[-1 1]);
8 xi = randn(2555,1);
9
10 % Simulate the process
11 y = sim([u xi],process_mod);
12
13 % Plot y as a function of u and xi
14 subplot(3,1,1), plot(y(1:500)),
15 label('',18,'','y',18)
16 subplot(3,1,2), plot(u(1:500))
17 label('',18,'','u',18)
18 subplot(3,1,3), plot(var*xi(1:500))
19 label('',18,'n','xi',18)
20
21 % Build iddata objects
22 z = iddata(y,u,1);
23
24 % Compute impulse response using
25 % CRA after removal of means
26 figure; [ir,r,cl] = cra(detrend(z,'constant'));
27 hold on
28
29 % Compare the first 10 impulse response
30 % computed from C(q)
```

```
31 ir_act = filter([0 0.6 -0.2],[1 -0.5],...
32            [1 zeros(1,9)]);
33
34 % Plot the actual IR
35 set(gca,'XLim',[0 9]); grid on;
36 h_act = stem((0:9),ir_act,'ro','filled');
37
38 % Add legend
39 ch_f = get(gcf,'Children');
40 ch_f2 = get(ch_f,'Children');
41 legend([ch_f2(5) h_act(1)],...
42        {'Estimated'; 'Actual'});
```

Matlab Code 6.15 Determination of ARX parameters as described in Example 6.25 on page 203. This code is available at `HOME/ident/matlab/arx_est.m`. This requires the identification toolbox.

```
1  % Create the plant and noise model objects
2  process_arx = idpoly([1 -0.5],[0 0 0.6 -0.2],...
3                 1,1,1,'Noisevariance',0.05,'Ts',1);
4
5  % Create input sequence and simulate
6  u = idinput(2555,'prbs',[0 0.2],[-1 1]);
7  xi = randn(2555,1);
8  y = sim([u xi],process_arx);
9
10 % Build iddata objects and remove means
11 z = iddata(y,u,1); zd = detrend(z,'constant');
12
13 % Compute IR for time-delay estimation
14 figure; [ir,r,cl] = cra(zd);
15
16 % Time-delay = 2 samples
17 % Estimate ARX model (assume known orders)
18 na = 1; nb = 2; nk = 2;
19 theta_arx = arx(zd,[na nb nk])
20
21 % Present the model
22 present(theta_arx)
23
24 % Check the residual plot
25 figure; resid(theta_arx,zd);
```

Matlab Code 6.16 Determination of ARMAX parameters as described in Example 6.27 on page 206. This code is available at `HOME/ident/matlab/armax_est.m`. This requires the identification toolbox.

```
1  % Create the plant and noise model objects
2  process_armax = idpoly([1 -0.5],[0 0 0.6 -0.2],...
3          [1 -0.3],1,1,'Noisevariance',0.05,'Ts',1);
4
5  % Create input sequence
6  u = idinput(2555,'prbs',[0 0.2],[-1 1]);
7  xi = randn(2555,1);
8
9  % Simulate the process
10 y = sim([u xi],process_armax);
11
12 % Build iddata objects and remove means
13 z = iddata(y,u,1); zd = detrend(z,'constant');
14
15 % Compute IR for time-delay estimation
16 figure; [ir,r,cl] = cra(zd);
17
18 % Time-delay = 2 samples
19 % Estimate ARMAX model (assume known orders)
20 na = 1; nb = 2; nc = 1; nk = 2;
21 theta_armax = armax(zd,[na nb nc nk]);
22
23 % Present the model
24 present(theta_armax)
25
26 % Check the residual plot
27 figure; resid(theta_armax,zd);
```

Matlab Code 6.17 Determination of OE parameters as described in Example 6.28 on page 209. This code is available at `HOME/ident/matlab/oe_est.m`. This requires the identification toolbox.

```
1  % Create the plant and noise model objects
2  process_oe = idpoly(1,[0 0 0.6 -0.2],1,1,[1 0.5],...
3                     'Noisevariance',0.05,'Ts',1);
4
5  % Create input sequence and simulate
6  u = idinput(2555,'prbs',[0 0.2],[-1 1]);
7  xi = randn(2555,1);
8  y = sim([u xi],process_oe);
9
10 % Build iddata objects and remove means
11 z = iddata(y,u,1); zd = detrend(z,'constant');
12
13 % Compute IR for time-delay estimation
```

```
14 figure; [ir,r,cl] = cra(zd);
15
16 % Time-delay = 2 samples
17 % Estimate OE model (assume known orders)
18 nb = 2; nf = 1; nk = 2;
19 theta_oe = oe(zd,[nb nf nk]);
20
21 % Present the model
22 present(theta_oe)
23
24 % Check the residual plot
25 figure; resid(theta_oe,zd);
```

Matlab Code 6.18 Determination of OE parameters as described in Example 6.29 on page 210. This code is available at `HOME/ident/matlab/bj_est.m`. This requires the identification toolbox.

```
1 % Create the plant and noise model objects
2 process_bj = idpoly(1,[0 0 0.6 -0.2],[1 -0.3],...
3     [1 -0.5],[1 -0.7],'Noisevariance',0.05,'Ts',1);
4
5 % Create input sequence and simulate
6 u = idinput(2555,'prbs',[0 0.2],[-1 1]);
7 xi = randn(2555,1);
8 y = sim([u xi],process_bj);
9
10 % Build iddata objects and remove means
11 z = iddata(y,u,1); zd = detrend(z,'constant');
12
13 % Compute IR for time-delay estimation
14 figure; [ir,r,cl] = cra(zd);
15
16 % Time-delay = 2 samples
17 % Estimate BJ model (assume known orders)
18 nb = 2; nc = 1; nd = 1; nf = 1; nk = 2;
19 theta_bj = bj(zd,[nb nc nd nf nk])
20
21 % Present the model
22 present(theta_bj)
23
24 % Check the residual plot
25 figure; resid(theta_bj,zd);
```

Matlab Code 6.19 Impact of frequency content of input on plant model mismatch, as discussed in Example 6.34 on page 223. This code is available at
`HOME/ident/matlab/input_freq.m`

```
1 m1 = idpoly([1 -0.9 0.08],[0 0.1 0.1],1);
2 u = idinput(400,'rbs');
```

```
3  e = 0.1*randn(400,1); y=sim(m1,[u e]);
4  z2=[y u]; idplot(z2), figure
5  m2=arx(z2,[1, 1, 2]);
6  u = idinput(400,'rbs',[0 0.2]);
7  e = 0.1*randn(400,1); y=sim(m1,[u e]);
8  z3=[y u]; idplot(z3), figure
9  m3=arx(z3,[1, 1, 2]);
10 u4 = idinput(400,'rbs',[0 0.05]);
11 e = 0.1*randn(400,1); y=sim(m1,[u4 e]);
12 z4=[y u4]; idplot(z4), figure
13 m4=arx(z4,[1, 1, 2]);
14 nyquist(m1,'k',m2,'r--',m3,'g*',m4,'b-.',{0.0001,3.14});
```

Matlab Code 6.20 Identifying a plant with a drifting noise model, discussed in Sec. 6.6.8. This code is available at `HOME/ident/matlab/case1.m`

```
1  % Generating the input sequence
2  u = idinput(2000,'rbs',[0 0.1],[-1 1]);
3  simin = [(0:1999)' u(:)];
4  open_system('drift_ex1.mdl')
5  R = input('Now execute drift_ex1.mdl and hit the return key '
     );
6
7  % Execute simulink code and generate output data
8  dataexp = iddata(simout(:),simin(:,2),1);
9  dataexp.Tstart = 0;
10 datatrain = dataexp(1:1000);
11 datatest = dataexp(1001:2000);
12
13 % Plots
14 plot(datatrain), cra(datatrain); grid
15 figure(2), step(datatrain); grid
16
17 % Estimating the OE Model
18 theta_oe1 = oe(datatrain,[1 1 1]);
19 present(theta_oe1);
20 figure(3), resid(theta_oe1,datatrain);
21 figure(4), compare(theta_oe1,datatrain);
22 figure(5); compare(theta_oe1,datatest);
23
24 % Estimating the Noise Model
25 err_oe1 = pe(theta_oe1,datatrain);
26 figure(6), plotacf(err_oe1.y,1e-03,10,1);
27 figure(7), pacf(err_oe1.y,10);
28
29 theta_n1 = ar(err_oe1.y,1);
30 present(theta_n1);
31 err_n1 = pe(theta_n1,err_oe1);
32 figure(8), plotacf(err_n1.y,1e-03,10,1);
```

```
33
34 % Building the BJ Model from OE and Noise Models
35 B = theta_oe1.b; F = theta_oe1.f; D = theta_n1.a;
36 gtotal = idpoly(1,B,1,D,F);
37 theta_bj1 = bj(datatrain,gtotal);
38 present(theta_bj1);
39 figure(9), resid(theta_bj1,datatrain);
40
41 dB = length(B)-1; dD = length(D)-1; dF = length(F)-1+1;
42 theta_bj2 = bj(datatrain,[dB 0 dD dF 1]);
43 present(theta_bj2);
44 figure(10), resid(theta_bj2,datatrain);
45 roots(theta_bj2.f); % Calculate Poles
46 figure(11), compare(theta_bj2,datatrain);
47 figure(12), compare(theta_bj2,datatest);
48
49 % Estimating the OE model on differenced data
50 datadifftrain = iddata(diff(datatrain.y),diff(datatrain.u),1)
     ;
51 theta_oediff = oe(datadifftrain,[1 1 1]);
52 present(theta_oediff);
53 figure(13), resid(theta_oediff,datadifftrain);
54
55 theta_oediff = oe(datadifftrain,[1 2 1]);
56 present(theta_oediff);
57 figure(14), resid(theta_oediff,datadifftrain);
58
59 theta_oediff = oe(datadifftrain,[2 2 1]);
60 present(theta_oediff);
61 figure(15), resid(theta_oediff,datadifftrain);
62
63 % Presenting the two models
64 m_oe = idpoly(1,theta_oediff.B,1,[1 -1],theta_oediff.F);
65 present(m_oe);
66 present(theta_bj2);
```

6.10 Problems

6.1. An LTI system can be modelled using the following relationship:

$$y(n) = a_2 y(n-2) + a_1 y(n-1) + b_3 u(n-3) + b_5 u(n-5) + e(n)$$

where $e(n)$ can be taken as white noise. Set up the data matrix Φ for the least squares problem. Comment on the number of rows of Φ. Write down the expression for the least squares estimate of the parameters. (Do *not* calculate.)

6.2. Run M 6.1 for M values of 1, 2, 5 and 10. How do the results compare with those presented in Table 6.2?

6.3. Calculate the theoretical ACF $\{\gamma_{yy}(k)\}$ for

$$y(n) = \xi(n) - 0.5\xi(n-1) + 2\xi(n-2) \tag{6.205}$$

6.4. In this problem, we will illustrate another method to arrive at Eq. 6.59 on page 175. Show that Eq. 6.57 is equivalent to

$$\begin{bmatrix} a & b \end{bmatrix} \begin{bmatrix} r_{uu}(0) & r_{uy}(k) \\ r_{uy}(k) & r_{yy}(0) \end{bmatrix} \begin{bmatrix} a \\ b \end{bmatrix} \geq 0, \text{ hence } \begin{bmatrix} r_{uu}(0) & r_{uy}(k) \\ r_{uy}(k) & r_{yy}(0) \end{bmatrix}$$

is positive definite. From this, arrive at Eq. 6.58 to Eq. 6.60.

6.5. This problem demonstrates that care should be taken while calculating the means in the time delay detection problem.

(a) Using the `xcov` function of Matlab, determine the CCF, *i.e.*, $r_{uy}(n)$, between two signals $\{u(n)\} = \{1, 2\}$ and the delayed signal $\{y(n)\} = \{0, 1, 2\}$. Do these results agree with hand calculations? Explain.

(b) Repeat the above steps with $\{u(n)\} = \{-0.5, 0.5\}$ and $\{y(n)\} = \{0, -0.5, 0.5\}$. What do you observe now? Why?

6.6. Although not useful computationally, Eq. 6.84 on page 182 helps identify the unknowns and the equations required to solve them, in the case of MA(q). Extend this approach to AR(p) problems. That is, identify the equations required to solve the unknowns that arise in AR(p) processes, modelled by Eq. 6.73 on page 179, repeated here for convenience:

$$v(n) + a_1 v(n-1) + \cdots + a_p v(n-p) = \xi(n)$$

Do not carry out the PACF calculations – the model order is given as p.

6.7. Using the method of theoretical prediction, determine the sequence of calculations required in the estimation of the parameters a_1 and c_1 in the following ARMA model:

$$y_1 + a_1 y(n-1) = \xi(n) + c_1 \xi(n-1)$$

6.8. Consider the system

$$y(n) + a_1 y(n-1) = b_1 u(n-1) + \xi(n) + c_1 \xi(n-1) \tag{6.206}$$

with $\gamma_{uu}(l) = \delta(l)\sigma_u^2$, $\gamma_{\xi\xi}(l) = \delta(l)\sigma_\xi^2$ with u and ξ uncorrelated.

(a) Show that

$$\gamma_{y\xi}(k) + a_1\gamma_{y\xi}(k-1) = \gamma_{\xi\xi}(k) + c_1\gamma_{\xi\xi}(k-1)$$

$$\gamma_{y\xi}(0) = \sigma_\xi^2$$

$$\gamma_{y\xi}(1) = (c_1 - a_1)\sigma_\xi^2$$

$$\gamma_{yu}(k) + a_1\gamma_{yu}(k-1) = b_1\gamma_{uu}(k-1)$$

$$\gamma_{yu}(0) = 0 \tag{6.207}$$

$$\gamma_{yu}(1) = b_1\sigma_u^2 \tag{6.208}$$

$$\gamma_{yy}(k) + a_1\gamma_{yy}(k-1) = b_1\gamma_{yu}(-k+1) + \gamma_{y\xi}(-k) + c_1\gamma_{y\xi}(-k+1)$$

$$\gamma_{yy}(0) = \frac{b_1^2\sigma_u^2 + (1 + c_1^2 - 2a_1c_1)\sigma_\xi^2}{1 - a_1^2} \tag{6.209}$$

$$\gamma_{yy}(1) = \frac{-a_1b_1^2\sigma_u^2 + (c_1 - a_1)(1 - a_1c_1)\sigma_\xi^2}{1 - a_1^2} \tag{6.210}$$

[Hint: Follow the method of Example 6.6 on page 170; derive the above equations in the same order as given.]

(b) How would you experimentally determine a_1, b_1 and c_1?

6.9. A simple ARX model was studied in Example 6.23 on page 202. Show that the values of ϕ and θ in Eq. 6.6 on page 162 for the general ARX model are given by

$$\phi = \begin{bmatrix} -y(k-1) & -y(k-2) & \cdots & u(k-1) & u(k-2) & \cdots \end{bmatrix}$$
$$\theta = \begin{bmatrix} a_1 & a_2 & \cdots & b_1 & b_2 & \cdots \end{bmatrix}^T$$

6.10. Arrive at Eq. 6.149 on page 203 using the following procedure. Show that the (1,1) term of the left-hand side of Eq. 6.148 on page 202 is

$$\frac{1}{N}\sum_{k=1}^{N} y^2(k-1) = r_{yy}(0) \tag{6.211}$$

Similarly evaluate all entries of both sides of Eq. 6.148 to arrive at Eq. 6.149.

6.11. Show that the values of ϕ and θ in Eq. 6.6 on page 162 for the ARMAX model are given by

$$\phi = [-y(k-1) \quad -y(k-2) \quad \cdots \quad u(k-1) \quad u(k-2) \quad \cdots$$
$$\varepsilon(k-1) \quad \varepsilon(k-2) \quad \cdots]$$
$$\theta = \begin{bmatrix} a_1 & a_2 & \cdots & b_1 & b_2 & \cdots & c_1 & c_2 & \cdots \end{bmatrix}^T$$

6.12. Explain how you would determine the parameters of the following ARMAX model, using pseudolinear regression:

$$(1 + a_1z^{-1} + a_2z^{-2})y(n) = (b_1z^{-1} + b_2z^{-2})u(n)$$
$$+ (1 + c_1z^{-1} + c_2z^{\ 2})\xi(n)$$

6.13. Explain the steps required to determine the model parameters in the OE model, given by Eq. 6.161 on page 208.

6.14. This problem shows the condition for a linear recursive estimator to be unbiased. Let the recursive estimator be given by

$$\hat{\theta}(n+1) = A(n+1)\hat{\theta}(n) + b(n+1)y(n+1) \tag{6.212}$$

for the model

$$y(n+1) = \phi(n+1)^T\theta + \xi(n+1) \tag{6.213}$$

$$\mathscr{E}\left[\xi(n+1)\right] = 0 \tag{6.214}$$

Show that when

$$\mathscr{E}\left[\hat{\theta}(n+1)\right] = \mathscr{E}\left[\hat{\theta}(n)\right] \tag{6.215}$$

for any value of n, $\hat{\theta}(n+1)$ given by Eq. 6.212 is an unbiased estimator of θ if

$$A(n+1) = I - b(n+1)\phi^T(n+1) \tag{6.216}$$

where $A(n+1)$ and $b(n+1)$ are deterministic [36].

Part III

Transfer Function Approach to Controller Design

Chapter 7

Structures and Specifications

There are many reasons why we use controllers. The important ones are: to account for model uncertainties, to negate the effect of disturbances and to improve the performance of the plant. In this section, we will look at a few strategies that are useful in this regard.

In this chapter, we present briefly the concepts of feedback and feed forward controllers. We introduce popular controllers, such as proportional, integral, derivative and lead–lag. We explain the concepts of internal stability and internal model principle. We introduce the topic of limits of performance. We translate the performance requirements into desired pole locations in the z plane.

7.1 Control Structures

We can classify the controllers into feedback and feed forward controllers. The feedback controllers can be further classified into one degree of freedom (abbreviated as 1-DOF) and two degree of freedom (2-DOF) controllers. This section is devoted to a study of these control structures.

7.1.1 Feed Forward Controller

Consider a model of a plant, given in the mixed notation of Sec. 6.4.1:

$$y = Gu + v \tag{7.1}$$

where u, y are, respectively, input and output signals. The variable v could denote disturbance or noise; it could be deterministic or stochastic. A schematic of this model is given on the left-hand side of Fig. 7.1. Compare this with Fig. 6.5 on page 177. Suppose that we want to design a controller so as to keep the plant at the operating point, known as the *regulation* or *disturbance rejection* problem. In other words, we would like to choose u so as to compensate for the effects of v and to keep the deviation variable y at zero.

The schematic on the right-hand side of Fig. 7.1 shows a feed forward control structure in which we measure the disturbance before it can upset the plant. Note that the negative of the disturbance v becomes the input to the controller block F.

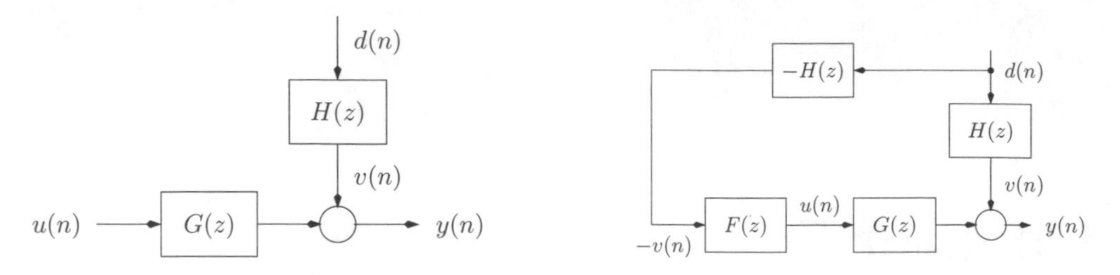

Figure 7.1: Transfer function model of a typical plant (left) and control of it using a feed forward technique (right)

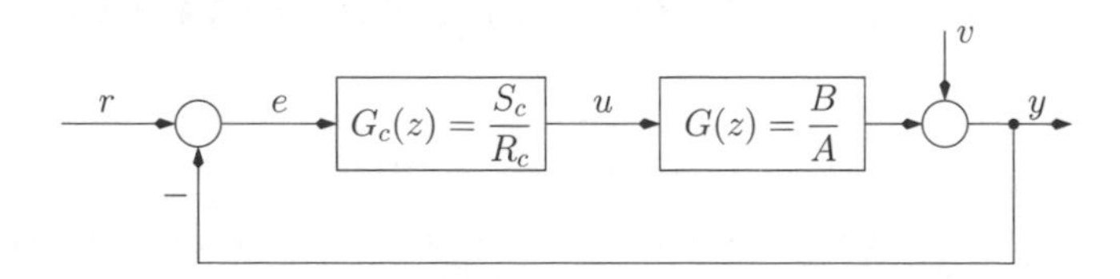

Figure 7.2: Schematic of a one degree of freedom feedback controller

The output y is given by

$$y = -GFv + v \tag{7.2}$$

We achieve the objective of regulation or $y = 0$, if $F = G^{-1}$, which implies that we should have a good knowledge of G. For this strategy to work, we also need to be able to measure v. We observe the control strategy to be

$$u = -Fv \tag{7.3}$$

If G is minimum phase, its inverse is stable and we can implement this controller.

Because this strategy has the capability to reject disturbances before they affect the plant, one should use the feed forward controller, if possible. Unfortunately, however, often we don't have good knowledge of the plant. Even if we know the plant, it may not be possible to get an exact inverse. All sampled data systems have at least one delay and hence the inverse of the plant model would result in a noncausal system. If G is nonminimum phase, the inverse would be unstable. It is possible to get an approximate inverse of the plant that is causal and stable, to be discussed in Sec. 10.2.1. Nevertheless, because of the restrictions posed by this scheme, the feed forward controller is rarely used alone. In Sec. 7.1.3, we show how the feed forward control action is possible in a feedback scheme.

7.1.2 One Degree of Freedom Feedback Controller

The feedback control strategy involves measurement, comparison with the required value and employing a suitable correction. The simplest feedback control strategy for the plant in Fig. 7.1 is given in Fig. 7.2, where G_c denotes the control block and r stands for the reference or the setpoint signal.

If the actual output y is different from the required value r, the error signal, $e = r - y$, acts on the controller to come up with an appropriate value of the control effort u. Exact knowledge of G and H is not required for this strategy to work and this is the main attraction of feedback over feed forward control strategy, presented in the previous section.

Through the algebraic method presented in Sec. 4.3.2, we obtain the following expression for y:

$$y(n) = \frac{G(z)G_c(z)}{1+G(z)G_c(z)}r(n) + \frac{1}{1+G(z)G_c(z)}v(n) \tag{7.4}$$

where we have once again used the mixed notation of Sec. 6.4.1. Recall the transfer functions T and S, defined in Eq. 4.23 and Eq. 4.24 on page 84, respectively:

$$T(z) = \frac{G(z)G_c(z)}{1+G(z)G_c(z)} \tag{7.5}$$

$$S(z) = \frac{1}{1+G(z)G_c(z)} \tag{7.6}$$

Substituting these, Eq. 7.4 becomes

$$y(n) = T(z)r(n) + S(z)v(n) \tag{7.7}$$

The feedback controller has to meet two requirements:

1. To make $y(n)$ follow $r(n)$ in an acceptable manner, explored further in Sec. 7.7. As mentioned earlier, this is known as the tracking or *servo* control problem.

2. To remove the effect of $v(n)$ on $y(n)$. This is known as the *regulatory* or *disturbance rejection* problem.

In order to address the above two issues independently, we should be able to modify S and T to our liking. Unfortunately, however, S and T defined as above satisfy the condition

$$S + T = 1 \tag{7.8}$$

which is easy to verify by straightforward substitution. As a result, once S is specified, T is fixed and vice versa. This is known as the *one degree of freedom* controller, which is abbreviated as *1-DOF* controller.

7.1.3 Two Degrees of Freedom Feedback Controller

We have seen in Sec. 7.1.2 that, using the 1-DOF controller, it is not possible to simultaneously shape the responses to both reference and disturbance signals. The *two degrees of freedom* (*2-DOF*) control structure addresses this problem. There are many different 2-DOF structures. We will discuss one of them in this section.

In the control structure of Fig. 7.3, G_b and G_f together constitute the controller. G_b is in the feedback path, while G_f is in the forward path. All other variables are as in Fig. 7.2. G_b is used to stabilize the system as well as to remove the effect of the disturbance. G_f is used to help y track r. We will look for a control law of the form

$$R_c(z)u(n) = T_c(z)r(n) - S_c(z)y(n) \tag{7.9}$$

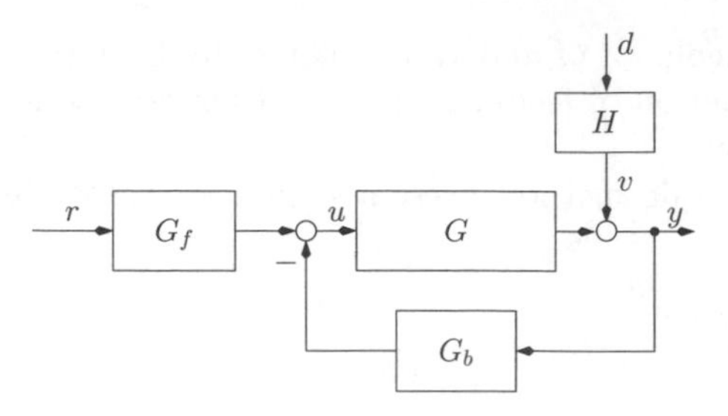

Figure 7.3: Two degrees of freedom feedback control structure

where R_c, S_c and T_c are polynomials in z^{-1}, see Footnote 5 on page 100 for the notation. It is easy to see that

$$\begin{aligned} G_f &= \frac{T_c}{R_c} \\ G_b &= \frac{S_c}{R_c} \end{aligned} \tag{7.10}$$

We will now illustrate the use of this controller for a plant with the model

$$A(z)y(n) = z^{-k}B(z)u(n) + v(n) \tag{7.11}$$

A and B are also polynomials in powers of z^{-1}. Substituting the controller defined in Eq. 7.9 in the plant model given in Eq. 7.11, we obtain

$$Ay(n) = z^{-k}\frac{B}{R_c}[T_c r(n) - S_c y(n)] + v(n)$$

Bringing the y terms to one side and simplifying, we obtain

$$\left(\frac{R_c A + z^{-k}BS_c}{R_c}\right) y(n) = z^{-k}\frac{BT_c}{R_c}r(n) + v(n)$$

This can be written as

$$y(n) = z^{-k}\frac{BT_c}{\phi_{cl}}r(n) + \frac{R_c}{\phi_{cl}}v(n) \tag{7.12}$$

where ϕ_{cl} is the closed loop characteristic polynomial

$$\phi_{cl}(z) = A(z)R_c(z) + z^{-k}B(z)S_c(z) \tag{7.13}$$

We see that the closed loop transfer function between $y(n)$ and $r(n)$ also has a delay of k samples. We want:

1. The zeros of ϕ_{cl} to be inside the unit circle, so that the closed loop system is stable.
2. R_c/ϕ_{cl} to be made small, so that we achieve disturbance rejection.
3. $z^{-k}BT_c/\phi_{cl}$ to be made close to 1, so that we achieve setpoint tracking.

We will use this approach in the subsequent chapters to design 2-DOF controllers. Although the 2-DOF controller has more capabilities, the 1-DOF controller is simpler to design, as it has fewer parameters. We now present the simplest 1-DOF controller.

7.2 Proportional Control

A proportional controller is one in which the control action is proportional to an error signal. The proportional controller is one of the most fundamental controllers.

Proportional control is the first strategy that a designer tries out while designing controllers. If this does not work, the designer looks at other options. Even in this case, the insights obtained while designing the proportional controller are extremely useful in the design of more sophisticated controllers. Because of these reasons, the design of proportional controllers becomes an extremely important approach in control system design.

One way to design a controller is through shifting the poles of a system to a better place. For example, if the plant to be controlled is unstable, we would like to move its poles to within the unit circle. Performance requirements could restrict the pole locations further, the topic of discussion of Sec. 7.7. We will illustrate these ideas with a simple example.

Example 7.1 Suppose that a plant with a transfer function

$$G(z) = \frac{1}{z(z-1)}$$

is put in a closed loop with a proportional controller K. How do the closed loop poles vary as K increases from 0?

The closed loop transfer function is

$$T = \frac{KG}{1+KG} = \frac{K}{z^2 - z + K}$$

The characteristic polynomial is $\phi_{cl}(z) = z^2 - z + K$ and the characteristic equation is $\phi_{cl}(z) = 0$, the roots of which are the closed loop poles, denoted as λ. We have

$$\lambda_{1,2} = \frac{1 \pm \sqrt{1-4K}}{2}$$

Notice that $K = 0$ gives $\lambda_{1,2} = 0, 1$, which are the same as the open loop poles. As K increases, the discriminant becomes smaller. For $K = 0.25$, $\lambda_{1,2} = 0.5$, *i.e.*, the two poles coincide. For $K > 0.25$, the poles become imaginary with the real part being 0.5. A plot of the evolution of the roots for this example is given in Fig. 7.4.

Note that for $K = 1$, the closed loop poles are at the unit circle, *i.e.*, the system is on the verge of instability. We refer to this as the *ultimate gain* and use the symbol K_u to denote it. The concept of ultimate gain is used in the tuning of PID controllers, in Sec. 8.3.2.

The plot of loci of the closed loop poles as the controller gain K is varied from 0 to ∞ is known as the *root locus* plot [16]. M 7.1 shows how to draw the root locus plot in Matlab. ∎

Proportional control is an important part of the popular PID controller, to be discussed in Sec. 7.3 and Chapter 8. We will discuss the related problem of pole placement in Chapters 9 and 14. The root locus method is a time domain technique. We will present a frequency domain technique in the next section.

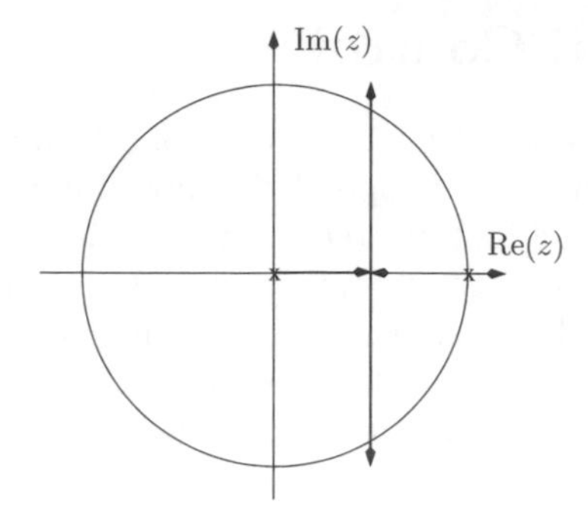

Figure 7.4: A plot of evolution of closed loop poles as K is varied from 0 to ∞ for $G(z) = 1/[z(z-1)]$, see Example 7.1

7.2.1 Nyquist Plot for Control Design

We will begin this section with the application of *Cauchy's principle.* Given a function $F(z)$, draw a closed contour C_1 in the z plane so that no zeros or poles of $F(z)$ lie on C_1. Let z zeros and p poles of $F(z)$ lie within the closed contour. Let us evaluate $F(z)$ at all points on the curve C_1 in the clockwise direction. Plot the evaluated values in another plane, called the F plane, in which $\text{Im}[F(z)]$ and $\text{Re}[F(z)]$ form y and x axes, respectively, see Footnote 2 on page 65. The new curve also will be a closed contour; call it C_2, see Fig. 7.5. Then Cauchy's principle states that C_2 will encircle the origin of the F plane $z - p$ times in the clockwise direction. This can be summarized as follows:

$$N = z - p \tag{7.14}$$

$N =$ Number of encirclements of origin in clockwise direction by curve C_2. N is positive if the encirclement is in the clockwise direction. It is negative if the encirclement is in the counterclockwise direction.

$z =$ Number of zeros of $F(z)$ that lie within the closed contour C_1.

$p =$ Number of poles of $F(z)$ that lie within the closed contour C_1.

Examples of N calculations in the $F(z)$ plane are given in Fig. 7.6. The direction of C_1 is assumed to be clockwise.

We will now explain how this approach can be used to design proportional controllers. Given $G(z)$, the open loop transfer function, the closed loop transfer function is given by $T(z) = KG(z)/(1 + KG(z))$, see Eq. 7.5. The closed loop system is unstable if the poles of $T(z)$ are outside the unit circle. That is, if the zeros of $1 + KG(z)$ are outside the unit circle.

We will choose C_1 to be the unit circle, taken in the clockwise direction. Suppose that $1 + KG(z)$ has n poles and n zeros, including those at infinity. Let

$Z =$ number of zeros of $1 + KG(z)$ *outside* the unit circle C_1
$P =$ number of poles of $1 + KG(z)$ *outside* the unit circle C_1

It follows that $1 + KG(z)$ will have $n - Z$ zeros and $n - P$ poles *inside* the unit circle.

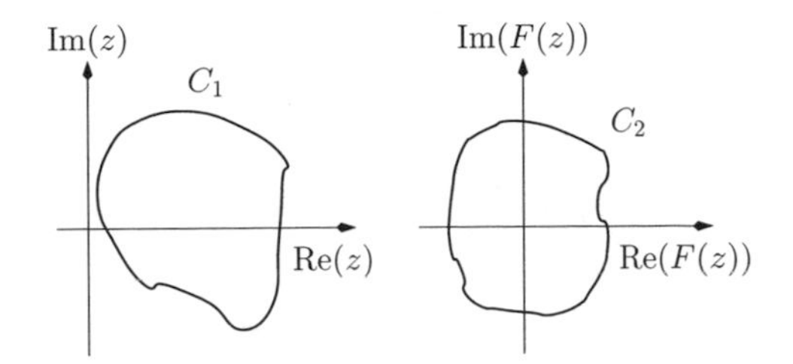

Figure 7.5: Evaluation of $F(z)$ along a closed contour C_1 (left) and plotting of imaginary part of $F(z)$ *vs.* real part of $F(z)$ results in a closed contour C_2 in the F plane

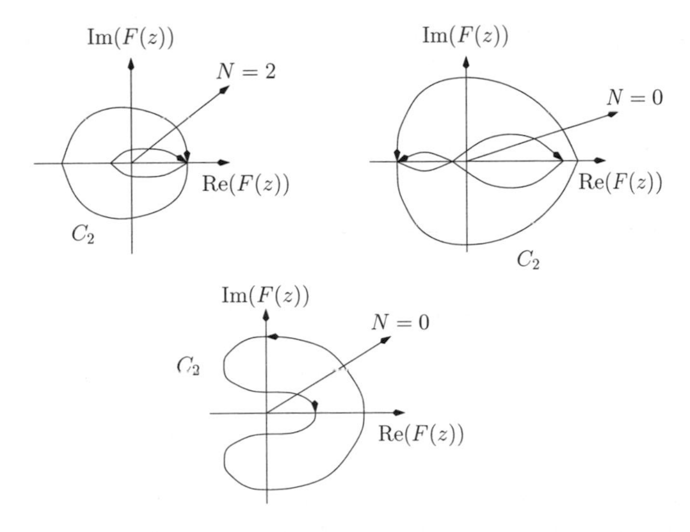

Figure 7.6: Calculation of number of encirclements in Cauchy's principle

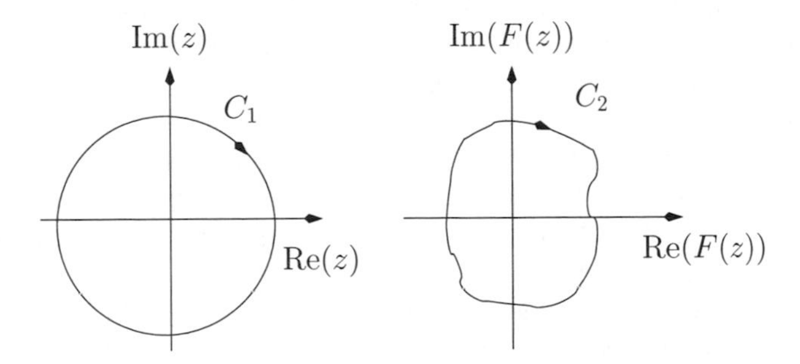

Figure 7.7: An approach to generation of Nyquist plot. $1 + KG(z)$ is evaluated along C_1 and the imaginary part plotted *vs.* the real part, to produce C_2.

Let us evaluate $1 + KG(z)$ along C_1 and plot its imaginary part *vs.* the real part and call it C_2, see Fig. 7.7. Then using Cauchy's principle, C_2 will encircle the origin of the $1 + KG$ plane

$$N = (n - Z) - (n - P) = P - Z \tag{7.15}$$

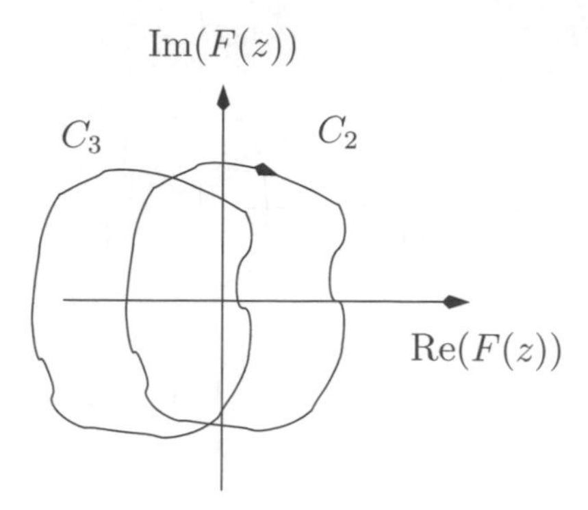

Figure 7.8: C_2 is a plot of $1 + KG(z)$ and C_3 is a plot of $KG(z)$. If the critical point of C_2 is $(0, 0)$, that of C_3 is $(-1, 0)$.

times in the clockwise direction. Since

$$1 + KG(z) = 1 + K\frac{b(z)}{a(z)} = \frac{a(z) + Kb(z)}{a(z)}$$

the poles of $1 + KG(z)$ = zeros of $a(z)$ = poles of $G(z)$. Thus we can interpret P as the number of poles of $G(z)$ outside C_1, *i.e.*, the number of open loop unstable poles. From $N = P - Z$, since we want $Z = 0$, we arrive at the stability condition:

$$N = P \tag{7.16}$$

We will refer to the origin of the F plane as the *critical point.*

The difficulty with the above approach is that we need to know K to draw the C_2 curve. To overcome this difficulty, instead of plotting $1+KG(z)$, we plot $1+KG(z)-1 = KG(z)$, and call the result the C_3 curve, see Fig. 7.8. Since C_3 is obtained by subtracting one from every point of C_2, the critical point also gets shifted by one to the left. In other words, N is then determined by the encirclements of the $(-1, 0)$ point by the C_3 curve.

As a further improvement, we can plot $G(z)$, call it C_4, and now the critical point becomes $(-1/K, 0)$. First we determine N required to satisfy the stability condition. Then we choose K such that the $(-1/K, 0)$ point is encircled the desired number (N) of times. The curve C_4 is known as the *Nyquist plot.*

We will summarize our findings now. For closed loop stability, the Nyquist plot C_4 should encircle the critical point P times, where P denotes the number of open loop unstable poles. We now illustrate this approach with an example.

Example 7.2 Draw the Nyquist plot of

$$G(z) = \frac{1}{z(z-1)} \tag{7.17}$$

and find out for what values of K the system becomes unstable.

To use the Nyquist approach, C_1 should not go through a pole or zero. So indent it with a semicircle of radius $\to 0$ near $z = 1$, as shown in the left-hand side plot of

Fig. 7.9. It is also possible to do this indentation so as to exclude the pole from the unit circle, see Problem 7.2. The number of poles outside C_1 is zero, *i.e.*, $P = 0$. First we evaluate $b(z)/a(z)$ along the main C_1 curve as follows. On substitution of $z = e^{j\omega}$, Eq. 7.17 becomes

$$G(e^{j\omega}) = \frac{1}{e^{j\omega}(e^{j\omega} - 1)} = \frac{1}{e^{j3\omega/2}\left(e^{j\omega/2} - e^{-j\omega/2}\right)}$$
$$= \frac{e^{-j3\omega/2}}{2j\sin\omega/2} = -\frac{j\left(\cos 3\omega/2 - j\sin 3\omega/2\right)}{2\sin\omega/2}$$

We write this in the standard form:

$$= -\frac{\sin 3\omega/2}{2\sin\omega/2} - j\frac{\cos 3\omega/2}{2\sin\omega/2} \tag{7.18}$$

On C_1, we mark A at $\omega = 180°$, B at $\omega = 120°$, D at $\omega = 60°$ and E at $\omega \to 0$ and evaluate G at these points. At point A, because $\omega = 180°$, from Eq. 7.18, we obtain $G = 0.5$. For $\omega = 120°$ at point B, Eq. 7.18 gives

$$G = -\frac{\sin 180°}{2\sin 60°} - j\frac{\cos 180°}{2\sin 60°} = j0.5774$$

At point D, for $\omega = 60°$, Eq. 7.18 gives $G = -1$. At point E, with $\omega \to 0$, we obtain

$$G = -\frac{0}{0} - j\infty$$

Because the real part is in $0/0$ form, we apply L'Hospital's rule to it. We obtain

$$G = -\frac{(3/2)\cos(3/2)\omega}{\cos(1/2)\omega} - j\infty - -\frac{3}{2} - j\infty$$

These points have been marked in the Nyquist plot, shown on the right-hand side of Fig. 7.9. We next evaluate G along the indentation next to the pole at 1. This indentation is in the form of a semicircle with an equation $z = 1 + \varepsilon e^{j\phi}$ with $\varepsilon \to 0$. When G is evaluated along this semicircle, we obtain

$$G(1 + \varepsilon e^{j\phi}) = \frac{1}{(1 + \varepsilon e^{j\phi})\varepsilon e^{j\phi}} = \frac{\infty e^{-j\phi}}{1 + \varepsilon e^{j\phi}} = \infty e^{-j\phi}$$

as $\varepsilon \to 0$. As ϕ starts at $+90°$, goes to $0°$ and then to $-90°$, G of radius ∞ starts at $-90°$, goes to $0°$ and then to $90°$. This is shown as a semicircle of infinite radius in the Nyquist plot of Fig. 7.9. M 7.2 shows the procedure to obtain the Nyquist plot in Matlab.

Using this Nyquist plot, we proceed to determine the range of proportional control gains for which the closed loop system is stable. Recall from Eq. 7.16 that the condition for stability is $N = P$. Because there are no open loop unstable poles, we have $P = 0$. As a result, we look for points that are not encircled by the Nyquist plot.

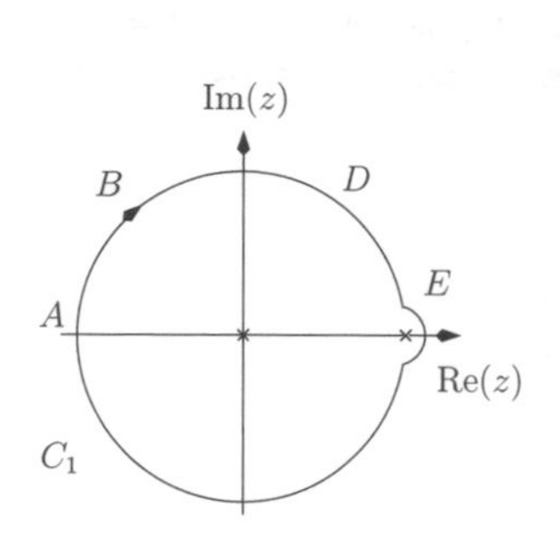

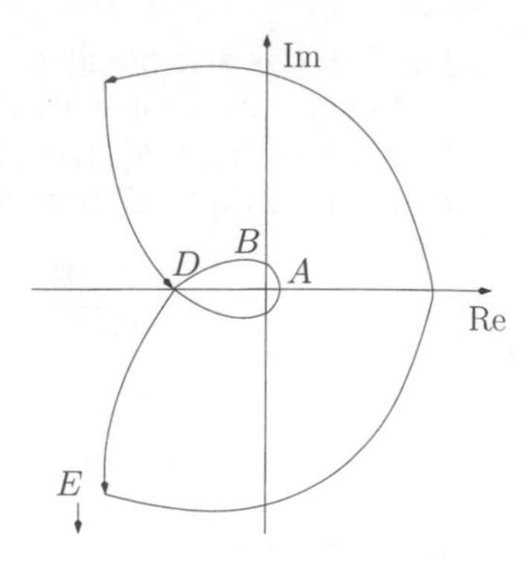

Figure 7.9: Contour C_1 and the corresponding Nyquist plot C_4, for the system studied in Example 7.2

Now we explore where the critical point $(-1/K, 0)$ has to be placed. If it is placed between D and A in the Nyquist plot shown in Fig. 7.9, we will have $N = -2$. This corresponds to the condition $0.5 > -1/K > -1$. As we are interested in positive values of K only, we take this as $0 > -1/K > -1$, which is equivalent to $0 < 1/K < 1$ or $K > 1$. Thus for $K > 1$, the closed loop system is unstable. Indeed, when $K > 1$, there will be two unstable poles in the closed loop: we substitute $N = -2$ in Eq. 7.15 and obtain $Z = 2$, because $P = 0$. This is in agreement with the results of Example 7.1.

If, on the other hand, the critical point $(-1/K, 0)$ is placed to the left of D in the Nyquist plot of Fig. 7.9, because there are no encirclements, we obtain $N = 0$, satisfying the condition for closed loop stability, given in Eq. 7.16. This corresponds to $-\infty < -1/K < -1$, or equivalently, for $0 < K < 1$. This result also is in agreement with that of Example 7.1. ∎

Although the Nyquist approach can be used to design other kinds of controllers also, we will not discuss this procedure any further. Using the Nyquist plot, it is possible to explain the concept of stability margins, an important metric that a controller should satisfy. We discuss this topic in the next section.

7.2.2 Stability Margins

Suppose that the plant is open loop stable. Then, the point $(-1, 0)$ should not be encircled by the Nyquist curve for stability. It is general practice to keep a good distance between this point and the Nyquist curve. The idea is that if this distance is large, it will act as a safety factor in case the transfer function and the Nyquist curve of the plant change.

The transfer function of the plant can change for many reasons. For example, wear and tear can change the behaviour of the system. We may not have a perfect understanding of the plant and hence there could be differences between the actual and the assumed transfer functions of the plant.

If the distance between the Nyquist curve and the point $(-1, 0)$ is large, hopefully the closed loop system will remain stable even if the transfer function is different from

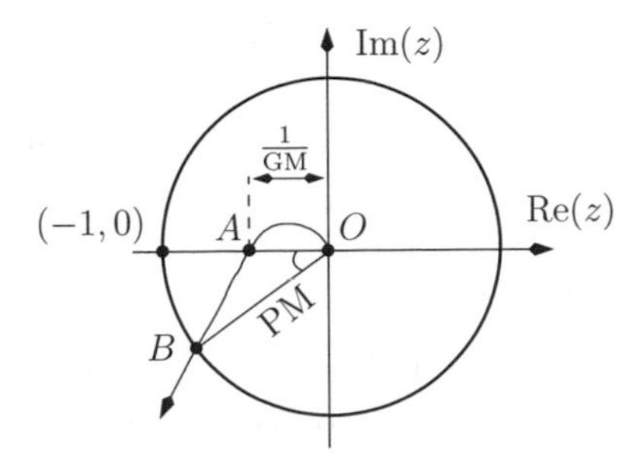

Figure 7.10: Definition of gain and phase margins. Margins act as a safety factor against uncertainties in the model of the plant.

what is used in control design. This distance is characterized in terms of gain margin and phase margin.

Consider the Nyquist curve of a stable plant, as in Fig. 7.10. Let B, A and O be three points on the Nyquist curve. At point B, it cuts the unit circle, centred at the origin. At point A, it cuts the real axis. If point A were to coincide with the point $(-1, 0)$, the system would become unstable. This will happen when the transfer function is multiplied by the scalar $1/OA$, where we have denoted the distance between O and A by OA. This scalar is defined as the *gain margin* of the system. We obtain

$$\text{gain margin} = \frac{1}{OA}$$

The Nyquist curve will go through the point $(-1, 0)$ also when it is rotated in the clockwise direction. This could happen when the plant has, for example, a larger delay than assumed.

With clockwise rotation, the Nyquist plot, such as the one in Fig. 7.10, could go through the critical point $(-1, 0)$, making the system unstable. In other words, unmodelled dynamics and delays could make the system unstable. It is to safeguard against such difficulties, that the notion of *phase margin* has been introduced: it is the angle through which the Nyquist plot can be rotated in the clockwise direction, so as to just reach the point of instability. From the Nyquist curve drawn in Fig. 7.10, we obtain

$$\text{phase margin} = \angle AOB$$

which is the same as $180° - \angle G(e^{j\omega_c})$, where ω_c is the *crossover frequency*, at which the magnitude of G is 1.

A gain margin of ≥ 2 and a phase margin of $\geq 45°$ are generally considered as a safe design. Nevertheless, in some situations even after satisfying these two conditions, the Nyquist plot can get arbitrarily close to the critical point of $(-1, 0)$. In view of this, it is recommended that the Nyquist curve be drawn to ensure that these two margins are a reasonable metric for the plant under consideration. The concept of stability margins is not restricted to proportional controllers alone. In fact, any controller that does not have good stability margins will not be of any use.

7.3 Other Popular Controllers

In this section, we briefly discuss two other popular controllers, namely lead–lag and PID controllers.

7.3.1 Lead–Lag Controller

When a proportional controller is inadequate, we consider more elaborate controllers. The lead–lag controllers are more advanced than the proportional controllers. When the phase margin is not sufficient, we may like to rotate the Nyquist curve in the counterclockwise direction. In other words, we would like to increase the phase angle of the system, thereby improving the stability margins of the closed loop system. The *lead controller* does precisely this.

A *lag controller*, on the other hand, is used to decrease the phase of the system. A lag controller is generally used to improve the performance properties at steady state. A lead–lag controller has both of these modes. We will discuss lead controllers in detail in this section. A more popular version of the lead–lag controller is the ubiquitous PID controller.

It is extremely easy to construct lead controllers directly in the discrete time domain [40]. Fig. 7.11 shows the Bode plot of two transfer functions, $G_1(z^{-1}) = 1 - 0.9z^{-1}$, plotted using solid lines and $G_2(z^{-1}) = 1 - 0.8z^{-1}$, plotted using dashed lines. These have been obtained using M 7.3. The poles of both of these have been chosen to lie in the same radial line from the centre of the circle [49]. Note that the phase of G_1 is greater than that of G_2. This difference goes through a maximum. This suggests that a filter of the form

$$G = \frac{1 - 0.9z^{-1}}{1 - 0.8z^{-1}} \tag{7.19}$$

will have a positive phase angle. This is indeed true, as can be seen from Fig. 7.11, where we have drawn a Bode plot of G.

We can generalize this result. Consider the transfer function

$$G(z^{-1}) = \frac{1 - bz^{-1}}{1 - az^{-1}} = \frac{B(z^{-1})}{A(z^{-1})} \tag{7.20}$$

with

$$1 > b > a > 0 \tag{7.21}$$

Substituting $e^{j\omega}$ for z, we obtain $B(e^{j\omega}) = 1 - be^{-j\omega}$. Using Euler's formula and simplifying, $B(e^{j\omega}) = (1 - b\cos\omega) + jb\sin\omega$. We obtain the phase angle as

$$\angle B(e^{j\omega}) = \tan^{-1}\frac{b\sin\omega}{1 - b\cos\omega} \tag{7.22}$$

Similarly, we obtain

$$\angle A(e^{j\omega}) = \tan^{-1}\frac{a\sin\omega}{1 - a\cos\omega} \tag{7.23}$$

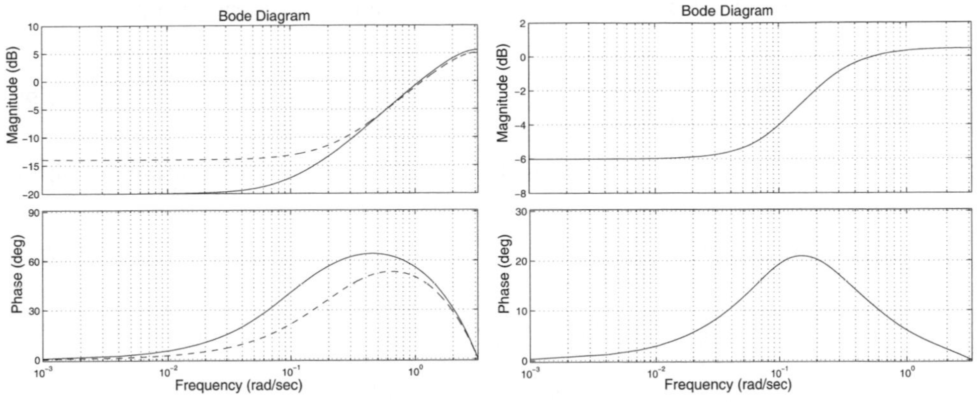

(a) Bode plots of $1 - 0.9z^{-1}$ (solid lines) and $1 - 0.8z^{-1}$ (dashed lines)

(b) Bode plot of $G_1 = (1 - 0.9z^{-1})/(1 - 0.8z^{-1})$

Figure 7.11: The phase angle of G_1 is positive and goes through a maximum, confirming lead nature

and hence,

$$\angle G(e^{j\omega}) = \angle B(e^{j\omega}) - \angle A(e^{j\omega}) \tag{7.24}$$

As $b > a$, we obtain for $\omega \in [0, \pi]$,

$$b \sin\omega > a \sin\omega \tag{7.25}$$

from which it follows that

$$\frac{b \sin\omega}{1 - b\cos\omega} > \frac{a \sin\omega}{1 - a\cos\omega} \tag{7.26}$$

which can be verified by cross multiplication. We conclude that $\angle G(e^{j\omega})$, defined in Eq. 7.22–7.24, is positive.

The transfer function G defined in Eq. 7.19 has a maximum phase of approximately 21° at 0.105 rad/s, as can be seen in Fig. 7.11. We next derive an expression for the maximum phase reached by G, defined in Eq. 7.24. Differentiating Eq. 7.22 with respect to ω, we obtain

$$\begin{aligned}\frac{d\angle B(e^{j\omega})}{d\omega} &= \frac{1}{1 + \left(\frac{b\sin\omega}{1 - b\cos\omega}\right)^2} \frac{(1 - b\cos\omega)b\cos\omega - b^2\sin^2\omega}{(1 - b\cos\omega)^2} \\ &= \frac{1 - b^2}{1 - 2b\cos\omega + b^2}\end{aligned}$$

Using a similar expression for $d\angle A(e^{j\omega})/d\omega$ and substituting in Eq. 7.24, we obtain the condition for the maximum in $\angle G$ as

$$\frac{1 - b^2}{1 - 2b\cos\omega_m + b^2} - \frac{1 - a^2}{1 - 2a\cos\omega_m + a^2} = 0$$

where ω_m denotes that it corresponds to the maximum phase lead. Simplifying this, we observe that

$$\cos\omega_m = \frac{a+b}{ab+1} \tag{7.27}$$

Next, we calculate the phase lead obtained at ω_m. From the above equation, we obtain

$$\sin\omega_m = \frac{\sqrt{(a^2-1)(b^2-1)}}{ab+1} \tag{7.28}$$

Substituting these in Eq. 7.22–7.24, we obtain the phase lead at ω_m as

$$\angle G\left(e^{j\omega_m}\right) = \tan^{-1} b\sqrt{\frac{1-a^2}{1-b^2}} - \tan^{-1} a\sqrt{\frac{1-b^2}{1-a^2}} \tag{7.29}$$

Taking tan of both sides,

$$\tan\angle G\left(e^{j\omega_m}\right) = \frac{b-a}{\sqrt{(1-a^2)(1-b^2)}} \tag{7.30}$$

Thus, we obtain the following expression for the maximum phase achieved as

$$\angle G\left(e^{j\omega_m}\right) = \tan^{-1}\frac{b-a}{\sqrt{(1-a^2)(1-b^2)}} \tag{7.31}$$

Thus, the transfer function defined by Eq. 7.20–7.21 introduces a phase lead, as in Eq. 7.31, at the frequency ω_m defined as in Eq. 7.27.

The lead control problem typically involves a reverse procedure: what transfer function should we use to achieve a specified lead $\angle G$ at a given frequency ω_m? There are two unknowns, a and b, and two constraints, Eq. 7.27 and Eq. 7.31. We can solve these and obtain values for a and b.

A graphical procedure to quickly arrive at approximate values of a and b is now proposed. Let a be a fraction of b, as given by

$$a = fb, \quad 0 < f < 1 \tag{7.32}$$

Fig. 7.12 shows a plot of maximum frequency ω_m and the zero location b as functions of maximum lead $\angle G(e^{j\omega})$, for different values of f. M 7.4 shows how to obtain one set of curves corresponding to $f = 0.9$.

Suppose that we want to design a lead controller that introduces a maximum lead of 45° at $\omega = 0.5$ rad/sample (see Sec. 5.1.2 for a discussion on the units used for discrete time sinusoids). We start in the top figure, with the ordinate value at 0.5 rad/sample. We choose the curve corresponding to $f = 0.3$, as it gives a maximum phase lead of 45°. From the bottom figure, we see that a maximum lead of 45° will be achieved for the zero location $b = 0.8$ when $f = 0.3$. The arrows show the direction of the calculation. Using the definition of f in Eq. 7.32, we obtain $a = 0.24$, and hence arrive at the required controller as

$$G = \frac{1-0.8z^{-1}}{1-0.24z^{-1}}$$

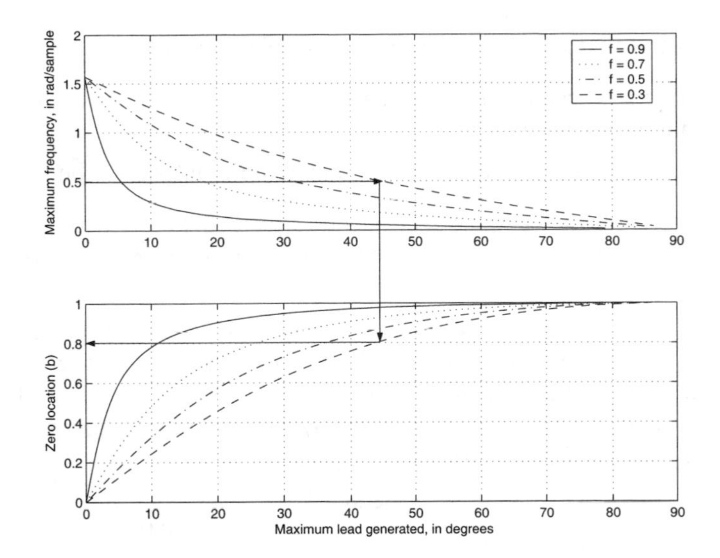

Figure 7.12: A graphical procedure to design a lead controller. Zero location (b) of 0.8 and a multiplying factor f of 0.3 to achieve a maximum phase lead ω_m of 45° at frequency $\omega = 0.5$ rad/sample: follow the arrows.

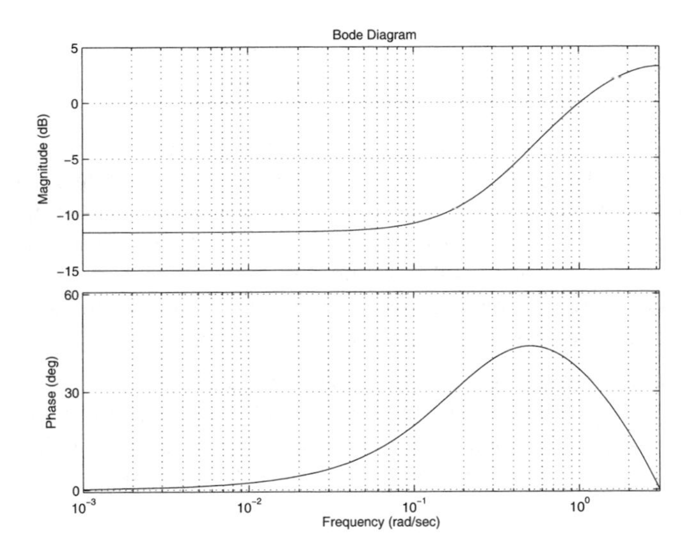

Figure 7.13: Bode plot of $(1 - 0.8z^{-1})/(1 - 0.24z^{-1})$

It is easy to verify that this controller indeed offers the required lead. Using M 7.5, we obtain the Bode plot as in Fig. 7.13. We now demonstrate the utility of lead controllers with an example.

Example 7.3 Recall the discrete time transfer function of the antenna control system, obtained in Example 4.18, for a sampling time $T_s = 0.2$ s:

$$G(z) = \frac{0.001987(z + 0.9931)}{(z - 1)(z - 0.9802)} \tag{7.33}$$

Design a controller that will help meet the following requirements:

1. For a unit step change in the reference signal, rise time should be less than 2.3 s and overshoot less than 0.25.
2. The steady state tracking error to a unit ramp input should be less than 1.

Although the technique of pole placement, discussed in Chapter 9, is a more appropriate method to solve this problem, we will now explore the usefulness of lead controllers.

From a root locus plot of this example, it is easy to see that the closed loop poles will lie between the poles at 1 and 0.9802, see Example 7.1 on page 247. These poles, when stable, result in a sluggish response, see Fig. 5.1 on page 114. For a faster response, the closed loop poles of the system should be closer to the origin. Suppose that we use a lead controller of the form

$$G_c(z) = \frac{1-\beta}{1-0.9802}\frac{z-0.9802}{z-\beta}, \quad 0 < \beta < 0.9802 \tag{7.34}$$

which has a steady state gain of 1, *i.e.*, $G_c(1) = 1$. The lead controller cancels the pole of the plant at 0.9802 and replaces it with another pole at β, closer to the origin. The loop transfer function becomes

$$G(z)G_c(z) = 0.001987\frac{1-\beta}{1-0.9802}\frac{z+0.9931}{(z-1)(z-\beta)} \tag{7.35}$$

It is instructive to study this system also from the frequency domain perspective. Fig. 7.14 shows a Bode plot of the antenna system with solid lines. It is easy to see that this plant has a phase angle of $165°$ at a gain of 1, resulting in a phase margin of about $15°$. When the above mentioned lead controller is used with an arbitrarily chosen value of β as 0.8, we obtain the Bode plot, drawn with dashed lines in the same figure. From this, we see that the phase margin has increased to about $50°$. Moreover, the bandwidth also has increased.

We will now explore whether it is possible to use the above discussed lead controller, possibly in series with a proportional controller, to meet the steady state requirement. It is given that the steady state error to a unit ramp input should be less than 1. The Z-transform of error to ramp is given by

$$E(z) = \frac{1}{1+KG(z)G_c(z)}\frac{T_s z}{(z-1)^2} \tag{7.36}$$

We need to meet the requirement of $e(\infty) < 1$. Applying the final value theorem, we obtain

$$\lim_{z\to 1}\frac{T_s}{(z-1)[1+KG(z)G_c(z)]} \le 1$$

As $G(1)G_c(1) = \infty$, the above inequality becomes

$$\lim_{z\to 1}\frac{T_s}{(z-1)[KG(z)G_c(z)]} \le 1$$

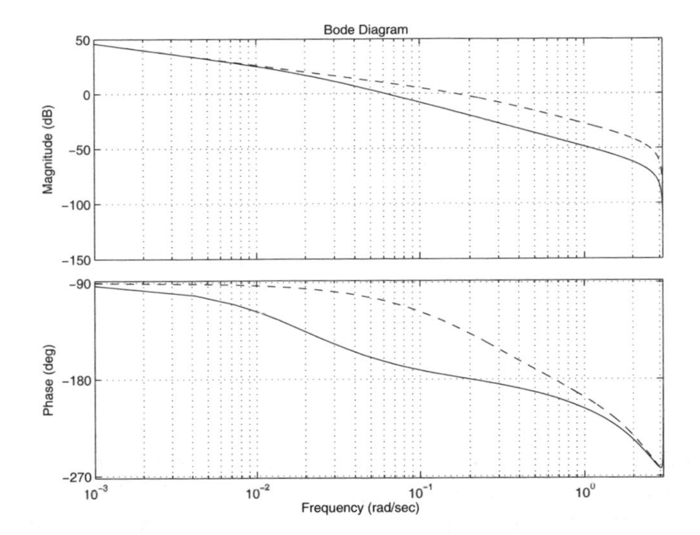

Figure 7.14: Bode plot of antenna tracking system, with (dashed lines) and without (solid lines) the lead controller G_c

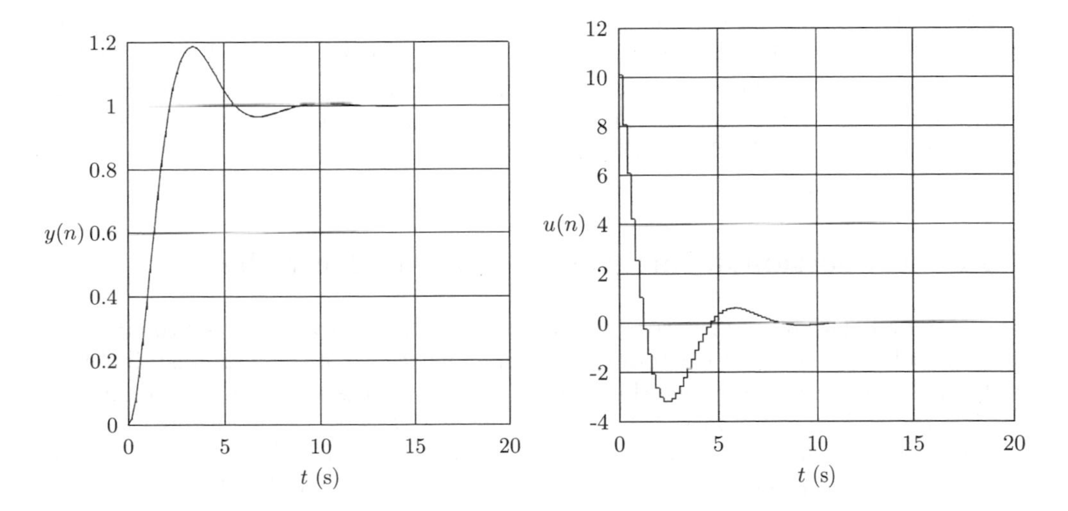

Figure 7.15: Output and input profiles of the antenna system with the lead controller

On substituting for G from Eq. 7.33 and simplifying, we obtain $KG_c(1) \geq 1$. We can take $KG_c(1) = 1$. Because $G_c(1) = 1$, we have $K = 1$.

We have just seen that the lead controller that we selected earlier meets the steady state error requirement. In addition, as mentioned earlier, this controller improves the transient performance. We will now carry out simulations and check whether this controller is adequate. The efficacy of this scheme is evaluated through M 7.6 and simulated using the Simulink code given in Fig. A.5. The resulting input–output profiles for a step input are in Fig. 7.15.

∎

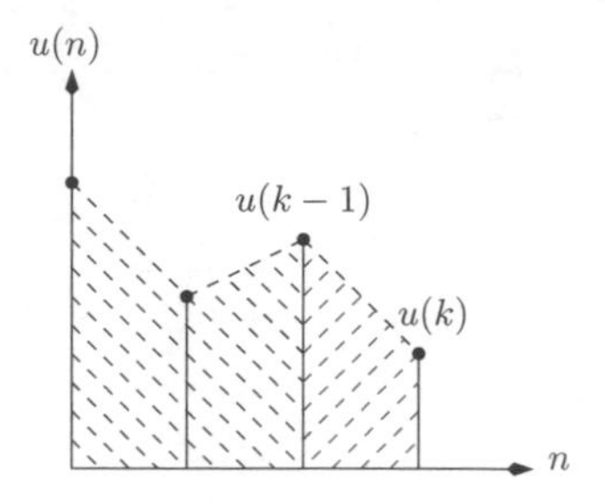

Figure 7.16: Graphical representation of discrete time integration

Now that we have given a detailed presentation of the lead controller, we will explain the lag controllers briefly. A lag controller also is given by an expression identical to Eq. 7.19 with the difference that now $b < a$. As a result, the phase is negative – hence the name. Supposing that we are generally happy with the dynamic response of the system but that we want to increase the controller gain so as to reduce the steady state offset, we include the lag controller.

By a combination of a lead and a lag controller, it is possible to achieve both requirements: improving dynamic performance and reducing the steady state gain.

The lead action is similar to the derivative action and the lag mode is similar to the integral mode in the popular controller of the industry, namely the PID controller, to be discussed briefly in the next section.

7.3.2 Proportional, Integral, Derivative Controller

The proportional, integral, derivative, or more popularly, the PID, is probably the most popular controller in use today. It is popular because of the large installed base and simplicity in its use: it has three parameters to be tuned, one corresponding to each of the P, I and D modes. While the proportional controller is always used, one has the option of choosing either one or both of the other two modes. As a result, one has the option of using any combination of P, PI, PD or PID. All these are loosely referred to as the PID controllers. We will discuss each of the modes separately in this section.

Suppose that our process is an integrator. Let the input to the process be denoted by u. Let the output, the integral of u with respect to time, be denoted by y. This can be graphically represented as in Fig. 7.16. A discrete time way to integrate the curve is through the trapezoidal scheme, as given below:

$$\begin{aligned} y(k) &= \text{shaded area with right slanting lines} \\ &\quad + \text{shaded area with left slanting lines} \\ y(k) &= y(k-1) + \text{shaded area with left slanting lines} \\ &= y(k-1) + \frac{T_s}{2}\left[u(k) + u(k-1)\right] \end{aligned} \tag{7.37}$$

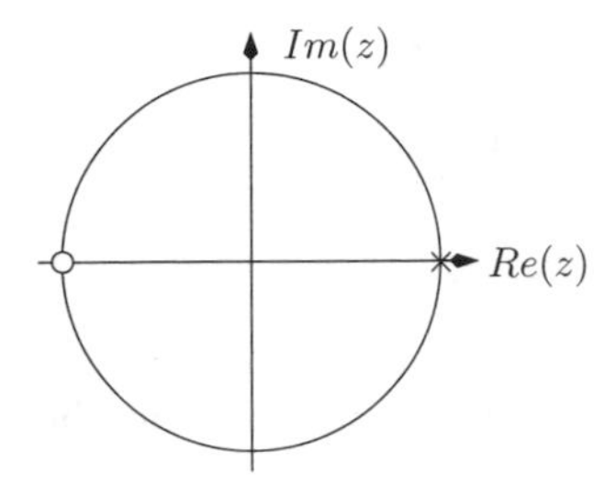

Figure 7.17: Pole–zero location of the integral mode

Taking the Z-transform of both sides, we obtain

$$Y(z) = z^{-1}Y(z) + \frac{T_s}{2}\left[U(z) + z^{-1}U(z)\right]$$

Simplifying this, we obtain

$$Y(z) = \frac{T_s}{2}\frac{1+z^{-1}}{1-z^{-1}}U(z) \tag{7.38}$$

If we denote the transfer function of the integrator by G_i, we see that it is given by

$$G_i(z) = \frac{T_s}{2}\frac{z+1}{z-1}. \tag{7.39}$$

This is known as the *trapezoidal approximation* or *Tustin approximation* or *bilinear approximation*. The transfer function G_i is low pass as it has a pole at $\omega = 0$ and a zero at $\omega = \pi$, see Fig. 7.17. Thus we see that integration, a smoothing operation, is low pass. We explain the role of the integral mode with an example.

Example 7.4 Evaluate the effect of an integrating controller

$$G_i(z) = \frac{z+1}{z-1}$$

when used with a nonoscillating plant given by

$$G(z) = \frac{z}{z-a}$$

where $a > 0$. The closed loop transfer function becomes

$$T(z) = \frac{\frac{z}{z-a}\frac{z+1}{z-1}}{1 + \frac{z}{z-a}\frac{z+1}{z-1}} = \frac{z(z+1)}{2z^2 - az + a}$$

The poles are at $(a \pm \sqrt{a^2 - 8a})/4$. For all $a < 8$, the closed loop system is oscillatory. When a PI controller of the following form is used

$$G_c(z) = K\left(1 + \frac{1}{\tau_i}\frac{z+1}{z-1}\right)$$

the overall transfer function becomes

$$T(z) = \frac{K\left(1 + \frac{1}{\tau_i}\frac{z+1}{z-1}\right)\frac{z}{z-a}}{1 + K\left(1 + \frac{1}{\tau_i}\frac{z+1}{z-1}\right)\frac{z}{z-a}}$$

The steady state output for a step input is given by $\lim_{n\to\infty} y(n)$, which is equal to $\lim_{z\to 1} T(z) = 1$, see Eq. 4.21 on page 83. This shows that there is no steady state offset.

∎

Because the reciprocal of integration is differentiation, we obtain the transfer function of discrete time differentiation G_d as the reciprocal of G_i, given in Eq. 7.39. We obtain

$$G_d(z) = \frac{1}{G_i(z)} = \frac{2}{T_s}\frac{z-1}{z+1} \tag{7.40}$$

This form, however, has a problem: $G_d(z)$ has a pole at $z = -1$ and hence it produces, in partial fraction expansion, a term of the form

$$\frac{z}{z+1} \leftrightarrow (-1)^n \tag{7.41}$$

which results in a wildly oscillating control effort, see Sec. 5.1.1. Because of this, we consider other ways of calculating the area given in Fig. 7.16. For example, we approximate the area under the curve with the following *backward difference* formula:

$$y(k) = y(k-1) + T_s u(k) \tag{7.42}$$

Taking the Z-transform and simplifying it, we obtain

$$Y(z) = T_s \frac{z}{z-1} U(z)$$

We arrive at the following transfer functions of the integrator and differentiator as

$$\begin{aligned} G_i(z) &= T_s \frac{z}{z-1} \\ G_d(z) &= \frac{1}{T_s}\frac{z-1}{z} \end{aligned} \tag{7.43}$$

It is also possible to use the *forward difference* approximation for integration

$$y(k) = y(k-1) + T_s u(k-1) \tag{7.44}$$

This results in the following transfer functions of the integrator and differentiator as

$$\begin{aligned} G_i(z) &= \frac{T_s}{z-1} \\ G_d(z) &= \frac{z-1}{T_s} \end{aligned} \tag{7.45}$$

It is easy to check that all the definitions of the differentiator given above are high pass.

Example 7.5 Examine the effect of using the derivative controller

$$G_d(z) = \tau_d \frac{z-1}{z}$$

on the oscillating plant

$$G(z) = \frac{z}{z+a}, \quad a > 0$$

As the pole is on the negative real axis, it has maximum number of oscillations. The overall transfer function T is given by

$$T(z) = \frac{G(z)G_d(z)}{1+G(z)G_d(z)} = \frac{\tau_d(z-1)}{(1+\tau_d)z + (a-\tau_d)}$$

which is less oscillatory than the plant $G(z)$. We next see the effect of using a proportional derivative controller on the steady state offset to a step input. Let the proportional derivative controller be given by

$$G_d(z) = K\left(1 + \tau_d \frac{z-1}{z}\right)$$

The closed loop transfer function $T(z)$ is given by

$$T(z) = \frac{K\left(1+\tau_d \frac{z-1}{z}\right)\frac{z}{z+a}}{1+K\left(1+\tau_d \frac{z-1}{z}\right)\frac{z}{z+a}}$$

If the input to this is a step function, The steady state value $y(n)$ is given by the final value theorem:

$$\lim_{n\to\infty} y(n) = \lim_{z\to 1} Y(z)\frac{z-1}{z} = \lim_{z\to 1} T(z) = \frac{K}{1+a+K}$$

Note that the steady state value is the same as that in the case of the proportional controller.

∎

We now summarize the properties of the three modes of the PID controller. The proportional mode is the most popular control mode. Increase in the proportional mode generally results in decreased steady state offset and increased oscillations. The integral mode is used to remove steady state offset. Increase in the integral mode generally results in zero steady state offset and increased oscillations. The derivative mode is mainly used for prediction purposes. Increase in the derivative mode generally results in decreased oscillations and improved stability. The derivative mode is sensitive to noise, however.

7.4 Internal Stability and Realizability

Recall that in Sec. 3.3.7, we have presented the concept of BIBO or external stability. In this section, we show that unstable poles should not be cancelled with zeros at the same location to achieve internal stability. We also show that internally stable sampled data systems are realizable.

$$E(z) \rightarrow \boxed{\frac{z+2}{z+0.5}} \xrightarrow{U(z)} \boxed{\frac{1}{z+2}} \rightarrow Y(z)$$

Figure 7.18: Unstable pole–zero cancellation

7.4.1 Forbid Unstable Pole–Zero Cancellation

Cancellation of a pole and a zero, although it helps simplify a transfer function, could lead to difficulties. By cancellation, essentially we ignore some dynamics due to the initial conditions. As a result, if the dynamics cancelled away is unstable, we could get into difficulties, see Problem 7.3. We now present an example to illustrate this difficulty.

Example 7.6 Study the effects of stabilizing the following system

$$G(z) = \frac{1}{z+2}$$

with a controller that has a zero at $z = -2$, as shown in Fig. 7.18.

We will now derive the state space realization of this system without cancellation. First we will do this for the first block:

$$\begin{aligned}
U(z) &= \frac{z+2}{z+0.5}E(z) = \left(1 + \frac{1.5}{z+0.5}\right)E(z) \\
X_1(z) &\triangleq \frac{1.5}{z+0.5}E(z) \\
U(z) &= X_1(z) + E(z) \\
x_1(k+1) &= -0.5x_1(k) + 1.5e(k) \\
u(k) &= x_1(k) + e(k)
\end{aligned}$$

For the second block, this becomes

$$\begin{aligned}
X_2(z) &\triangleq \frac{1}{z+2}U(z) \\
x_2(k+1) &= -2x_2(k) + u(k) \\
y(k) &= x_2(k)
\end{aligned}$$

Substituting for $u(k)$ from above

$$x_2(k+1) = -2x_2(k) + x_1(k) + e(k)$$

Writing these equations together in matrix form, we obtain

$$\begin{aligned}
\begin{bmatrix} x_1(k+1) \\ x_2(k+1) \end{bmatrix} &= \begin{bmatrix} -0.5 & 0 \\ 1 & -2 \end{bmatrix} \begin{bmatrix} x_1(k) \\ x_2(k) \end{bmatrix} + \begin{bmatrix} 1.5 \\ 1 \end{bmatrix} e(k) \\
y(k) &= \begin{bmatrix} 0 & 1 \end{bmatrix} \begin{bmatrix} x_1(k) \\ x_2(k) \end{bmatrix}
\end{aligned}$$

These are of the form

$$
\begin{aligned}
x(k+1) &= Ax(k) + bu(k) \\
y(k) &= Cx(k)
\end{aligned}
$$

where

$$
A = \begin{bmatrix} -0.5 & 0 \\ 1 & -2 \end{bmatrix}, \quad b = \begin{bmatrix} 1.5 \\ 1 \end{bmatrix}, \quad C = \begin{bmatrix} 0 & 1 \end{bmatrix}
$$

Using Eq. 3.46 on page 56, we find the solution to this system as

$$
x(k) = A^k x(0) + \sum_{i=0}^{k-1} A^{k-(i+1)} be(i) = A^k x(0) + \sum_{m=0}^{k-1} A^m be(k-m-1) \tag{7.46}
$$

Using the diagonalization procedure of Sec. A.1.2, we obtain

$$
\begin{aligned}
A &= S\Lambda S^{-1} \\
A^k &= S\Lambda^k S^{-1}
\end{aligned}
$$

For this problem, we can calculate the following:

$$
S = \begin{bmatrix} 1.5 & 0 \\ 1 & 1 \end{bmatrix}, \quad \Lambda = \begin{bmatrix} -0.5 & 0 \\ 0 & -2 \end{bmatrix}, \quad S^{-1} = \frac{1}{1.5} \begin{bmatrix} 1 & 0 \\ -1 & 1.5 \end{bmatrix}
$$

Then we can calculate $A^m b$ as $S\Lambda^m S^{-1} b$. The calculations are a bit easier if we calculate from right to left. In this case, we obtain

$$
A^m b = S\Lambda^m S^{-1} b = \begin{bmatrix} 1.5 \\ 1 \end{bmatrix} (-0.5)^m
$$

Substituting this in Eq. 7.46, we obtain

$$
x(k) = A^k x(0) + \begin{bmatrix} 1.5 \\ 1 \end{bmatrix} \sum_{m=0}^{k-1} (-0.5)^m e(k-m-1)
$$

Because $y = Cx$, we obtain

$$
y(k) = CA^k x(0) + C \begin{bmatrix} 1.5 \\ 1 \end{bmatrix} \sum_{m=0}^{k-1} (-0.5)^m e(k-m-1) \tag{7.47}
$$

The first term is

$$
\begin{aligned}
CA^k x(0) &= \frac{1}{1.5} \begin{bmatrix} 0 & 1 \end{bmatrix} \begin{bmatrix} 1.5 & 0 \\ 1 & 1 \end{bmatrix} \begin{bmatrix} (-0.5)^k & 0 \\ 0 & (-2)^k \end{bmatrix} \begin{bmatrix} 1 & 0 \\ -1 & 1.5 \end{bmatrix} \begin{bmatrix} x_1(0) \\ x_2(0) \end{bmatrix} \\
&= \frac{1}{1.5} \begin{bmatrix} (-0.5)^k & (-2)^k \end{bmatrix} \begin{bmatrix} x_1(0) \\ -x_1(0) + 1.5x_2(0) \end{bmatrix}
\end{aligned}
$$

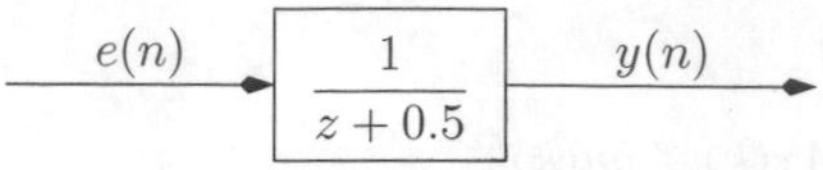

Figure 7.19: Same as Fig. 7.18, but after cancellation of pole and zero

If $x(0)$ is not identically zero, this term grows unbounded. The second term in Eq. 7.47 decays to zero. This term is identical to what we will obtain if we realize the simplified transfer function obtained after cancellation, with block diagram as shown in Fig. 7.19. Suppose that the state of this reduced system is indicated by x_s. Then we obtain

$$\begin{aligned} X_s(z) &\triangleq \frac{1}{z+0.5}E(z) \\ x_s(k+1) &= -0.5x_s(k) + e(k) \\ y_s(k) &= x_s(k) \end{aligned}$$

Iteratively solving this equation, we obtain

$$y_s(k) = x_s(k) = (-0.5)^m x_s(0) + \sum_{m=0}^{k-1} (-0.5)^m e(k-m-1) \tag{7.48}$$

When the initial values are zero, the outputs are identical, as can be seen from Eq. 7.47 and Eq. 7.48. In case, however, $x(0)$ consists of small but nonzero entries, like 10^{-15}, the first element of Eq. 7.47 grows unbounded, while Eq. 7.48 is stable. Such small but nonzero entries can very well be due to the way a computer system stores zero or due to noise. ■

This example illustrates that an unstable pole and zero should not be cancelled. That is, the adverse effect of an unstable pole cannot be wished away by placing a zero at that exact location. We now look at the issue of unstable pole–zero cancellation in a closed loop system.

7.4.2 Internal Stability

In the last section we showed that unless we have perfect zero initial condition, cancellation of an unstable pole with a zero would result in instability. We say that a closed loop system, in which the transfer function between any two points is stable, has the property of *internal stability*. It is instructive to compare this with the property of external or BIBO stability, defined in Sec. 3.3.7. In this section, using a feedback loop, we show that if unstable pole–zero cancellation is permitted, some signals in the loop will become unbounded. In other words, the system will not have the property of internal stability.

Further discussion in this section will centre around the closed loop system, given in Fig. 7.20. Let

$$G = \frac{n_1}{d_1} \tag{7.49a}$$

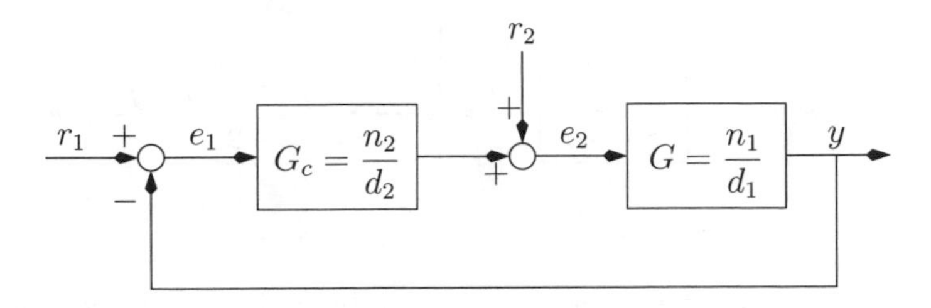

Figure 7.20: Internal stability

be the plant transfer function and let

$$G_c = \frac{n_2}{d_2} \tag{7.49b}$$

be the controller transfer function. At first sight, it may seem that if the transfer function between the output of the system and the reference signal is stable, everything is fine. Unfortunately, however, this is insufficient. We claim that every signal in the loop has to be bounded if stability is to be ensured. As any signal inside the loop may be influenced by noise, we have the following requirement of stability: the transfer function between any two points in the loop should be stable. Equivalently, for any bounded signal injected at any part of the loop, all signals in the loop must also be bounded. We obtain the following matrix relation.

$$\begin{bmatrix} e_1 \\ e_2 \end{bmatrix} = \begin{bmatrix} \dfrac{1}{1+GG_c} & -\dfrac{G}{1+GG_c} \\ \dfrac{G_c}{1+G_cG} & \dfrac{1}{1+G_cG} \end{bmatrix} \begin{bmatrix} r_1 \\ r_2 \end{bmatrix} \tag{7.50}$$

Substituting for G and G_c from Eq. 7.49, we obtain

$$\begin{bmatrix} e_1 \\ e_2 \end{bmatrix} = \begin{bmatrix} \dfrac{d_1 d_2}{n_1 n_2 + d_1 d_2} & -\dfrac{n_1 d_2}{n_1 n_2 + d_1 d_2} \\ \dfrac{n_2 d_1}{n_1 n_2 + d_1 d_2} & \dfrac{d_1 d_2}{n_1 n_2 + d_1 d_2} \end{bmatrix} \begin{bmatrix} r_1 \\ r_2 \end{bmatrix} \tag{7.51}$$

First we will illustrate the effect of pole–zero cancellation. Suppose that d_1 and n_2 have a common factor $(z + a)$. That is, let

$$d_1 = (z + a)d_1' \tag{7.52a}$$

$$n_2 = (z + a)n_2' \tag{7.52b}$$

where d_1' and n_2' are polynomials that do not have $z + a$ as a factor. Thus we obtain

$$G(z) = \frac{n_1(z)}{(z + a)d_1'(z)}$$

and

$$G_c(z) = \frac{(z + a)n_2'(z)}{d_2(z)}$$

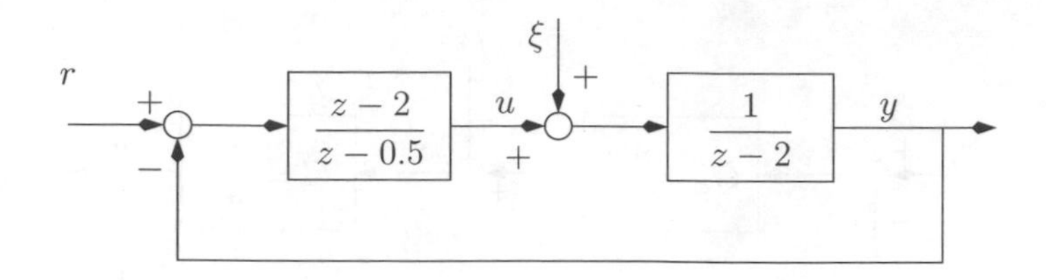

Figure 7.21: A feedback loop with unstable pole–zero cancellation

with $|a| > 1$. Let us suppose also that the transfer function between r_1 and e_1, namely $T = 1/(1 + GG_c) = d_1' d_2/(d_1' d_2 + n_1' n_2)$, is stable. By direct calculation, the transfer function between r_2 and y can be shown to be unstable. Let $r_1 = 0$.

$$\frac{y}{r_2} = \frac{G}{1 + GG_c} = \frac{n_1(z) d_2(z)}{(d_1'(z) d_2(z) + n_1(z) n_2'(z))(z + a)}$$

Thus this transfer function is unstable and a bounded signal injected at r_2 will produce an unbounded signal at y.

Because of the symmetric nature of the problem, the unstable pole–zero cancellation is to be avoided – it does not matter whether the pole comes in G or in G_c:

1. Suppose G has an unstable pole cancelling a zero of G_c so that the zeros of $1 + GG_c$ are inside the unit circle. We see that $1/(1 + GG_c)$ and $G_c/(1 + GG_c)$ are stable but not $G/(1 + GG_c)$.

2. Suppose now G_c has an unstable pole cancelling a zero of G such that the zeros of $1 + GG_c$ are inside the unit circle. We see that $G_c/(1 + G_c G)$ is unstable.

We will now present an example to explain this idea.

Example 7.7 Verify whether the feedback system given in Fig. 7.21, in which a controller is designed with unstable pole–zero cancellation, is internally stable.

It is easy to see that the transfer function between r and y is $1/(z - 0.5)$. This shows that for bounded changes in r, the changes in y will remain bounded. Let ξ denote the actuator noise. The closed loop transfer function between ξ and y can easily be calculated as $(z - 0.5)/[(z + 0.5)(z - 2)]$. Thus if even an extremely small noise is introduced in the actuator, the plant output y will become unbounded. Thus this system is not internally stable.

∎

We have seen that if there is an unstable pole–zero cancellation, the system is not internally stable. We would now like to know the converse: if the transfer function between r_1 and e_1 is stable and if unstable pole–zero cancellation does not take place, will we have internal stability? The answer is yes and such a system is said to be *internally stable.*

7.4.3 Internal Stability Ensures Controller Realizability

The next question we have to address is whether the discrete time controller that guarantees internal stability is realizable, *i.e.*, whether it can be implemented. Suppose that the controller given by Eq. 7.49b is written as a polynomial in z^{-1}:

$$G_c = \frac{n_0 + n_1 z^{-1} + \cdots}{d_0 + d_1 z^{-1} + \cdots}$$

If $d_0 = 0$ and $d_1 \neq 1$, G_c will have a factor of z^{-1} in the denominator and hence can be written as

$$G_c = \frac{G_c'}{z^{-1}} \tag{7.53}$$

where G_c' is causal. G_c is not realizable as it is not causal. Next, note that all realistic discrete time systems have at least one sample time delay.[1] As a result, plant transfer functions can in general be written as

$$G = z^{-d} \frac{b_1(z^{-1})}{a_1(z^{-1})} \tag{7.54}$$

where b_1 and a_1 are polynomials in z^{-1} and $d \geq 1$.

If a noncausal controller, as in Eq. 7.53, is obtained, it would imply that there is a cancellation of z^{-1} in the denominator of G_c and that in the numerator of G, in Eq. 7.54. This is equivalent to cancelling a factor with a root at $z^{-1} = 0$ or at $z = \infty$. It follows that the unrealizable controller of the form given in Eq. 7.53 could result if unstable pole–zero cancellation is allowed. Conversely, if this cancellation is disallowed, the controller G_c cannot have z^{-1} as a factor in its denominator and hence will be realizable. A procedure to avoid unstable pole–zero cancellation during controller design is discussed in Sec. 9.2.

7.4.4 Closed Loop Delay Specification and Realizability

In this section, we show that the delays in the closed loop system have to be at least as large as that of the open loop system, if a controller is to be realizable. Suppose that the open loop transfer function of a system is given by

$$G(z) = z^{-k} \frac{B(z)}{A(z)} = z^{-k} \frac{b_0 + b_1 z^{-1} + \cdots}{1 + a_1 z^{-1} + \cdots}$$

with $b_0 \neq 0$. Suppose that we use a feedback controller of the form

$$G_c(z) = z^{-d} \frac{S(z)}{R(z)} = z^{-d} \frac{s_0 + s_1 z^{-1} + \cdots}{1 + r_1 z^{-1} + \cdots} \tag{7.55}$$

with $s_0 \neq 0$. Note that $d \geq 0$ for the controller to be realizable. Closing the loop, we obtain the closed loop transfer function as

$$T = \frac{GG_c}{1 + GG_c} \tag{7.56}$$

[1]The reader may wish to look at the examples presented in Sec. 8.2.2 at this juncture.

Substituting the values, we obtain

$$T = z^{-k-d} \frac{b_0 s_0 + (b_0 s_1 + b_1 s_0) z^{-1} + \cdots}{1 + (a_1 + r_1) z^{-1} + \cdots} \tag{7.57}$$

Thus, the delay in the closed loop system is $k + d$, which is greater than k, the open loop delay, for all $d > 0$. The only way to make $k + d < k$ is to choose $d < 0$. But this will make the controller unrealizable, see Eq. 7.55. We illustrate this idea with an example.

Example 7.8 Suppose that

$$G = z^{-2} \frac{1}{1 - 0.5z^{-1}}$$

and that we want the closed loop transfer function as

$$T = z^{-1} \frac{1}{1 - az^{-1}}$$

Determine the controller that is required for this purpose.

Solving Eq. 7.56 for G_c, we obtain

$$G_c = \frac{1}{G} \frac{T}{1 - T}$$

Substituting the values of T and G, and simplifying, we obtain

$$G_c = \frac{1}{z^{-1}} \frac{1 - 0.5z^{-1}}{1 - (a + 1)z^{-1}}$$

We see that this controller is unrealizable, no matter what a is. ∎

We will see in the subsequent chapters that it is possible to improve the dynamic response and stability, but not the delay. The delay of the closed loop system has to be at least as large as that of the open loop system.

7.5 Internal Model Principle and System Type

We are interested in finding out the conditions to be satisfied to have the plant output follow the reference trajectories and to reject disturbances. We will show that the concept of internal model principle helps answer this question. We will also show that the popular topic of *system type* is a consequence of this principle.

7.5.1 Internal Model Principle

Consider the problem of controlling a system with transfer function G with a controller G_c, as in the standard feedback loop of Fig. 7.22. The variables R, Y, U and V, respectively, refer to the Z-transforms of the reference, output, control input and disturbance variables, respectively.

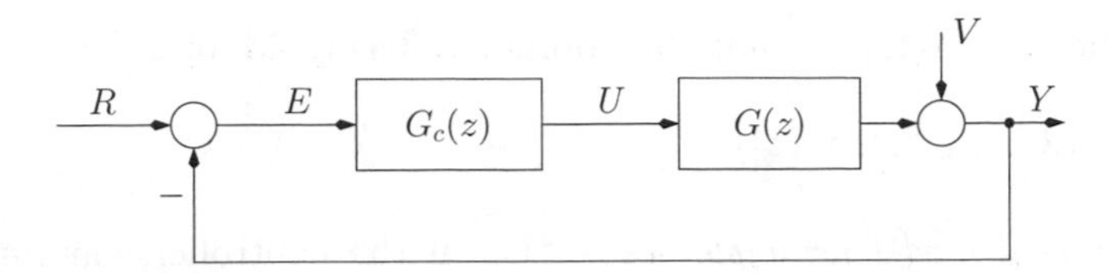

Figure 7.22: Servo and regulation problem in a feedback loop

The objectives of the controller are two fold: stabilization of the feedback loop and rejection of the effect of unbounded inputs. The first issue of stabilization of the closed loop is the topic of discussion in the subsequent chapters. In the rest of this discussion, we will assume that this requirement is fulfilled by a suitable controller. We will now look at the condition that needs to be fulfilled to satisfy the second requirement of rejecting the effect of unbounded inputs. We will consider two cases:

1. Make the output Y follow the reference signal R, or equivalently, make the error E zero. When we study this, we take the disturbance variable V to be zero and call it the servo problem, as mentioned earlier. Thus, we need to consider the relation between E and R, given by

$$E(z) = \frac{1}{1 + GG_c} R(z) \triangleq S(z)R(z) \tag{7.58}$$

where $S(z)$ is $1/(1 + GG_c)$. We need to make $E(z)$ well behaved.

2. Eliminate the effect of the disturbance variable V on the output Y. When we study this case, we take R to be zero and call it the regulation problem, as mentioned earlier. We study the relation between Y and V, given by

$$Y(z) = \frac{1}{1 + GG_c} V(z) = S(z)V(z) \tag{7.59}$$

We need to make Y well behaved.

From Eq. 7.58 and 7.59, we see that both the requirements will be met if we can make the transfer function $1/(1 + GG_c)$ small. To study this, suppose that we have the following functional relationship for R and V:

$$\begin{aligned} R(z) &= \frac{N_r}{D_r \alpha_r} \\ V(z) &= \frac{N_v}{D_v \alpha_v} \end{aligned} \tag{7.60}$$

The unstable poles of the reference and the disturbance signals have been grouped, respectively, in α_r and α_v. For example, these could contain the factors:

- $1 - z^{-1}$, if constant change is intended
- $(1 - z^{-1})^2$, if a ramp-like change is intended
- $1 - 2\cos\omega z^{-1} + z^{-2}$, if sinusoidal change is present
- $1 - z^{-N}$, if the change is periodic, with a period N

Let $\alpha(z)$ be the least common multiple, abbreviated as LCM, of $\alpha_r(z)$ and $\alpha_v(z)$, *i.e.*,

$$\alpha(z) = \text{LCM}(\alpha_r(z), \alpha_v(z)) \tag{7.61}$$

Then, the *internal model principle* states that if the controller contains the factor $1/\alpha$, *i.e.*, if the controller is of the form

$$G_c(z) = \frac{S_c(z)}{\alpha(z)R_1(z)} \tag{7.62}$$

the effect of unbounded signals will be eliminated. Suppose that the plant transfer function is given by

$$G(z) = \frac{B(z)}{A(z)} \tag{7.63}$$

where B and A are polynomials in powers of z^{-1}. In order to ensure internal stability, there should not be any unstable pole–zero cancellation in G, G_c and between them. The transfer functions of Eq. 7.58 and 7.59 are given by

$$S \triangleq \frac{1}{1+GG_c} = \frac{1}{1+\frac{B}{A}\frac{S_c}{\alpha R_1}} = \frac{A\alpha R_1}{A\alpha R_1 + BS_c} \tag{7.64}$$

Substituting this relation and that for $R(z)$ from Eq. 7.60 in Eq. 7.58, we obtain

$$E(z) = \frac{A\alpha R_1}{A\alpha R_1 + BS_c}\frac{N_r}{D_r\alpha_r} \tag{7.65}$$

Because of Eq. 7.61, α contains α_r and hence the unbounded inputs, if any, are rejected. Thus, the servo problem has been addressed by this α. Because the requirement for the regulation problem is identical, see Eq. 7.58 and 7.59, the unbounded inputs in the disturbance variable also are rejected. Recall that we have assumed the closed loop system to be stable and hence $A\alpha R_1 + BS_c$ has zeros inside the unit circle only.

We need to address the issue of unstable pole–zero cancellation once again. If such a cancellation is to be forbidden, as explained in detail in Sec. 7.4.1, how can we allow this in Eq. 7.64? The answer is that we have no option but to cancel them in this case. To reduce the impact of the inherent difficulties in such a cancellation, we need exact knowledge of the unstable nature of the external inputs. If there is at all some uncertainty in the external inputs, which are taken as unbounded, the uncertainty should be bounded. Moreover, the unstable pole–zero cancellation indicated in Eq. 7.64 is between a signal and a system, whereas the one presented in Sec. 7.4.1 is between two systems. The latter gives rise to other difficulties, such as the one in Sec. 7.4.2.

We have shown that when the unstable parts of the external inputs explicitly appear as a part of the loop, they are rejected. Hence, this is known as the internal model principle.

Example 7.9 If $\alpha = 1 - z^{-1}$, step changes in both R and V will be rejected if the closed loop is stable. If, on the other hand, r is a step change, *i.e.*, $R = 1/(1-z^{-1})$, and v is a sine function, *i.e.*,

$$V = \frac{z^{-1}\sin\omega}{1 - 2\cos\omega z^{-1} + z^{-2}}$$

or a cosine function,

$$V = \frac{1 - z^{-1}\cos\omega}{1 - 2\cos\omega z^{-1} + z^{-2}}$$

then α is given by

$$\alpha = (1 - z^{-1})(1 - 2\cos\omega z^{-1} + z^{-2})$$

As before, we assume that the controller is designed so as to make the closed loop system stable. ∎

We now discuss the concept of type of a system, which is a direct consequence of the internal model principle.

7.5.2 System Type

Consider a plant G with a controller G_c in our standard 1-DOF feedback control configuration, Fig. 7.22. The system is said to be of type N if GG_c has N integrators at $z = 1$. That is, if

$$G(z)G_c(z) = \frac{1}{(z-1)^N}\frac{G_n(z)}{G_d(z)} \tag{7.66}$$

where G_n and G_d, polynomials in z, don't have any zero at $z = 1$.

From the internal model principle presented in the previous section, this system is capable of completely rejecting the effect of exogeneous signals, whose Z-transform could have $(z-1)^N$ in the denominator. What happens if the order of the poles at $z = 1$ in the exogeneous signal is different from that of GG_c? In other words, let the loop transfer function GG_c be given by

$$GG_c = \frac{1}{\alpha}\frac{BS_c}{AR_1}, \quad \alpha = (z-1)^N \tag{7.67}$$

where we have used Eq. 7.62 and 7.63, with $\alpha = (z-1)^N$ and N, a nonnegative integer. Let the reference signal be given by

$$R = \frac{N_r}{D_r\alpha_r}, \quad \alpha_r = (z-1)^M \tag{7.68}$$

see Eq. 7.60. Using Eq. 7.65, $E(z)$ becomes

$$E(z) = \frac{A(z-1)^N R_1}{A(z-1)^N R_1 + BS_c}\frac{N_r}{D_r(z-1)^M} \tag{7.69}$$

Table 7.1: Value of error $e(\infty)$ for different M and N values in Eq. 7.70

Condition on M and N	$e(\infty)$
$M = N$	0
$M = N + 1$, $N = 0$	$\left.\frac{AR_1}{AR_1 + BS_c}\frac{N_r}{D_r}\right\|_{z=1}$
$M = N + 1$, $N > 0$	$\left.\frac{AR_1}{BS_c}\frac{N_r}{D_r}\right\|_{z=1}$
$M > N + 1$	∞
$M < N$	0

Table 7.2: Values of $e(\infty)$, given in Table 7.1, for popular systems and signals

	Step, $1(n)$	Ramp, $nT_s1(n)$	Parabola, $\frac{1}{2}n^2T_s^21(n)$
	$\frac{z}{z-1}$, $M = 1$	$\frac{zT_s}{(z-1)^2}$, $M = 2$	$\frac{(z^2+z)T_s^2}{(z-1)^3}$, $M = 3$
Type 0, $N = 0$	$\left.\frac{AR_1}{AR_1 + BS_c}\right\|_{z=1}$	∞	∞
Type 1, $N = 1$	0	$\left.T_s\frac{AR_1}{BS_c}\right\|_{z=1}$	∞
Type 2, $N = 2$	0	0	$\left.T_s^2\frac{AR_1}{BS_c}\right\|_{z=1}$

Under the assumption that the closed loop system is stable, we can apply the final value theorem:

$$\lim_{n\to\infty} e(n) = \lim_{z\to 1}\frac{z-1}{z}E(z)$$

Substituting for $E(z)$ from Eq. 7.69, we obtain

$$\lim_{n\to\infty} e(n) = \lim_{z\to 1}\frac{A(z-1)^{N+1}R_1}{A(z-1)^N R_1 + BS_c}\frac{N_r}{D_r(z-1)^M} \tag{7.70}$$

By substituting different values for M and N, we obtain the values given in Table 7.1 for $e(\infty)$. For N values of 0, 1 and 2, the system is known as *type 0*, *type 1* and *type 2*, respectively. These systems give rise to different errors for step, ramp and parabola signals, with corresponding models $1(n)$, $nT_s1(n)$ and $\frac{1}{2}n^2T_s^21(n)$, respectively. The matrix of error values is given in Table 7.2.

It is customary to obtain the diagonal values in Table 7.2 in terms of the values of GG_c. It is easy to show that the $(1,1)$ entry becomes

$$\left.\frac{AR_1}{AR_1 + BS_c}\right|_{z=1} = \frac{1}{1+K_p} \tag{7.71}$$

where

$$K_p \triangleq \left. \frac{BS_c}{AR_1} \right|_{z=1} \tag{7.72}$$

is known as the *position constant*. The $(2, 2)$ entry becomes

$$T_s \left. \frac{AR_1}{BS_c} \right|_{z=1} = \frac{1}{K_v} \tag{7.73}$$

where

$$K_v \triangleq \lim_{z \to 1} (z-1) \frac{GG_c}{T_s} \tag{7.74}$$

is known as the *velocity constant*. Finally, the $(3, 3)$ entry in Table 7.2 becomes

$$T_s^2 \left. \frac{AR_1}{BS_c} \right|_{z=1} = \frac{1}{K_a} \tag{7.75}$$

where

$$K_a \triangleq \lim_{z \to 1} (z-1)^2 \frac{GG_c}{T_s^2} \tag{7.76}$$

is known as the *acceleration constant*. It is also customary to state the performance specifications in terms of the constants defined above. We will now explain these ideas with an example.

Example 7.10 Illustrate the ideas of this section with the plant

$$G(z) = \frac{z^{-1}}{1 - z^{-1}} = \frac{1}{z - 1}$$

Rejecting step inputs: Suppose that we want to reject step inputs. Because the plant already has an integral term, we do not need to include it in the controller. Note that a precondition for discussing steady state errors is the closed loop stability. Suppose that we want the closed loop characteristic polynomial to be

$$\phi_{cl} = z - 0.5 \tag{7.77}$$

The controller $G_c = 1/2$ helps achieve this requirement. A procedure to design such a controller will be discussed in detail in Example 9.14 on page 351. M 7.7 implements this control design procedure.

Note that this controller results in a Type 1 system, because it results in GG_c of the form given by Eq. 7.66, with $N = 1$. This should reject step inputs and give a constant steady state error to ramps. To verify these observations, we determine the closed loop transfer function between the input and the error, given by Eq. 7.64. We obtain

$$S = \frac{1}{1 + GG_c} = \frac{z - 1}{z - 0.5} \tag{7.78}$$

The closed loop characteristic polynomial agrees with our requirement, stated by Eq. 7.77. The steady state error is given by the final value theorem.

$$\lim_{n\to\infty} e(n) = \lim_{z\to 1} \frac{z-1}{z} E(z) \tag{7.79}$$

Expressing $E(z)$ as a product of reference signal and the error transfer function S,

$$\lim_{n\to\infty} e(n) = \lim_{z\to 1} \frac{z-1}{z} S(z)R(z) \tag{7.80}$$

where $R(z)$ is the reference signal. For step inputs, $R(z) = z/(z-1)$. Substituting this and the expression for S from Eq. 7.78 in the above equation, we obtain

$$\lim_{n\to\infty} e(n) = \lim_{z\to 1} \frac{z-1}{z}\frac{z-1}{z-0.5}\frac{z}{z-1} = 0 \tag{7.81}$$

Thus, the steady state error to step inputs is zero.

How does this controller handle ramp inputs? To evaluate this, we take $R(z) = z/(z-1)^2$, where we have taken the sampling time to be unity. Substituting this and the expression for S from Eq. 7.78 in Eq. 7.80, we obtain

$$\lim_{n\to\infty} e(n) = \lim_{z\to 1} \frac{z-1}{z}\frac{z-1}{z-0.5}\frac{z}{(z-1)^2} = 2$$

These results can be verified by executing M 7.7 and the Simulink program, given in Fig. A.1. To calculate the controller, we assign a value of 0 to the variable `reject_ramps` in Line 8 of M 7.7. By double clicking the manual switch in the Simulink program of Fig. A.1, one can toggle between step and ramp inputs. From the response in Simulink, one can see that this controller rejects step inputs. A ramp input, on the other hand, produces a steady state error of 2. For example, at $n = 10$, $y(n) = 8$, while $r(n) = 10$.

Rejecting ramp inputs: Suppose that now we want to reject ramp inputs as well. Because the plant already has one integral term, we need to include just one more in the controller. A procedure for designing the controller will be given in Chapter 9. Using M 7.7, we obtain the controller

$$G_c = \frac{1.5 - z^{-1}}{1 - z^{-1}} = \frac{1.5z - 1}{z - 1} \tag{7.82}$$

From Eq. 7.66, it is easy to see that this is a Type 2 system. Substituting this controller and the plant expression in Eq. 7.64, we obtain

$$S = \frac{1}{1 + GG_c} = \frac{(z-1)^2}{z(z-0.5)} \tag{7.83}$$

Substituting this in Eq. 7.80 with $R(z) = z/(z-1)$, we obtain the steady state error to steps:

$$\lim_{n\to\infty} e(n) = \lim_{z\to 1} \frac{z-1}{z}\frac{(z-1)^2}{z(z-0.5)}\frac{z}{z-1} = 0 \tag{7.84}$$

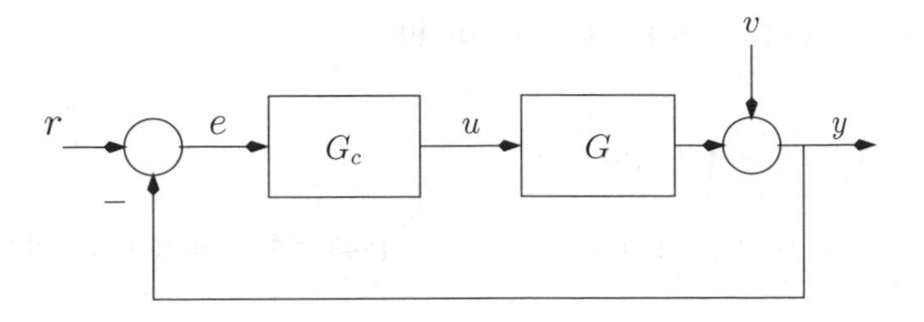

Figure 7.23: 1-DOF feedback control configuration

Substituting the expression for S from Eq. 7.83 in Eq. 7.80 with $R(z) = z/(z-1)^2$, we obtain the steady state error to ramp inputs:

$$\lim_{n\to\infty} e(n) = \lim_{z\to 1} \frac{z-1}{z}\frac{(z-1)^2}{z(z-0.5)}\frac{z}{(z-1)^2} = 0 \tag{7.85}$$

These results can be verified by executing M 7.7 once again, but with `reject_ramps` in Line 8 of M 7.7 assigned as 1. The closed loop response to steps and ramps may be verified with the help of the Simulinkprogram, given in Fig. A.1. Before leaving this example, let us point out that there will be a nonzero offset to a parabola input. We find that the findings of this example are in perfect agreement with the observations made earlier in this section. ∎

In the next section, we present the important concept of performance limits.

7.6 Introduction to Limits of Performance

Is it possible to achieve a given performance for any given plant? One clue is already available in Sec. 7.4.4: closed loop delay should be at least as large as that of the open loop. In this section, we will present a few other limits on the achievable performance. We will begin the discussion with time domain limits.

7.6.1 Time Domain Limits

In this section, we discuss some limits on achievable time domain performance [20]. Consider the 1-DOF control feedback given in Fig. 7.23. Let us suppose that the closed loop system is stable. Let us also suppose that GG_c has a pole at $z = 1$ and hence y follows step changes in r without offset, as discussed in Sec. 7.5. For the same reason, step changes in disturbance are also rejected. We will examine how the transient responses in y and e depend on pole–zero locations of G.

Suppose that we introduce a unit step change in r. We obtain the Z-transform of e as

$$E(z) = \frac{1}{1+G(z)G_c(z)}\frac{z}{z-1} \tag{7.86}$$

Evaluating this at z_0, the zero of G, we obtain

$$E(z_0) = \frac{z_0}{z_0 - 1} = \frac{1}{1 - z_0^{-1}} \tag{7.87}$$

Suppose that $z_0 \in$ ROC, see Footnote 1 on page 64. Using the definition of the Z-transform, we obtain

$$\sum_{k=0}^{\infty} e(k) z_0^{-k} = E(z_0) = \frac{1}{1 - z_0^{-1}} \tag{7.88}$$

Suppose that, in addition, z_0 is real and less than 1. For all causal systems, the ROC is outside a circle of some radius a, see Sec. 4.1.5. In other words, we have $a < z_0 < 1$. Using this value of z_0 in the above equation, we obtain

$$\sum_{k=0}^{\infty} e(k) z_0^{-k} < 0 \tag{7.89}$$

From $e(k) = r(k) - y(k)$ we obtain $e(0) = 1 - 0 = 1$. As a result, $e(k)$ has to be negative for some $k \geq 1$, so as to satisfy Eq. 7.89. Because $e(k) = 1 - y(k)$, this implies an overshoot in $y(k)$, for $k \geq 1$. We also see that as z_0 approaches the unit circle (*i.e.*, $z_0 \to 1$), $\sum_{k=0}^{\infty} e(k) z_0^{-k}$ becomes a large negative number, see Eq. 7.88. In other words, as z_0 approaches the unit circle, the overshoot becomes larger. We will now illustrate this with an example.

Example 7.11 Design controllers for two plant transfer functions

$$G_1 = z^{-1} \frac{1 - 0.85z^{-1}}{1 - 0.8z^{-1}}, \quad G_2 = z^{-1} \frac{1 - 0.9z^{-1}}{1 - 0.8z^{-1}}$$

with integral mode such that the closed loop characteristic polynomial in both cases is equal to $(1 - 0.7z^{-1})^2$. Calculate the step responses and verify the validity of the observations made in this section.

Because of the presence of the integral mode, the controller is of the form

$$G_c(z) = \frac{S(z)}{\Delta(z) R(z)} \tag{7.90}$$

where $\Delta = 1 - z^{-1}$ and S and R are polynomials in z^{-1}. Let the plant be denoted by

$$G(z) = z^{-k} \frac{B(z)}{A(z)} \tag{7.91}$$

The characteristic equation of the closed loop is given by $1 + G(z)G_c(z) = 0$. Substituting for G_c and G from Eq. 7.90–7.91, we obtain

$$A \Delta R + z^{-k} B S = \phi$$

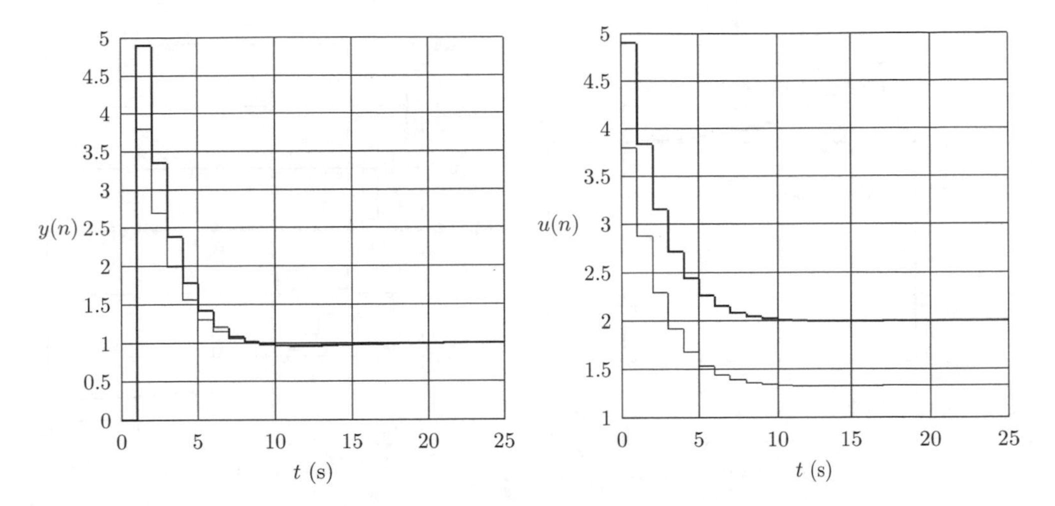

Figure 7.24: Output and input profiles of the closed loop system with G_1 (thin line) and G_2 (thick line), as discussed in Example 7.11. Presence of slow open loop zeros has resulted in overshoots of the output. Overshoot of G_2 is larger than that of G_1.

where ϕ is the desired characteristic polynomial, which we require to be $(1 - 0.7z^{-1})^2$. For the first plant, we need to solve

$$(1 - 0.8z^{-1})(1 - z^{-1})R + z^{-1}(1 - 0.85z^{-1})S = 1 - 1.4z^{-1} + 0.49z^{-2}$$

A procedure to solve such equations will be outlined in Sec. 7.8. It is easy to verify that $R = 1 - 3.4z^{-1}$ and $S = 3.8 - 3.2z^{-1}$ satisfy this equation and hence the controller for the first plant is given by

$$G_{c_1} = \frac{3.8 - 3.2z^{-1}}{(1 - z^{-1})(1 - 3.4z^{-1})}$$

In a similar way, the controller to the second plant can be verified as

$$G_{c_2} = \frac{4.9 - 4z^{-1}}{(1 - z^{-1})(1 - 4.5z^{-1})}$$

Thus, we have two plants with different zeros but the same closed loop pole positions. The step responses of these two are shown in Fig. 7.24. Observe that both responses have overshoot, as predicted. Also notice that the response of the second plant has a larger overshoot compared to the first. This agrees with the observation that zeros close to the unit circle result in a large overshoot. From the accompanying plot of the control effort u, we see that larger overshoot is accompanied by larger initial control effort. ∎

Conclusions similar to the ones obtained in this section can be arrived at for the pole locations of G. Let p_0 be a real pole of G, such that $p_0 < a$. Recall that a defines

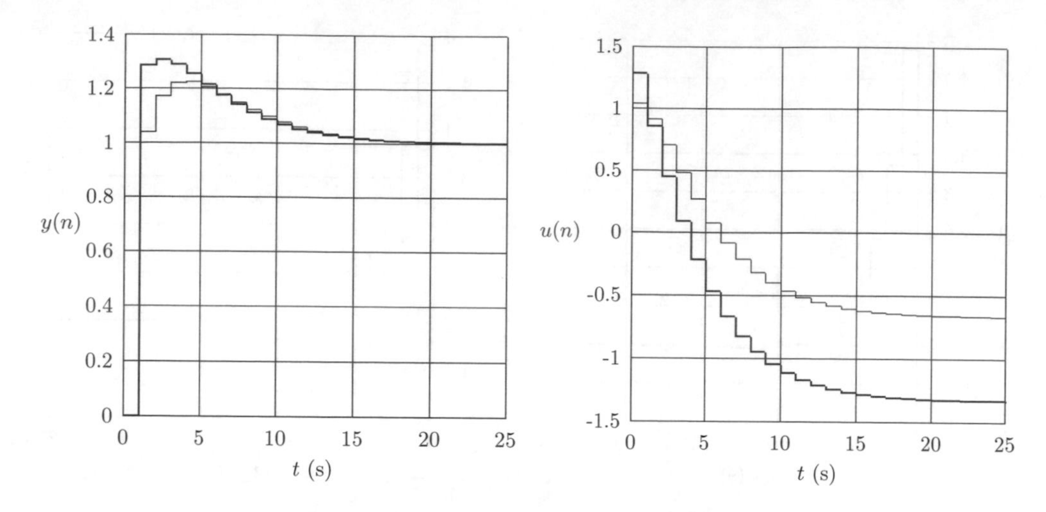

Figure 7.25: Output and input profiles of the closed loop system with G_3 (thin line) and G_4 (thick line) of Eq. 7.93. Presence of slow open loop poles has resulted in overshoots of the output. Overshoot of G_4 is larger than that of G_3.

ROC. Evaluating Eq. 7.86 at p_0, we obtain $E(z = p_0) = 0$. Using the definition of the Z-transform of $E(z)$, we conclude that

$$\sum_{k=0}^{\infty} e(k) p_0^{-k} = 0 \tag{7.92}$$

This shows once again that the error to a unit step change in r has to become negative for $k \geq 1$, because $e(0) = 1$. Note that this conclusion is true also when $p_0 > 1$. If p_0 becomes larger, the weighting on $e(k)$ in Eq. 7.92 is smaller and hence $e(k)$ should attain a larger negative value, or equivalently, y should assume a larger overshoot.

To verify these, Example 7.11 is repeated for two transfer functions

$$G_3 = z^{-1} \frac{1 - 0.85z^{-1}}{1 - 1.1z^{-1}}, \quad G_4 = z^{-1} \frac{1 - 0.85z^{-1}}{1 - 1.2z^{-1}} \tag{7.93}$$

The controllers are once again designed with the same ϕ. The resulting input–output profiles are shown in Fig. 7.25. From these plots, we can observe that the outputs have overshoots. We also see that the overshoot corresponding to G_4 is larger than that of G_3, as predicted above.

We conclude this section with a study of nonminimum phase zeros of G. Let $y(n)$ be the plant response for a unit step change in r. We obtain

$$Y = \frac{GG_c}{1 + GG_c} \frac{z}{z - 1} \tag{7.94}$$

where Y is the Z-transform of y. From the definition of Y, we also have

$$Y = \sum_{k=0}^{\infty} y(k) z^{-k}, \ \forall z \in \text{ROC} \tag{7.95}$$

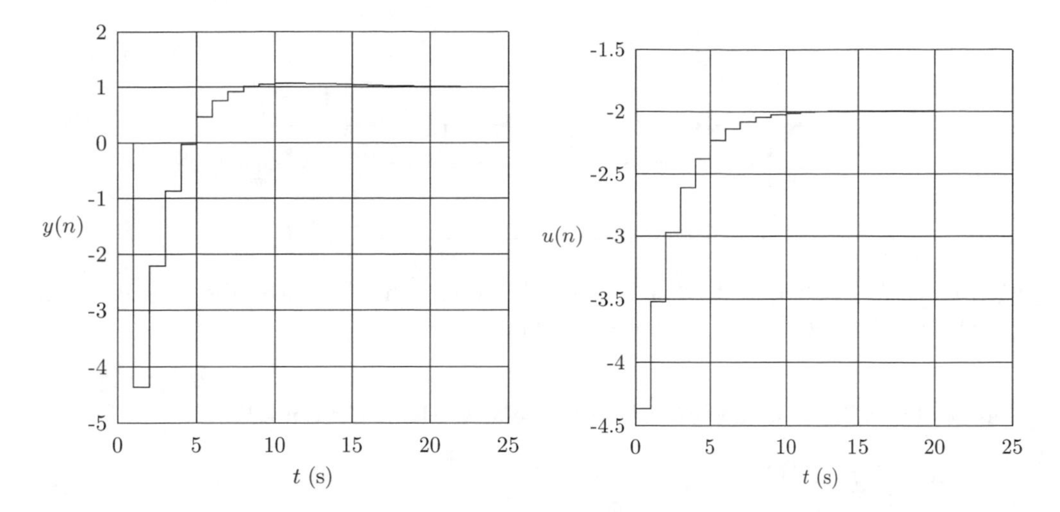

Figure 7.26: Output and input profiles of the closed loop system with $G_5 = z^{-1}(1 - 1.1z^{-1})/(1 - 0.8z^{-1})$. Presence of nonminimum phase zero has resulted in an undershoot of the output.

where ROC is the region of convergence, see Footnote 1 on page 64. Let G have a real zero z_0 with $z_0 > 1$. Thus, $z_0 \in$ ROC. Evaluating Eq. 7.94 at $z = z_0$, we obtain $Y(z_0) = 0$. Comparing with Eq. 7.95, we arrive at

$$\sum_{k=0}^{\infty} y(k) z_0^{-k} = 0 \tag{7.96}$$

Because z_0 is positive and $y(k)$ will eventually reach a value of 1, $y(k)$ should assume a negative value for some k if the above equality is to be satisfied. This is demonstrated in Fig. 7.26, where we have shown the input–output profiles for $G_5 = z^{-1}(1 - 1.1z^{-1})/(1 - 0.8z^{-1})$, when controlled in the same manner as in Example 7.11, so that the closed loop poles are at 0.7. This study shows that if the open loop system is nonminimum phase, there will be overshoots in the negative direction for a positive step change, known as *inverse response*.

In this section, we have studied in detail what happens when the closed loop poles are chosen smaller than the open loop poles and zeros in magnitude. When a pole is chosen close to the origin, the signals change quickly. For example, a^k goes to zero quickly as k increases when a is small. Thus, when we make the closed loop system faster than the open loop poles, there is a penalty we have to pay through overshoots, etc.

7.6.2 Sensitivity Functions

One of the main objectives of using feedback controllers is to cope with the problem of not knowing the plant exactly. Suppose that the plant transfer function is G and that the closed loop transfer function is denoted by T. Suppose also that when G

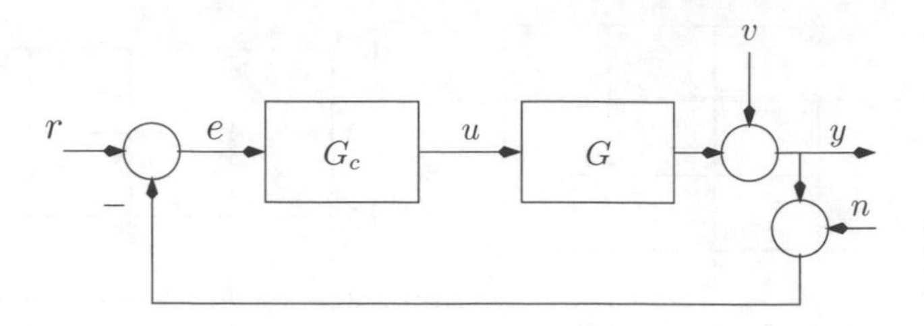

Figure 7.27: A simple feedback control loop with reference, disturbance and measurement noise signals as external inputs

changes by ΔG, T changes by ΔT. If the relative change, given by

$$\frac{\Delta T/T}{\Delta G/G} = \frac{\Delta T}{\Delta G}\frac{G}{T} \tag{7.97}$$

is small, we say that the closed loop is not sensitive to perturbations in plant parameters. When the changes are infinitesimal, this ratio, denoted by S, becomes

$$S = \frac{\partial T}{\partial G}\frac{G}{T} \tag{7.98}$$

S is known as the *sensitivity function.* We will now apply this formula to the closed loop system presented in the previous section. Let us evaluate the fractional change in T, defined in Eq. 7.5 on page 245, to that in G:

$$S = \frac{\partial T}{\partial G}\frac{G}{T} = \frac{(1+GG_c)G_c - GG_c^2}{(1+GG_c)^2}\frac{G(1+GG_c)}{GG_c}$$

Simplifying, we obtain

$$S = \frac{1}{1+GG_c}$$

which turns out to be identical to the expression in Eq. 7.6. Thus, if $1/(1+GG_c)$ is made small, the plant becomes less sensitive to perturbations in the plant. Observe from Eq. 7.4 and 7.6 that this fraction is also the transfer function between $v(n)$ and $y(n)$. Thus, by making S small, the effect of v on y also is reduced.

In view of the fact that $S + T = 1$ (see Eq. 7.8 on page 245), T is known as the *complementary sensitivity function.* This constraint restricts the achievable performance, as we will see in the next section.

7.6.3 Frequency Domain Limits

In this section, we will look at the achievable performance from a frequency domain perspective. For this discussion, we will refer to Fig. 7.27. Notice that measurement noise has been included in this, as compared to Fig. 7.23. Using straightforward algebra and Eq. 7.6 and 7.5 on page 245, we arrive at the following equations:

$$y = \frac{GG_c}{1+GG_c}(r-n) + \frac{1}{1+GG_c}v = T(r-n) + Sv \tag{7.99}$$

$$e = r - y - n = \frac{1}{1+GG_c}(r-v-n) = S(r-v-n) \tag{7.100}$$

$$u = \frac{G_c}{1+GG_c}(r-v-n) \tag{7.101}$$

We assume that the system is internally stable. We will look at the restrictions on S and T for tracking, disturbance rejection and noise suppression.

In tracking or servo control, the objective is to make e small while assuming v and n to be zero. To make e small at any frequency ω, we obtain from Eq. 7.99,

$$\max |S(e^{j\omega})| \ll 1 \tag{7.102}$$

Another condition that is of interest in control applications is that the disturbance v should not greatly affect y, known as the disturbance rejection problem. For this to happen, we see from Eq. 7.99 that $|S|$ has to be small, *i.e.*, the above equation has to be satisfied.

From these conditions, one may be tempted to keep $S(e^{j\omega})$ close to zero and $T(e^{j\omega})$ close to one at all frequencies. Unfortunately, there are problems in implementing this. One of them is due to the mismatch between the model G and the actual plant it represents. Suppose that the exact model of the plant, indicated by $G_L(e^{j\omega})$ at a frequency ω, is given by

$$G_L(e^{j\omega}) = \left[1 + L(e^{j\omega})\right] G(e^{j\omega}) \tag{7.103}$$

with

$$\max |L(e^{j\omega})| < l(\omega) \tag{7.104}$$

Generally the uncertainty is large at high frequencies and, as a result, l is large at high frequencies and small at low frequencies. In other words, the mismatch between the model and the actual plant is generally large at high frequencies.

Suppose that we design a controller G_c, assuming the plant transfer function is G. If the actual transfer function is G_L, the following condition has to be satisfied for stability [1]:

$$\max |T(e^{j\omega})| < \frac{1}{l(\omega)} \tag{7.105}$$

This implies that $|T(e^{j\omega})|$ should be small at high frequencies. In other words, the bandwidth of T should be only as large as the frequency range over which we have good knowledge of the plant. Thus, we revise the condition on S and T: keep $|S(e^{j\omega})| \simeq 0$ and $|T(e^{j\omega})| \simeq 1$ in the range $0 \le \omega < \omega_k$ and $|S(e^{j\omega})| \simeq 1$ and $|T(e^{j\omega})| \simeq 0$ in the range $\omega_k \le \omega < \pi$, where ω_k is the frequency until which we have good knowledge of the plant. Thus, the robustness condition restricts the range of frequencies over which $|S|$ can be made small.

The magnitude of control effort u also poses a restriction on S. When $|S|$ is small, $|1/S| = |1 + GG_c|$ is large. This implies $|GG_c| \gg 1$ and hence from Eq. 7.101, we obtain the following approximate expression:

$$u \simeq \frac{G_c}{GG_c}(r - n - u) = G^{-1}(r - n - u) \tag{7.106}$$

If $|G|$ is small in this frequency range, the control effort u will become large and the actuator may saturate. It follows that S can be made small only in the frequency range corresponding to the bandwidth of G.

We end this discussion by looking at the requirement for noise suppression from Eq. 7.101. The condition

$$|T(e^{j\omega})| \ll 1 \tag{7.107}$$

has to be satisfied if the effect of n is not to be felt on y. Hopefully, the measurement noise n is large only at high frequencies. Thus, we need to satisfy the above inequality only at high frequencies.

Gathering all the requirements stated above, we can summarize that $|S(e^{j\omega})|$ could be made small only over the smallest of the following frequency ranges:

1. The frequency range in which we have good knowledge of the plant.

2. The bandwidth of the plant is large.

3. The frequency range in which clean measurements are available.

We conclude this section by briefly presenting the so called *water-bed effect.* The following restriction on the sensitivity function can be proved [20]:

$$\frac{1}{\pi}\int_0^{\pi} \ln |S(e^{j\omega})| d\omega = \sum_{i=1}^{q} \ln |p_i| \tag{7.108}$$

where p_i, $q \geq i \geq 1$, refer to the unstable poles of G. At frequencies where $|S| \ll 1$, the contribution to the integral is negative. If the integral has to be greater than or equal to zero (will be equal to zero for stable G), it is clear that S has to become large at some other frequencies. In other words, if we want a good performance at some frequency ω_0 (*i.e.*, $|S(\omega_0)| \ll 1$), we have no option but to accept a poor performance at some other frequency ω_L (*i.e.*, $|S(\omega_L)|$ is large). This poor performance becomes worse if the plant under consideration is open loop unstable.

7.7 Well Behaved Signals

The objective of a controller is to make a plant behave better, both from transient and from steady state viewpoints. There are two ways of specifying this requirement. One approach is to specify the closed loop transfer function and to look for a controller that will help achieve this. The second approach is to specify the requirements on the plant response to some reference signal, such as a step signal. In this section, we will follow the second approach. It turns out that this objective may be met by placing the poles of the transfer function in some locations. This approach is well known for

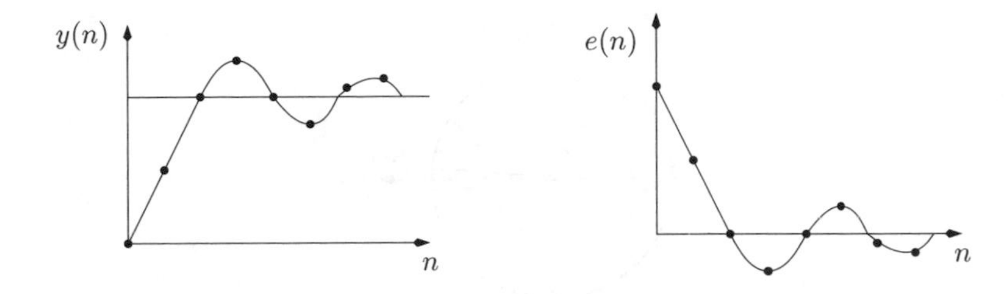

Figure 7.28: Typical $y(n)$ (left) and $e(n)$ (right) profiles

continuous time systems [16]. In this section, we will be concerned with discrete time systems.

Let us first look at the transient response to a unit step change in r. We would want y to have:

1. Small *rise time*. This is an indication of how fast the system responds to a stimulus. For oscillatory or underdamped systems, it is the time when the response reaches the final value, at the first instance. For overdamped systems, it is the time required for the response to go from 10% to 90% of the final value.

2. Small *overshoot*. This is the amount by which the response exceeds the final value. It is usually specified as a fraction and it is applicable only for underdamped systems.

3. Small *settling time*. From this time onwards, the response comes within a small band of the final value and stays within it. Some of the typical values of the band are ±5% and ±2%.

If, instead, we specify the requirements in terms of the error $e(n)$ between the reference signal $r(n)$ and the actual response $y(n)$, the problem becomes easier to handle. A plot of a step response $y(n)$ and the error $e(n)$ are shown in Fig. 7.28.

We first address the issue of using a suitable functional form for the error $e(n)$. The following expression

$$e(n) = \rho^n \cos \omega n, \quad 0 < \rho < 1 \tag{7.109}$$

is a suitable and a convenient form for error, as its initial error is one, it oscillates about zero with decaying amplitude and its steady state value is zero. Translating the requirements on the response $y(n)$ in terms of the error, it is easy to see that the error should meet the following requirements: it should have a small fall time, a small undershoot and a large decay ratio.

The first step in designing the controller G_c is to determine suitable ρ and ω values from the step response requirements. This will be the topic of discussion in this section. In Chapter 9, we will explain a method to design the controller, given ρ and ω.

7.7.1 Small Rise Time in Response

We will first determine the values of ρ and ω that satisfy the rise time requirement. From Eq. 7.109, we see that when $\omega n = \pi/2$, $e(n) = 0$ for the first time, the condition

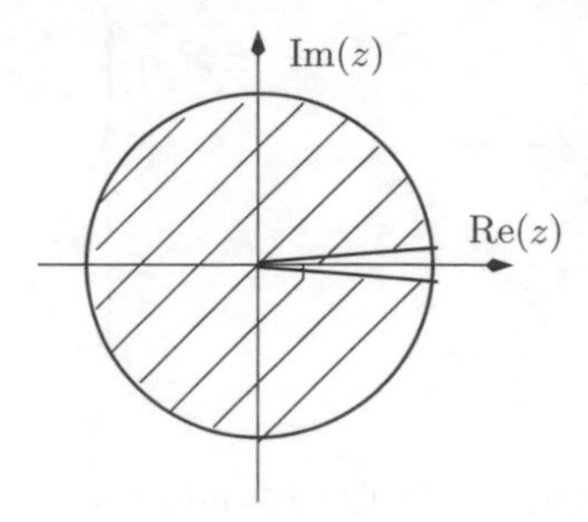

Figure 7.29: Desired location of poles for small fall time is shown shaded

for rise time. That is, this happens at the sampling instant given by

$$n = \frac{\pi}{2\omega}$$

Suppose that we want the rise time to be less than a specified value. If we divide this specification in time units by the sampling time, we get it as the number of samples in one rise time, which we will denote by the symbol N_r. It is clear that n should be less than N_r. In other words,

$$\frac{\pi}{2\omega} < N_r$$

Rewriting this, we obtain the expression for small rise time in the response or small fall time in error as

$$\omega > \frac{\pi}{2N_r} \tag{7.110}$$

The desired region is shown shaded in Fig. 7.29. All points in this region satisfy the condition given by Eq. 7.110. Note that there is no strict requirement that N_r should be an integer.

7.7.2 Small Overshoot in Response

In this section, we will derive an expression for the maximum overshoot in $y(n)$. Because y is not a continuous function of n, we will not take the differentiation route.

Suppose that the maximum undershoot occurs approximately when $\omega n = \pi$. Substituting this in Eq. 7.109, we obtain

$$e(n)|_{\omega n=\pi} = \rho^n \cos \omega n|_{\omega n=\pi} = -\rho^n|_{\omega n=\pi} = -\rho^{\pi/\omega}$$

This is the value of the first undershoot in $e(n)$, which occurs at $n = \pi/\omega$. This has to be less than the user specified overshoot of ε. That is, $\rho^{\pi/\omega} < \varepsilon$. Simplifying, we obtain the condition for small overshoot in response or small undershoot in error as

$$\rho < \varepsilon^{\omega/\pi} \tag{7.111}$$

The values of ρ and ω satisfying the above requirement for different values of ε are listed in Table 7.3. The desired region is shown shaded in Fig. 7.30.

Table 7.3: Values of ρ and ω satisfying the *small fall* requirement

ε	15	30	45	60	75	90	105	120	135	150	165	180
0.05	0.78	0.61	0.47	0.37	0.28	0.22	0.17	0.14	0.11	0.08	0.06	0.05
0.1	0.83	0.68	0.56	0.46	0.38	0.32	0.26	0.22	0.18	0.15	0.12	0.1

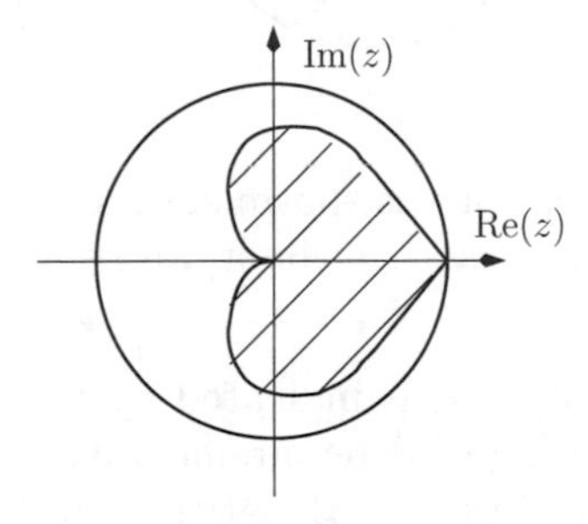

Figure 7.30: Desired location of poles for small undershoot is shown shaded

Although we have referred to the expression given by Eq. 7.111 as approximate, it turns out to be exact, as can be seen in examples of Chapter 9, starting from 9.3 on page 331. This is because ρ^n is a decaying function.

7.7.3 Large Decay Ratio

The time that it takes for a step response of a system to come within a specified band is known as the settling time. One of the objectives of control design is to achieve a small settling time. This will be achieved if the ratio of two successive peaks or troughs in a step response is small. In control parlance, this ratio is known as the decay ratio. In this section, we consider the ratio of a peak and the immediate trough; the reason will become clear shortly.

Suppose that we make the approximation that the first undershoot in $e(n)$ occurs at $\omega n = \pi$ and that the second overshoot occurs at $\omega n = 2\pi$. Then we would want $|e(n)|_{\omega n=2\pi}|/|e(n)|_{\omega n=\pi}|$ to be small. In other words, we would want

$$\frac{\rho^n|_{\omega n=2\pi}}{\rho^n|_{\omega n=\pi}} < \delta$$

where $\delta = 0.5$ and 0.25 roughly work out to one-quarter and one-eighth decay, approximately. Simplifying, we obtain

$$\rho < \delta^{\omega/\pi}$$

This is in the same form as the condition for small undershoot. Normally, as $\varepsilon < \delta$, the small undershoot condition is sufficient to take care of small decay as well.

We now summarize the procedure to select the desired area where the closed loop poles should lie. Choose the angle of the pole location ω, satisfying $\omega > \pi/2N_r$, as given in Eq. 7.110. Here N_r is the rise time in number of samples. This is indicated by the shaded region in Fig. 7.29. Choose the distance of the pole location from the origin, ρ, satisfying $\rho = \varepsilon^{\omega/\pi}$, as given in Eq. 7.111. Here, ε is the overshoot specified as the

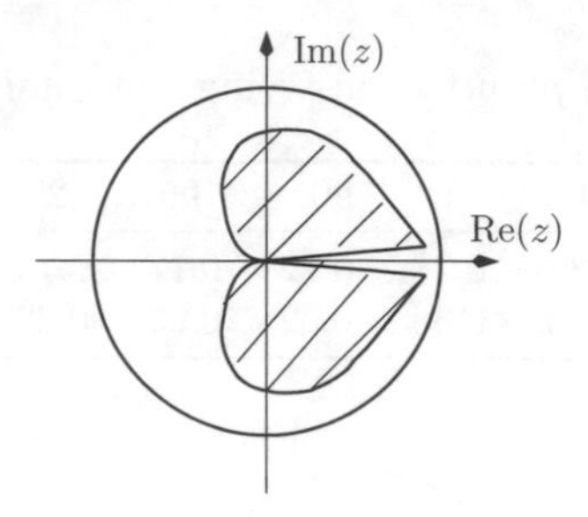

Figure 7.31: Desired location of poles is shown shaded. The poles in the shaded area would result in small rise time, small overshoot, and small settling time.

fraction of the setpoint change. This is indicated by the shaded region in Fig. 7.30. This is likely to satisfy the settling time requirement as well. The intersection of these two shaded regions, as indicated in Fig. 7.31, satisfies all the three requirements stated on page 285.

Using the above approach, we obtain the closed loop characteristic polynomial as

$$\phi_{cl}(z^{-1}) = (1 - \rho e^{j\omega} z^{-1})(1 - \rho e^{-j\omega} z^{-1}) = 1 - \rho z^{-1}(e^{j\omega} + e^{-j\omega}) + \rho^2 z^{-2}$$

Using Euler's formula, this can be simplified as

$$\phi_{cl}(z^{-1}) = 1 - 2\rho \cos\omega \; z^{-1} + \rho^2 z^{-2} \tag{7.112}$$

How do we use this approach to design controllers? Given the procedure detailed above, we can first find ρ and ω and then obtain the expression for error to a step input as $e(n) = \rho^n \cos\omega n$. It is possible to back-calculate the desired transfer function between r and y to be in the form of $z^{-1}\psi(z^{-1})/\phi_{cl}(z^{-1})$. Unfortunately, however, there are a few difficulties in implementing this approach. First of all, there are some restrictions imposed by the nonminimum phase zeros of the plant, to be discussed in detail in Sec. 9.2. In the presence of nonminimum phase zeros of the plant, the zeros introduced by ψ could cause undesirable overshoots, as discussed in Sec. 7.6.1. Finally, we have made approximations to arrive at the desired region. In view of these observations, we will use only the closed loop characteristic polynomial, ϕ_{cl}, to design controllers. A methodology based on this procedure is explained in detail in Chapter 9.

7.8 Solving Aryabhatta's Identity[2]

In several control applications, there is a need to solve polynomial equations. A standard software package for this purpose is available [28]. In this section, we present the approach we have adopted to solve them. We begin with Euclid's algorithm.

7.8.1 Euclid's Algorithm for GCD of Two Polynomials

We will now present Euclid's well known algorithm to compute the greatest common divisor (GCD) of two polynomials. Let the two polynomials be denoted by $D(z)$

[2]This section may be skipped during a first reading.

and $N(z)$. Without loss of generality, let

$$\mathrm{d}D > \mathrm{d}N \tag{7.113}$$

Recall from Footnote 3 on page 68 the notation to refer to the degree of polynomials. Carry out the following steps. Divide $D(z)$ by $N(z)$ and call the reminder $R_1(z)$. Divide $N(z)$ by $R_1(z)$ and obtain $R_2(z)$ as the reminder. Divide $R_1(z)$ by $R_2(z)$ and obtain $R_3(z)$ as the reminder. As we continue this procedure, the degree of the reminder keeps decreasing, eventually going to zero. We arrive at the following relations:

$$D(z) = Q_1(z)N(z) + R_1(z) \tag{7.114a}$$
$$N(z) = Q_2(z)R_1(z) + R_2(z) \tag{7.114b}$$
$$R_1(z) = Q_3(z)R_2(z) + R_3(z) \tag{7.114c}$$
$$\vdots$$
$$R_{k-3}(z) = Q_{k-1}(z)R_{k-2}(z) + R_{k-1}(z) \tag{7.114d}$$
$$R_{k-2}(z) = Q_k(z)R_{k-1}(z) + R_k(z) \tag{7.114e}$$
$$R_{k-1}(z) = Q_{k+1}(z)R_k(z) \tag{7.114f}$$

Note that in the last step, the reminder is taken as zero. We will show in two steps that $R_k(z)$ in Eq. 7.114f is the GCD of $D(z)$ and $N(z)$:

1. We first show that R_k divides both D and N. For this, we start from the bottom.
   - The last equation shows that $R_k(z)$ divides $R_{k-1}(z)$.
   - When this fact is applied to the second last equation, we see that $R_k(z)$ divides $R_{k-2}(z)$.
   - As it divides $R_{k-1}(z)$ and $R_{k-2}(z)$, from the third last equation, we see that $R_k(z)$ divides $R_{k-3}(z)$.
   - Continuing upwards, we see that $R_k(z)$ divides both $D(z)$ and $N(z)$.
2. Next we show that the $R_k(z)$ computed above is the GCD. Suppose that there exists another divisor of $N(z)$ and $D(z)$; call it $S(z)$.
   - From the first equation, we see that S divides R_1, denoted as $S|R_1$.
   - As it divides N and R_1, $S_2|R_2$ – see the second equation.
   - Continuing downwards, we see that $S|R_k$.
   - As an arbitrary divisor $S(z)$ of $D(z)$ and $N(z)$ divides it, $R_k(z)$ is indeed the GCD of $D(z)$ and $N(z)$.

The GCD is unique up to a constant. If we demand that it be chosen monic, *i.e.*, the coefficient of the constant term is one, then the GCD becomes unique.

Now, we will state an important result. If $R(z)$ is the GCD of $D(z)$ and $N(z)$, it is possible to find polynomials $X(z)$ and $Y(z)$ such that

$$X(z)D(z) + Y(z)N(z) = R(z) \tag{7.115}$$

If $\mathrm{d}D(z)$ and $\mathrm{d}N(z) > 0$, we can take it that

$$\mathrm{d}X(z) < \mathrm{d}N(z) \tag{7.116}$$

$$\mathrm{d}Y(z) < \mathrm{d}D(z) \tag{7.117}$$

We will prove this claim now. It is given that $R(z)$ is the same as $R_k(z)$ of Eq. 7.114f. From Eq. 7.114e, we obtain

$$R(z) = R_{k-2}(z) - Q_k R_{k-1}(z)$$

Substituting for $R_{k-1}(z)$ in terms of $R_{k-2}(z)$ and $R_{k-3}(z)$ using Eq. 7.114d, we arrive at

$$R(z) = (1 - Q_k Q_{k-1}) R_{k-2}(z) - Q_k(z) R_{k-3}(z)$$

Continuing this procedure upwards, we arrive at Eq. 7.115. The degree condition can be proved easily, as shown in Sec. 7.8.2.

Two polynomials are said to be *coprime* if they do not have a polynomial of degree one or higher as a common factor. From the above result, it is easy to see that if the polynomials $D(z)$ and $N(z)$ are coprime, there exist polynomials $X(z)$ and $Y(z)$ that satisfy

$$X(z)D(z) + Y(z)N(z) = 1 \tag{7.118}$$

7.8.2 Aryabhatta's Identity

In many control problems, we will be interested in solving the polynomial equation

$$X(z)D(z) + Y(z)N(z) = C(z) \tag{7.119}$$

for X and Y, with D, N and C specified. This equation is variously known as the *Bezout identity*, *Diophantine equation* [24] and *Aryabhatta's identity* [59]. In this book, we will refer to it as Aryabhatta's identity. In this section, we will first present the condition under which there is a solution to Eq. 7.119. Aryabhatta's identity, given by Eq. 7.119, has a solution if and only if the GCD of $D(z)$ and $N(z)$ divides $C(z)$.

We will first prove the *only if* part. Let the GCD of $D(z)$ and $N(z)$ be $R(z)$ and

$$D(z) = D_0(z)R(z)$$

$$N(z) = N_0(z)R(z)$$

Substituting the above in Eq. 7.119, we obtain

$$X(z)D_0(z)R(z) + Y(z)N_0(z)R(z) = C(z)$$

or

$$(XD_0 + YN_0)R = C$$

Thus, we see that $R|C$.

Next, we will show the *if* part. Let the GCD of $D(z)$ and $N(z)$ be $R(z)$ and that $R|C$; that is, $C_0(z)R(z) = C(z)$. Multiplying Eq. 7.115 by $C_0(z)$, we obtain

$$C_0(z)X(z)D(z) + C_0(z)Y(z)N(z) = C_0(z)R(z) = C(z)$$

Comparing with Eq. 7.119, we see that $C_0(z)X(z)$, $C_0(z)Y(z)$ is the required solution.

There are infinitely many solutions to Aryabhatta's identity, given by Eq. 7.119. But we can obtain a *unique solution* under special conditions, as we see now. Suppose that D and N are coprime with respective degrees $\mathrm{d}D > 0$ and $\mathrm{d}N > 0$. If the degree of C satisfies

$$0 \leq \mathrm{d}C < \mathrm{d}D + \mathrm{d}N \tag{7.120}$$

Eq. 7.119 has a unique least degree solution given by

$$\begin{aligned} \mathrm{d}X(z) &< \mathrm{d}N(z) \\ \mathrm{d}Y(z) &< \mathrm{d}D(z) \end{aligned} \tag{7.121}$$

Note that this implies that $\mathrm{d}X$ is at least one less than the degree of the known quantity, in this case N, in the other term, in this case YN. The second inequality in Eq. 7.121 may also be interpreted in a similar way. We will refer to this as the *degree condition* of unique minimum degree solution. Suppose that, as per Eq. 7.118, we determine polynomials $a(z)$ and $b(z)$ satisfying

$$a(z)D(z) + b(z)N(z) = 1 \tag{7.122}$$

We multiply throughout by $C(z)$. If we substitute $X(z)$ and $Y(z)$, respectively for $a(z)C(z)$ and $b(z)C(z)$, we arrive at Eq. 7.119. We will now show the degree condition. Suppose that we solve for X and Y satisfying Eq. 7.119 with $\mathrm{d}X \geq \mathrm{d}N$. We divide X by N to obtain

$$X(z) = Q(z)N(z) + R(z), \quad \mathrm{d}R(z) < \mathrm{d}N \tag{7.123}$$

where Q is the quotient and R is the remainder. Substituting in Eq. 7.119 and dropping the dependence on z, we obtain

$$(QN + R)D + YN = C$$

We rearrange the terms to arrive at

$$RD + (Q + Y)N = C$$

We see that $\mathrm{d}(Q + Y) < \mathrm{d}D$, because $\mathrm{d}C < \mathrm{d}N + \mathrm{d}D$.

7.8.3 Algorithm to Solve Aryabhatta's Identity

In this section, we will present an outline of the approach we have taken to solve Aryabhatta's identity, Eq. 7.119. We will illustrate our approach with the following specific example:

$$X(z)(1 - 5z^{-1} + 4z^{-2}) + Y(z)(z^{-1} + z^{-2}) = 1 - z^{-1} + 0.5z^{-2} \tag{7.124}$$

Eq. 7.119 is equivalent to solving

$$\begin{bmatrix} X(z) & Y(z) \end{bmatrix} \begin{bmatrix} D(z) \\ N(z) \end{bmatrix} = C(z) \tag{7.125}$$

We solve this equation by comparing the coefficients of powers of z^{-1}. Let

$$V(z) = \begin{bmatrix} X(z) & Y(z) \end{bmatrix} \tag{7.126}$$

$$F(z) = \begin{bmatrix} D(z) \\ N(z) \end{bmatrix} \tag{7.127}$$

Let $V(z)$ be of the form

$$V(z) = V_0 + V_1 z^{-1} + \cdots + V_v z^{-v} \tag{7.128}$$

where v is the order of the polynomial V, an unknown at this point. We need to determine v, as well as the coefficients V_k, $v \geq k \geq 0$. Let F and C be of the form

$$F(z) = F_0 + F_1 z^{-1} + \cdots + F_{\mathrm{d}F} z^{-\mathrm{d}F} \tag{7.129}$$

$$C(z) = C_0 + C_1 z^{-1} + \cdots + C_{\mathrm{d}C} z^{-\mathrm{d}C} \tag{7.130}$$

Thus, Eq. 7.125 becomes

$$\begin{aligned} &[V_0 + V_1 + \cdots + V_v z^{-v}][F_0 + F_1 + \cdots + F_{\mathrm{d}F} z^{-\mathrm{d}F}] \\ &\qquad = C_0 + C_1 z^{-1} + \cdots + C_{\mathrm{d}C} z^{-\mathrm{d}C} \end{aligned} \tag{7.131}$$

The resulting equations, in powers of z^{-1}, can be arranged in the following matrix form:

$$\begin{aligned} \begin{bmatrix} V_0 & V_1 & \cdots & V_v \end{bmatrix} \begin{bmatrix} F_0 & F_1 & \cdots & F_{\mathrm{d}F} & 0 & \cdots & 0 \\ 0 & F_0 & F_1 & \cdots & F_{\mathrm{d}F} & \cdots & 0 \\ \vdots & & & & & & \\ 0 & \cdots & 0 & F_0 & F_1 & \cdots & F_{\mathrm{d}F} \end{bmatrix} \\ = \begin{bmatrix} C_0 & C_1 & \cdots & C_{\mathrm{d}C} \end{bmatrix} \end{aligned} \tag{7.132}$$

For example, multiplying these out, the first two equations are obtained as

$$\begin{aligned} V_0 F_0 &= C_0 \\ V_0 F_1 + V_1 F_0 &= C_1 \end{aligned}$$

which agree with Eq. 7.131. The matrix consisting of the elements of F_k has v rows, to be consistent with the matrix of elements of V_l. Thus, solving Aryabhatta's identity boils down to determining v and the elements V_l, $v \geq l \geq 0$. We write Eq. 7.132 as

$$\mathscr{V}\mathscr{F} = \mathscr{C} \tag{7.133}$$

where the new symbols are defined in an obvious way. If $\mathscr{F}$ is right invertible, we can find the solution to Eq. 7.133 as

$$\mathscr{V} = \mathscr{C}\mathscr{F}^{-1} \tag{7.134}$$

For the inverse $\mathscr{F}^{-1}$ to exist, the rows of $\mathscr{F}$ have to be linearly independent. We choose v to be the largest possible integer, satisfying this requirement. Once $\mathscr{V}$ is known, $V(z)$ and then $X(z)$, $Y(z)$ can be determined. We outline this approach with an example.

Example 7.12 Explain the steps involved in solving Aryabhatta's identity, Eq. 7.124.

Comparing Eq. 7.124 with Eq. 7.119, we obtain

$$\begin{aligned} D(z) &= 1 - 5z^{-1} + 4z^{-2} \\ N(z) &= z^{-1} + z^{-2} \end{aligned} \tag{7.135}$$

and hence, using Eq. 7.127,

$$F(z) = \begin{bmatrix} 1 - 5z^{-1} + 4z^{-2} \\ z^{-1} + z^{-2} \end{bmatrix} = \begin{bmatrix} 1 \\ 0 \end{bmatrix} + \begin{bmatrix} -5 \\ 1 \end{bmatrix} z^{-1} + \begin{bmatrix} 4 \\ 1 \end{bmatrix} z^{-2} \tag{7.136}$$

which is of the form $F_0 + F_1 z^{-1} + F_2 z^{-2}$. First we explore the possibility of $v = 1$. We obtain

$$\mathscr{F} = \begin{bmatrix} F_0 & F_1 & F_2 \end{bmatrix} = \begin{bmatrix} 1 & -5 & 4 \\ 0 & 1 & 1 \end{bmatrix} \tag{7.137}$$

Although the rows of $\mathscr{F}$ are linearly independent, v is not the largest integer. In view of this, we explore $v = 2$. We obtain

$$\mathscr{F} = \begin{bmatrix} F_0 & F_1 & F_2 & 0 \\ 0 & F_0 & F_1 & F_2 \end{bmatrix} = \begin{bmatrix} 1 & -5 & 4 & 0 \\ 0 & 1 & 1 & 0 \\ 0 & 1 & -5 & 4 \\ 0 & 0 & 1 & 1 \end{bmatrix}$$

As these rows are independent, we observe that v is still not the maximum. We explore the possibility of $v = 3$ and verify once again. We obtain

$$\mathscr{F} = \begin{bmatrix} F_0 & F_1 & F_2 & 0 & 0 \\ 0 & F_0 & F_1 & F_2 & 0 \\ 0 & 0 & F_0 & F_1 & F_2 \end{bmatrix} = \begin{bmatrix} 1 & -5 & 4 & 0 & 0 \\ 0 & 1 & 1 & 0 & 0 \\ 0 & 1 & -5 & 4 & 0 \\ 0 & 0 & 1 & 1 & 0 \\ 0 & 0 & 1 & -5 & 4 \\ 0 & 0 & 0 & 1 & 1 \end{bmatrix} \tag{7.138}$$

These rows are no longer independent. As a matter of fact, we can verify

$$b\mathscr{F} = 0 \tag{7.139}$$

where b can be chosen as

$$b = \begin{bmatrix} 0 & 0.25 & -0.25 & -1.25 & -0.25 & 1 \end{bmatrix} \tag{7.140}$$

We can also see that while all the rows of $\mathscr{F}$ in Eq. 7.138 form a dependent set, the first five rows are independent. Thus, Eq. 7.134 becomes

$$\mathscr{V} = \begin{bmatrix} 1 & -1 & 0.5 & 0 & 0 \end{bmatrix} \begin{bmatrix} 1 & -5 & 4 & 0 & 0 \\ 0 & 1 & 1 & 0 & 0 \\ 0 & 1 & -5 & 4 & 0 \\ 0 & 0 & 1 & 1 & 0 \\ 0 & 0 & 1 & -5 & 4 \end{bmatrix}^{-1}$$

From the above, we see that $\mathscr{V}$ is a 1×5 vector. To account for the row removed from $\mathscr{F}$, we add a zero to $\mathscr{V}$. We obtain

$$\mathscr{V} = \begin{bmatrix} 1 & 3.25 \,|\, 0.75 & -3 \,|\, 0 & 0 \end{bmatrix}$$

where we have separated coefficients of powers of z^{-1} with vertical lines. That is, we have $V_0 = \begin{bmatrix} 1 & 3.25 \end{bmatrix}$ and $V_1 = \begin{bmatrix} 0.75 & -3 \end{bmatrix}$. Using Eq. 7.128 and Eq. 7.126, we obtain

$$\begin{aligned} X(z) &= 1 + 0.75z^{-1} \\ Y(z) &= 3.25 - 3z^{-1} \end{aligned} \tag{7.141}$$

These form the solution to Aryabhatta's identity, given in Eq. 7.124, which can be verified by substitution. M 7.8 shows how to solve this problem using Matlab. This code calculates a few other variables as well, which will be explained next. ∎

Eq. 7.139 is equivalent to

$$E(z)F(z) = 0 \tag{7.142}$$

which can be written as

$$\begin{bmatrix} -B(z) & A(z) \end{bmatrix} \begin{bmatrix} D(z) \\ N(z) \end{bmatrix} = 0 \tag{7.143}$$

which is equivalent to $B(z)D(z) = A(z)N(z)$ or

$$A^{-1}(z)B(z) = N(z)D^{-1}(z) \tag{7.144}$$

Comparing Eq. 7.139, Eq. 7.140, Eq. 7.142 and Eq. 7.124, we obtain

$$\begin{aligned} B(z) &= 0.25z^{-1} + 0.25z^{-2} \\ A(z) &= 0.25z^{-1} - 1.25z^{-1} + z^{-2} \end{aligned} \tag{7.145}$$

It is easy to verify that the values of A, B given above and $N(z)$, $D(z)$ given in Eq. 7.135 satisfy Eq. 7.144. Of course, when N and D are scalars, it is trivial to obtain A and B. The proposed approach works also when N and D are matrices. The algorithm given above can be used solve a matrix version of Aryabhatta's identity as well, which we will encounter in Chapter 13.

The way we have chosen b in Eq. 7.140 ensures that the coefficients of highest powers of A form a nonsingular matrix, which ensures that A is invertible. Construction of the maximal independent row set out of $\mathscr{F}$ involves a little more book-keeping. These topics are beyond the scope of this book. The interested reader is referred to [24], [9], [8] and [39]. The reader can also go through the Matlab code available with this book for solving polynomial equations.

Recall that there is a unique least degree solution to Aryabhatta's identity when Eq. 7.120 is satisfied. In case this condition is violated, there is no longer a unique solution. Nevertheless, depending on the implementation, it may be possible to find some solution. We have seen an occurrence of this situation in Example 9.9 on page 341.

The solver for Aryabhatta's identity is available at `HOME/matlab/xdync.m`, see Footnote 3 on the following page. In this implementation, we have used Kwakernaak's [27] convention of writing the coefficient matrices in ascending powers of z^{-1}. This convention is nothing but storing a polynomial matrix, given in Eq. 7.136, with its constant matrix equivalent of Eq. 7.137. We conclude this section with two matrix versions of examples.

Example 7.13 Starting with a left factorization N and D given below, find a right coprime factorization satisfying Eq. 7.144.

$$N(z) = \begin{bmatrix} 1 & 0 & 0 \\ 0 & 1 & 0 \\ 0 & 0 & 1 \end{bmatrix}, \quad D(z) = \begin{bmatrix} 1 & 0 & 0 \\ z^{-1} & 1 & 0 \\ z^{-1} & z^{-1} & 1 \end{bmatrix} \tag{7.146}$$

M 7.9 carries out the indicated operations. Notice that there are only four input arguments to the function-call `left_prm`: the polynomial and the degree, for both D and N. With the help of this code, we obtain left coprime factorization B and A satisfying Eq. 7.144:

$$B(z) = \begin{bmatrix} 1 & 0 & 0 \\ -z^{-1} & 1 & 0 \\ z^{-1} - z^{-2} & z^{-1} & -1 \end{bmatrix}, \quad A(z) = \begin{bmatrix} 1 & 0 & 0 \\ 0 & 1 & 0 \\ 0 & 0 & -1 \end{bmatrix}$$

∎

Example 7.14 Solve Aryabhatta's identity, Eq. 7.119, when

$$N = \begin{bmatrix} 0 & 4 + z^{-1} \\ -1 & 3 + 3z^{-1} \end{bmatrix}, \quad D = \begin{bmatrix} z^{-1} & 4z^{-1} + z^{-2} \\ -z^{-1} & 0 \end{bmatrix}$$

$$C = \begin{bmatrix} 1 + z^{-1} & z^{-1} \\ 0 & 2 + z^{-1} \end{bmatrix}$$

This is solved by invoking M 7.10. We obtain the following result:

$$Y = \begin{bmatrix} 2 & -1 - 0.5z^{-1} \\ 0.5 & -0.125z^{-1} \end{bmatrix}, \quad X = \begin{bmatrix} 1.5 & 1 \\ 0.375 & 0.5 \end{bmatrix}$$

∎

7.9 Matlab Code

Matlab Code 7.1 Procedure to draw root locus in Matlab for the problem discussed in Example 7.1 on page 247. This code is available at HOME/specs/matlab/rlocus_ex1.m[3]

```
1 H = tf(1,[1 -1 0],-1);
2 rlocus(H)
```

Matlab Code 7.2 Procedure to draw the Nyquist plot, as discussed in Example 7.2 on page 250. This code is available at HOME/specs/matlab/nyquist_ex1.m

```
1 H = tf(1,[1 -1 0],-1);
2 nyquist(H)
```

Matlab Code 7.3 Procedure to draw Bode plots in Fig. 7.11 on page 255. This code is available at HOME/freq/matlab/lead_exp.m

```
1 pol1 = [1 -0.9]; pol2 = [1 -0.8];
2 G1 = tf(pol1,[1 0],-1);
3 G2 = tf(pol2,[1 0],-1);
4 w = linspace(0.001,pi,1000);
5 bode(G1,'-',G2,'--',w), grid
6 figure
7 G = tf(pol1,pol2,-1);
8 bode(G,w), grid
```

Matlab Code 7.4 A procedure to design lead controllers, as explained in Fig. 7.12 on page 257. This code is available at HOME/freq/matlab/lead_lag.m

```
1 w = linspace(0.001,pi,1000);
2 a = linspace(0.001,0.999,100);
3 lena = length(a);
4 omega = []; lead = [];
5 for i = 1:lena,
6     zero = a(i);
7     pole = 0.9*zero;
8     sys = tf([1 -zero],[1 -pole],-1);
9     [mag,phase] = bode(sys,w);
10    [y,j] = max(phase);
11    omega = [omega w(j)];
12    lead = [lead y];
13    comega = (pole+zero)/(pole*zero+1);
14    clead = zero-pole;
15    clead1 = sqrt((1-zero^2)*(1-pole^2));
16    clead = clead/clead1;
17 %     [w(j) acos(comega) y atan(clead)*180/pi]
```

[3] HOME stands for http://www.moudgalya.org/dc/ – first see the software installation directions, given in Appendix A.2.

```
18 end
19 subplot(2,1,1), plot(lead,omega)
20 ylabel('Frequency, in radians'), grid
21 pause
22 subplot(2,1,2), plot(lead,a)
23 ylabel('Zero location')
24 xlabel('Lead generated, in degrees'), grid
```

Matlab Code 7.5 Bode plot of a lead controller, as shown in Fig. 7.13 on page 257. This code is available at HOME/freq/matlab/lead_vfy.m

```
1 w = linspace(0.001,pi,1000);
2 G = tf([1 -0.8],[1 -0.24],-1);
3 bode(G,w), grid
```

Matlab Code 7.6 Verification of performance of lead controller on antenna system, as discussed in Example 7.3. This code is available at HOME/specs/matlab/ant_lead.m

```
1 % continuous time antenna model
2 a = 0.1;
3 F = [0 1;0 -a]; g = [0; a]; c = [1 0]; d = 0;
4 Ga = ss(F,g,c,d); [num,den] = tfdata(Ga,'v');
5 Ts = 0.2;
6 G = c2d(Ga,Ts);
7 
8 % lead controller
9 beta = 0.8;
10 N = [1 -0.9802]*(1-beta)/(1-0.9802); Rc = [1 -beta];
11 
12 % simulation parameters using g_s_cl2.mdl
13 gamma = 1; Sc = 1; Tc = 1; C = 0; D = 1;
14 st = 1; st1 = 0;
15 t_init = 0; t_final = 20;
```

Matlab Code 7.7 Illustration of system type, as explained in Example 7.10 on page 275. This code is available at HOME/specs/matlab/type_test.m

```
1 % Plant
2 B = 1; A = [1 -1]; zk = [0 1]; Ts = 1;
3 
4 % Specify closed loop characteristic polynomial
5 phi = [1 -0.5];
6 
7 % Design the controller
8 reject_ramps = 1;
9 if reject_ramps == 1,
10    Delta = [1 -1]; % to reject ramps another Delta
11 else
12    Delta = 1; % steps can be rejected by plant itself
```

```
13 end
14 [Rc,Sc] = pp_pid(B,A,k,phi,Delta);
15
16 % parameters for simulation using stb_disc.mdl
17 Tc = Sc; gamma = 1; N = 1;
18 C = 0; D = 1; N_var = 0;
19 st = 1; t_init = 0; t_final = 20;
20 open_system('stb_disc.mdl')
```

Matlab Code 7.8 Solution to Aryabhatta's identity, presented in Example 7.12 on page 293. This code is available at HOME/specs/matlab/abex.m

```
1 N = conv([0 1],[1 1]);
2 D = conv([1 -4],[1 -1]);
3 dN = 2; dD = 2;
4 C = [1 -1 0.5];
5 dC = 2;
6 [Y,dY,X,dX,B,dB,A,dA] = xdync(N,dN,D,dD,C,dC)
```

Matlab Code 7.9 Left coprime factorization, as discussed in Example 7.13 on page 295. This code is available at HOME/specs/matlab/data01.m

```
1 D = [
2 1 0 0 0 0 0
3 0 1 0 1 0 0
4 0 0 1 1 1 0]
5 N = [
6 1 0 0
7 0 1 0
8 0 0 1]
9 dD = 1
10 dN = 0
11 [B,dB,A,dA] = left_prm(N,dN,D,dD)
```

Matlab Code 7.10 Solution to polynomial equation, as discussed in Example 7.14 on page 295. This code is available at HOME/specs/matlab/data05.m

```
1 N = [0    4 0 1
2        -1 8 0 3]
3 dN = 1
4 D = [0 0   1 4 0 1
5        0 0 -1 0 0 0]
6 dD = 2
7 C = [1 0 1 1
8      0 2 0 1]
9 dC = 1
10 [Y,dY,X,dX,B,dB,A,dA] = xdync(N,dN,D,dD,C,dC)
```

7.10 Problems

7.1. If the system of Example 7.1 on page 247 is operated in a closed loop, with a proportional controller of gain of K_u, what will be the frequency of oscillation for impulse and step inputs?

7.2. In Example 7.2 on page 250, we designed a proportional controller for the plant $H = 1/[z(z-1)]$ using a *particular* contour C_1 that encircled *both* the poles. In this problem you indent the contour in such a way that the pole at 1 is *excluded*, as in Fig. 7.32. Using the Nyquist plot for this C_1, determine the

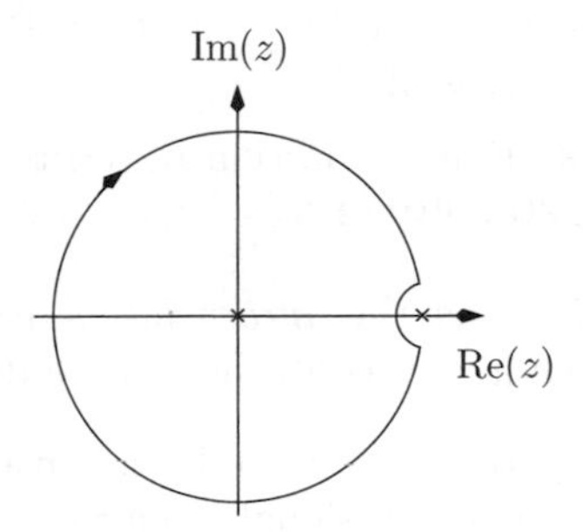

Figure 7.32: An alternative contour C_1

range of proportional controller for which the closed loop system is stable.

7.3. Is it acceptable to cancel the factor $z - a$ when the following system is Z-transformed?

$$\begin{aligned} x(k+1) - ax(k) &= u(k+1) - au(k) \\ y(k) &= x(k) \end{aligned}$$

[Hint: Show that the solution to this system is given by $y(k) = a^k x(0) + u(k)$.]

7.4. This problem is concerned with the demonstration that internal stability of the closed loop system shown below can be expressed in terms of a few conditions on the sensitivity function (to be described below).

(a) Write down the 2×2 transfer function between $\begin{bmatrix} r & d \end{bmatrix}^T$ (treated as input) and $\begin{bmatrix} e & u \end{bmatrix}^T$ (treated as output) in the following figure. [Hint: See Eq. 7.50 on page 267 in the text.]

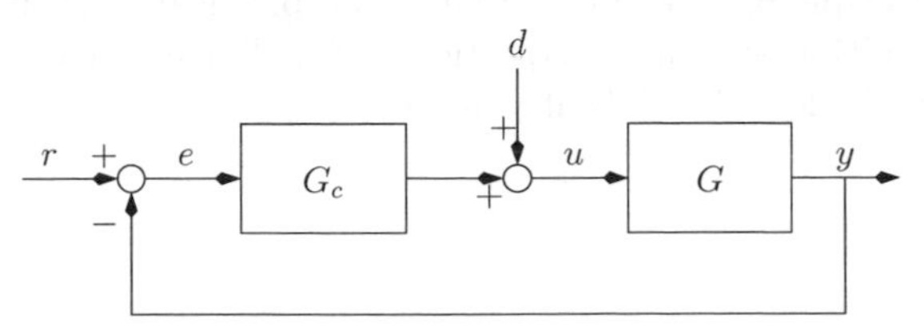

(b) State the internal stability condition, in terms of the stability of the entries of the 2×2 transfer function matrix, obtained in part (a).

(c) State the internal stability condition in terms of the unstable pole–zero cancellation and external stability of the closed loop system.

(d) Let the sensitivity function S be defined as

$$S = \frac{1}{1 + GG_c}$$

Show that the unstable poles of G will not be cancelled by the zeros of G_c, if $S(p_i) = 0$, $n \geq i \geq 1$, where p_i are the unstable poles of G.

(e) Show that the nonminimum phase zeros of $G(z)$ will not be cancelled by the unstable poles of G_c, if $S(z_j) = 1$, $m \geq j \geq 1$, where z_j are the nonminimum phase zeros of G.

(f) State the internal stability condition in terms of conditions on S, derived in (d) and (e), and stability of S.

By the way, (f) is known as the *interpolation condition* on the sensitivity function, the use of which allows controllers to be designed [15].

7.5. Consider a closed loop system that includes a controller so that the error to a unit step change in the reference signal behaves as in Eq. 7.109 on page 285. Show that if the steady state error of this system to a ramp of unit slope has to be less than a small value, say μ, the following relation should be satisfied:

$$T_s \frac{1 - \rho\cos\omega}{1 - 2\rho\cos\omega + \rho^2} < \mu \tag{7.147}$$

7.6. Find the minimum degree polynomials x and y in the following equation:

$$xD + yN = C$$

where

$$\begin{aligned} D &= 1 - z^{-1} + 2z^{-2} \\ N &= 1 + z^{-1} + z^{-2} \\ C &= 2 + 2z^{-1} - 3z^{-2} + 4z^{-3} \end{aligned}$$

by the following two methods:

(a) Using the routine `xdync.m`.

(b) Multiplying the polynomials, comparing the coefficients and solving the resulting system of equations by a linear solver – you can use the one available in Matlab, if you wish.

Chapter 8

Proportional, Integral, Derivative Controllers

Proportional, integral and derivative controllers are the most popular in industrial practice. In Sec. 7.3.2, we looked at the effect of using each of the three modes present in these controllers. But we did not discuss how much of each of these modes should be present in the implementation. This is decided by the tuning rule of the PID controllers. There are many rules to tune the PID controllers. All the tuning rules, however, are based on continuous time analysis. In view of this, to understand this chapter, a background on analog control is required.

Because all the PID tuning rules are in the continuous time domain and because this is the most popular controller, the PID implementations are achieved by discretizing the analog counterpart. In view of this, we revisit the topic of sampling rate selection. In Sec. 2.5, we presented the procedure to arrive at the ZOH equivalent discrete time system. In this chapter, we present a few other discretization techniques.

We present different ways of implementing the PID controllers. We also present discretized versions of the PID controllers. We eschew the topic of tuning the continuous time PID controllers and present only one popular technique of tuning them.

8.1 Sampling Revisited

In Sec. 5.3, we determined the minimum rate at which continuous time functions have to be sampled so as to not lose any information. The minimum rate should be greater than twice the highest frequency component present in the continuous time signal. Unfortunately, this sampling rate is not useful in real time applications, because the reconstruction procedure is not causal.

In reality, we need to sample much faster than the minimum rate suggested by the sampling theorem. In Sec. 5.3.3, we have seen that the reconstruction procedure using ZOH gets better if we increase the sampling rate. Unfortunately, however, we don't know the sampling rate that should be used. As a matter of fact, there is no rigorous answer to this question.

If we use a small sampling rate, we may lose some crucial information. If we use a large sampling rate, we will end up increasing the computational load. Based on

Digital Control Kannan M. Moudgalya

the experience, many different ad hoc rules have been proposed for sampling rate selection. We list them now:

- Use ten times the rate suggested by Shannon's sampling theorem.
- Make sure that there are four to ten samples per rise time of the closed loop system.
- Make sure that there are 15 to 45 samples per period of the oscillating closed loop system.
- Sampling frequency should be 10 to 30 times the bandwidth.
- Choose $\omega_c T_s$ to be 0.15 to 0.5, where ω_c is the crossover frequency and T_s is the sampling time.
- Choose the sampling time in such a manner that the decrease in phase margin of the discretized system is not more than 5° to 15° degrees of the margin in the continuous time system.

Whatever sampling rate is used, it has to be validated through simulations. The efficacy of the discrete time controller with ZOH should be tested on the continuous time model of the plant. Note that validation of the controller on the discrete time model of the plant is not sufficient in sampled data systems.

8.2 Discretization Techniques

We presented the method of ZOH equivalent discretization in Sec. 2.5.2. This approach gives discrete time models that are exact at sampling instants. We also presented a simple approximation technique in Sec. 2.5.3. These methods gave rise to discrete time state space models. There are also methods for converting the continuous time transfer functions into discrete time equivalents. We present some of these techniques in this section. This section assumes prior knowledge of Laplace transforms. These approximation methods can be classified into three broad categories: area based approximation, equivalence in response to inputs, such as step and ramp, and pole–zero mapping. We restrict ourselves to area based approximation and step response equivalent techniques.

8.2.1 Area Based Approximation

Recall that we derived the transfer function of the discrete time integrator in Eq. 7.39 on page 261. As the output y is the integral of the input u, and as $1/s$ is the transfer function in the Laplace domain, we obtain the following correspondence between the Laplace domain variable s and the Z-domain variable z:

$$\frac{1}{s} \leftrightarrow \frac{T_s}{2}\frac{z+1}{z-1} = \frac{T_s}{2}\frac{1+z^{-1}}{1-z^{-1}} \tag{8.1}$$

This result says that one way to discretize a continuous transfer function is to replace all occurrences of $1/s$ by the right-hand side of Eq. 8.1. In Sec. 7.3.2, we explained why we need to consider other mapping approaches. Using the backward difference approximation of Eq. 7.43 on page 262, we obtain

$$\frac{1}{s} \leftrightarrow T_s \frac{z}{z-1} = \frac{T_s}{1-z^{-1}} \tag{8.2}$$

Using the forward difference approximation of Eq. 7.45 on page 262, we obtain

$$\frac{1}{s} \leftrightarrow \frac{T_s}{z-1} = T_s \frac{z^{-1}}{1-z^{-1}} \tag{8.3}$$

8.2.2 Step Response Equivalence Approximation

We will now present a four step method to determine the ZOH equivalent Z-transform transfer function of a Laplace domain transfer function of a continuous system:

1. Determine the step response of the continuous transfer function $y_s(t)$.
2. Discretize the step response to arrive at $y_s(nT_s)$.
3. Z-transform the step response to obtain $Y_s(z)$.
4. Divide the function obtained in the above step by the Z-transform of a step input, namely $z/(z-1)$.

To summarize, if $G(s)$ is the continuous domain transfer function of a system, its discrete time transfer function $G(z)$ is given by

$$G(z) = \frac{z-1}{z} Z\left[\mathcal{L}^{-1}\frac{G(s)}{s}\right] \tag{8.4}$$

We will now present a few examples to illustrate this idea.

Example 8.1 Find the ZOH equivalent of $1/s$.

The step response of $1/s$ is given by $1/s^2$. In the time domain, it is given by

$$y_s(t) = \mathcal{L}^{-1}\frac{1}{s^2} = t$$

Sampling it with a period of T_s, $y_s(nT_s) = nT_s$. Taking Z-transforms and using Problem 4.9,

$$Y_s(z) = \frac{T_s z}{(z-1)^2}$$

Dividing by $z/(z-1)$, we obtain the ZOH equivalent discrete time transfer function $G(z)$ as

$$G(z) = \frac{T_s}{z-1} = T_s \frac{z^{-1}}{1-z^{-1}}$$

When written as a function of z^{-1}, the presence of delay in the transfer function becomes clear. This fact has been used to arrive at the realizability condition in Sec. 7.4.3.

∎

Example 8.2 Find the ZOH equivalent of $1/s^2$.

The step response of $1/s^2$ is given by $1/s^3$. In the time domain, it is given by

$$y_s(t) = \mathcal{L}^{-1}\frac{1}{s^3} = \frac{1}{2}t^2$$

Sampling it with a period of T_s,

$$y_s(nT_s) = \frac{1}{2}n^2T_s^2$$

It is easy to check (see Problem 4.9) that

$$Y_s(z) = \frac{T_s^2 z(z+1)}{2(z-1)^3}$$

Dividing by $z/(z-1)$, we obtain

$$G(z) = \frac{T_s^2(z+1)}{2(z-1)^2} = \frac{T_s^2}{2}\frac{(1+z^{-1})z^{-1}}{(1-z^{-1})^2}$$

This example also shows that there is a delay that is introduced due to sampling.

∎

Example 8.3 Find the ZOH equivalent of $K/(\tau s + 1)$.

$$Y_s(s) = \frac{1}{s}\frac{K}{\tau s + 1} = K\left[\frac{1}{s} - \frac{1}{s + \frac{1}{\tau}}\right]$$

Inverting this,

$$y_s(t) = K\left[1 - e^{-t/\tau}\right], \quad t > 0$$

Sampling this, we obtain

$$y_s(n) = K\left[1 - e^{-nT_s/\tau}\right]1(n) = K\left[1(n) - e^{-nT_s/\tau}1(n)\right]$$

The Z-transform of $1(n)$ is $z/(z-1)$. Using Example 4.5 on page 71, we obtain

$$Y_s(z) = K\left[\frac{z}{z-1} - \frac{z}{z - e^{-T_s/\tau}}\right] = \frac{Kz(1 - e^{-T_s/\tau})}{(z-1)(z - e^{-T_s/\tau})}$$

Dividing by $z/(z-1)$, we obtain

$$G(z) = \frac{K(1 - e^{-T_s/\tau})}{z - e^{-T_s/\tau}}$$

∎

Matlab can be used to find the ZOH equivalent transfer functions. We illustrate this with an example.

Example 8.4 Find the ZOH equivalent transfer function of $10/(5s+1)$ obtained with sampling period $T_s = 0.5$ s.

Using M 8.1, we see that

$$G(z) = \frac{0.9546}{z - 0.9048} = \frac{10(1 - e^{-0.1})}{z - e^{-0.1}}$$

which is in agreement with the above example. ∎

Example 8.5 Calculate the step response discrete time equivalent of the Laplace domain transfer function

$$G(s) = \frac{e^{-Ds}}{\tau s + 1} \tag{8.5}$$

where $0 < D < T_s$ and T_s is the sampling time.

We cannot work with the above form directly because

$$e^{-Ds} = \frac{1}{e^{Ds}} = \frac{1}{1 + Ds + Ds^2 + \cdots}$$

will contribute to an infinite number of poles, making the problem difficult. In view of this, we first write $G(s) = e^{-T_s s} G_1(s)$, where

$$G_1(s) = \frac{e^{(T_s - D)s}}{\tau s + 1}$$

We apply a continuous unit step input to $G_1(s)$ and write down the partial fraction expansion of the resulting output, call it $Y_1(s)$, to arrive at

$$Y_1(s) = \frac{e^{(T_s - D)s}}{\tau s + 1}\frac{1}{s} = \frac{A}{s} + \frac{B}{\tau s + 1}$$

Multiplying by s and letting $s = 0$, we obtain $A = 1$. Multiplying by $\tau s + 1$ and letting $s = -1/\tau$, we obtain

$$B = \left.\frac{e^{(T_s - D)s}}{s}\right|_{s=-1/\tau} = -\tau e^{-(T_s - D)/\tau}$$

Thus we arrive at the partial fraction expansion for $Y_1(s)$ as

$$Y_1(s) = \frac{1}{s} - \frac{e^{-(T_s - D)/\tau}}{s + 1/\tau}$$

Inverting this, we obtain

$$y_1(t) = \left(1 - e^{-(T_s - D)/\tau} e^{-t/\tau}\right) 1(t)$$

where $1(t)$ is the continuous time unit step function. Discretizing it, we obtain

$$y_1(nT_s) = \left(1 - e^{-(T_s - D)/\tau} e^{-nT_s/\tau}\right) 1(n)$$

Z-transforming it, we obtain

$$\begin{aligned} Y_1(z) &= \frac{z}{z-1} - e^{-(T_s - D)/\tau} \frac{z}{z - e^{-T_s/\tau}} \\ &= z \frac{z\left[1 - e^{-(T_s - D)/\tau}\right] + \left[e^{-(T_s - D)/\tau} - e^{-T_s/\tau}\right]}{(z-1)(z - e^{-T_s/\tau})} \end{aligned}$$

As this is the response due to a unit step function, we divide by $z/(z-1)$ to arrive at the transfer function

$$G_1(z) = \frac{z\left[1 - e^{-(T_s - D)/\tau}\right] + \left[e^{-(T_s - D)/\tau} - e^{-T_s/\tau}\right]}{z - e^{-T_s/\tau}}$$

To account for the $e^{-T_s s}$ that we pulled out, we need to multiply $G_1(z)$ by z^{-1} to obtain

$$G(z) = \frac{G_1(z)}{z} = \frac{z\left[1 - e^{-(T_s - D)/\tau}\right] + \left[e^{-(T_s - D)/\tau} - e^{-T_s/\tau}\right]}{z(z - e^{-T_s/\tau})} \tag{8.6}$$

∎

In the above example, we had assumed D to be less than T_s. For a general D, *i.e.*, $D = kT_s + D'$, where k is an integer, $k > 0$, the discrete time equivalent of Eq. 8.5 is

$$G(z) = \frac{z\left[1 - e^{-(T_s - D')/\tau}\right] + \left[e^{-(T_s - D')/\tau} - e^{-T_s/\tau}\right]}{z^{k+1}(z - e^{-T_s/\tau})} \tag{8.7}$$

We will next solve the problem discussed in Example 8.5 through state space techniques.

Example 8.6 Using the state space approach, discretize the following transfer function

$$G(s) = \frac{e^{-Ds}}{\tau s + 1}$$

where $0 < D < T_s$ and T_s is the sampling time.

It is easy to check that the following state space equations result:

$$\begin{aligned} \dot{x}(t) &= -\frac{1}{\tau} x(t) + \frac{1}{\tau} u(t - D) \\ y(t) &= x(t) \end{aligned}$$

The discrete time equivalent of the state equation has been obtained in Example 2.5. In particular, we arrive at

$$\begin{aligned} x(n+1) &= Ax(n) + Bu(n) \\ y(n) &= Cx(n) + Du(n) \end{aligned}$$

with

$$A = \begin{bmatrix} \Phi & B_1 \\ 0 & 0 \end{bmatrix}, \quad B = \begin{bmatrix} B_0 \\ 1 \end{bmatrix}, \quad C = \begin{bmatrix} 1 & 0 \end{bmatrix}, \quad D = 0$$

Taking the Z-transform, we obtain the transfer function as

$$G(z) = \frac{B_0 z + B_1}{z(z - \Phi)}$$

Substituting the values of Φ, B_0 and B_1 from Eq. 2.54 on page 28, we once again obtain Eq. 8.7. Thus both the transfer function and state space approaches give rise to identical results.

∎

We have seen several approximations to continuous time transfer functions: step response or ZOH equivalent, trapezoidal or Tustin or bilinear equivalent, forward difference approximation and backward difference approximation. There are several other ways to approximate the continuous time transfer functions, the popular ones being *impulse response approximation* and *ramp response approximation*, with the former being popular in DSP. There is no theory that categorically states the preferred order of these approximations. In control applications, the ZOH or step response and Tustin approximations are most preferred, although backward and forward difference approximations are also used to approximate continuous time transfer functions.

In this section, we discretized several continuous time transfer functions. The inverse problem of determining the continuous time transfer functions from discrete time transfer functions is a more difficult one. Only under special conditions is it possible to obtain a unique solution to this problem. Problem 8.3 presents one such instance. This inverse problem is beyond the scope of this book and hence will not be considered further.

8.3 Discretization of PID Controllers

As mentioned earlier, PID controllers are the most popular form of controllers used in industry. These controllers have proportional, integral and derivative modes. Arriving at the extent of presence of these modes by trial and error is known as tuning. Most tuning techniques for PID controllers are in the continuous time domain. In this section, we discuss a few tuning methods and structures of PID controllers, and some implementation issues.

8.3.1 Basic Design

Let the input to the controller be $e(t)$ and the output from it be $u(t)$. The PID controller in continuous time is given by

$$u(t) = K\left[e(t) + \frac{1}{\tau_i}\int_0^t e(t)dt + \tau_d\frac{de(t)}{dt}\right] \tag{8.8}$$

where K is the gain, τ_i is the integral time and τ_d is the derivative time or lead time. On taking Laplace transforms, we obtain the continuous controller to be

$$u(t) = K\left(1 + \frac{1}{\tau_i s} + \tau_d s\right)e(t) \tag{8.9}$$

where we have used the mixed notation, as in the discrete time control. The above equation can be written as

$$u(t) = \frac{S_c(s)}{R_c(s)}e(t) \tag{8.10}$$

where

$$\begin{aligned} S_c(s) &= \tau_i s + 1 + \tau_i\tau_d s^2 \\ R_c(s) &= \tau_i s \end{aligned} \tag{8.11}$$

From the above equation, we arrive at the important property of controllers with integral modes:

$$R_c(0) = 0 \tag{8.12}$$

In the above equations, the derivative mode is difficult to implement. It is made implementable by converting into a lead term. The modified PID control law is given by

$$u(t) = K\left(1 + \frac{1}{\tau_i s} + \frac{\tau_d s}{1 + \tau_d s/N}\right)e(t) \tag{8.13}$$

where N is a large number, of the order of 100. It is easy to see that at low frequencies, the last term behaves like a derivative, becoming equal to N at large frequencies. For all practical purposes, the number of tuning parameters goes up to four with the introduction of N. It is important to note that if this controller is written in fractional form S_c/R_c, as above, Eq. 8.12 will still be satisfied. It is easy to verify that Eq. 8.13 reduces to Eq. 8.9 in the limiting case of $N \to \infty$. This PID control law is said to be in *filtered form*.

8.3.2 Ziegler–Nichols Method of Tuning

There are a large number of methods of tuning a PID controller. The most popular ones amongst them are the *reaction curve method* and *instability method*. Both are referred to as the *Ziegler–Nichols tuning* method.

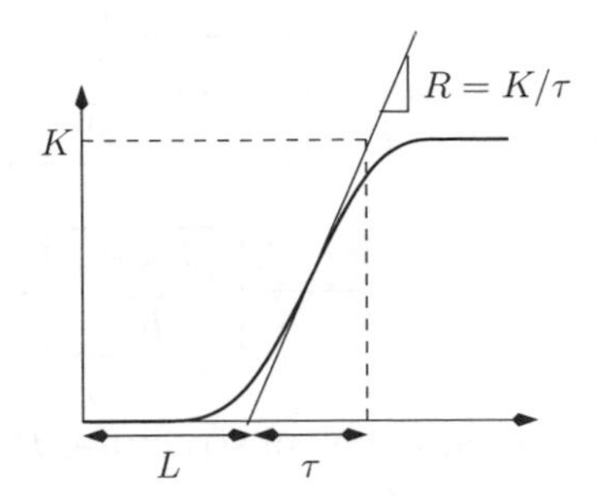

Figure 8.1: Reaction curve method of Ziegler–Nichols tuning

Table 8.1: Ziegler–Nichols settings using reaction curve method

	K	τ_i	τ_d
P	$1/RL$		
PI	$0.9/RL$	$3L$	
PID	$1.2/RL$	$2L$	$0.5L$

Table 8.2: Ziegler–Nichols settings using instability method

	K	τ_i	τ_d
P	$0.5K_u$		
PI	$0.45K_u$	$P_u/1.2$	
PID	$0.6K_u$	$P_u/2$	$P_u/8$

We will first present the reaction curve method. Give a unit step input to a stable system (see Fig. 8.1) and obtain the time lag after which the system starts responding (L), the steady state gain (K) and the time the output takes to reach the steady state, after it starts responding (τ). This method is known as the *transient response method* or reaction curve method.

We will now summarize the instability Ziegler–Nichols tuning method. Put the system in a feedback loop with a proportional controller, whose gain is increased until the closed loop system becomes unstable. At the verge of instability, note down the gain of the controller (K_u) and the period of oscillation (P_u). Then the PID settings are as given in Table 8.2. This is known as the *instability* method of Ziegler–Nichols.

These tuning methods are based on the specifications imposed on model transfer functions. The reader may refer to [45, p. 683] for more details.

Ziegler–Nichols methods are the most popular tuning methods. Although there are several other conventional methods of tuning the PID controllers, we will not devote any more time to this topic. Nevertheless, several tuning methods that have their origin in the discrete time domain will be presented throughout this book.

8.3.3 2-DOF Controller with Integral Action at Steady State

We have seen the advantages of the 2-DOF control structure in Sec. 7.1.3. In view of this, PID controllers are often implemented in the 2-DOF framework. There are many

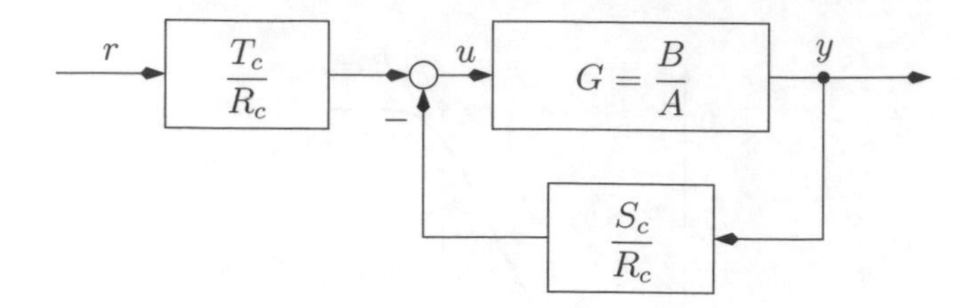

Figure 8.2: A 2-DOF feedback control structure. B includes time delay.

different 2-DOF implementations of PID controllers, some of which will be presented in this chapter. A necessary condition that these implementations have to satisfy is derived in this section. The presentation is general in the sense that this development is applicable to any controller with an integral action.

Consider the 2-DOF control structure presented in Fig. 8.2. The controller transfer functions S_c/R_c and T_c/R_c are assumed to have the integral mode. Because this structure is applicable to both continuous time and discrete time systems, the dependence on s or z is omitted. The term B in this figure includes time delay. The control law implemented in Fig. 8.2 is

$$u = \frac{T_c}{R_c} r - \frac{S_c}{R_c} y \tag{8.14}$$

It is easy to arrive at the following relation between r and y:

$$y = \frac{T_c}{R_c} \frac{B/A}{1 + BS_c/AR_c} r = \frac{BT_c}{AR_c + BS_c} r$$

The error between the reference signal and the actual value is given by

$$e = r - y = \left(1 - \frac{BT_c}{AR_c + BS_c}\right) r$$

Simplifying, we arrive at

$$e = \frac{AR_c + BS_c - BT_c}{AR_c + BS_c} r \tag{8.15}$$

We will now use this expression to arrive at a necessary condition that the 2-DOF controllers have to satisfy at steady state. We restrict our attention to step inputs in the reference signal, r.

Continuous time systems: The continuous time form of the error expression in Eq. 8.15 is given by

$$E(s) = \frac{A(s)R_c(s) + B(s)S_c(s) - B(s)T_c(s)}{A(s)R_c(s) + B(s)S_c(s)} R(s) \tag{8.16}$$

where $E(s)$ and $R(s)$ are, respectively, the Laplace transforms of $e(t)$ and $r(t)$. Using the final value theorem for continuous time systems, we obtain

$$\lim_{t \to \infty} e(t) = \lim_{s \to 0} sE(s) \tag{8.17}$$

Substituting the expression for $E(s)$ from Eq. 8.16, the above equation becomes

$$\lim_{t \to \infty} e(t) = \lim_{s \to 0} s \frac{A(s)R_c(s) + B(s)S_c(s) - B(s)T_c(s)}{A(s)R_c(s) + B(s)S_c(s)} \frac{1}{s}$$

where we have made use of the fact that $R(s)$ is the transfer function of unit step. Because the controller has an integral action, $R_c(0) = 0$, as in Eq. 8.12. Using this, the above equation reduces to

$$e(\infty) = \left. \frac{S_c(s) - T_c(s)}{S_c(s)} \right|_{s=0} = \frac{S_c(0) - T_c(0)}{S_c(0)} \tag{8.18}$$

The above condition can be satisfied if one of the following conditions is met:

$$\begin{aligned} T_c &= S_c \\ T_c &= S_c(0) \\ T_c(0) &= S_c(0) \end{aligned} \tag{8.19}$$

We have assumed that $S_c(0)$ is nonzero while arriving at the above conditions. We will make use of these to arrive at 2-DOF continuous time PID controllers in subsequent sections.

Discrete time systems: The discrete time form of the error expression in Eq. 8.15 is given by

$$E(z) = \frac{A(z)R_c(z) + B(z)S_c(z) - B(z)T_c(z)}{A(z)R_c(z) + B(z)S_c(z)} R(z) \tag{8.20}$$

where $E(z)$ and $R(z)$ are, respectively, the Z-transforms of $e(n)$ and $r(n)$. Using the final value theorem for discrete time systems, we obtain

$$\lim_{n \to \infty} e(n) = \lim_{z \to 1} \frac{z-1}{z} E(z) \tag{8.21}$$

Substituting the expression for $E(z)$ from Eq. 8.20, the above equation becomes

$$\lim_{n \to \infty} e(n) = \lim_{z \to 1} \frac{z-1}{z} \frac{A(z)R_c(z) + B(z)S_c(z) - B(z)T_c(z)}{A(z)R_c(z) + B(z)S_c(z)} \frac{z}{z-1}$$

where we have made use of the fact that $R(z)$ is the transfer function of unit step. Because the controller has an integral action, $R_c(1) = 0$. Using this, the above equation reduces to

$$e(\infty) = \left. \frac{S_c(z) - T_c(z)}{S_c(z)} \right|_{z=1} = \frac{S_c(1) - T_c(1)}{S_c(1)} \tag{8.22}$$

The above condition can be satisfied if one of the following conditions is met:

$$\begin{aligned} S_c &= T_c \\ S_c &= T_c(1) \\ S_c(1) &= T_c(1) \end{aligned} \tag{8.23}$$

We have assumed that $S_c(1)$ is nonzero. If any of these conditions is satisfied, Fig. 8.2 is reduced, at steady state, to a 1-DOF control structure, as in Fig. 7.2 on page 244. Because R_c has an integral term, by the internal model principle, steps in the reference signal are rejected, verifying the offset free tracking condition derived above.

It should be remembered that although we have used the same symbols, the actual functional forms of R_c, S_c and T_c above will be different from those in the continuous time case. This is explained in detail in the subsequent sections.

8.3.4 Bumpless PID Controller with $T_c = S_c$

Using the relations given by Eq. 8.1 and Eq. 8.2, we can map the controller given by Eq. 8.9 to the discrete time domain:

$$u(n) = K\left[1 + \frac{1}{\tau_i}\frac{T_s}{2}\frac{z+1}{z-1} + \frac{\tau_d}{T_s}\frac{z-1}{z}\right]e(n) \tag{8.24}$$

where we have used the trapezoidal approximation for the integral mode and backward difference formula for the derivative mode. On cross multiplying, we obtain

$$\left(z^2 - z\right)u(n) = K\left[\left(z^2 - z\right) + \frac{T_s}{2\tau_i}\left(z^2 + z\right) + \frac{\tau_d}{T_s}(z-1)^2\right]e(n)$$

We divide by z^2 and invert, to obtain

$$\begin{aligned} u(n) - u(n-1) = K\Big[&e(n) - e(n-1) + \frac{T_s}{2\tau_i}\{e(n) + e(n-1)\} \\ &+ \frac{\tau_d}{T_s}\{e(n) - 2e(n-1) + e(n-2)\}\Big] \end{aligned} \tag{8.25}$$

This formula can also be derived starting from the continuous version of the PID controller given by Eq. 8.8. Let us approximate this expression by discretization:

$$\begin{aligned} u(n) = K\Big[&e(n) + \frac{1}{\tau_i}\left\{T_s\frac{e(0)+e(1)}{2} + \cdots + T_s\frac{e(n-1)+e(n)}{2}\right\} \\ &+ \tau_D\frac{e(n) - e(n-1)}{T_s}\Big] \end{aligned}$$

One can easily check that by writing $u(n-1)$ using the above and finding the difference between these two expressions, the PID expression given above is once again obtained. The PID expression is usually written as

$$u(n) - u(n-1) = s_0 e(n) + s_1 e(n-1) + s_2 e(n-2) \tag{8.26}$$

where

$$\begin{aligned} s_0 &= K\left[1 + \frac{T_s}{2\tau_i} + \frac{\tau_d}{T_s}\right] \\ s_1 &= K\left[-1 + \frac{T_s}{2\tau_i} - 2\frac{\tau_d}{T_s}\right] \\ s_2 &= K\frac{\tau_d}{T_s} \end{aligned} \tag{8.27}$$

Although the continuous time formulation is not implementable, because of the presence of the derivative term, its discrete time equivalent does not have this difficulty.

The control law, as given by Eq. 8.26, is in the difference or incremental formulation: the control law at the current instant is the control law at the previous time instant plus an increment based on the error signals. This formulation helps while switching a plant from the manual or open loop mode to automatic or closed loop mode even when the control level used in the manual mode is not exactly known to the controller. If, on the other hand, the controller specifies only the *total* control action to be implemented, there could be problems if the previous open loop state of the plant is not known exactly. In this case, the mismatch in the control levels at the previous and the current time instants could result in the plant experiencing a bump. For this reason, Eq. 8.26 is supposed to implement a *bumpless control law.* We will refer to this controller as PID-1.

Example 8.7 Determine the discrete time PID controller if we have the following continuous time PID settings: $K = 2$, $\tau_d = 2.5$ s, $\tau_i = 40$ s and $T_s = 1$ s.

Substituting these values in Eq. 8.27, we obtain $s_0 = 7.03$, $s_1 = -11.98$, $s_2 = 5$. Thus,

$$u(t) = 2\left[e(t) + \frac{1}{40}\int_0^t e(\tau)d\tau + 2.5\frac{de(t)}{dt}\right]$$

is approximated as

$$u(n) = u(n-1) + 7.03e(n) - 11.98e(n-1) + 5e(n-2)$$

We see that the control action is an increment over the previous control action, as explained above. ∎

We will now write the controller obtained in this section in standard notation. Writing Eq. 8.26 in the following form,

$$(1 - z^{-1})u(n) = (s_0 + s_1 z^{-1} + s_2 z^{-2})e(n) \tag{8.28}$$

and substituting for $e(n)$ using $e(n) = r(n) - y(n)$, we obtain the controller in the form of Eq. 7.9 on page 245 with

$$\begin{aligned} R_c &= 1 - z^{-1} \\ S_c &= s_0 + s_1 z^{-1} + s_2 z^{-2} \\ T_c &= S_c \end{aligned} \tag{8.29}$$

8.3.5 PID Controller with Filtering and $T_c = S_c$

In this section, we will discretize the control law of Eq. 8.13 with the backward difference formula of Eq. 8.2 on page 303 for all occurrences of s. The derivative term becomes

$$\frac{1}{1 + \tau_d s/N} \leftrightarrow -\frac{NT_s}{\tau_d}\frac{r_1}{1 + r_1 z^{-1}} \tag{8.30}$$

where

$$r_1 = -\frac{\tau_d/N}{\tau_d/N + T_s} \tag{8.31}$$

Using the same rule for approximating integration as well, Eq. 8.13 becomes

$$\frac{S_c}{R_c} = K\left[1 + \frac{T_s}{\tau_i}\frac{1}{1-z^{-1}} - \frac{Nr_1(1-z^{-1})}{1+r_1z^{-1}}\right] \tag{8.32}$$

Simplifying, we obtain

$$\frac{S_c}{R_c} = K\frac{(1-z^{-1})(1+r_1z^{-1}) + T_s/\tau_i(1+r_1z^{-1}) - Nr_1(1-z^{-1})^2}{(1-z^{-1})(1+r_1z^{-1})}$$

On comparing the denominator and the numerator, we obtain

$$R_c(z) = (1-z^{-1})(1+r_1z^{-1}) \tag{8.33}$$

$$S_c(z) = s_0 + s_1z^{-1} + s_2z^{-2} \tag{8.34}$$

where

$$\begin{aligned} s_0 &= K\left(1 + \frac{T_s}{\tau_i} - Nr_1\right) \\ s_1 &= K\left[r_1\left(1 + \frac{T_s}{\tau_i} + 2N\right) - 1\right] \\ s_2 &= -Kr_1(1+N) \end{aligned} \tag{8.35}$$

Thus, we obtain the control law,

$$(1-z^{-1})(1+r_1z^{-1})u(n) = (s_0 + s_1z^{-1} + s_2z^{-2})e(n) \tag{8.36}$$

We will refer to this controller as PID-2.

We observe that on introduction of the filtering action in the derivative mode, we lose the useful property of bumpless control action.

Recall that the filtering action was introduced to make the derivative mode implementable. We will now point out another advantage in using this action. The closed loop characteristic polynomial is given by $AR_c + BS_c$, see Eq. 8.20. We could equate this to the desired polynomial and determine R_c and S_c.[1] Comparing Eq. 8.33 with Eq. 8.29, we find that filtering increases the degree of R_c. The larger degree of R_c allows plants of larger degree to be accommodated by this procedure, because of the condition $\mathrm{d}B = \mathrm{d}R_c + 1$.

Example 8.8 Using the discretization method presented in this section, determine the polynomial coefficients that correspond to the continuous time PID controller presented in Example 8.7. The filter constant N may be taken as 10.

[1]This procedure is illustrated in Sec. 9.2.

Using Eq. 8.31, we obtain

$$r_1 = -\frac{0.25}{0.25+1} = -0.2$$

Using Eq. 8.35, we obtain

$$\begin{aligned} s_0 &= 2(1 + 1/40 + 2) = 6.05 \\ s_1 &= 2[-0.2(1 + 1/40 + 20) - 1] = -10.41 \\ s_2 &= 2 \times 0.2(11) = 4.4 \end{aligned}$$

Using Eq. 8.36, the control law is given by

$$(1 - z^{-1})(1 - 0.2z^{-1})u(n) = (6.05 - 10.41z^{-1} + 4.4z^{-2})e(n)$$

As mentioned above, the control action is not in incremental form. ∎

There are instances when one may not want to use the integral mode of control. For example, if the plant already has an integrator, one may not wish to introduce an integral mode through the controller. That is, one may want to implement a proportional, derivative (PD) controller. This is achieved by setting the second term of Eq. 8.13 and Eq. 8.32 equal to zero to obtain continuous time and discrete time PD controllers, respectively. The discrete time PD controller is

$$\frac{S_c}{R_c} = K\left[1 - \frac{Nr_1(1 - z^{-1})}{1 + r_1 z^{-1}}\right] = K\frac{(1 - Nr_1) + r_1(1 + N)z^{-1}}{1 + r_1 z^{-1}} \tag{8.37}$$

In other words, the discrete time PD control law is given by

$$(1 + r_1 z^{-1})u(n) = (s_0 + s_1 z^{-1})e(n) \tag{8.38}$$

where

$$\begin{aligned} s_0 &= K(1 - Nr_1) \\ s_1 &= Kr_1(1 + N) \end{aligned} \tag{8.39}$$

Compare this control law with that of Eq. 8.36. This control law also is not in incremental form. We will refer to this as PID-3.

Example 8.9 Using the discretization method presented above, determine the polynomial coefficients that correspond to the continuous time PID controller with $K = 2$, $\tau_d = 2.5$, $N = 10$, and compare with the values in Example 8.7.

It is easy to see that r_1 remains as -0.2. Using Eq. 8.39, we calculate s_0 and s_1:

$$\begin{aligned} s_0 &= K(1 - Nr_1) = 2(1 + 2) = 6 \\ s_1 &= Kr_1(1 + N) = -2 \times 0.2(11) = -4.4 \end{aligned}$$

The control law is given by

$$(1 - 0.2z^{-1})u(n) = (6 - 4.4z^{-1})e(n)$$

∎

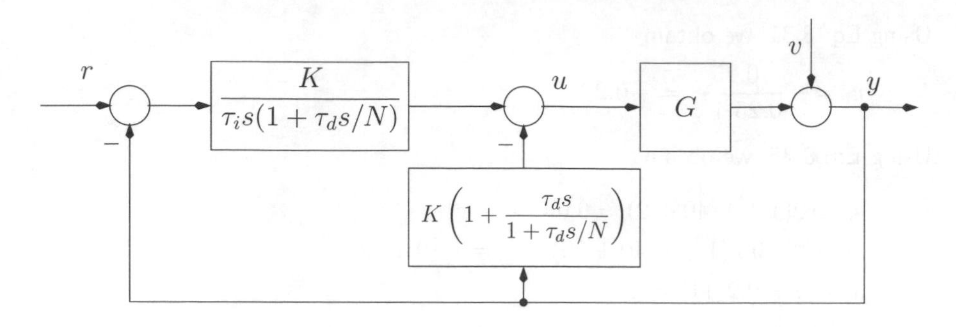

Figure 8.3: A 2-DOF PID control structure without proportional and derivative actions on a reference signal. It satisfies second condition of Eq. 8.19 for tracking step changes without offset.

8.3.6 2-DOF PID Controller with $T_c = S_c(1)$

In this section, we present a control structure that satisfies the second condition in Eq. 8.19.

The control effort given by Eq. 8.8 has the following shortcoming: if there is a sudden change in the setpoint, both the proportional and the derivative modes will introduce large jumps in the control effort, known as *setpoint kick*. The large change introduced by the derivative mode is known as *derivative kick*. Although the proportional mode may introduce a smaller change than the derivative mode, it may still give rise to a larger than acceptable control effort, known as *proportional kick*. Both derivative and proportional kicks are generally not acceptable. The control law implemented in Fig. 8.3 addresses these difficulties [30]. Note that the feed forward path also has the low pass filter from the derivative term. The control law can be expressed as

$$u(t) = \frac{K}{\tau_i s\left(1+\frac{\tau_d s}{N}\right)}(r(t)-y(t)) - K\left[1+\frac{\tau_d s}{1+\frac{\tau_d s}{N}}\right]y(t) \tag{8.40}$$

Grouping the terms involving $y(t)$, we obtain

$$u(t) = \frac{K}{\tau_i s\left(1+\frac{\tau_d s}{N}\right)}r(t) - K\left[1+\frac{\tau_d s}{1+\frac{\tau_d s}{N}}+\frac{1}{\tau_i s\left(1+\frac{\tau_d s}{N}\right)}\right]y(t)$$

which can be simplified as

$$u = \frac{K}{\tau_i s(1+\tau_d s/N)}r - K\left[1+\frac{\tau_i\tau_d s^2+1}{\tau_i s(1+\tau_d s/N)}\right]y \tag{8.41}$$

Comparing this with Eq. 8.14, we see that the first term is $T_c r/R_c$ and the second term is $S_c y/R_c$. It is easy to see that this equation implies $T_c = S_c(s=0) = K$, the continuous time condition for offset free tracking of step inputs, as given by Eq. 8.19.

We proceed to discretize Eq. 8.41. Because the second term is $S_c y/R_c$, we can write it as follows:

$$\frac{S_c}{R_c} = K\left[1 + \frac{\tau_i\tau_d s^2 + 1}{\tau_i s}\frac{1}{1+\tau_d s/N}\right]$$

Using Eq. 8.30 and the backward difference rule, the discrete approximation of this expression is

$$\frac{S_c}{R_c} = K\left[1 - \frac{\tau_i\tau_d(1-z^{-1})^2/T_s^2 + 1}{\tau_i(1-z^{-1})/T_s}\frac{NT_s}{\tau_d}\frac{r_1}{1+r_1 z^{-1}}\right]$$

which can be simplified as

$$\frac{S_c}{R_c} = K\left[1 - \frac{(1-z^{-1})^2 + T_s^2/(\tau_i\tau_d)}{1-z^{-1}}\frac{Nr_1}{1+r_1 z^{-1}}\right]$$

On equating the terms in the numerator and denominator, we obtain

$$R_c = (1-z^{-1})(1+r_1 z^{-1}) \tag{8.42}$$

and

$$S_c = K\left[(1-z^{-1})(1+r_1 z^{-1}) - \left(1 - 2z^{-1} + z^{-2} + \frac{T_s^2}{\tau_i\tau_d}\right)Nr_1\right]$$

Simplifying, we obtain

$$S_c = K\left[1 - Nr_1 - \frac{Nr_1 T_s^2}{\tau_i\tau_d} + (r_1 - 1 + 2Nr_1)z^{-1} - (r_1 + Nr_1)z^{-2}\right]$$

which is equivalent to the following:

$$\begin{aligned} S_c(z) &= s_0 + s_1 z^{-1} + s_2 z^{-2} \\ s_0 &= K\left[1 - Nr_1 - \frac{Nr_1 T_s^2}{\tau_i\tau_d}\right] \\ s_1 &= K[r_1(1+2N) - 1] \\ s_2 &= -Kr_1(1+N) \end{aligned} \tag{8.43}$$

with r_1 given by Eq. 8.31. Following the same procedure as above, the first term in Eq. 8.41 becomes

$$\frac{T_c}{R_c} = -\frac{KNT_s^2 r_1/(\tau_i\tau_d)}{(1-z^{-1})(1+r_1 z^{-1})} \tag{8.44}$$

Comparing the numerator and the denominator, we obtain

$$T_c = -\frac{KNT_s^2 r_1}{\tau_i\tau_d} \tag{8.45}$$

and R_c is as in Eq. 8.42. We will refer to this controller as PID-4. Substituting from Eq. 8.43, and simplifying, we obtain

$$S_c(1) = s_0 + s_1 + s_2 = -\frac{KNT_s^2 r_1}{\tau_i \tau_d} \tag{8.46}$$

We see that $T_c = S_c(1)$, the discrete time condition for offset free tracking of step changes in the reference signal $r(n)$, see Eq. 8.23. Of course, this is only expected because the continuous time version of this controller has this property.

Example 8.10 Using the 2-DOF formulation given in this section, determine the discrete time PID controller for the continuous controller parameters as in Example 8.8.

As in the previous example, we obtain

$$r_1 = -0.2$$

Using Eq. 8.43, we obtain

$$\begin{aligned} s_0 &= 2\left[1 + 10 \times 0.2 + \frac{10 \times 0.2}{40 \times 2.5}\right] = 6.04 \\ s_1 &= 2[-0.2(1+20) - 1] = -10.4 \\ s_2 &= 2 \times 0.2 \times 11 = 4.4 \end{aligned}$$

Using the fact that $T_c = S_c(1)$, we obtain

$$T_c = s_0 + s_1 + s_2 = 0.04$$

The control law is given by

$$(1 - z^{-1})(1 - 0.2z^{-1})u(n) = 0.04r(n) - (6.04 - 10.4z^{-1} + 4.4z^{-2})e(n)$$

As mentioned above, the control action is not in the incremental form. ∎

Several variations of the control law given in Eq. 8.40 are possible. Substituting $N \to \infty$ in this equation, we obtain

$$u(t) = K\left[-y(t) + \frac{1}{\tau_i s}e(t) - \tau_d s y(t)\right] \tag{8.47}$$

where $e(t) = r(t) - y(t)$. Equivalently, this can be obtained also from Eq. 8.9 on page 308 by replacing $e(t)$ with $-y(t)$ in proportional and derivative modes. In other words, $r(t)$ has been replaced with zero in these modes. The usual implementations with a large proportional gain result in large control actions for step changes in the reference signal. The current formulation of not sending the reference signal through the derivative mode addresses this problem. Nevertheless, setpoint tracking is not affected by this change, because the integral term is unchanged: so long as there is a nonzero error, the integral term will try to correct it. Because only the integral term takes care of offset free tracking, it cannot be made zero. In other words, proportional or proportional derivative controllers cannot be used in this configuration.

We conclude the discussion in this section with another discretization. Using the backward difference formula, given by Eq. 8.2, we obtain the discrete time equivalent of this control law:

$$u(n) = K\left[-y(n) + \frac{T_s}{\tau_i \Delta}e(n) - \tau_d \frac{\Delta}{T_s}y(n)\right]$$

where $\Delta = 1 - z^{-1}$. Multiplying throughout by Δ and rearranging the terms, we obtain

$$\Delta u(n) = \frac{KT_s}{\tau_i}e(n) - K\left[\Delta + \frac{\tau_d}{T_s}\Delta^2\right]y(n) \tag{8.48}$$

Substituting for Δ as $1 - z^{-1}$ on the right-hand side, this equation becomes

$$\Delta u(n) = \frac{KT_s}{\tau_i}(r(n) - y(n)) - K\left[(1 - z^{-1}) + \frac{\tau_d}{T_s}(1 - 2z^{-1} + z^{-2})\right]y(n)$$

Rearranging the terms, this becomes

$$\Delta u(n) = \frac{KT_s}{\tau_i}r(n) - K\left[\left(1 + \frac{T_s}{\tau_i} + \frac{\tau_d}{T_s}\right) - \left(1 + 2\frac{\tau_d}{T_s}\right)z^{-1} + \frac{\tau_d}{T_s}z^{-2}\right]y(n) \tag{8.49}$$

Defining

$$L(z) = K\left[\left(1 + \frac{T_s}{\tau_i} + \frac{\tau_d}{T_s}\right) - \left(1 + 2\frac{\tau_d}{T_s}\right)z^{-1} + \frac{\tau_d}{T_s}z^{-2}\right], \tag{8.50}$$

Eq. 8.49 becomes

$$\Delta u(n) = L(1)r(n) - L(z)y(n) \tag{8.51}$$

This is in the form of Eq. 8.14 with $S_c(z) = L(z)$ and $T_c(z) = L(1)$. We see that the second condition in Eq. 8.23 on page 311 is satisfied. The controller is once again in incremental form, useful for bumpless implementation. We will refer to this controller as PID-5.

Example 8.11 Using the relations obtained above, repeat Example 8.10 without the filtering action.

Let us evaluate L using Eq. 8.50:

$$\begin{aligned} L(z) &= 2\left[\left(1 + \frac{1}{40} + 2.5\right) - (1 + 2 \times 2.5)z^{-1} + 2.5z^{-2}\right] \\ &= 7.05 - 12z^{-1} + 5z^{-2} \end{aligned}$$

Substituting z as 1, we obtain $L(1) = 0.05$. Thus, the control law given by Eq. 8.51 becomes

$$\Delta u(n) = 0.05r(n) - (7.05 - 12z^{-1} + 5z^{-2})e(n)$$

It is easy to check that this controller satisfies the offset free tracking condition, $T_c = S_c(1)$, each of which is equal to 0.05. ∎

8.3.7 2-DOF PID Controller with $T_c(1) = S_c(1)$

The 2-DOF PID controller presented in the previous section does not have dynamic elements in T_c. Moreover, because only the integral mode acts on the measurement $y(t)$, the system could be sluggish in tracking the reference signal. The response can be made faster by feeding back a fraction of y through the proportional mode. We modify Eq. 8.13 on page 308 to arrive at the following control law [2]:

$$u(t) = K\left[br(t) - y(t) + \frac{1}{s\tau_i}(r(t) - y(t)) - \frac{s\tau_d}{1 + \frac{s\tau_d}{N}}y(t)\right] \tag{8.52}$$

where we have substituted for $e(t)$ using $r(t) - y(t)$ and b is selected to be in the range of zero to one, *i.e.*, $0 < b < 1$. Many different ways of discretizing the continuous controller of Eq. 8.52 exist. To compare with the result of [2], however, we use forward difference approximation of Eq. 8.3 to discretize the integral mode and backward difference approximation of Eq. 8.2 to approximate the derivative mode. Recall that there is no hard and fast rule about which approximation of continuous time transfer function is the best. The above choice illustrates that there are many ways of approximating continuous time transfer functions. Because there is no difference in the proportional mode between the continuous and discrete time domains, Eq. 8.52 becomes the following:

$$u(n) = K\left[br(n) - y(n) + \frac{1}{\tau_i}\frac{T_s}{z-1}(r(n) - y(n)) - \frac{\frac{z-1}{zT_s}\tau_d}{1 + \frac{z-1}{zT_s}\frac{\tau_d}{N}}y(n)\right] \tag{8.53}$$

Simplifying and dropping the explicit dependence on n, we obtain

$$u = K\left[br - y + \frac{T_s}{\tau_i}\frac{1}{z-1}(r - y) - \frac{N\tau_d}{NT_s + \tau_d}\frac{z-1}{z - \frac{\tau_d}{NT_s+\tau_d}}y\right]$$

With a_d defined as follows

$$a_d = \frac{\tau_d}{NT_s + \tau_d} \tag{8.54}$$

the above equation becomes

$$u = K\left[br - y + \frac{T_s}{\tau_i}\frac{1}{z-1}(r - y) - \frac{Na_d(z-1)}{z - a_d}y\right]$$

After multiplying by $(z-1)(z-a_d)$ on both sides, we arrive at

$$\begin{aligned}(z-1)(z-a_d)u = {} & K(br-y)(z-1)(z-a_d) \\ & + Kb_i(r-y)(z-a_d) - Kb_d(z-1)^2 y\end{aligned} \tag{8.55}$$

where

$$\begin{aligned}b_i &= \frac{T_s}{\tau_i} \\ b_d &= N a_d\end{aligned} \tag{8.56}$$

Dividing by z^2 and rearranging, we arrive at the following controller expression:

$$\begin{aligned}(1-z^{-1})(1-a_d z^{-1})u = {} & K[b + (b_i - b(1+a_d))z^{-1} + (ba_d - b_i a_d)z^{-2}]r \\ & - K[(1+b_d) + (b_i - 2b_d - (1+a_d))z^{-1} + (a_d - b_i a_d + b_d)z^{-2}]y\end{aligned} \tag{8.57}$$

This is of the form

$$R_c u(n) = T_c r(n) - S_c y(n)$$

where the definition of these variables follows from the previous equation. It is easy to check that

$$T_c(1) = S_c(1) = Kb_i(1-a_d)$$

satisfying the requirement for offset free tracking of step change in R, namely Eq. 8.23 on page 311. We will refer to this controller as PID-6.

The use of PID controllers will be a continuing theme in this book. The next occurrence of this will be in Sec. 9.8, where we will take up the implementation issues of PID controllers.

8.4 Matlab Code

Matlab Code 8.1 Continuous to discrete time transfer function. Available at `HOME/Z-trans/matlab/disc2.m`[2]

```
1 sys = tf(10,[5 1]);
2 sysd = c2d(sys,0.5);
```

[2]`HOME` stands for `http://www.moudgalya.org/dc/` – first see the software installation directions, given in Appendix A.2.

8.5 Problems

8.1. This problem presents a state space approach to Example 8.3 on page 304. Show that the state space equivalent of this problem is

$$\dot{x} = -\frac{1}{\tau}x + \frac{K}{\tau}u$$
$$y = x$$

Show that this can be written in the standard discrete time state space formulation of Eq. 2.2 on page 6, with

$$A = e^{-T_s/\tau}, \quad B = K\left[1 - e^{-T_s/\tau}\right]$$

Show that this discrete time state space system has a transfer function identical to the one presented in Example 8.3.

8.2. This problem is concerned with the determination of where the stable region in the s plane gets mapped to the z plane under the trapezoidal approximation.

(a) Show that the trapezoidal or the Tustin approximation, given by Eq. 8.3 on page 303, is equivalent to

$$z = \frac{1 + sT_s/2}{1 - sT_c/2}$$

(b) Find out where the left half of the s plane will be mapped in the z plane using the above transformation [Hint: Substitute $s = a + jb$ in the above equation for $a < 0$ and find out what z you get. Repeat this for $a = 0$.] Does this agree with the notion of the z domain stability region discussed earlier?

8.3. Consider the following discrete time transfer function

$$G(z) = \frac{z^{-1}}{1 - 0.9z^{-1}} \tag{8.58}$$

obtained by sampling a continuous system with a period $T_s = 0.1$ s. If the discrete time transfer function is obtained by the ZOH equivalent approximation of a first order system, determine the continuous transfer function.

8.4. Find the ZOH equivalent of $G_a(s) = 1/s^3$ in the following manner.

(a) Show that

$$F = \begin{bmatrix} 0 & 1 & 0 \\ 0 & 0 & 1 \\ 0 & 0 & 0 \end{bmatrix}, \quad G = \begin{bmatrix} 0 \\ 0 \\ 1 \end{bmatrix}, \quad C = \begin{bmatrix} 1 & 0 & 0 \end{bmatrix}, \quad D = 0$$

is a state space realization of $1/s^3$.

(b) Show that the discrete time state space matrices are

$$A = \begin{bmatrix} 1 & T_s & T_s^2/2 \\ 0 & 1 & T_s \\ 0 & 0 & 1 \end{bmatrix}, \quad B = \begin{bmatrix} T_s^3/3! \\ T_s^2/2! \\ T_s \end{bmatrix}$$

(c) Using

$$\begin{bmatrix} A - zI & B \\ C & D \end{bmatrix} = \begin{bmatrix} A - zI & 0 \\ C & I \end{bmatrix} \begin{bmatrix} I & (A - zI)^{-1}B \\ 0 & D + C(zI - A)^{-1}B \end{bmatrix}$$

show that the required discrete time transfer function $G_d(z)$ is given by

$$\begin{aligned} G_d(z) &= \det \begin{bmatrix} A - zI & B \\ C & D \end{bmatrix} / \det(A - zI) \\ &= (-1)^3 \frac{T_s^3}{3!} \frac{1 + 4z + z^2}{(1 - z)^3} \end{aligned}$$

(d) Check that $G_d(z)$ is a nonminimum phase transfer function, even though $G_a(s) = 1/s^3$ is not [10].

8.5. With the derivative mode controller

$$D(z) = \tau_d \frac{z - 1}{z + 1}$$

in a feedback loop with the oscillating plant

$$H(z) = \frac{z}{z + a}, \qquad a > 0$$

show that the closed loop transfer function is given by

$$T(z) = \frac{\tau_d z(z - 1)}{(\tau_d + 1)z^2 + (a + 1 - \tau_d)z + a}$$

With $a = 1$, show that the closed loop poles are at the following locations:

(a) $-0.25 \pm j0.6614 = 0.707e^{\pm j111^\circ}$ when $\tau_d = 1$

(b) $\pm \frac{1}{\sqrt{3}} j$ when $\tau_d = 2$

(c) 0.565 or 0.161 when $\tau_d = 10$

Justify that the effect of this derivative mode is to reduce the oscillations.

8.6. When the plant in the above problem is put in a feedback loop with the following PD controller,

$$D = K \left(1 + \tau_d \frac{z - 1}{z + 1} \right)$$

show that the steady state value of the output is given by

$$\lim_{n \to \infty} y(n) = \frac{K}{1 + a + K}$$

8.7. Using the root locus technique, or any other method, determine the proportional controller (K_u) that will bring the open loop transfer function

$$G(z) = \frac{1}{z(z-1)}$$

to the verge of instability. If the Ziegler–Nichols settings for a PI controller are $K_p = 0.45K_u$ and $\tau_I = P_u/1.2$, determine K_p and τ_I. A sampling time of 0.1 second is used to arrive at this transfer function.

8.8. This question is concerned with discretization of the PID controller given in Eq. 8.52 using the trapezoidal approximation, with $0 < b < 1$, and N is of the order of 10 [2]. As usual, R, Y and U, respectively, refer to setpoint, output and input.

(a) Use the trapezoidal approximation for both derivative and integral terms, *i.e.*, substitute for s as

$$s \leftrightarrow \frac{2}{T_s}\frac{z-1}{z+1}$$

and arrive at the following result:

$$U(z) = K\left[b + b_i\frac{z+1}{z-1}\right]R(z) - K\left[1 + b_i\frac{z+1}{z-1} + \frac{(z-1)b_d}{z-a_d}\right]Y(z)$$

where

$$b_i = \frac{T_s}{2\tau_i}, \quad b_d = \frac{2N\tau_d}{2\tau_d + NT_s}, \quad a_d = \frac{2\tau_d - NT_s}{2\tau_d + NT_s}$$

(b) Simplify the above expressions to arrive at a controller in the usual R_c, S_c, T_c form with $S_c \neq T_c$:

$$\begin{aligned}(1 - z^{-1})(1 - a_d z^{-1})U(z) = {} & [t_0 + t_1 z^{-1} + t_2 z^{-2}]R(z) \\ & - [s_0 + s_1 z^{-1} + s_2 z^{-2}]Y(z)\end{aligned}$$

where

$$\begin{aligned}
t_0 &= K(b + b_i) \\
t_1 &= -K(b(1 + a_d) - b_i(1 - a_d)) \\
t_2 &= Ka_d(b - b_i) \\
s_0 &= K(1 + b_i + b_d) \\
s_1 &= -K(1 + a_d + 2b_d - b_i(1 - a_d)) \\
s_2 &= K(a_d + b_d - b_i a_d)
\end{aligned}$$

We will refer to this controller as PID-7.

(c) Check that $T_c(1) = S_c(1) = 2Kb_i(1 - a_d)$ and hence that this controller satisfies the condition required for the plant output to track the setpoint, namely Eq. 8.23 on page 311.

8.9. In this problem, we study the 2-DOF PID controller.

(a) Argue that the PID controller of Fig. 8.3 on page 316 can be redrawn as in the following figure:

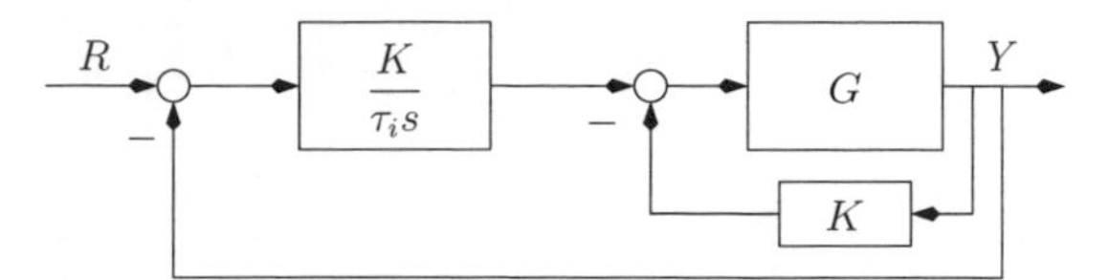

when $N = \infty$, $\tau_d = 0$ and disturbance $= 0$. Show that this is equivalent to the following block diagram:

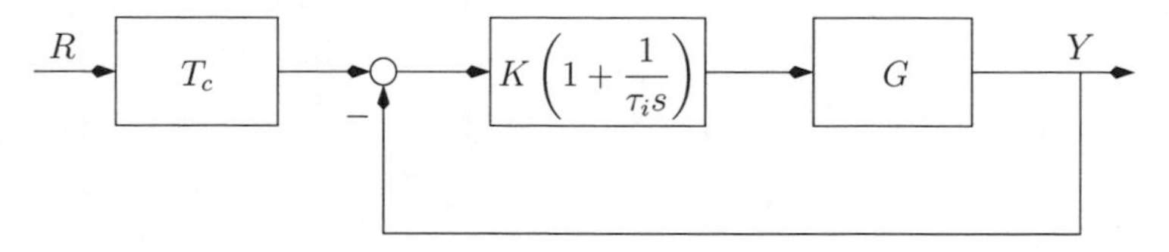

for a specific T_c. Determine T_c.

(b) Suppose that the normal 1-DOF PI controller has a control saturation (input saturation) problem. Will the 2-DOF PI controller discussed in this problem be able to reject disturbances better, perhaps for different tuning parameters? If so, explain what you will do. If not, explain why not.

(c) If the normal 1-DOF PI controller configuration does not have a saturation problem, will the response to a step change in R be faster than that in the 2-DOF PI controller, discussed here? Make suitable assumptions.

Chapter 9

Pole Placement Controllers

Pole placement controllers can be considered as the mother of all controllers in the sense that all controllers can be thought of as belonging to this category in one way or another. In this chapter, we will present mainly two approaches: one that uses a model system and one that is based on performance specification. The first section is based on a first order model as a target system. The remaining sections are devoted to treating the closed loop as a second order system. This forms the major part of this chapter.

9.1 Dead-Beat and Dahlin Control

In this section, we first design a dead-beat control, which is essentially a zero order system with possibly a time delay. We also explain Dahlin control, whose objective is to produce a first order overall system.

One of the advantages in working with discrete time systems is that digital controllers can be designed through direct techniques, as we show in this section. First we present dead-beat control, then Dahlin control. Consider the feedback system presented in Fig. 4.6.

Supposing we want the transfer function between the setpoint and the output to be $G_m(z)$, we have

$$\frac{G_D(z)G(z)}{1+G_D(z)G(z)} = G_m(z)$$

the solution of which is

$$G_D(z) = \frac{G_m(z)}{[1-G_m(z)]G(z)}$$

In dead-beat control, an effort is made to reach the setpoint as quickly as possible and stay there. The minimum time required to reach the setpoint, however, is the dead time of the system. As a result, we get the condition

$$G_m(z) = z^{-k}, \quad k \geq 1$$

where $k = D/T_s$, D is the system dead time and T_s the sampling time. The resulting dead-beat controller is given by

$$G_D(z) = \frac{z^{-k}}{(1 - z^{-k})G(z)}$$

Example 9.1 Design a dead-beat controller for

$$G(z) = \frac{z^{-2}}{1 - z^{-1}}$$

Using the above formula, we get the dead-beat controller to be

$$G_D(z) = \frac{z^{-2}}{(1 - z^{-2})z^{-2}/(1 - z^{-1})} = \frac{1 - z^{-1}}{1 - z^{-2}}$$

∎

The Dahlin controller is similar to the dead-beat controller in the sense that the desired closed loop transfer function $G_m(z)$ is specified, from which the discrete time controller is found using the formula given above. The only difference is that now the desired transfer function is no longer dead-beat, but a first order transfer function with a time delay.

Example 9.2 For the open loop transfer function in the continuous domain

$$G(s) = \frac{10e^{-s}}{5s + 1}$$

a choice of a desired closed loop transfer function is

$$G_m(s) = \frac{10e^{-s}}{2s + 1}$$

Note that the closed loop transfer function has the same dead time, but a smaller time constant, indicating a faster response.

∎

9.2 Pole Placement Controller with Performance Specifications

In this section, we explore the possibility of making the closed loop system a second order system, with possibly a time delay. In this setting, one can naturally specify requirements, such as rise time, overshoot and settling time, in addition to ensuring internal stability. After introducing a basic design, we present a design that has an internal model of the disturbance in the loop. We also explain the steps to be taken to prevent oscillations in the loop. We will work exclusively with the 2-DOF control

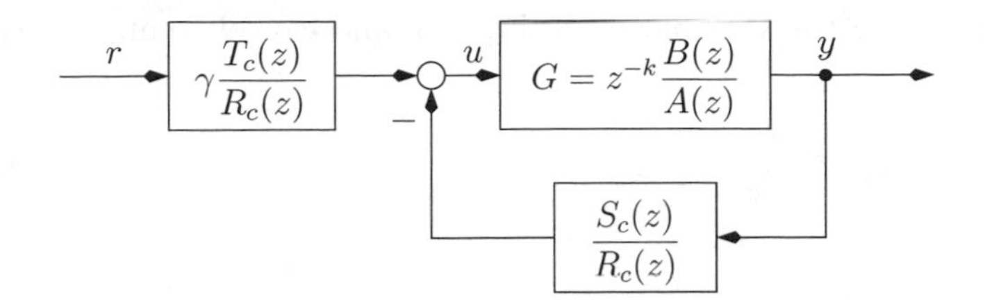

Figure 9.1: Schematic of 2-DOF pole placement controller

structure presented in Sec. 7.1.3. Although we ignore the effects of noise for now, we study them in detail from Chapter 11 onwards. Let us begin with the plant

$$Y(z) = G(z)U(z) \tag{9.1}$$

with transfer function

$$G(z) = z^{-k}\frac{B(z)}{A(z)} \tag{9.2}$$

where $B(z)$ and $A(z)$ are coprime. Recall that this means that A and B do not have a common factor. We would like to design a controller such that the plant output y is related to the setpoint or command signal r in the following manner:

$$Y_m(z) = \gamma z^{-k}\frac{B_r}{\phi_{cl}}R(z) \tag{9.3}$$

where ϕ_{cl} is the characteristic polynomial obtained by the desired location analysis and

$$\gamma = \frac{\phi_{cl}(1)}{B_r(1)} \tag{9.4}$$

so that at steady state $Y = R$. Towards this end, we look for a controller of the form

$$R_c(z)U(z) = \gamma T_c(z)R(z) - S_c(z)Y(z) \tag{9.5}$$

where $R_c(z)$, $S_c(z)$ and $T_c(z)$ are polynomials in z^{-1}, to be determined, see Fig. 9.1. Note that this structure is identical to the one in Eq. 7.9 on page 245, but for the introduction of γ now. The controller has two components:

1. A feedback component S_c/R_c that helps ensure internal stability and reject disturbances.

2. A feed forward component T_c/R_c that helps Y track R.

Because of these two objectives, as mentioned earlier, it is known as the 2-DOF controller.

Simplifying the block diagram given in Fig. 9.1, or equivalently, substituting for $Y(z)$ from Eq. 9.1 and Eq. 9.2 in Eq. 9.5, and dropping the argument z for convenience, we obtain

$$Y = \gamma z^{-k}\frac{T_c B}{AR_c + z^{-k}BS_c}R \tag{9.6}$$

We equate this Y to the variable Y_m in Eq. 9.3, and cancel common terms to arrive at

$$\frac{BT_c}{AR_c + z^{-k}BS_c} = \frac{B_r}{\phi_{cl}} \tag{9.7}$$

In general,

$$\deg B_r < \deg B \tag{9.8}$$

so that the desired closed loop transfer function is of lower order than that of BT_c. This is achieved by cancelling common terms between the numerator and denominator. But we know that such cancellations, if any, should be between factors that have zeros inside the unit circle, or in our notation, between good factors only. In view of this, we factorize B as good and bad factors:

$$B = B^g B^b \tag{9.9}$$

Similarly, we factorize A into good and bad factors to arrive at

$$A = A^g A^b \tag{9.10}$$

If we let

$$R_c = B^g R_1 \tag{9.11a}$$

$$S_c = A^g S_1 \tag{9.11b}$$

$$T_c = A^g T_1 \tag{9.11c}$$

Eq. 9.7 becomes

$$\frac{B^g B^b A^g T_1}{A^g A^b B^g R_1 + z^{-k} B^g B^b A^g S_1} = \frac{B_r}{\phi_{cl}}$$

which can be simplified by cancelling the good common factors. We obtain

$$\frac{B^b T_1}{A^b R_1 + z^{-k} B^b S_1} = \frac{B_r}{\phi_{cl}} \tag{9.12}$$

We can equate the numerator and denominator parts. On equating the numerator, we obtain

$$B_r = B^b T_1 \tag{9.13}$$

and equating the denominator results in the following Aryabhatta's identity

$$A^b R_1 + z^{-k} B^b S_1 = \phi_{cl} \tag{9.14}$$

which can be solved for R_1 and S_1. There are many options to choose T_1. We can shape the transfer function between the reference signal and the plant output by a suitable choice of T_1 [30]. By choosing T_1 to be equal to S_1, the 2-DOF controller is reduced to the 1-DOF configuration. Another simple choice is to make T_1 equal to one. We compare the last two choices in Sec. 9.6. For most of this chapter, however, we will make the following choice:

$$T_1 = 1 \tag{9.15}$$

In view of Eq. 9.13 and 9.15, the expression for γ obtained in Eq. 9.4 becomes

$$\gamma = \frac{\phi_{cl}(1)}{B^b(1)} \tag{9.16}$$

and the closed loop transfer function, as desired in Eq. 9.3, is now obtained as

$$G_Y = \gamma z^{-k} \frac{B^b}{\phi_{cl}} \tag{9.17}$$

We see that the bad factors in the numerator of the original transfer function appear in the closed loop transfer function as well. In other words, the bad zeros of the original transfer function cannot be changed by feedback.

Systems with zeros outside the unit circle are known as nonminimum phase systems, see the discussion in Sec. 5.4.2. Recall that the zeros outside the unit circle are known as the nonminimum phase zeros. Because these zeros cannot be altered by feedback control, the performance of a controller will be lower than that achievable with the corresponding minimum phase systems.

Example 9.3 Let us control the magnetically suspended ball presented in detail in Sec. 2.2.1. Let the distance between the ball and the armature h be the only measured variable. That is, the output is modelled as

$$y = Cx$$

with

$$C = \begin{bmatrix} 1 & 0 & 0 \end{bmatrix}$$

The controller should help track step changes in the reference signal, with the following specifications:

$$\begin{aligned} \text{Steady state error} &\le 2\% \\ \text{Overshoot} = \varepsilon &\le 5\% \\ \text{Settling time} &\le 0.5 \text{ s} \end{aligned}$$

The Matlab code in M 9.1 carries out controller design, as we now explain:

1. The continuous time transfer function can be obtained as

$$G(s) = \frac{-280.14}{s^3 + 100s^2 - 981s - 98100}$$

which has poles at 31.32, −31.32 and −100 and hence is unstable. Sample at

$$T_s = 0.01 \text{ s}$$

Using `myc2d.m` listed in M 9.2, obtain

$$\begin{aligned} G(z) &= z^{-1}\frac{(-3.7209 \times 10^{-5} - 1.1873 \times 10^{-4}z^{-1} - 2.2597 \times 10^{-5}z^{-2})}{1 - 2.4668z^{-1} + 1.7721z^{-2} - 0.3679z^{-3}} \\ &= z^{-1}\frac{-3.7209 \times 10^{-5}(1 + 2.9877z^{-1})(1 + 0.2033z^{-1})}{(1 - 1.3678z^{-1})(1 - 0.7311z^{-1})(1 - 0.3679z^{-1})} = z^{-k}\frac{B}{A} \end{aligned}$$

where $k = 1$. We factor A and B into good and bad parts:

$$
\begin{aligned}
A &= A^g A^b \\
A^g &= (1 - 0.7311z^{-1})(1 - 0.3679z^{-1}) \\
A^b &= (1 - 1.3678z^{-1}) \\
B &= B^g B^b \\
B^g &= -3.7209 \times 10^{-5}(1 + 0.2033z^{-1}) \\
B^b &= (1 + 2.9877z^{-1})
\end{aligned}
$$

This is carried out using the function given in M 9.3.

2. Next, we proceed to determine ρ and ω required to satisfy the transient conditions, using the method presented in Sec. 7.7. As the rise time is not specified, we guess it to be about one third of the settling time. Thus the rise time has to be 0.15 s. Let us choose the sampling time to be 0.01 s. Let us first apply the rise time constraint:

$$N_r \leq \text{rise time}/T_s = 15$$

Choose $N_r = 15$

$$\omega = \frac{\pi}{2N_r} = \frac{\pi}{30} = 0.1047$$

Overshoot constraint:

$$\rho \leq \varepsilon^{\omega/\pi} = 0.05^{0.1047/\pi} = 0.905$$

Choose $\rho = 0.905$

Desired closed loop poles:

$$z = \rho e^{\pm j\omega}$$

3. We calculate the desired closed loop polynomials as

$$\phi_{cl}(z) = 1 - 2z^{-1}\rho\cos\omega + \rho^2 z^{-2} = 1 - 1.8z^{-1} + 0.819z^{-2}$$

M 9.4 carries out these calculations.

4. Next, solve Eq. 9.14. That is, solve

$$(1 - 1.3678z^{-1})R_1 + z^{-1}(1 + 2.9877z^{-1})S_1 = 1 - 1.8z^{-1} + 0.819z^{-2} \tag{9.18}$$

Using `xdync.m`, we obtain

$$
\begin{aligned}
S_1 &= 0.0523 \\
R_1 &= 1 - 0.4845z^{-1}
\end{aligned} \tag{9.19}
$$

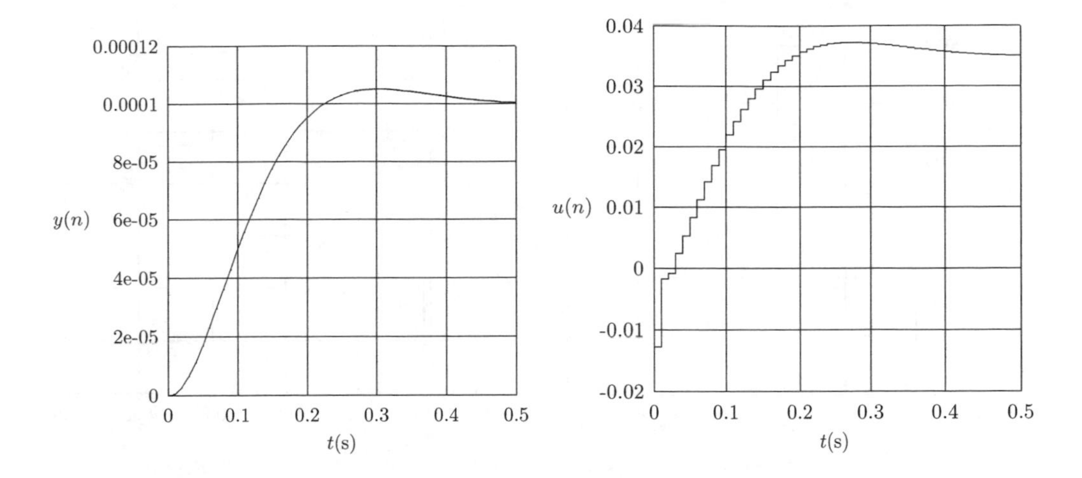

Figure 9.2: Output and input profiles for ball suspension example obtained by executing the Simulink block diagram in Fig. A.8. The controller used is CTRL-1, derived in Example 9.3. The overshoot and the settling time constraints of $y(n)$ have been met. The auxiliary constraint in rise time is not met, however.

5. Using Eq. 9.11, we calculate the control parameters:

$$R_c = -3.7209 \times 10^{-5}(1 - 0.2812z^{-1} - 0.0985z^{-2})$$
$$S_c = 0.0523 - 0.0575z^{-1} + 0.0141z^{-2}$$
$$T_c = A^g = (1 - 0.7311z^{-1})(1 - 0.3679z^{-1})$$

   These calculations are carried out in M 9.5. We will refer to this controller as CTRL-1.

6. The function in M 9.1 calculates the 2-DOF controller, discussed above. After that, simulation is carried out using the Simulink program in Fig. A.8. The resulting profiles are shown in Fig. 9.2. Note that the output y is measured in m. The equilibrium value is 1 cm or 0.01 m. The simulations have been carried out for 1% change in that value.

It is easy to see from the above profiles that while the overshoot and settling time requirements are met, the auxiliary condition of rise time constraint is not met. ∎

The controller designed above (CTRL-1) works well, but for the auxiliary condition of rise time. What if the rise time requirement also has to be fulfilled? It is possible to use N_r as a tuning parameter to meet this requirement, as we show in the next example.

Example 9.4 Use the rise time N_r as a tuning parameter and achieve the required rise time.

The rise time achieved is about one and a half times the specified rise time. In view of this, we carry out the control design procedure outlined in Example 9.3

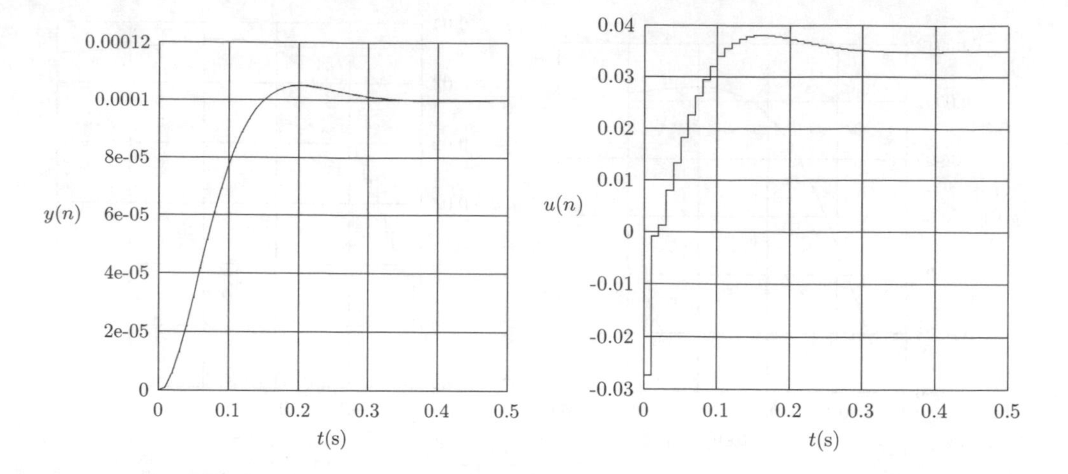

Figure 9.3: Output and input profiles for ball suspension example obtained by executing the Simulink block diagram in Fig. A.8 with $T_1 = 1$. The controller used is CTRL-2, derived in Example 9.4. All specifications on $y(n)$ have been met. The tradeoff is that now the control effort is larger than that in Fig. 9.2.

with rise time = 0.1 s or equivalently, $N_r = 10$. With this reduced rise time, we obtain the following:

$$\begin{aligned}
\rho &= 0.8609 \\
\omega &= 0.1571 \\
\phi_{cl} &= 1 - 1.7006z^{-1} + 0.7411z^{-2} \\
\gamma &= 0.0102 \\
R_1 &= 1 - 0.3984z^{-1} \\
S_1 &= 0.0657 \\
R_c &= -3.7209 \times 10^{-5}(1 - 0.1952z^{-1} - 0.081z^{-2}) \\
S_c &= 0.0657 - 0.0722z^{-1} + 0.0177z^{-2} \\
T_c &= 1 - 1.099z^{-1} + 0.269z^{-2}
\end{aligned}$$

We will refer to this controller as CTRL-2. As before, the Simulink code in Fig. A.8 is executed to evaluate the efficacy of CTRL-2. The resulting $y(n)$ and $u(n)$ profiles are shown in Fig. 9.3. With this controller, it is easy to see that we also satisfy the auxiliary condition of rise time. The price we have to pay for this is the slightly increased control effort at initial time, as compared to the one achieved using CTRL-1, see Fig. 9.2.

∎

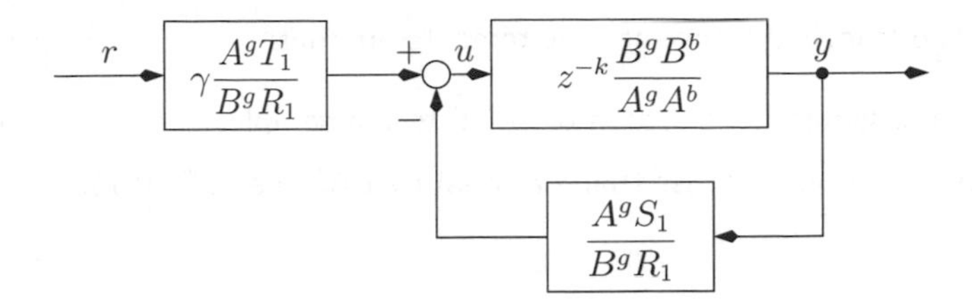

Figure 9.4: 2-DOF pole placement controller in factored form

9.3 Implementation of Unstable Controllers

In this section, we present a case that requires the denominator of the controller, namely R_c, to be taken inside the loop.

Let us begin by calculating the closed loop transfer function for the system with a pole placement controller. Let us first redraw the closed loop block diagram given in Fig. 9.1 with Eq. 9.11 substituted. The resulting diagram is given in Fig. 9.4. The closed loop transfer function is given by

$$T = \gamma\frac{A^gT_1}{B^gR_1}\frac{z^{-k}\dfrac{B^gB^b}{A^gA^b}}{1+z^{-k}\dfrac{B^gB^b}{A^gA^b}\dfrac{A^gS_1}{B^gR_1}} = \gamma\frac{T_1}{R_1}\frac{z^{-k}\dfrac{B^b}{A^b}}{1+z^{-k}\dfrac{B^bS_1}{A^bR_1}}$$

Simplifying, we obtain

$$T = \gamma\frac{T_1}{R_1}\frac{z^{-k}B^bR_1}{A^bR_1+z^{-k}B^bS_1} \tag{9.20}$$

Cancelling common factors, we obtain

$$= \gamma\frac{z^{-k}B^bT_1}{A^bR_1+z^{-k}B^bS_1} = \gamma z^{-k}\frac{B_b}{\phi_{cl}} \tag{9.21}$$

First observe that the closed loop transfer function, given by Eq. 9.21, is in the desired form, specified by Eq. 9.3 on page 329.

It is important to note that this desired form is obtained by cancelling R_1 in Eq. 9.20. On re-examining Fig. 9.4, we see that R_1 in the denominator is from the feed forward term while that in the numerator comes from the feedback loop calculation. Recall that R_1 is calculated by solving Aryabhatta's identity given by Eq. 9.14 on page 330. There is no constraint that R_1 has to be stable. In case R_1 is unstable, and if there are differences between the numerically obtained R_1 in the feedback loop and that in the feed forward term, the closed loop system will become unstable.

We will now illustrate this with an example, in which R_c has a zero outside the unit circle.

Example 9.5 Design a 2-DOF pole placement controller for the plant with transfer function

$$G(z) = z^{-1}\frac{1-3z^{-1}}{(1-2z^{-1})(1+4z^{-1})}$$

such that the following transient requirements are met:

1. Rise time should be less than or equal to ten samples (*i.e.*, $N = 10$).
2. Overshoot should be less than or equal to 10% (*i.e.*, $\varepsilon = 0.1$).

We have

$$B^b = 1 - 3z^{-1}$$
$$A^b = (1 - 2z^{-1})(1 + 4z^{-1}) = 1 + 2z^{-1} - 8z^{-2}$$
$$A^g = 1,\ B^g = 1$$

Let us first consider the rise time condition:

$$\omega \geq \frac{\pi}{2N} = \frac{\pi}{2 \times 10} = 0.1571$$

We will choose $\omega = 0.1571$. Let us next consider the overshoot condition:

$$\rho \leq \varepsilon^{\omega/\pi} = 0.1^{\pi/20/\pi} = 0.1^{0.05} = 0.8913$$

We will let $\rho = 0.8913$. The characteristic polynomial is obtained as

$$\phi_{cl} = 1 - 2z^{-1}\rho\cos\omega + \rho^2 z^{-2} = 1 - 1.7607z^{-1} + 0.7944z^{-2}$$

We obtain the following Aryabhatta's identity:

$$A^b R_1 + z^{-1} B^b S_1 = \phi_{cl}$$

Solving this, we obtain

$$S_1 = -2.6945 + 2.8431z^{-1}$$
$$R_1 = 1 - 1.0662z^{-1}$$

M 9.7 may be used to design this controller. Implementation of this controller as in Fig. 9.4 makes this system unstable. Depending on the accuracy with which the implementation is carried out, it could take more or less time. For example, the above mentioned closed loop becomes unstable after about 1,000 time steps when simulated using Simulink code, given in Fig. A.3. In real implementations, it could become unstable a lot sooner.

∎

The problem associated with the implementation as in Fig. 9.1 on page 329 or as in Fig. 9.4 is that there is unstable pole–zero cancellation between the feed forward element and the closed loop transfer function. The obvious solution to this problem is to shift R_c inside the loop, as in Fig. 9.5. Note that it is easy to shift R_c inside the loop as we are using z^{-1} as the independent variable. If, instead, we work with polynomials in powers of z, some extra work is required.

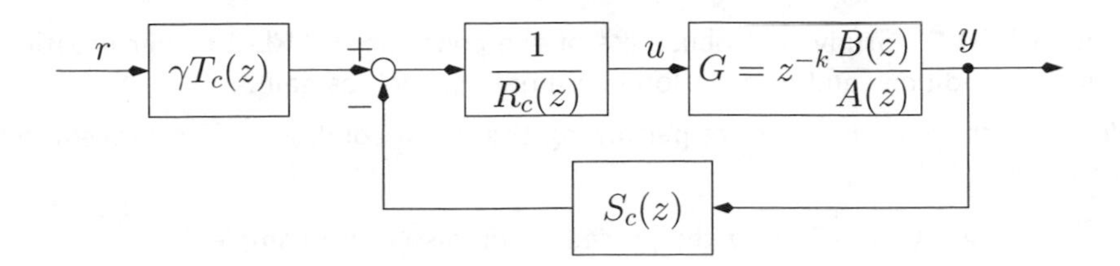

Figure 9.5: 2-DOF pole placement controller with unstable R_c taken inside the feedback loop

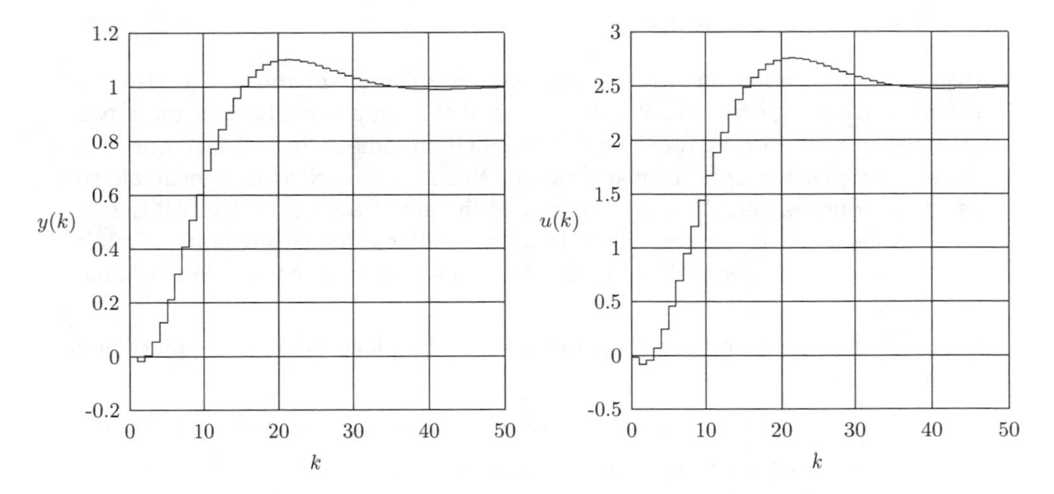

Figure 9.6: Output and input profiles for closed loop system with unstable controller, discussed in Example 9.6

Example 9.6 Solve Example 9.5 once again, with R_c taken inside the loop.

Note that there is no change in the control design procedure. The only change is in simulation: now, it has to be carried out as in Fig. 9.5. M 9.7, in addition to designing the controller, sets up parameters for simulation using Simulink code, given in Fig. A.1 as well.

The system remains stable even after 10,000 steps of iteration. The output and input profiles have been plotted in Fig. 9.6.

Because of the limitations introduced by the nonminimum phase zeros, the controller does not perform as well as the design specifications. ∎

9.4 Internal Model Principle for Robustness

We would like to evaluate the efficacy of CTRL-2 designed in Example 9.4 for handling any mismatches between the model and reality for the magnetically suspended ball problem. This is the topic of discussion in the next example.

Example 9.7 Study the robustness of the controller CTRL-2 to perturbations in initial conditions and in open loop parameters, such as gain.

We will first study the effect of perturbing the initial conditions. The procedure is as follows:

1. Calculate CTRL-2 using the procedure discussed in Example 9.4.
2. Invoke Simulink code presented in Fig. A.8.
3. Perturb the initial conditions using the vector `xInitial` in Matlab work space.
4. Carry out the simulation.

Recall that x_1 denotes the distance between the ball and armature and x_3 denotes the current through the circuit. We perturbed the initial conditions in these two variables in the following range: $x_1(0) = \pm 0.005$ m and $x_3(0) = \pm 1$ A, one at a time. Note that the order of magnitude of these perturbations is comparable to the equilibrium values. We can assume that the initial velocity of the ball is zero and, hence, $x_2(0)$ is not perturbed. In all simulations, the steady state offset is nil. Thus the controller CTRL-2 is able to handle the perturbations in the initial conditions.

We study the effect of perturbing the gain of the plant through the procedure given next:

1. The controller (CTRL-2) is designed with a nominal value of $c = \begin{bmatrix} 1 & 0 & 0 \end{bmatrix}$.
2. After the design is completed, the value of c is perturbed to $\begin{bmatrix} 1.1 & 0 & 0 \end{bmatrix}$ and $\begin{bmatrix} 0.9 & 0 & 0 \end{bmatrix}$ in Matlab work space.
3. For each perturbation, simulation is carried out through the Simulink block diagram given in Fig. A.8.

The resulting profiles are plotted, respectively, as solid and dotted lines, in Fig. 9.7. It is clear that there is a big offset in y for both perturbations. ∎

It is easy to explain the reason for the offset in the above example: output following is proposed to be achieved by choosing γ as the reciprocal of steady state gain. If the gain of the plant used in simulation is different from that used for controller design, naturally γ will be calculated wrongly. We can conclude that the above procedure has a shortcoming whenever there is uncertainty in the steady state gain. We will next describe a method that overcomes this difficulty.

We know from Sec. 7.5 that if an internal model of a step is present in the loop, step disturbances can be rejected. One way to realize this is to ensure that the denominator R_c has this component, see Eq. 7.62 on page 272. If we denote the denominator of the step input as Δ, *i.e.*,

$$\Delta = 1 - z^{-1} \tag{9.22}$$

Eq. 9.11a will have to be replaced by

$$R_c = B^g \Delta R_1 \tag{9.23}$$

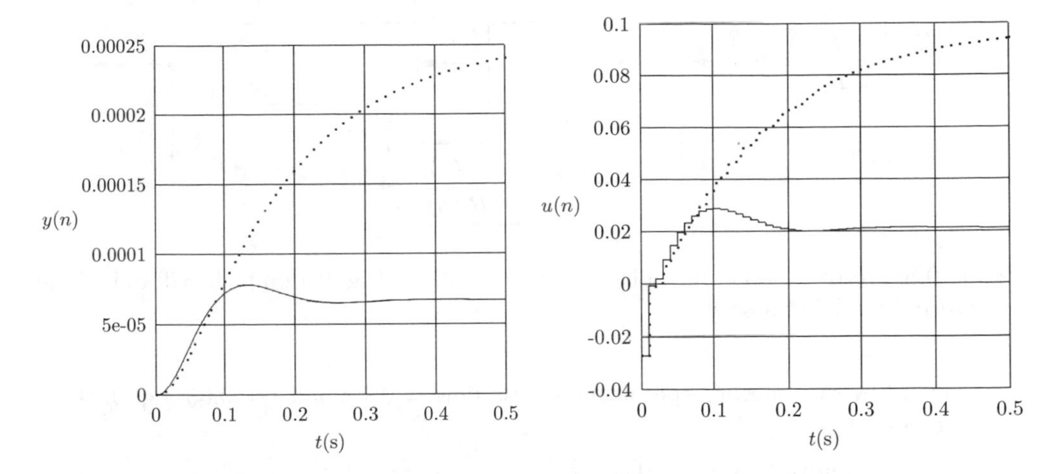

Figure 9.7: Output $y(n)$ and input $u(n)$ profiles for perturbations in plant gain, as discussed in Example 9.7 using the controller designed in Example 9.4 (CTRL-2). Solid lines correspond to increasing the gain ($c = [1.1\ 0\ 0]$) and dotted lines denote decrease of gain ($c = [0.9\ 0\ 0]$). There is steady state offset in y for both perturbations.

As Δ has a root on the unit circle, it has to be treated as the bad part and not cancelled. This implies that all earlier occurrences of R_1 have to be replaced by ΔR_1. For example, we now have to solve the following Aryabhatta's identity, as opposed to the one in Eq. 9.14 on page 330:

$$A^b \Delta R_1 + z^{-k} B^b S_1 = \phi_{cl} \tag{9.24}$$

Comparing Eq. 9.5 on page 329 with Fig. 8.2 on page 310, we see that we have an extra factor of γ in T_c now. In view of the condition for offset free tracking of step inputs derived in Eq. 8.23 on page 311, we see that a sufficient condition for offset free tracking is now obtained as

$$S_c(1) = \gamma T_c(1) \tag{9.25}$$

Using Eq. 9.11 on page 330, we see that this condition is equivalent to $S_1(1) = \gamma T_1(1)$. Because of Eq. 9.15, this condition becomes

$$S_1(1) = \gamma \tag{9.26}$$

Fig. 9.8 shows a schematic of this approach. We now illustrate this procedure and its efficacy with a simple example.

Example 9.8 Explore the effect of incorporating the internal model of a step in the closed loop of the suspended ball problem.

The calculation begins as in Example 9.3 on page 331. Nevertheless, as there is now an extra factor of Δ, Eq. 9.14 on page 330 has to be changed. Recall that we require rise time = 0.15 s, overshoot = 0.05 and settling time = 0.5 s. This is

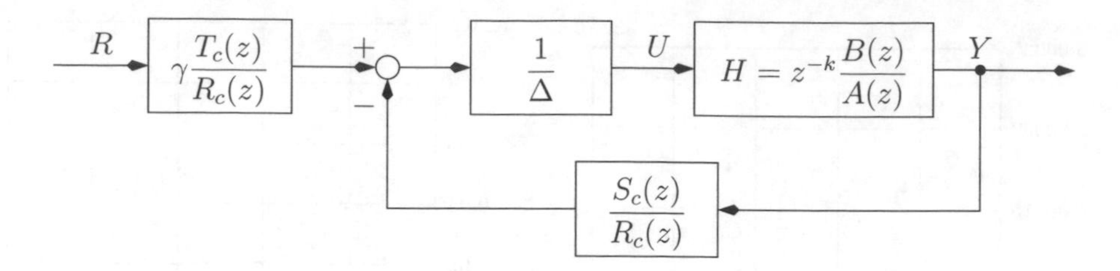

Figure 9.8: Modification of pole placement controller of Fig. 9.1 on page 329 to include an internal model of a step

achieved by the following specifications: rise time = 0.1 s and $\varepsilon = 0.05$. Eq. 9.24 becomes

$$(1-1.3678z^{-1})(1-z^{-1})R_1 + z^{-1}(1+2.9877z^{-1})S_1 = 1-1.7006z^{-1}+0.7411z^{-2} \tag{9.27}$$

Solving this equation for R_1 and S_1 using xdync.m, we obtain

$$\begin{aligned} R_1 &= 1+0.4507z^{-1} \\ S_1 &= 0.2165-0.2063z^{-1} \end{aligned} \tag{9.28}$$

Using Eq. 9.11,

$$\begin{aligned} R_c &= -3.7209\times 10^{-5}(1+0.654z^{-1}+0.0916z^{-2})(1-z^{-1}) \\ S_c &= 0.2165-0.4443z^{-1}+0.285z^{-2}-0.0555z^{-3} \\ T_c &= A^g = 1-1.099z^{-1}+0.269z^{-2} \end{aligned}$$

Using Eq. 9.16 on page 331, we obtain $\gamma = 0.0102$, which can be shown to be equal to $S_1(1)$, satisfying the offset free tracking condition of Eq. 9.26. We will refer to this controller as CTRL-3.

By incorporating the following changes, M 9.8 generalizes M 9.5:

1. There is now one more input parameter. The polynomial, whose internal model has to be introduced in the loop, has to be passed through the argument `Delta`. For example, if `Delta` is set as [1 -1], an internal model of $\Delta = 1 - z^{-1}$ is introduced in the loop. Note that if no internal model is to be introduced, Δ has to be chosen as 1.
2. The call to `xdync.m` is now changed: we now pass $A^b\Delta$ in place of A^b.

The controller described above, namely CTRL-3, is obtained by setting `rise=0.1`, `Delta=[1 -1]` in M 9.9, which also sets up the simulation parameters for the magnetically suspended ball problem. If instead we let `rise=0.15`, `Delta=1`, we obtain CTRL-1, discussed in Example 9.3.

After execution of M 9.9, Simulink code given in Fig. A.8 is executed. The resulting profiles for nominal as well as positive and negative perturbations in c are reported in Fig. 9.9. It is easy to see that the presence of an internal model of the step in the control loop has removed the offsets seen in Fig. 9.7.

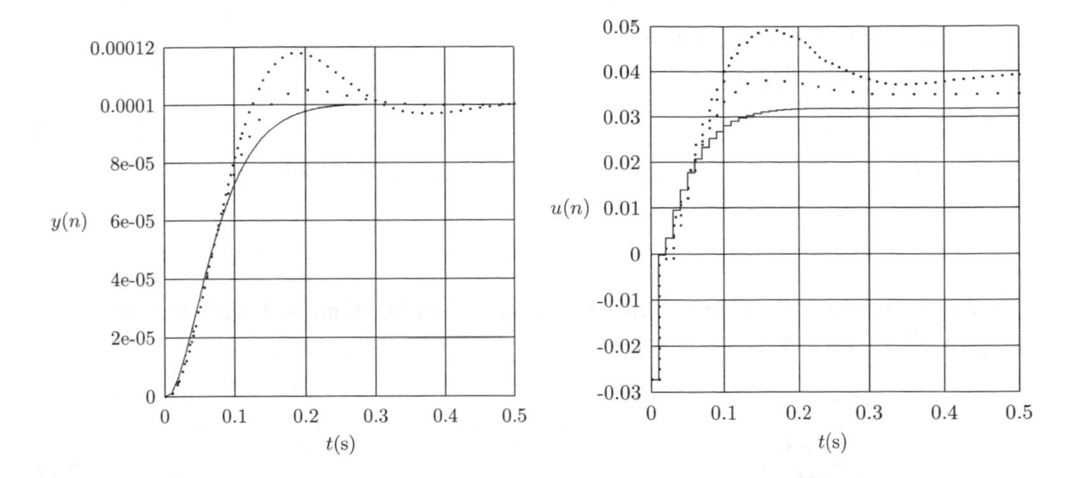

Figure 9.9: Output $y(n)$ and input $u(n)$ profiles for perturbations in plant gain, as discussed in Example 9.7 using the controller designed in Example 9.8 (CTRL-4). Solid lines correspond to increasing the gain ($c = [1.1\ 0\ 0]$), dense dotted lines denote decrease of gain ($c = [0.9\ 0\ 0]$) and sparse dotted lines correspond to nominal state. Offsets seen in Fig. 9.7 have been eliminated.

Although the steady state offsets in y have been removed, the transient requirements are no longer met. This is because the internal model principle provides relief only at steady state, see Sec. 7.5. The control efforts required for perturbed conditions are different from that of the nominal model.

Simulations have shown that CTRL-3 can reject the perturbations in initial conditions as well.

Note that the code presented in this example can generate and evaluate the performance of all controllers derived so far, namely CTRL-1 to CTRL-3.

∎

We conclude this section with an example that has to do with the control of an IBM Lotus Domino server. Through this example, we illustrate another utility of the integral mode.

Example 9.9 Determine a 2-DOF pole placement control for an IBM Lotus Domino server [22] with the transfer function

$$G(z) = \frac{0.47z^{-1}}{1 - 0.43z^{-1}}$$

such that the following transient specifications in tracking a step input are met: rise time ≤ 10 and overshoot condition, $\varepsilon \leq 0.01$.

We obtain the following factors:

$$\begin{aligned} A^g &= 1 - 0.43z^{-1} \\ A^b &= 1 \\ B^g &= 0.47 \\ B^b &= 1 \\ k &= 1 \end{aligned}$$

In all these naturally discrete time problems, T_s can be taken as 1 without loss of generality. We obtain

$$\begin{aligned} N_r &= 10 \\ \omega &= 0.1571 \\ \rho &= 0.7943 \\ \phi_{cl} &= 1 - 1.5691z^{-1} + 0.6310z^{-2} \end{aligned}$$

Because step signals are to be followed, we will assume an internal model of steps in the loop. We need to solve Aryabhatta's identity, given by Eq. 9.14, reproduced here for convenience:

$$A^b \Delta R_1 + z^{-k} B^b S_1 = \phi_{cl}$$

For the current problem, the above equation becomes

$$(1 - z^{-1})R_1 + z^{-1}S_1 = 1 - 1.5692z^{-1} + 0.6310z^{-2}$$

Because the condition on C, namely Eq. 7.120 on page 291, is not satisfied, there is no unique least degree solution. Nevertheless, our implementation of this solver gives the following solution,

$$\begin{aligned} R_1 &= 1 - 0.6310z^{-1} \\ S_1 &= 0.0619 \end{aligned}$$

the correctness of which can be easily verified. The 2-DOF pole placement controller is given by

$$\begin{aligned} R_c &= 0.47 - 0.7665z^{-1} + 0.2965z^{-2} \\ S_c &= 0.0619 - 0.0266z^{-1} \\ T_c &= 1 - 0.43z^{-1} \\ \gamma &= 0.0619 \end{aligned}$$

M 9.10 implements this solution. The resulting output and input profiles are shown in Fig. 9.10. As in the previous examples, all requirements, except the rise time, are met. The rise time requirement can also be met by specifying a tighter requirement.

We will take this up example once again in Example 9.16 on page 357 to discuss performance in the presence of input limits.

∎

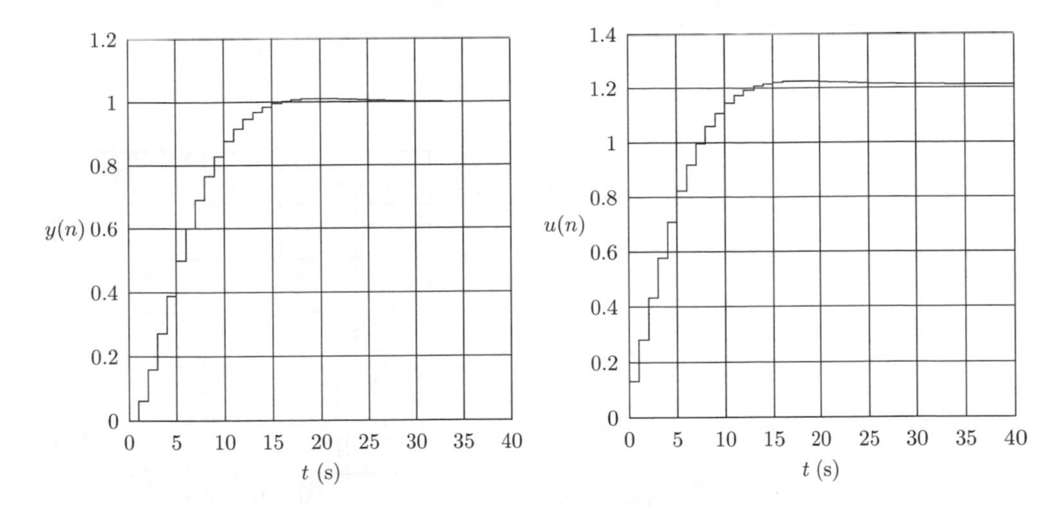

Figure 9.10: Step response of the IBM Lotus Domino server, discussed in Example 9.9

9.5 Redefining Good and Bad Polynomials

In the previous section, we had taken the polynomial with roots inside the unit circle to be good. In this section, through an example, we illustrate that we need to redefine the concept of what is meant by good and bad polynomials.

Example 9.10 We will now design a controller for a DC motor described by [2]. The system is described by the following state space equations:

$$\frac{d}{dt}\begin{bmatrix}x_1\\x_2\end{bmatrix} = \begin{bmatrix}-1 & 0\\1 & 0\end{bmatrix}\begin{bmatrix}x_1\\x_2\end{bmatrix} + \begin{bmatrix}1\\0\end{bmatrix}u$$

$$y = \begin{bmatrix}0 & 1\end{bmatrix}\begin{bmatrix}x_1\\x_2\end{bmatrix}$$

The objective of this exercise is to design a pole placement controller that has a rise time of 3 seconds and overshoot of not more than 0.05 for a step change in the command signal.

The sampling time T_s is chosen as 0.25 s. The transfer function of the system is

$$G(z) = z^{-1}\frac{0.0288 + 0.0265z^{-1}}{1 - 1.7788z^{-1} + 0.7788z^{-2}}$$

We obtain the following factorizations:

$$\begin{aligned}
A^g &= 1 - 0.7788z^{-1}\\
A^b &= 1 - z^{-1}\\
B^g &= 0.0288 + 0.0265z^{-1} = 0.0288(1 + 0.9201z^{-1})\\
B^b &= 1
\end{aligned}$$

The controller design proceeds as in the ball system, explained in Example 9.3. The transient specifications result in the following relations, leading to the desired

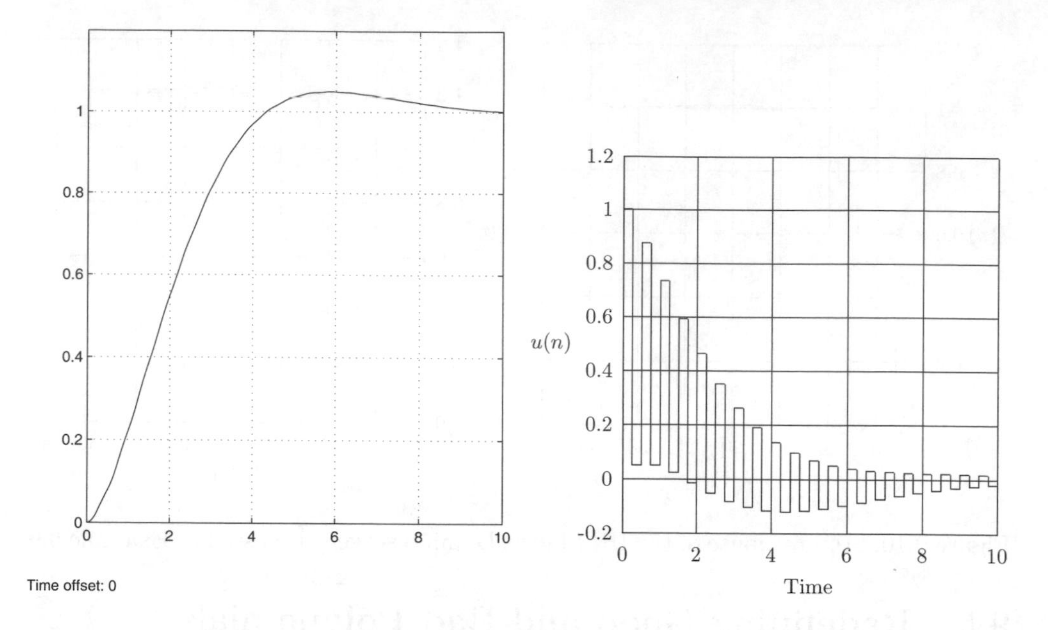

Figure 9.11: Output (left) and input (right) profiles for the motor example obtained with $T_1 = 1$. Control effort $u(n)$ is oscillatory and, hence, not acceptable. The rise time constraint is not met in $y(n)$.

closed loop characteristic polynomial:

$$\begin{aligned}
N &= 12 \\
\omega &= 0.1309 \\
\rho &= 0.8827 \\
\phi_{cl} &= 1 - 1.7502z^{-1} + 0.7791z^{-2}
\end{aligned}$$

Note that as A has $1 - z^{-1}$ as a factor, an internal model of the step is already present. By solving Aryabhatta's identity, of the form in Eq. 9.14, we obtain the following control parameters:

$$\begin{aligned}
R_1 &= 1 - 0.7791z^{-1} \\
S_1 &= 0.0289 \\
R_c &= 0.0288 + 0.0041z^{-1} - 0.0206z^{-2} \\
S_c &= 0.0289 - 0.0225z^{-1} \\
T_c &= 1 - 0.7788z^{-1}
\end{aligned}$$

M 9.11 carries out these calculations.

The efficacy of this controller is checked by simulation through Simulink code in Fig. A.8. The resulting profiles are shown in Fig. 9.11.

The control effort is highly oscillatory. In view of this, this controller is rejected. As in the ball problem (see Example 9.3), while the overshoot and the settling time constraints have been met, the rise time is larger than what is specified. ∎

It is easy to explain why the control effort is oscillatory. Substituting the expressions for R_c, S_c and T_c from Eq. 9.11 on page 330 into the control law given by Eq. 9.5, we obtain

$$u(n) = \gamma \frac{A^g T_1}{B^g R_1} r(n) - \frac{A^g S_1}{B^g R_1} y(n)$$

Because B^g has a zero at -0.9201, the control variable u is oscillatory. Recall from Fig. 5.1 on page 114 that poles on the negative real axis result in oscillations. Recall also that the very definition of good factors has been introduced earlier in this chapter to effect cancellations. And recall from Sec. 7.4.2 that the pole–zero cancellation effected between the plant and the controller appears in other transfer functions. When such a cancellation occurs, the output also oscillates in every sampling interval, even though it may appear to be nonoscillatory if one observes it at the sampling instants only. These are known as hidden oscillations.

One way to prevent this from happening in the current design is to define the factor with a root in the left half of the plane as a bad one, whether or not it lies within the unit circle. As a matter of fact, we can make use of the desired region analysis carried out earlier. For example, we could say that the entire area outside the shaded region in Fig. 7.31 is bad. Unfortunately, however, it is not easy to calculate this region. As a way out, we can take a root whose real part is negative to be bad. We can take this opportunity to define also as bad the factors that are inside the unit circle, but close to it. The reason is that if the factors close to the unit circle are cancelled, this would show up as poles of the controller transfer function and make the control action sluggish. In view of this, we would define the stable poles inside the unit circle also as bad and thus ensure that they don't get cancelled. These two changes have been implemented in M 9.12. Compare this with M 9.3 that defines all the factors inside the unit circle to be good.

We will now apply this new definition of good and bad polynomials to the motor control problem and check whether the oscillations are removed.

Example 9.11 Redo Example 9.10 with good and bad factors defined as in M 9.12.

In Lines 7 and 8 in M 9.8, let the function `polsplit3` be called instead of `polsplit2`. B is factored as

$$B^g = 0.0288$$
$$B^b = 1 + 0.9201z^{-1}$$

while the factoring of A is same as that in Example 9.10. As the transient specifications have not changed, we obtain the old ϕ_{cl}, namely

$$\phi_{cl} = 1 - 1.7502z^{-1} + 0.7791z^{-2}$$

Solving

$$(1 - z^{-1})R_1 + z^{-1}(1 + 0.9201z^{-1})S_1 = 1 - 1.7502z^{-1} + 0.7791z^{-2}$$

using `xdync.m`, we obtain

$$R_1 = 1 - 0.7652z^{-1}$$
$$S_1 = 0.015$$

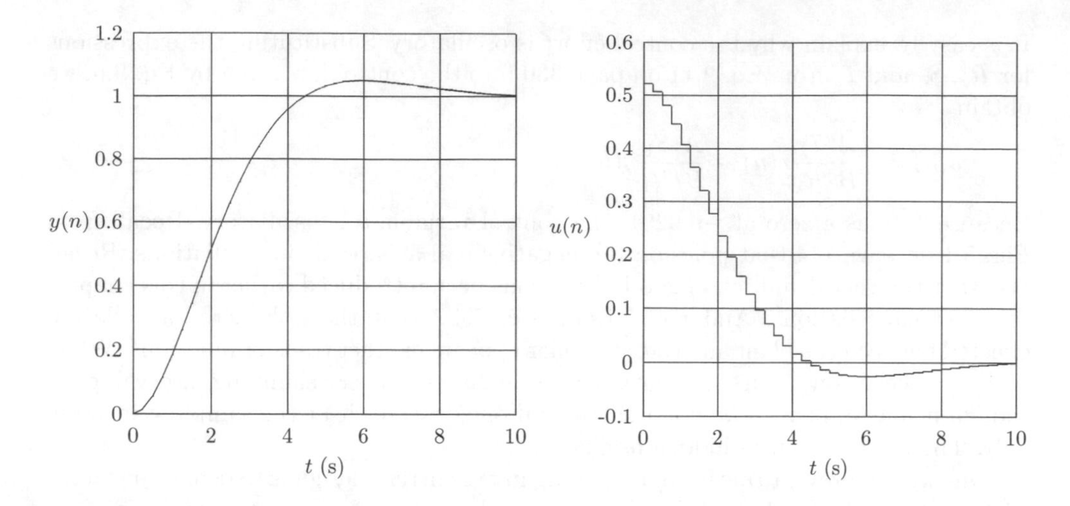

Figure 9.12: Output and input profiles for the motor problem using the new cancellation rule. Oscillations in the control effort have been removed. The rise time constraint is not met in $y(n)$.

Using Eq. 9.11 on page 330,

$$R_c = 0.0288 - 0.022z^{-1}$$
$$S_c = 0.015 - 0.0117z^{-1}$$
$$T_c = A^g = 1 - 0.7788z^{-1}$$

The performance of this controller is checked with the Simulink code given in Fig. A.8. We obtain the profiles as in Fig. 9.12. Observe that the oscillations have been completely removed. In addition, the initial control effort is smaller. Unfortunately, however, the rise time condition is not met.

As in the ball suspension problem, the best control effort is achieved by making the required rise time less than that required. Fig. 9.13 shows the profiles obtained with rise time specified as 2 s. The required conditions of rise time = 3 s and overshoot = 0.05 have been met simultaneously. The controller parameters are

$$\phi_{cl} = 1 - 1.6266z^{-1} + 0.6877z^{-2}$$
$$R_1 = 1 - 0.6584z^{-1}$$
$$S_1 = 0.0318$$

As before, using Eq. 9.11, we obtain

$$R_c = 0.0288 - 0.019z^{-1}$$
$$S_c = 0.0318 - 0.0248z^{-1}$$
$$T_c = A^g$$

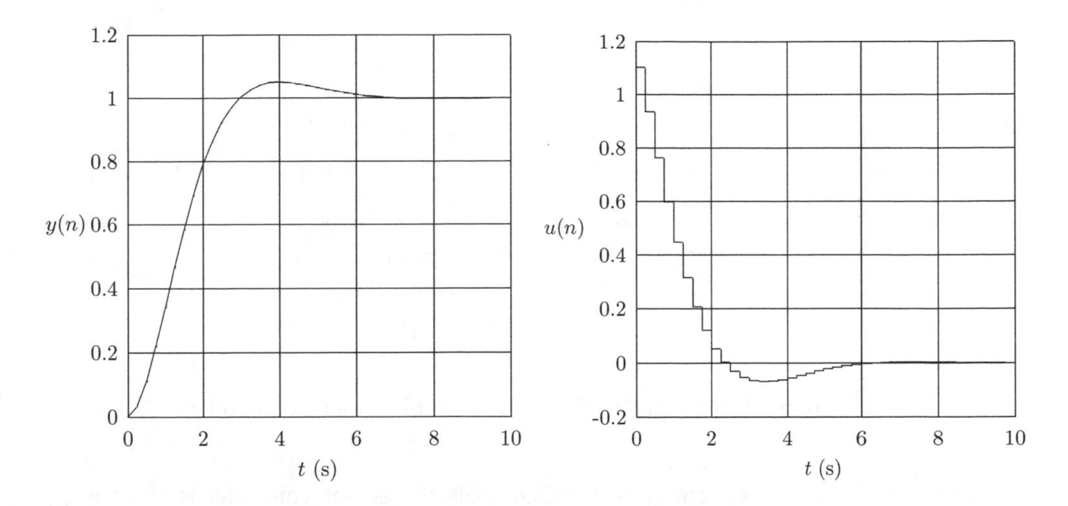

Figure 9.13: The profiles in Fig. 9.12 are improved by requiring the rise time to be less than required. With rise time specified as 2 s, we meet the required conditions of rise time = 3 s and overshoot = 0.05, simultaneously with a reasonable control effort.

Because the plant has Δ as a factor, the controller is capable of handling setpoint changes even if there are some changes in the model parameters. This controller is found to track the perturbations in the initial conditions as well. M 9.13 helps design this controller.

∎

We illustrate the efficacy of the proposed method [40] with a difficult control problem, presented by [52, p. 47].

Example 9.12 Suppose that we have a system with plant ($G(s)$) and disturbance ($H(s)$) transfer functions, as given below,

$$G(s) = \frac{200}{10s+1}\frac{1}{(0.05s+1)^2}$$
$$H(s) = \frac{100}{10s+1}$$

with time in seconds. Design a 2-DOF pole placement controller, so as to satisfy the following requirements:

1. The 90% rise time to a step change in r should be less than 0.3 s and the overshoot should be less than 5%.
2. The output in response to a unit step disturbance should remain within $[-1, 1]$. It should satisfy $y(t) < 0.1$ after 3 s.
3. $u(t)$ should remain within $[-1, 1]$ at all times.

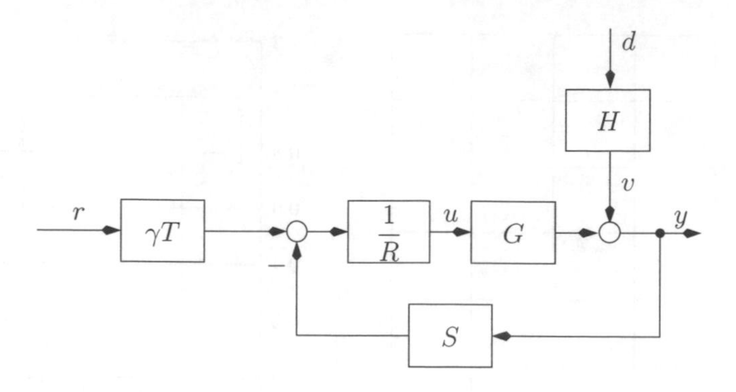

Figure 9.14: Schematic of 2-DOF pole placement controller

A schematic of this system with a 2-DOF pole placement controller is given in Fig. 9.14. The zero order hold equivalent of this system with sampling period $T_s = 0.025$ s yields

$$
\begin{aligned}
G(z) &= \frac{0.0163(1 + 2.9256z^{-1})(1 + 0.2071z^{-1})z^{-1}}{(1 - 0.9975z^{-1})(1 - 1.2131z^{-1} + 0.3679z^{-2})} \\
H(z) &= \frac{0.2497z^{-1}}{1 - 0.9975z^{-1}}
\end{aligned}
$$

where we have used the same symbols G and H. If we define only the roots outside the unit circle as bad, the transfer function between d and y will have a sluggish pole at 0.9975. In view of this, we define this factor as bad. In order to avoid oscillations in the control effort, we take the factor $1 - 0.2071z^{-1}$, which has a negative zero, as in Example 9.11, also as bad. We obtain

$$
\begin{aligned}
B^g &= 0.0163 \\
B^b &= (1 + 2.9256z^{-1})(1 + 0.2071z^{-1}) \\
A^g &= 1 - 1.2131z^{-1} + 0.3679z^{-2} \\
A^b &= 1 - 0.9975z^{-1}
\end{aligned}
$$

The controller designed with rise time as 0.3 s and overshoot as 0.05 results in the first condition being violated: we obtain a rise time of 0.37 s. The is because of the presence of $B_r = B^b$, the nonminimum phase part of the plant numerator. We handle this by overspecifying the rise time requirement. With the required rise time as 0.24 s, we obtain the following:

$$
\begin{aligned}
\phi_{cl} &= 1 - 1.6882z^{-1} + 0.7319z^{-2} \\
R &= 0.0163(1 + 0.2357z^{\;1} + 0.0391z^{\;2})(1 - z^{\;1}) \\
S &= 0.0736(1 - 2.0877z^{-1} + 1.4288z^{-2} - 0.3218z^{-3}) \\
T &= 1 - 1.2131z^{-1} + 0.3679z^{-2}
\end{aligned}
$$

M 9.14 implements these designs. By choosing `st = 1` and `st1 = 0` in Lines 31 and 32 of this code, we can simulate the tracking problem. If instead we choose

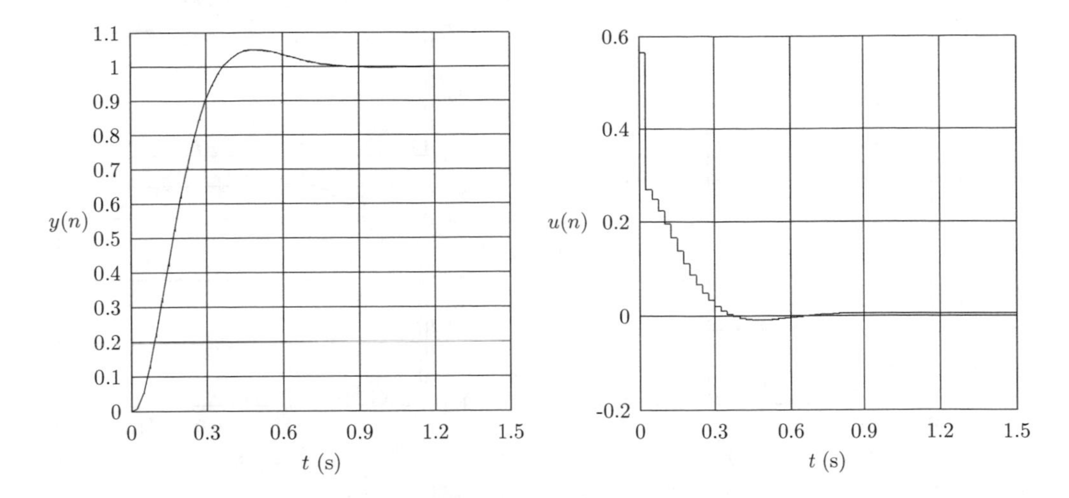

Figure 9.15: Profiles for tracking problem

`st = 0` and `st1 = 1`, we can verify the efficacy of the controller in rejecting the unit step disturbance.

The plots of the output y and the control effort u for the tracking problem are given in Fig. 9.15. One can see that all the tracking requirements are fulfilled. The corresponding plots for the regulation problem are given in Fig. 9.16. The response y returns to within 0.1 in 0.4 s itself, as opposed to the requirement to do this within 3 s. On the flip side, the control effort exceeds the maximum permissible value by 12%. M 9.14 calculates also the gain and phase margins as `Gm = 4.7` and `Pm = 46°`, respectively. We see that the simple 2-DOF pole placement technique has been able to provide a reasonable controller for a difficult plant, characterized by two time constants that vary by two orders of magnitude. Considering that the transient requirements are also quite stiff, we see the efficacy of the proposed control technique.

We have used two tuning parameters in this design: sampling time and rise time. We note that the sampling time T_s has to be smaller than the smallest time constant of the plant, which is 0.05 s. A smaller T_s, of the order of 0.01 s, results in large control efforts. We could quickly arrive at a value of 0.025 s. Tuning of the rise time specification also is straightforward: we have made it small enough so that the required rise time of 0.3 s is met.

Thus, we see that a fairly simple control design technique, posed entirely in the discrete time domain, has been able to provide a reasonably good controller for a difficult problem.

∎

In the next example, we will present the original controller, designed entirely in the continuous time domain, and discuss the implementation issues.

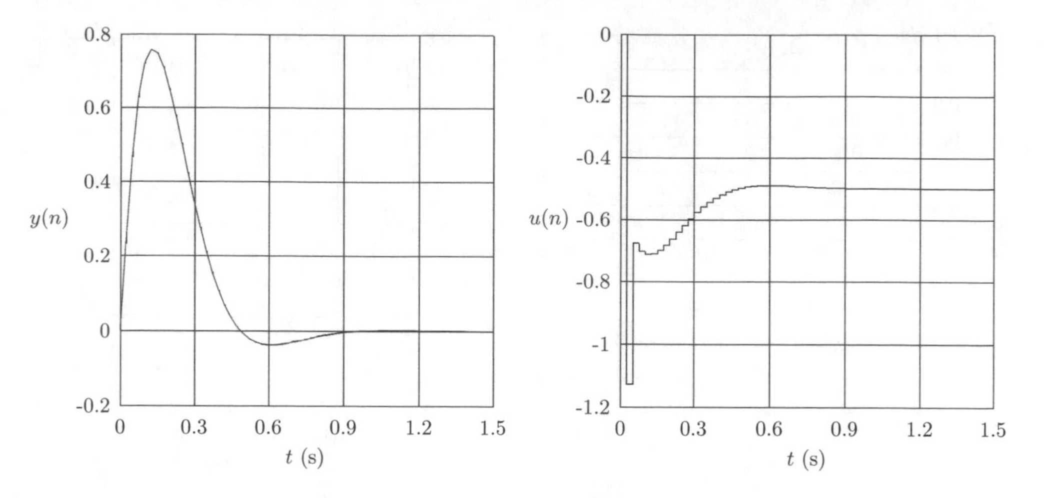

Figure 9.16: Profiles for regulation problem

Example 9.13 For the plant studied in Example 9.12, the controller has been designed by [52, p. 52], in the form of two control blocks:

$$K_y(s) = 0.5\frac{s+2}{s}\frac{0.05s+1}{0.005s+1}$$
$$K_r(s) = \frac{0.5s+1}{0.65s+1}\frac{1}{0.03s+1}$$

Discuss the efficacy of this controller.

Because continuous time domain techniques have been used to arrive at this controller, we do not present them here. The interested reader is referred to the original reference for details. We would, however, like to focus on the implementation issues of this controller.

M 9.15 sets up the plant and the controller blocks. The Simulink block diagram, given in Fig. A.6, carries out the simulations entirely in the continuous time domain. Through execution of these programs, one can easily see that the problem specifications are achieved. M 9.15 calculates also the gain and phase margins as `Gm = 19.7` and `Pm = 51°`, respectively.

Unfortunately, however, it requires some more work to implement this controller, the reason being that one has to discretize the controller first. M 9.15 can be used to study these issues as well. The Simulink block diagram, given in Fig. A.7, may be used to carry out the indicated simulations. If T_s is chosen to be equal to 0.025 s, as in Example 9.12, the controller performs extremely poorly in meeting the setpoint tracking requirements. The degradation in performance is expected because the sampling time used is larger than the time constants of the controller.

The performance of the controller gets better as the sampling time is made smaller. If T_s is chosen to be equal to 0.01 s, the performance becomes comparable to that of the discrete time controller, designed in Example 9.12. In the setpoint tracking problem, the initial control effort exceeds the specified limits of $[-1, 1]$ by about 10%. The gain and phase margins for this implementation are `Gm = 6.7` and `Pm` = 68°, respectively. Indeed, to achieve the continuous time domain performance, sampling periods smaller than 0.001 s are required, as can be verified using the programs mentioned.

In summary, if small sampling times are difficult to achieve, one may consider using the discrete time controller, presented in Example 9.12. If smaller sampling times are permissible, the continuous time controller presented in this example is more suitable.

We would like to conclude this example with a few comments on the design of the continuous time controller. It is not obvious how to arrive at some of the transfer functions in the above expressions. The number of tuning parameters also is large. It is safe to say that the time required to tune the continuous time controller would be at least one order of magnitude more than that required for the discrete time controller of Example 9.12. ∎

9.6 Comparing 1-DOF and 2-DOF Controllers

The main benefit of the 2-DOF controller is that we can shape the response to reference and disturbance signals independently. We have seen in Sec. 8.3.6 that the PID controllers may be implemented in a 2-DOF configuration to overcome the derivative kick. This allows tight tuning for disturbance rejection, without worrying about input saturation, see Problem 8.9. The 2-DOF pole placement controller generalizes these ideas. Although we have chosen T_1 to be unity in Eq. 9.15 on page 330, it is possible to assign a suitable value so as to shape the response to reference signals [30].

In this section, we show that in 2-DOF controllers, the system type is dependent on the type of signal – reference, as opposed to disturbance signal. The reason is that S_c, which is present in the loop, helps achieve the desired system type for the disturbance signal. Because T_c is not in the loop, it does not have to satisfy any special condition. Because T_c could be different from S_c, the type of the system could be different from the reference signal. We illustrate these ideas with an example.

Example 9.14 For the system given in Example 7.10, design 1-DOF and 2-DOF controllers, with and without an additional integral term, and evaluate the system type.

In Example 7.10, for a closed loop characteristic polynomial of $\phi_{cl} = 1 - 0.5z^{-1}$, we have designed two controllers for the 1-DOF controller, for two cases: with and without an additional integral term. Let us refer to them as G_{c1} and G_{c2}. That is,

$$G_{c1} = \frac{1}{2}, \quad G_{c2} = \frac{1.5z - 1}{z - 1}$$

These controllers have been designed with the following two changes from the 2-DOF control design techniques. The first step is to take the entire A and B as bad. That is, we do not cancel the good parts. We end up solving Aryabhatta's identity,

$$AR_c + z^{-k}BS_c = \phi_{cl} \tag{9.29}$$

If the controller includes one integral term, we choose $R_c = \Delta R_1$, as explained in Sec. 9.4, and solve the following Aryabhatta's identity:

$$A\Delta R_1 + z^{-k} B S_c = \phi_{cl} \tag{9.30}$$

The second step is to take $\gamma T_c = S_c$, irrespective of whether or not we include the integral term. We have implemented these calculations through M 7.7 in Example 7.10 on page 275 and seen that G_{c1} and G_{c2} give rise to Type 1 and Type 2 systems, respectively.

We will now calculate the 2-DOF controller for the same ϕ_{cl}. By choosing `Delta` to be 1 in Line 4, we design a controller with no additional integral term. We will refer to the resulting controller as G_{c3}. It is given by

$$R_c = 1, \quad S_c = 0.5, \quad T_c = 1, \quad \gamma = 0.5$$

By choosing `Delta` to be `[1 -1]` in Line 4, we design a controller with an additional integral term. We will refer to the resulting controller as G_{c4}. It is given by

$$R_c = 1 - z^{-1}, \quad S_c = 1.5 - z^{-1}, \quad T_c = 1, \quad \gamma = 0.5$$

The transfer function between the reference signal r and the output signal y is given by Eq. 9.21. Because $\gamma = 0.5$ for both cases, we obtain the following transfer function T between the reference signal and the output signal:

$$T = \frac{0.5}{z - 0.5}$$

Thus, the sensitivity transfer function S is given by

$$S = 1 - T = \frac{z-1}{z-0.5}$$

for both controllers. It follows that the system type is the same for both the controllers. In other words, we obtain the same system type, whether or not the controller includes an additional integral term. Let us calculate the steady state offsets for a step signal first. Substituting the above S and $R = z/(z-1)$ in Eq. 7.80, we obtain the steady state error to step inputs as

$$\lim_{n\to\infty} e(n) = \lim_{z\to 1} \frac{z-1}{z}\frac{z-1}{z-0.5}\frac{z}{z-1} = 0$$

Thus, both controllers reject step changes completely. Now, let us see what happens to step inputs to ramps. Using the same procedure as above, the following steady state error results:

$$\lim_{n\to\infty} e(n) = \lim_{z\to 1} \frac{z-1}{z}\frac{z-1}{z-0.5}\frac{z}{(z-1)^2} = 2$$

This can be verified by executing the Simulink program `stb_disc.mdl` of Fig. A.1 on page 525, followed by M 9.16.

Although G_{c4} introduces an extra integral term, the system is only Type 1 with respect to the reference signal. It is easy to check that the system is of Type 2 with respect to the disturbance signal.

∎

In this example, we have taken ϕ_{cl} as $1 - 0.5z^{-1}$. Instead, we could have arrived at it through performance specifications, as in Sec. 7.7.

Now we address the question of cancellation of good factors. Why should we even cancel them? Why can't we just solve Eq. 9.29 to design a 1-DOF controller? The 2-DOF control design ensures that at least some part of the controller is good, see for example Eq. 9.11 on page 330. If, on the other hand, R_c and S_c are obtained by solving only Eq. 9.29, there is no guarantee that R_c and S_c have good factors. Of course, if we don't cancel the good factors, we also end up solving a higher degree Aryabhatta's identity. We illustrate these ideas with an example.

Example 9.15 Design a pole placement controller that places the poles of the plant

$$G = \frac{z^{-1}(1 + 0.9z^{-1})(1 - 0.8z^{-1})}{(1 - z^{-1})(1 - 0.5z^{-1})}$$

so as to give rise to a closed loop characteristic polynomial $\phi_{cl} = 1 - z^{-1} + 0.5z^{-2}$ and to have nonzero offset to step inputs.

Let us first design the 2-DOF controller. M 9.17 may be used for this purpose with the variable `control` in Line 10 assigned as 2. We obtain

$$B^g = 1 - 0.8z^{-1}, \quad B^b = 1 + 0.9z^{-1}, \quad k = 1$$
$$A^g = 1 - 0.5z^{-1}, \quad A^b = 1 - z^{-1}$$

Because the plant has an integral term, we need to solve Eq. 9.14 on page 330, which is

$$(1 - z^{-1})R_1 + z^{-1}(1 + 0.9z^{-1})S_c = 1 - z^{-1} + 0.5z^{-2}$$

Solving this, we obtain $R_1 = 1 - 0.2632z^{-1}$, $S_1 = 0.2632$. Notice that the zero of R_1 is in a good location. Using Eq. 9.11 and Eq. 9.16 on page 331, we obtain

$$R_c = 1 - 1.0632z^{-1} + 0.2105z^{-2}, \quad S_c = 0.2632 - 0.1316z^{-1}$$
$$T_c = 1 - 0.5z^{-1}, \quad \gamma = 0.2632$$

We now calculate a 1-DOF pole placement controller, by assigning a value of 1 to the variable `control` in Line 10 of M 9.17. We now solve Eq. 9.29, which is

$$(1 - z^{-1})(1 - 0.5z^{-1})R_c + z^{-1}(1 + 0.9z^{-1})(1 - 0.8z^{-1})S_c = \phi_{cl}$$

with ϕ_{cl} unchanged. The solution is obtained as

$$R_c = 1 - 2.4292z^{-1} - 2.3233z^{-2}, \quad S_c = 2.9292$$

The roots of R_c are 3.1636 and -0.7344. We see that this controller is unstable. Notice also that it has a negative pole, which should result in oscillatory control effort, see the discussion in Sec. 9.5. In contrast, the previous controller has only good poles. Our observations may be verified by executing Simulink code given in Fig. A.1. We obtain the plots given in Fig. 9.17. Note that, as expected, the controller obtained after cancellation of good factors is performing a lot better. ∎

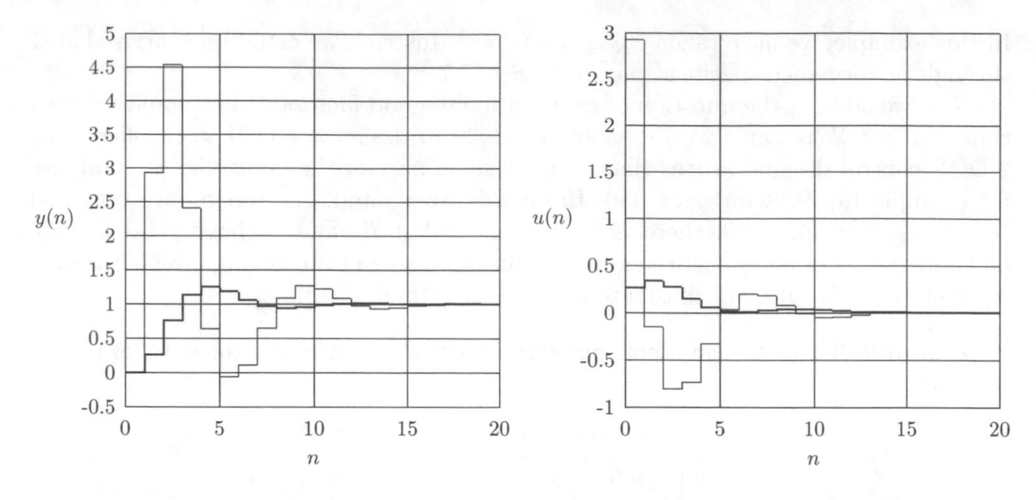

Figure 9.17: Comparison of pole placement controllers with cancellation (thick lines) and without cancellation (thin lines) of good factors, as discussed in Example 9.15

This example shows the benefits of cancelling good factors. Is it possible to cancel the good factors in the 1-DOF controller design so as to overcome the difficulties explained in the above example? We invite the reader to explore this option in the above example, as well as in Problem 9.5.

We now summarize the pole placement controller design:

1: Specify the constraints on rise time, overshoot and decay ratio, and determine ϕ_{cl}.
2: **if** 2-DOF controller **then**
3: Split A and B into good and bad factors: $A = A^g A^b$ and $B = B^g B^b$.
4: Solve Eq. 9.14 on page 330 or Eq. 9.24 on page 339 for R_1 and S_1. Assign a suitable value for T_1; a possible choice is 1.
5: R_c, S_c and T_c given by Eq. 9.11. In case the internal model of a step is used, R_c has to be calculated using Eq. 9.23 on page 338. In this case, check the calculations using Eq. 9.25 or Eq. 9.26.
6: **else if** 1-DOF controller **then**
7: Solve Eq. 9.29 for R_c and S_c. Let $T_c = S_c$ and choose appropriate γ.
8: **end if**
9: The controller is given by Eq. 9.5 on page 329.
10: Use the configuration of Fig. 9.5 on page 337 for implementation.

9.7 Anti Windup Controller

Sometimes there are limits put on the control action. This may be done from a safety point of view. This could also happen when the actuator saturates. A schematic of a 2-DOF controller with a limiter is shown in Fig. 9.18. It is clear that if the controller is not allowed to work freely, the performance could deteriorate. There could be additional difficulties, as we now explain.

If the output of a control block with integral mode is constrained by a limiter, the controller should be informed about it. If not, the controller would have no idea

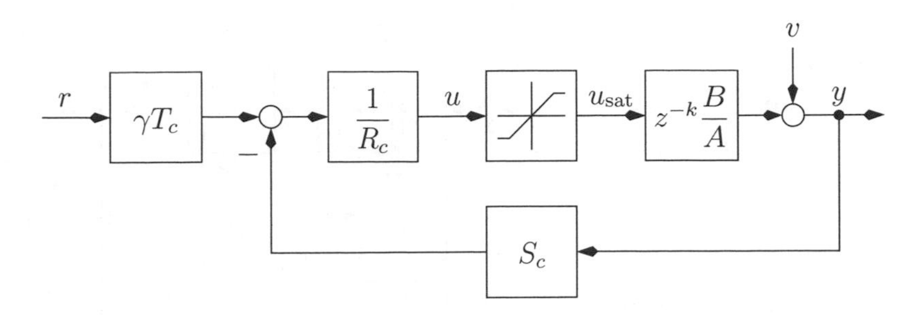

Figure 9.18: 2-DOF controller with limits on control effort. Up to some value, the controller output u is the same as the plant input u_{sat}. Beyond this, u_{sat} is different from u.

about the saturation and hence would increase its output further. This situation is known as *integrator windup*, an illustration of which is given in Example 9.16.

The problem with integrator windup is that, if not corrected, the controller output u could be quite different from the actual effort u_{sat} that goes into the plant. Now suppose that the sign of the control effort has to be changed. Because the magnitude of u is much larger than u_{sat}, it could take a lot of time to change the sign, as required. As a result, u_{sat} would continue to be at the wrong maximum value.

Note that the reduced performance due to the presence of limiters and the integrator windup are two different issues. The former is inevitable, as the performance in the presence of limits cannot be the same as the one without any limits. On the other hand, the integrator windup is an avoidable problem. An ideal situation is that u should not increase beyond u_{sat}. This will allow u to change sign quickly, as and when required.

The most obvious way to handle the integrator windup is to compare u and u_{sat} and when a difference occurs, hold u at u_{sat}. We now present a feedback approach [50] to solve this problem in our familiar 2-DOF controller framework. In particular, we will show that the control scheme presented in Fig. 9.19 can negate the effects of the saturating element, where we have introduced new polynomials P, E and F. Note that we use a positive feedback around the limiter.

First we will show that it is possible to choose E and F that will reduce Fig. 9.19 to Fig. 9.18 when the control effort stays within the limits. Because the gain of the limiting block is one under this assumption, the inner loop can be replaced by the transfer function $1/(1 - E/F) = F/(F - E)$. As a result, the schematic of Fig. 9.19 can be reduced to that in Fig. 9.20. Note that if we choose

$$F = E + PR_c \tag{9.31}$$

where P is stable, Fig. 9.20 reduces to Fig. 9.18. Thus, we have shown that when the control effort is within limits, the scheme of Fig. 9.19 implements the standard 2-DOF pole placement control.

We will next show that when the control effort exceeds the limits, its impact on y is eliminated, at least asymptotically. Thus, there will be a difference between the computed control effort u and the one actually used, u_{sat}. We will model the impact of this through a variable δ acting at the location of the nonlinear element, as in

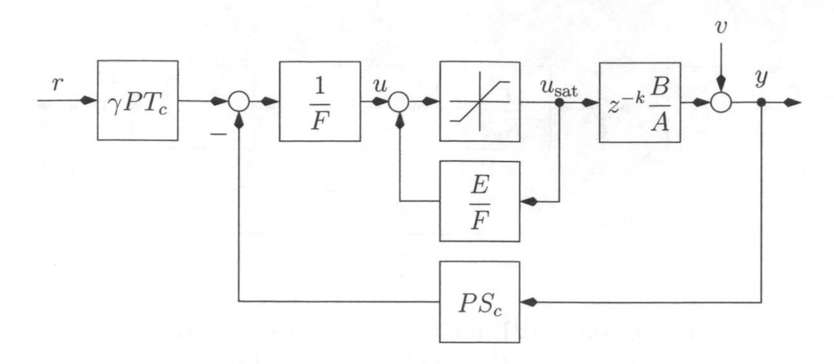

Figure 9.19: 2-DOF controller with feedback loop around the limiter. The controller output u is modified by comparing it with the actual control effort u_{sat} used.

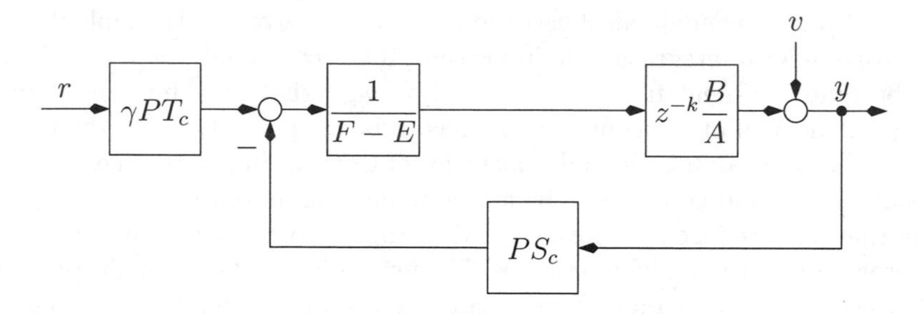

Figure 9.20: 2-DOF controller with feedback around limiter is reduced to 2-DOF controller without limiter when $F = E + PR_c$ and P is stable

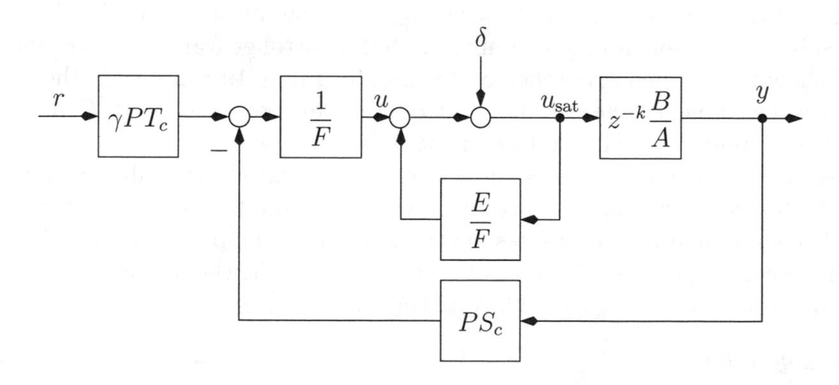

Figure 9.21: Modelling the mismatch between u and u_{sat} by an exogeneous variable δ

Fig. 9.21, where we have taken v to be zero as the focus is on δ. It is easy to see that Fig. 9.21 can be redrawn as in Fig. 9.22. Because the two feedback elements are in parallel, they can be added to give $PS_c - Ez^kA/B$ in the negative feedback path.

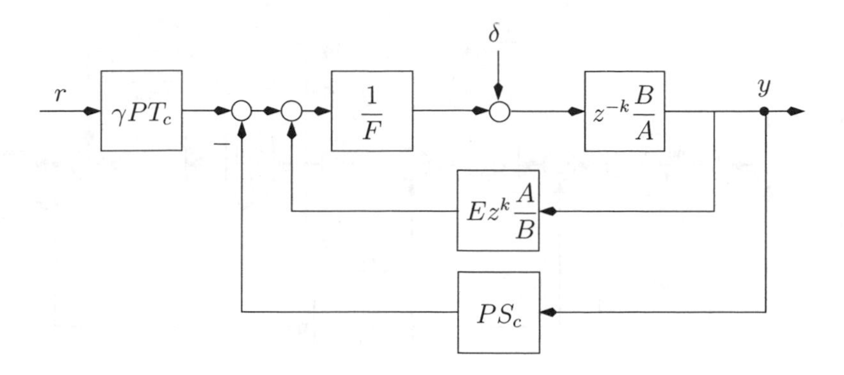

Figure 9.22: Simplified version of Fig. 9.21

The transfer function between δ and y is given by

$$T_{\delta y} = \frac{z^{-k}\frac{B}{A}}{1 + z^{-k}\frac{B}{A}\left(PS_c - Ez^k\frac{A}{B}\right)\frac{1}{F}} = \frac{z^{-k}BF}{FA + (z^{-k}BPS_c - EA)}$$

Using Eq. 9.31 in the denominator, we obtain

$$T_{\delta y} = \frac{z^{-k}BF}{PR_cA + z^{-k}BPS_c} \tag{9.32}$$

Note that $R_cA + z^{-k}BS_c$ is the closed loop characteristic polynomial; an expression for it can be obtained as $\phi_{cl}A^gB^g$, by multiplying both sides of Eq. 9.14 on page 330 with A^gB^g. If we choose

$$F = AR_c + z^{-k}BS_c = \phi_{cl}A^gB^g \tag{9.33}$$

Eq. 9.32 becomes

$$T_{\delta y} = z^{-k}\frac{B}{P} \tag{9.34}$$

If P is stable and well behaved, the effect of $T_{\delta y}$ will diminish with time. Many choices have been proposed in the literature for P. A popular choice is

$$P = A \tag{9.35}$$

if A is stable. If any factor of A is not stable, we replace this factor with its stable reflection. We illustrate this approach with an example.

Example 9.16 Explain the steps discussed in this section with the problem of control of an IBM Lotus Domino server, discussed in Example 9.9.

The transfer function of this system is given by

$$G(z) = \frac{0.47z^{-1}}{1 - 0.43z^{-1}}$$

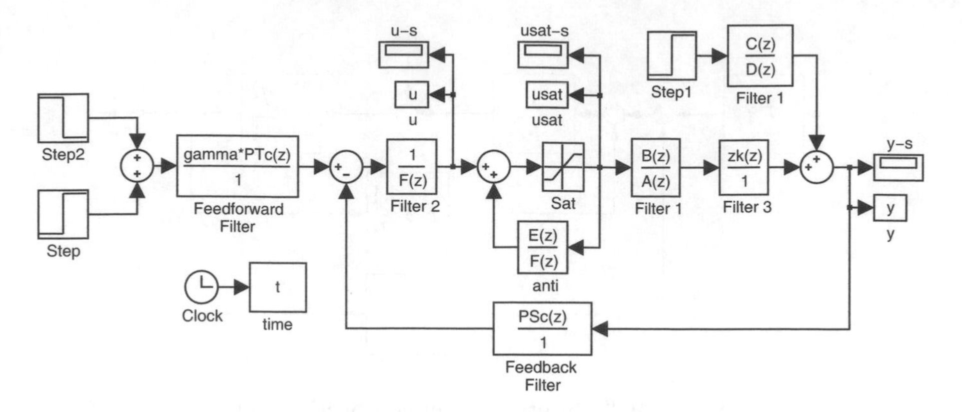

Figure 9.23: Anti windup control (AWC) in the 2-DOF framework. The Simulink code is available at HOME/matlab/stb_disc_sat.mdl, see Footnote 1 on page 367. Parameters for simulation may be established by executing a program, such as the one in M 9.18.

From Example 9.9, we obtain $A^g = 1 - 0.43z^{-1}$, $A^b = 1$, $B^g = 0.47$, $B^b = 1$ and $k = 1$. We calculate the desired characteristic polynomial now:

$$\begin{aligned}
\phi_{cl} &= 1 - 1.5691z^{-1} + 0.6310z^{-2} \\
R_c &= 0.47 - 0.7665z^{-1} + 0.2965z^{-2} \\
S_c &= 0.0619 - 0.0266z^{-1} \\
T_c &= 1 - 0.43z^{-1} \\
\gamma &= 0.0619
\end{aligned}$$

From Eq. 9.32–9.34, we obtain

$$\begin{aligned}
F &= 0.47 - 0.9396z^{-1} + 0.6137z^{-2} - 0.1275z^{-3} \\
P &= A^g = 1 - 0.43z^{-1}
\end{aligned}$$

Using M 9.18, and the Simulink code in Fig. 9.23, we design the anti windup controller and carry out the following different simulations. In each of them, we give a reference signal of $+1$ at the zeroth instant and -1 at the 500th instant. The resulting plots are shown in Fig. 9.24.

1. The first simulation is carried out assuming that there are no actuator constraints. No anti windup controller is used in this simulation. The resulting u, u_{sat} and y profiles are plotted with thick dotted lines in Fig. 9.24. Because there are no limits on the control effort, the output of the plant tracks the reference signal without any difficulty. Note that the magnitude of the control signal is greater than one. Also note that u and u_{sat} are identical, as there are no constraints.

   When M 9.18 is executed, three options come in a menu. The first option is to be selected for this design, followed by execution of the Simulink program in Fig. 9.23.

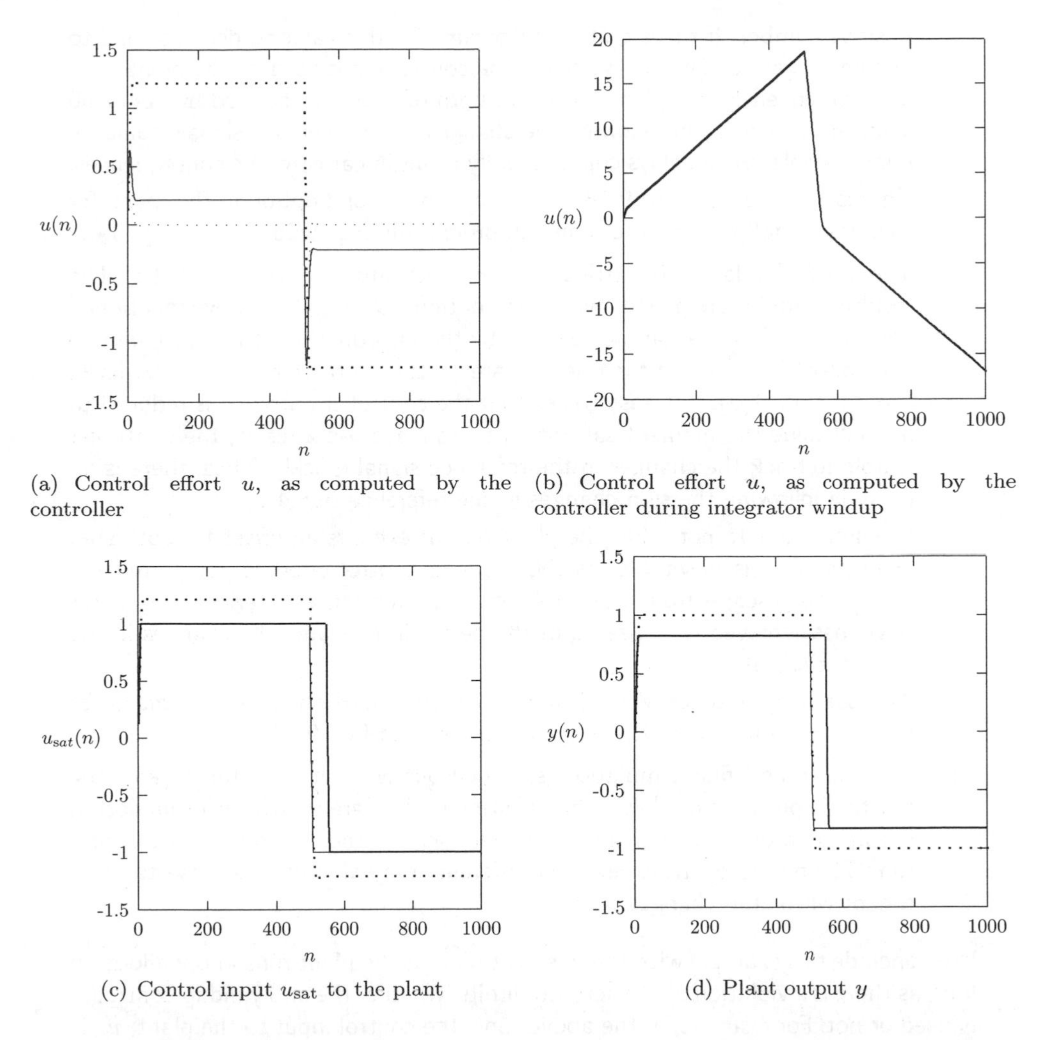

(a) Control effort u, as computed by the controller
(b) Control effort u, as computed by the controller during integrator windup
(c) Control input u_{sat} to the plant
(d) Plant output y

Figure 9.24: Controller output u, plant input u_{sat} and the plant output y for the IBM Lotus Domino server problem, discussed in Example 9.16: thick dotted lines, limits not imposed and AWC not used; thick continuous lines, limits imposed but AWC not used; thin continuous lines, limits imposed and AWC used; sparse continuous lines, limits not imposed but AWC used

2. The second simulation is carried out with actuator constraints of ± 1. The resulting profiles are plotted with thick continuous lines. Because the magnitude of the required control effort is larger than one and because the maximum allowable value is one, there is a difference between u and u_{sat}. As mentioned earlier, u goes on increasing, see Fig. 9.24(b). As a matter of fact, because the value of u for this run are a lot larger than that in other runs, it has been shown in a separate figure. The plots corresponding to this run are shown with thick lines.

   When the sign of r changes at $k = 500$, the required value of u is negative. But because of the integral windup, u is approximately 20, a large positive number. It takes quite a bit of time for this value to decrease and to become negative. This delay affects the control input used by the plant u_{sat}, and consequently, the plant output y: both of them are delayed by about 50 samples. This delay in following the changes in the reference signals could be unacceptable. Such delays could actually result in catastrophic consequences.

The user may execute M 9.18 and choose the second option in the menu for this design, followed by the Simulink program in Fig. 9.23.

3. The third simulation is carried out with actuator constraints of ± 1. The resulting profiles are plotted with thin continuous lines. The above mentioned bad effects of saturation are reduced by the introduction of the anti windup controller. The profiles for this run are plotted with thin continuous lines. From Fig. 9.24(a), it is easy to see that the controller output u is reduced to a small value the moment saturation takes place. As a result, the controller is able to track the changes in the reference signal quickly. Thus, there is no delay in following the sign changes in the reference signal.

   It is important to note that the plant output exhibits an offset for both step changes. This is inevitable, as the required control effort is not permitted. We cannot escape from this problem of performance deterioration. What AWC has achieved is to get rid of the delays in following the changes in the reference signal.

   The user may execute M 9.18 and choose the third option in the menu for this design, followed by the Simulink program in Fig. 9.23.

4. The fourth and final simulation is carried out without actuator constraints, but the in presence of AWC. The resulting profiles are shown with thin dotted lines. The profiles of u_{sat} and y are identical to those obtained in the first run. The profile of u, however, is different: the value of u is close to zero, except when step changes occur.

We conclude this example with the observation that the plant runs in open loop so long as there is a violation of the actuator limits, whether the anti windup controller is used or not. For example, in the above runs, the control input to the plant, u_{sat}, is constant at ± 1 during most of the simulation. If this open loop operation is not permissible, there is no option but to change the operational procedures or the actuator itself.

∎

In this section, we discussed in detail a possible way to incorporate anti windup controllers. A related issue is that of bumpless transfer. It was mentioned in Sec. 8.3.4 that if the control input is in difference form, switching from manual mode to automatic mode is somewhat easier than if we have the absolute control effort. There is a requirement for a bumpless transfer in other situations as well. It could be because of a change in the plant's operating region or substitution of an analog controller with a digital controller or switching to another controller simply to take care of maintenance.

The similarities between bumpless operation and AWC are striking. It is possible to think of a switch between two controllers of the former, as equivalent to the introduction of an extra disturbance (possibly nonlinear), as in Fig. 9.22. In this sense, the anti windup controller discussed in this section could help achieve bumpless transfer as well.

9.8 PID Tuning Through Pole Placement Control

PID controllers are extremely popular in industry. One of the difficulties associated with the use of these controllers, however, is the task of tuning: it can be time consuming, to say the least. Often the guidelines are inadequate and a lot of trial and error is required in tuning. In view of this, practitioners welcome any realistic method of tuning these controllers. In recent years, there has been a move to come up with initial tuning parameters through model based control techniques. This procedure involves design of a model based controller and then implementing it through a PID controller. In this section, we will concentrate on tuning the PID controllers through pole placement techniques.

In Sec. 8.3, we presented several methods of discretizing the PID controllers. We also showed how they can be presented in RST format. In this section, we will reverse the procedure: we will first solve Aryabhatta's identity for polynomials R_c and S_c and then back calculate the PID tuning parameters. As the PID framework allows only four parameters, including the filter constant N, the maximum number of polynomial coefficients we can work with is also four. Because the degrees of R_c and S_c depend on that of A and B, it is clear that this introduces a constraint on the type of plants for which this is possible. In the latter part of this section, we will discuss this topic in detail.

In continuous time PID controllers, we will not even think of assigning negative values to the parameters. Nevertheless, when they are assigned by solving simultaneous equations, there is no guarantee that the resulting solution will be positive all the time. We will now show that there could be advantages in assigning negative values to the PID parameters.

Recalling Eq. 8.31 on page 314, we see that r_1 has to be negative if both τ_d and N are to be positive. Unfortunately, as r_1 is obtained by solving Aryabhatta's identity, there is no guarantee of r_1 remaining negative.

We demonstrate this with an example.

Example 9.17 Design a PID control law for the plant with continuous time transfer function

$$G(s) = \frac{1}{2s+1}e^{-0.5s}$$

sampled at $T_s = 0.5$. Let the overshoot ε be 0.05. Solve Aryabhatta's identity for the rise time in the range of 4 to 9 s and determine the corresponding r_1, defined in Eq. 8.31– 8.33 on page 314.

We first calculate ϕ_{cl} for the given ε and a rise time in the given range. Eq. 9.37 is then solved for R_1 and S_c. R_1 is of the form $1 + r_1 z^{-1}$. M 9.19 carries out these calculations by invoking M 9.20. Closed loop simulations can be carried out using the Simulink code in Fig. A.4.

First we list r_1 as a function of rise time in Table 9.1. It is easy to see that if we demand a rise time of 6 seconds or less, r_1 is positive. From Eq. 8.31 on

Table 9.1: Parameter r_1 as a function of rise time

Rise time	r_1
4	0.1522
5	0.0782
6	0.0286
7	−0.0070
8	−0.0337
9	−0.0545

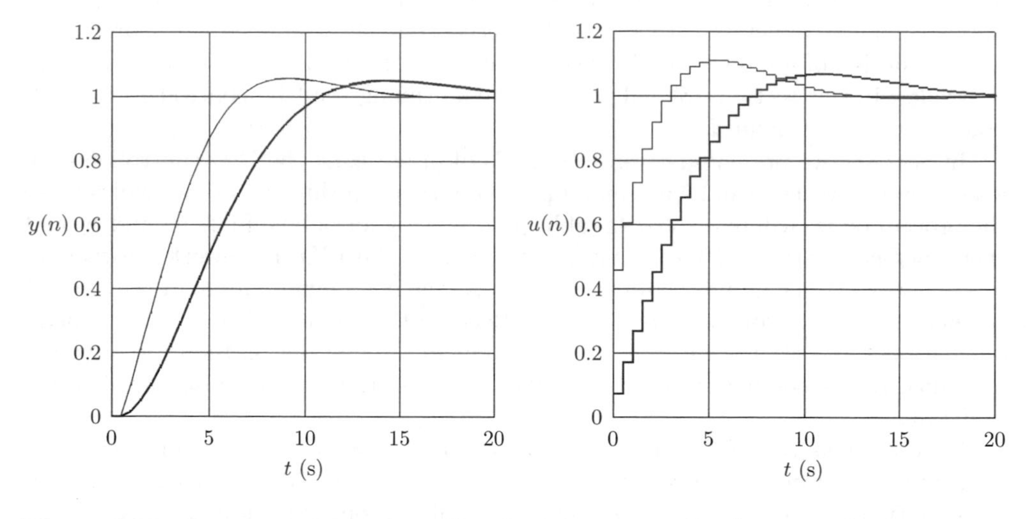

Figure 9.25: Output and input profiles for rise times of 5 (thick line) and 7 (thin line), in Example 9.17. The faster response is realized through a PID controller with negative tuning parameters.

page 314, we see that either τ_d or N has to be negative if we wish to implement this controller through the PID framework. In other words, the PID controller with positive parameters achieves only a slow response. If we wish to speed it up, we may have no option but to use negative values in the PID parameters. The output and the input profiles achieved with rise time specifications of 5 and 7 are shown using thin and thick lines, respectively, in Fig. 9.25. Actual rise times achieved are worse than what is designed for, as explained earlier.

If we worked with analog PID controllers, we would not even think about using negative tuning parameters. The digital controller, on the other hand, could actually promote such usage. In view of this, we remark that it may be possible to get better performance with digital controllers than with analog controllers. ∎

Having argued for the flexibility of assigning negative values to the PID parameters, we will proceed to derive explicit expressions for them and also the conditions on the plant.

The main reason for splitting the PID action into forward and backward paths was presented in Sec. 8.3.6 and 8.3.7. Here we approach this problem purely from the point of view of solvability of polynomial equations. Suppose that we wish to place the closed loop poles of the system at the roots of ϕ_{cl}, the desired closed loop characteristic polynomial. From Eq. 7.13, we have

$$A(z)R_c(z) + B(z)S_c(z) = \phi_{cl}(z) \tag{9.36}$$

This equation assumes that B includes time delays. If we wish to express the delay explicitly and if R has an integral action, *i.e.*, $R_c = \Delta R_1$, where $\Delta = 1 - z^{-1}$, the above equation becomes, after dropping the explicit dependence on z^{-1},

$$A\Delta R_1 + z^{-k}BS_c = \phi_{cl} \tag{9.37}$$

Comparing this with Eq. 9.24, we see that this is equivalent to setting A^b and B^b as A and B, respectively. That is, we don't cancel any factor of A or B to arrive at the control law. We now invoke the degree condition of the solution of Aryabhatta's identity, as in Eq. 7.121 on page 291. For a unique minimum degree solution, we see that $\mathrm{d}R_1$ has to be less than $\mathrm{d}z^{-k}B$, the degree of the known quantity, in the other term. As a result, we can equate $\mathrm{d}R_1$ to one less than $\mathrm{d}z^{-k}B$, with the possibility that some coefficients of R_1 are zero. Using a similar logic for the other unknown S_c, we obtain

$$\begin{aligned} \mathrm{d}R_1 &= \mathrm{d}B + k - 1 \\ \mathrm{d}S_c &= \mathrm{d}A + 1 - 1 = \mathrm{d}A \end{aligned} \tag{9.38}$$

where we have taken B to be a polynomial in z^{-1}, see Footnote 5 on page 100. Thus, we see that

$$\begin{aligned} \mathrm{d}B &\le \mathrm{d}R_1 - k + 1 \\ \mathrm{d}A &\le \mathrm{d}S_c \end{aligned} \tag{9.39}$$

Suppose that we use the control law given in Eq. 8.33– 8.35 on page 314. Because $\mathrm{d}R_1 = 1$ and $\mathrm{d}S_c = 2$, we obtain

$$\begin{aligned} \mathrm{d}B &\le 2 - k \\ \mathrm{d}A &\le 2 \end{aligned} \tag{9.40}$$

When the plant satisfies these conditions, it is possible to assign values to the four PID parameters, namely K, τ_i, τ_d and N from r_1, s_0, s_1 and s_2. We will now state the tuning rules for this PID structure [30]. From Eq. 8.31, it is easy to derive the following relation:

$$\frac{\tau_d}{N} = -T_s \frac{r_1}{1 + r_1} \tag{9.41}$$

In the following, we will use Eq. 8.35 for s_0, s_1 and s_2. By straightforward calculation, it is easy to show that $s_0 r_1 - s_1 - (2 + r_1)s_2 = K(1 + r_1)^2$. From this, we obtain

$$K = \frac{s_0 r_1 - s_1 - (2 + r_1)s_2}{(1 + r_1)^2} \tag{9.42}$$

By direct substitution, we can also obtain $s_0 r_1^2 - s_1 r_1 + s_2 = -KNr_1(1+r_1)^2$. Then, using Eq. 9.41, we obtain

$$\tau_d = T_s \frac{s_0 r_1^2 - s_1 r_1 + s_2}{K(1+r_1)^3} \tag{9.43}$$

By straightforward addition, we obtain $s_0 + s_1 + s_2 = K(1+r_1)T_s/\tau_i$, from which we obtain

$$\tau_i = T_s \frac{K(1+r_1)}{s_0 + s_1 + s_2} \tag{9.44}$$

It is easy to verify that if we substitute r_1, s_0, s_1 and s_2 values, respectively, as -0.2, 6.05, -10.41 and 4.4, in the above expressions, we obtain $K = 2$, $\tau_d = 2.5$, $\tau_i = 40$ and $N = 10$, consistent with Example 8.8.

We will next consider the control law that does not have an integral action, *i.e.*, PD controllers. It immediately follows that we cannot use the 2-DOF configuration presented in Sec. 8.3.6, in which only the integral mode is fed back. It is possible to use the configuration given in Sec. 8.3.7, in which a fraction of the reference signal goes through the proportional mode. In this configuration, even if the integral mode is excluded, fractional proportional feedback is available.

We will now demonstrate how to use the PD controller described in Sec. 8.3.5 to arrive at the tuning parameters. First, we will begin with the standard polynomial equation

$$AR_c + z^{-k} BS_c = \phi_{cl} \tag{9.45}$$

where R_c does not have Δ as a factor. Compare this with Eq. 9.37. The condition for unique minimum degree solution is

$$\begin{aligned} \mathrm{d}R_c &= \mathrm{d}B + k - 1 \\ \mathrm{d}S_c &= \mathrm{d}A - 1 \end{aligned} \tag{9.46}$$

or

$$\begin{aligned} \mathrm{d}B &\le \mathrm{d}R_c - k + 1 \\ \mathrm{d}A &\le \mathrm{d}S_c + 1 \end{aligned} \tag{9.47}$$

The control law presented in Eq. 8.37– 8.39 on page 315 has $\mathrm{d}R_c = 1$ and $\mathrm{d}S_c = 1$ and the above equations become

$$\begin{aligned} \mathrm{d}B &\le 2 - k \\ \mathrm{d}A &\le 2 \end{aligned} \tag{9.48}$$

We are now in a position to derive the tuning parameters for this PD control law. The expression for τ_d/N is the same as in Eq. 9.41. From Eq. 8.39, we obtain

$$\begin{aligned} K &= \frac{s_0 + s_1}{1 + r_1} \\ N &= \frac{s_1 - s_0 r_1}{r_1(s_0 + s_1)} \end{aligned} \tag{9.49}$$

It is easy to verify that if we substitute -0.2, 6 and -4.4, for r_1, s_0 and s_1, respectively, we obtain $K = 2$, $\tau_d = 2.5$ and $N = 10$, consistent with Example 8.9. We will present another example.

Example 9.18 Explore the possibility of arriving at a PID controller to achieve the performance specifications posed in Example 9.10 in the control of a DC motor.

This system has $B = 0.0288 + 0.0265z^{-1}$, $A = 1 - 1.7788z^{-1} + 0.7788z^{-2}$, $k = 1$, $\mathrm{d}A = 2$ and $\mathrm{d}B = 1$. Thus, the conditions of Eq. 9.48 are satisfied. We conclude that it is possible to tune a PD controller using the pole placement approach. From Example 9.10, we see the characteristic polynomial as

$$\phi_{cl} = 1 - 1.7502z^{-1} + 0.7791z^{-2}$$

Solving Eq. 9.45 for R_c and S_c, we obtain

$$R_c = 1 + 0.0073z^{-1}$$
$$S_c = 0.7379 - 0.2158z^{-1}$$

which is equivalent to

$$r_1 = 0.0073$$
$$s_0 = 0.7379$$
$$s_1 = -0.2158$$

We next calculate the PD controller parameters. Using Eq. 9.49, we obtain

$$K = 0.5183$$
$$N = -57.7038$$

Using Eq. 9.41, we obtain

$$\tau_d = 0.1052$$

M 9.21 implements these calculations. Observe that N has turned out to be negative.

Simulink block diagrams shown in Fig. A.5–A.6 on page 527 have been used to validate the performance of these controllers. The efficacy of the *RST* controller is demonstrated using thin lines in Fig. 9.26. Both the output variable y and the control effort u are shown.

In the same figure, the performance of the controller, implemented in the continuous time PD framework, is drawn using thick lines. The continuous time controller is of the form given in Eq. 8.13, with τ_i taken as infinity. The other tuning parameters are assigned the values given above.

It is clear from this figure that the control performance is better with the discrete time controller, although with a larger initial control effort. This is not surprising, because the specifications and the controller design have been carried out completely in the discrete time domain and there is no guarantee that implementation in the continuous time domain will give the same performance. ∎

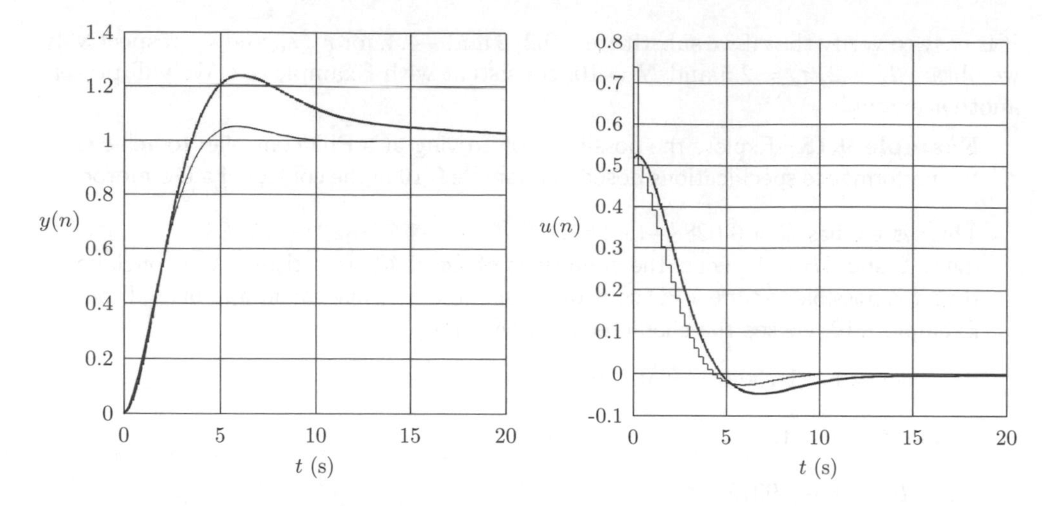

Figure 9.26: Output and input profiles of the DC motor control problem using discrete time (thin lines) and continuous time controllers, as explained in Example 9.18

We will conclude this section with a discussion on tuning the 2-DOF PID controller presented in Sec. 8.3.6. Because the controller has an integral mode, Eq. 9.37–9.39 are applicable now also. As the degree of R_1 is one and that of S_c is two, Eq. 9.40 also is applicable. For plants that satisfy these degree conditions, we can once again use the four parameters r_1, s_0, s_1 and s_2, to assign the values for K, τ_i, τ_d and N. The expression for r_1 is identical to the one in Eq. 9.41, reproduced here for convenience:

$$\frac{\tau_d}{N} = -T_s \frac{r_1}{1+r_1} \tag{9.50}$$

As the expressions for S_c are now different, the PID parameters also will be different, however. In the following, we will use the expressions for s_0, s_1 and s_2 from Eq. 8.43. By evaluating $s_1 + 2s_2$, it is easy to arrive at

$$K = -\frac{s_1 + 2s_2}{1+r_1} \tag{9.51}$$

By straightforward calculation, we can show $r_1 s_1 + (r_1 - 1)s_2 = KNr_1(1+r_1)$. Using the above expressions for K and τ_d/N, it is easy to arrive at

$$\tau_d = T_s \frac{r_1 s_1 + (r_1 - 1)s_2}{(1+r_1)(s_1 + 2s_2)} \tag{9.52}$$

Making use of Eq. 8.46 for $s_0 + s_1 + s_2$, Eq. 9.50–9.51 for τ_d/N and K, it is easy to arrive at the following expression for τ_i:

$$\tau_i = -T_s \frac{s_1 + 2s_2}{s_0 + s_1 + s_2} \tag{9.53}$$

The tuning of PID controllers using model based controllers will be a continuing theme throughout this book. We will take up this topic again in Sec. 10.2.3.

9.9 Matlab Code

Matlab Code 9.1 Pole placement controller for magnetically suspended ball problem, discussed in Example 9.3 on page 331. Available at
HOME/place/matlab/ball_basic.m[1]

```
1  % Magnetically suspended ball problem
2  % Operating conditions
3  M = 0.05; L = 0.01; R = 1; K = 0.0001; g = 9.81;
4  %
5  % Equilibrium conditions
6  hs = 0.01; is = sqrt(M*g*hs/K);
7  %
8  % State space matrices
9  a21 = K*is^2/M/hs^2; a23 = - 2*K*is/M/hs; a33 = - R/L;
10 b3 = 1/L;
11 a = [0 1 0; a21 0 a23; 0 0 a33];
12 b = [0; 0; b3]; c = [1 0 0]; d = 0;
13
14 % Transfer functions
15 G = ss(a,b,c,d); Ts = 0.01; [B,A,k] = myc2d(G,Ts);
16 [num,den] = tfdata(G,'v');
17
18 % Transient specifications
19 rise = 0.15; epsilon = 0.05;
20 phi = desired(Ts,rise,epsilon);
21
22 % Controller design
23 [Rc,Sc,Tc,gamma] = pp_basic(B,A,k,phi);
24
25 % Setting up simulation parameters for basic.mdl
26 st = 0.0001; % desired change in h, in m.
27 t_init = 0; % simulation start time
28 t_final = 0.5; % simulation end time
29
30 % Setting up simulation parameters for c_ss_cl
31 N_var = 0; xInitial = [0 0 0]; N = 1; C = 0; D = 1;
```

Matlab Code 9.2 Discretization of continuous transfer function. The result is numerator and denominator in powers of z^{-1} and the delay term k. For an example of how to use it, see M 9.1. This code is available at HOME/matlab/myc2d.m

```
1 % function [B,A,k] = myc2d(G,Ts)
2 % Produces numerator and denominator of discrete transfer
3 % function in powers of z^{-1}
4 % G is continuous transfer function, it can have time delays
5 % Ts is the sampling time, all in consistent time units
```

[1] HOME stands for http://www.moudgalya.org/dc/ – first see the software installation directions, given in Appendix A.2.

```
6
7  function [B,A,k] = myc2d(G,Ts)
8  H= c2d(G,Ts,'zoh');
9  [num,A] = tfdata(H,'v');
10 nonzero = find(num);
11 first_nz = nonzero(1);
12 k = first_nz -1 + H.ioDelay;
13 B = num(first_nz:length(num));
```

Matlab Code 9.3 Procedure to split a polynomial into good and bad factors, as discussed in Sec. 9.2. For an example of the usage, see M 9.1. This code is available at `HOME/matlab/polsplit2.m`

```
1  % function [goodpoly,badpoly] = polsplit2(fac,a)
2  % Splits a scalar polynomial of z^{-1} into good and bad
3  % factors.
4  % Input is a polynomial in increasing degree of z^{-1}
5  % Optional input is a, where a <= 1.
6  % Factor that has roots of z^{-1} outside a is called
7  % good and the rest bad.
8  % If a is not specified, it will be assumed as 1-1.0e-5
9
10 function [goodpoly,badpoly] = polsplit2(fac,a)
11 if nargin == 1, a = 1-1.0e-5; end
12 if a>1 error('good polynomial is unstable'); end
13 rts = roots(fac);
14 %
15 % extract good and bad roots
16 badindex = find(abs(rts)>=a);
17 badpoly = poly(rts(badindex));
18 goodindex = find(abs(rts)<a);
19 goodpoly = poly(rts(goodindex));
20 %
21 % scale by equating the largest terms
22 [m,index] = max(abs(fac));
23 goodbad = conv(goodpoly,badpoly);
24 factor = fac(index)/goodbad(index);
25 goodpoly = goodpoly * factor;
```

Matlab Code 9.4 Calculation of desired closed loop characteristic polynomial, as discussed in Sec. 7.7. This code is available at `HOME/matlab/desired.m`

```
1  % function [phi,dphi] = desired(Ts,rise,epsilon)
2  % Based on transient requirements,
3  % calculates closed loop characteristic polynomial
4  %
5  function [phi,dphi] = desired(Ts,rise,epsilon)
6
```

```
7  Nr = rise/Ts; omega = pi/2/Nr; rho = epsilon^(omega/pi);
8  phi = [1 -2*rho*cos(omega) rho^2]; dphi = length(phi)-1;
```

Matlab Code 9.5 Design of 2-DOF pole placement controller, as discussed in Sec. 9.2. For an example of the usage, see M 9.1. This code is available at HOME/matlab/pp_basic.m

```
1  % function [Rc,Sc,Tc,gamma] = pp_basic(B,A,k,phi)
2  % calculates pole placement controller
3
4  function [Rc,Sc,Tc,gamma] = pp_basic(B,A,k,phi)
5
6  % Setting up and solving Aryabhatta identity
7  [Ag,Ab] = polsplit2(A); dAb = length(Ab) - 1;
8  [Bg,Bb] = polsplit2(B); dBb = length(Bb) - 1;
9  [zk,dzk] = zpowk(k);
10 [N,dN] = polmul(Bb,dBb,zk,dzk);
11 dphi = length(phi) - 1;
12 [S1,dS1,R1,dR1] = xdync(N,dN,Ab,dAb,phi,dphi);
13
14 % Determination of control law
15 Rc = conv(Bg,R1); Sc = conv(Ag,S1);
16 Tc = Ag; gamma = sum(phi)/sum(Bb);
```

Matlab Code 9.6 Evaluates z^{-k}. This code is available at HOME/matlab/zpowk.m

```
1  function [zk,dzk] = zpowk(k)
2  zk = zeros(1,k+1); zk(1,k+1) = 1;
3  dzk = k;
```

Matlab Code 9.7 Simulation of closed loop system with an unstable controller, as discussed in Example 9.5 on page 335. This code is available at HOME/place/matlab/unstb.m

```
1  Ts = 1; B = [1 -3]; A = [1 2 -8]; k = 1;
2  [zk,dzk] = zpowk(k); int = 0;
3
4  % Transient specifications
5  rise = 10; epsilon = 0.1;
6  phi = desired(Ts,rise,epsilon);
7
8  % Controller design
9  [Rc,Sc,Tc,gamma] = pp_basic(B,A,k,phi);
10
11 % simulation parameters for basic_disc.mdl
12 st = 1.0; % desired change in setpoint
13 t_init = 0; % simulation start time
14 t_final = 1000; % simulation end time
15
```

```
16 % simulation parameters for stb_disc.mdl
17 N_var = 0; C = 0; D = 1; N = 1;
```

Matlab Code 9.8 Pole placement controller using internal model principle, as discussed in Sec. 9.4. It generalizes M 9.5. For an example of the usage, see M 9.9. This code is available at `HOME/matlab/pp_im.m`

```
1  % function [Rc,Sc,Tc,gamma,phit] = pp_im(B,A,k,phi,Delta)
2  % Calculates 2-DOF pole placement controller.
3
4  function [Rc,Sc,Tc,gamma] = pp_im(B,A,k,phi,Delta)
5
6  % Setting up and solving Aryabhatta identity
7  [Ag,Ab] = polsplit3(A); dAb = length(Ab) - 1;
8  [Bg,Bb] = polsplit3(B); dBb = length(Bb) - 1;
9  [zk,dzk] = zpowk(k);
10 [N,dN] = polmul(Bb,dBb,zk,dzk);
11 dDelta = length(Delta)-1;
12 [D,dD] = polmul(Ab,dAb,Delta,dDelta);
13 dphi = length(phi)-1;
14 [S1,dS1,R1,dR1] = xdync(N,dN,D,dD,phi,dphi);
15
16 % Determination of control law
17 Rc = conv(Bg,conv(R1,Delta)); Sc = conv(Ag,S1);
18 Tc = Ag; gamma = sum(phi)/sum(Bb);
```

Matlab Code 9.9 Pole placement controller, with internal model of a step, for the magnetically suspended ball problem, as discussed in Example 9.8 on page 339. This code is available at `HOME/place/matlab/ball_im.m` After executing this code, simulation can be carried out using the Simulink block diagram in Fig. A.8 on page 528. By choosing `rise=0.15, delta=0`, CTRL-1 of Example 9.3 on page 331 is realized. By choosing `rise=0.1, delta=1`, CTRL-3 of Example 9.8 is achieved.

```
1  % PP control with internal model for ball problem
2  % Operating conditions
3  M = 0.05; L = 0.01; R = 1; K = 0.0001; g = 9.81;
4
5  % Equilibrium conditions
6  hs = 0.01; is = sqrt(M*g*hs/K);
7
8  % State space matrices
9  a21 = K*is^2/M/hs^2; a23 = - 2*K*is/M/hs; a33 = - R/L;
10 b3 = 1/L;
11 a = [0 1 0; a21 0 a23; 0 0 a33];
12 b = [0; 0; b3]; c = [1 0 0]; d = 0;
13
14 % Transfer functions
15 G = ss(a,b,c,d); Ts = 0.01; [B,A,k] = myc2d(G,Ts);
16
```

```
17 % Transient specifications
18 rise = 0.1; epsilon = 0.05;
19 phi = desired(Ts,rise,epsilon);
20
21 % Controller design
22 Delta = [1 -1]; % internal model of step used
23 [Rc,Sc,Tc,gamma] = pp_im(B,A,k,phi,Delta);
24
25 % simulation parameters for c_ss_cl.mdl
26 st = 0.0001; % desired change in h, in m.
27 t_init = 0; % simulation start time
28 t_final = 0.5; % simulation end time
29 xInitial = [0 0 0];
30 N = 1; C = 0; D = 1; N_var = 0;
```

Matlab Code 9.10 Pole placement controller IBM Lotus Domino server, discussed in Example 9.9 on page 341. This code is available at HOME/place/matlab/ibm_pp.m

```
1 % Control of IBM lotus domino server
2 % Transfer function
3 B = 0.47; A = [1 -0.43]; k = 1;
4 [zk,dzk] = zpowk(k);
5
6 % Transient specifications
7 rise = 10; epsilon = 0.01; Ts = 1;
8 phi = desired(Ts,rise,epsilon);
9
10 % Controller design
11 Delta = [1 -1]; % internal model of step used
12 [Rc,Sc,Tc,gamma] = pp_im(B,A,k,phi,Delta);
13
14 % Simulation parameters for stb_disc
15 st = 1; % desired change
16 t_init = 0; % simulation start time
17 t_final = 40; % simulation end time
18 C = 0; D = 1; N_var = 0;
```

Matlab Code 9.11 Pole placement controller for motor problem, discussed in Example 9.10 on page 343. This code is available at HOME/place/matlab/motor.m

```
1 % Motor control problem
2 % Transfer function
3 a = [-1 0; 1 0]; b = [1; 0]; c = [0 1]; d = 0;
4 G = ss(a,b,c,d); Ts = 0.25; [B,A,k] = myc2d(G,Ts);
5
6 % Transient specifications
7 rise = 3; epsilon = 0.05;
8 phi = desired(Ts,rise,epsilon);
```

```
9
10 % Controller design
11 Delta = 1; % No internal model of step used
12 [Rc,Sc,Tc,gamma] = pp_im(B,A,k,phi,Delta);
13
14 % simulation parameters for c_ss_cl.mdl
15 st = 1; % desired change in position
16 t_init = 0; % simulation start time
17 t_final = 10; % simulation end time
18 xInitial = [0 0]; % initial conditions
19 N = 1; C = 0; D = 1; N_var = 0;
```

Matlab Code 9.12 Procedure to split a polynomial into good and bad factors, as discussed in Sec. 9.5. The factors that have roots outside unit circle or with negative real parts are defined as bad. This code is available at HOME/matlab/polsplit3.m

```
1 % function [goodpoly,badpoly] = polsplit3(fac,a)
2 % Splits a scalar polynomial of z^{-1} into good and bad
3 % factors. Input is a polynomial in increasing degree of
4 % z^{-1}. Optional input is a, where a <= 1.
5 % Factors that have roots outside a circle of radius a or
6 % with negative roots will be called bad and the rest
7 % good. If a is not specified, it will be assumed as 1.
8
9 function [goodpoly,badpoly] = polsplit3(fac,a)
10 if nargin == 1, a = 1; end
11 if a>1 error('good polynomial also is unstable'); end
12 rts = roots(fac);
13
14 % extract good and bad roots
15 badindex = find((abs(rts)>=a-1.0e-5)|(real(rts)<-0.05));
16 badpoly = poly(rts(badindex));
17 goodindex = find((abs(rts)<a-1.0e-5)&(real(rts)>=-0.05));
18 goodpoly = poly(rts(goodindex));
19
20 % scale by equating the largest terms
21 [m,index] = max(abs(fac));
22 goodbad = conv(goodpoly,badpoly);
23 factor = fac(index)/goodbad(index);
24 goodpoly = goodpoly * factor;
```

Matlab Code 9.13 Pole placement controller without intra sample oscillations, as discussed in Sec. 9.5. For an example of the usage, see M 9.14. This code is available at HOME/matlab/pp_im2.m

```
1 % function [Rc,Sc,Tc,gamma,phit] = pp_im2(B,A,k,phi,Delta,a)
2 % 2-DOF PP controller with internal model of Delta and
     without
3 % hidden oscillations
```

```
4
5 function [Rc,Sc,Tc,gamma,phit] = pp_im2(B,A,k,phi,Delta,a)
6
7 if nargin == 5, a = 1; end
8 dphi = length(phi)-1;
9
10 % Setting up and solving Aryabhatta identity
11 [Ag,Ab] = polsplit3(A,a); dAb = length(Ab) - 1;
12 [Bg,Bb] = polsplit3(B,a); dBb = length(Bb) - 1;
13 [zk,dzk] = zpowk(k);
14 [N,dN] = polmul(Bb,dBb,zk,dzk);
15 dDelta = length(Delta)-1;
16 [D,dD] = polmul(Ab,dAb,Delta,dDelta);
17 [S1,dS1,R1,dR1] = xdync(N,dN,D,dD,phi,dphi);
18
19 % Determination of control law
20 Rc = conv(Bg,conv(R1,Delta)); Sc = conv(Ag,S1);
21 Tc = Ag; gamma = sum(phi)/sum(Bb);
22
23 % Total characteristic polynomial
24 phit = conv(phi,conv(Ag,Bg));
```

Matlab Code 9.14 Controller design for the case study presented in Example 9.12 on page 347. This code is available at HOME/place/matlab/sigurd.m

```
1 clear
2 num = 200;
3 den = conv([0.05 1],[0.05 1]);
4 den = conv([10 1],den);
5 G = tf(num,den); Ts = 0.025;
6 [B,A,k] = myc2d(G,Ts);
7 [zk,dzk] = zpowk(k); int = 0;
8
9 % Transient specifications
10 a = 0.9; rise = 0.24; epsilon = 0.05;
11 phi = desired(Ts,rise,epsilon);
12
13 % Controller design
14 Delta = [1 -1]; % internal model of step is present
15 [Rc,Sc,Tc,gamma] = pp_im2(B,A,k,phi,Delta,a);
16
17 % margin calculation
18 Lnum = conv(Sc,conv(B,zk));
19 Lden = conv(Rc,A);
20 L = tf(Lnum,Lden,Ts);
21 [Gm,Pm] = margin(L);
22
23 num1 = 100; den1 = [10 1];
24 Gd = tf(num1,den1);
```

```
25 [C,D,k1] = myc2d(Gd,Ts);
26 [zk,dzk] = zpowk(k);
27 C = conv(C,zk);
28
29 % simulation parameters g_s_cl2.mdl
30 N = 1;
31 st = 1; % desired change in setpoint
32 st1 = 0; % magnitude of disturbance
33 t_init = 0; % simulation start time
34 t_final = 1.5; % simulation end time
```

Matlab Code 9.15 Evaluation of continuous time controller for the case study presented in Example 9.13 on page 349. This code is available at HOME/place/matlab/sigurd_his.m

```
1 clear
2 num = 200;
3 den = conv([0.05 1],[0.05 1]);
4 den = conv([10 1],den);
5 G = tf(num,den); Ts = 0.005;
6 [B,A,k] = myc2d(G,Ts);
7 [zk,dzk] = zpowk(k); int = 0;
8
9 % Sigurd's feedback controller'
10 numb = 0.5*conv([1 2],[0.05 1]);
11 denb = conv([1 0],[0.005 1]);
12 Gb = tf(numb,denb);
13 [Sb,Rb,kb] = myc2d(Gb,Ts);
14 [zkb,dzkb] = zpowk(kb);
15 Sb = conv(Sb,zkb);
16
17 % Sigurd's feed forward controller'
18 numf = [0.5 1];
19 denf = conv([0.65 1],[0.03 1]);
20 Gf = tf(numf,denf);
21 [Sf,Rf,kf] = myc2d(Gf,Ts);
22 [zkf,dzkf] = zpowk(kf);
23 Sf = conv(Sf,zkf);
24
25 % Margins
26 L = series(G,Gb);
27 [Gm,Pm] = margin(L);
28 Lnum = conv(Sb,conv(zk,B));
29 Lden = conv(Rb,A);
30 L = tf(Lnum,Lden,Ts);
31 [DGm,DPm] = margin(L);
32
33 % Noise
34 num1 = 100; den1 = [10 1];
```

```
35 % simulation parameters for
36 % entirely continuous simulation: g_s_cl3.mdl
37 % hybrid simulation: g_s_cl6.mdl
38 st = 1; % desired change in setpoint
39 st1 = 0;
40 t_init = 0; % simulation start time
41 t_final = 5; % simulation end time
```

Matlab Code 9.16 System type with 2-DOF controller. It is used to arrive at the results of Example 9.14. This code is available at
HOME/place/matlab/type_2DOF.m

```
1  B = 1; A = [1 -1]; k = 1; zk = zpowk(k); Ts = 1;
2  phi = [1 -0.5];
3
4  Delta = 1; % Choice of internal model of step
5  [Rc,Sc,Tc,gamma] = pp_im(B,A,k,phi,Delta);
6  %
7  % simulation parameters for stb_disc.mdl
8  st = 1; % desired step change
9  t_init = 0; % simulation start time
10 t_final = 20; % simulation end time
11 xInitial = [0 0];
12 C = 0; D = 1; N_var = 0;
13 open_system('stb_disc.mdl')
```

Matlab Code 9.17 Illustrating the benefits of cancellation. It is used to arrive at the results of Example 9.15. This code is available at
HOME/place/matlab/dof_choice.m

```
1  % test problem to demonstrate benefits of 2_dof
2  % Ts = 1; B = [1 0.9]; A = conv([1 -1],[1 -0.8]); k = 1;
3  Ts = 1; k = 1;
4  B = conv([1 0.9],[1 -0.8]); A = conv([1 -1],[1 -0.5]);
5  %
6  % closed loop characteristic polynomial
7  phi = [1 -1 0.5];
8
9  Delta = 1; % Choice of internal model of step
10 control = 1;
11 if control == 1, % 1-DOF with no cancellation
12     [Rc,Sc] = pp_pid(B,A,k,phi,Delta);
13     Tc = Sc; gamma = 1;
14 else % 2-DOF
15     [Rc,Sc,Tc,gamma] = pp_im(B,A,k,phi,Delta);
16 end
17 %
18 % simulation parameters for stb_disc.mdl
19 [zk,dzk] = zpowk(k);
```

```
20 st = 1; % desired step change
21 t_init = 0; % simulation start time
22 t_final = 20; % simulation end time
23 xInitial = [0 0];
24 C = 0; D = 1; N_var = 0;
25 open_system('stb_disc.mdl')
```

Matlab Code 9.18 Anti windup control (AWC) of IBM Lotus Domino server, studied in Example 9.16 on page 357. It can be used for the following situations: with and without saturation, and with and without AWC. This code is available at `HOME/place/matlab/ibm_pp_sat.m`

```
1  % Transfer function
2  B = 0.47; A = [1 -0.43]; k = 1;
3  [zk,dzk] = zpowk(k);
4
5  % Transient specifications
6  rise = 10; epsilon = 0.01; Ts = 1;
7  phi = desired(Ts,rise,epsilon);
8
9  % Controller design
10 delta = [1 -1]; % internal model of step used
11 [Rc,Sc,Tc,gamma,F] = pp_im2(B,A,k,phi,delta);
12
13 % Study of Antiwindup Controller
14
15 key = menu('Please choose one of the following', ...
16       'Simulate without any saturation limits', ...
17           'Simulate saturation, but do not use AWC', ...
18           'Simulate saturation with AWC in place', ...
19           'Simulate without any saturation limits, but with
             AWC');
20
21 if key == 1
22   U = 2; L = -2; P = 1; F = Rc; E = 0; PSc = Sc; PTc = Tc;
23 elseif key == 2
24   U = 1; L = -1; P = 1; F = Rc; E = 0; PSc = Sc; PTc = Tc;
25 else
26   if key == 3 % Antiwindup controller and with saturation
27     U = 1; L = -1;
28   elseif key == 4 % Antiwindup controller, but no saturation
29     U = 2; L = -2;
30   end
31   P = A;
32   dF = length(F) - 1;
33   PRc = conv(P,Rc); dPRc = length(PRc) - 1;
34   [E,dE] = poladd(F,dF,-PRc,dPRc);
35   PSc = conv(P,Sc); PTc = conv(P,Tc);
36 end
```

```
37 % Setting up simulation parameters for stb_disc_sat
38 t_init = 0; % first step begins
39 st = 1; % height of first step
40 t_init2 = 500; % second step begins
41 st2 = -2; % height of second step
42 t_final = 1000; % simulation end time
43 st1 = 0; % no disturbance input
44 C = 0; D = 1; N_var = 0;
```

Matlab Code 9.19 Demonstration of usefulness of negative PID parameters, discussed in Example 9.17 on page 361. This code is available at HOME/place/matlab/pid_neg.m

```
1 % Discretize the continuous plant
2 num = 1; den = [2 1]; tau = 0.5;
3 G = tf(num,den);
4 G.ioDelay = tau;
5 Ts = 0.5;
6 [B,A,k] = myc2d(G,Ts);
7
8 % Specify transient requirements
9 epsilon = 0.05; rise = 5;
10 phi = desired(Ts,rise,epsilon);
11
12 % Design the controller
13 Delta = [1 -1];
14 [Rc,Sc] = pp_pid(B,A,k,phi,Delta);
15 %
16 % parameters for simulation using g_s_cl
17 Tc = Sc; gamma = 1; N = 1;
18 C = 0; D = 1; N_var = 0;
19 st = 1; t_init = 0; t_final = 20;
```

Matlab Code 9.20 Solution to Aryabhatta's identity arising in PID controller design, namely Eq. 9.37 on page 363. For an example of the usage, see M 9.19. This code is available at HOME/matlab/pp_pid.m

```
1 function [Rc,Sc] = pp_pid(B,A,k,phi,Delta)
2
3 % Setting up and solving Aryabhatta identity
4 dB = length(B) - 1; dA = length(A) - 1;
5 [zk,dzk] = zpowk(k);
6 [N,dN] = polmul(B,dB,zk,dzk);
7 dDelta = length(Delta)-1;
8 [D,dD] = polmul(A,dA,Delta,dDelta);
9 dphi = length(phi)-1;
10 [Sc,dSc,R,dR] = xdync(N,dN,D,dD,phi,dphi);
11 Rc = conv(R,Delta);
```

Matlab Code 9.21 DC motor with PID control, tuned through pole placement technique, as in Example 9.18 on page 365. This code is available at HOME/place/matlab/motor_pd.m

```
1  % Motor control problem
2  % Transfer function
3  a = [-1 0; 1 0]; b = [1; 0]; c = [0 1]; d = 0;
4  G = ss(a,b,c,d); Ts = 0.25; [B,A,k] = myc2d(G,Ts);
5  [num,den] = tfdata(G,'v');
6
7  % Transient specifications
8  rise = 3; epsilon = 0.05;
9  phi = desired(Ts,rise,epsilon);
10
11 % Controller design
12 Delta = 1; % No internal model of step used
13 [Rc,Sc] = pp_pid(B,A,k,phi,Delta);
14
15 % continuous time controller
16 [K,taud,N] = pd(Rc,Sc,Ts);
17 numb = K*[1 taud*(1+1/N)]; denb = [1 taud/N];
18 numf = 1; denf = 1;
19
20 % simulation parameters
21 st = 1; % desired change in position
22 t_init = 0; % simulation start time
23 t_final = 20; % simulation end time
24 st1 = 0;
25
26 % continuous controller simulation: g_s_cl3.mdl
27 num1 = 0; den1 = 1;
28
29 % discrete controller simulation: g_s_cl2.mdl
30 C = 0; D = 1; N = 1; gamma = 1; Tc = Sc;
```

Matlab Code 9.22 PD control law from polynomial coefficients, as explained in Sec. 9.8. For an example of the usage, see M 9.21. This code is available at HOME/place/matlab/pd.m

```
1  function [K,taud,N] = pd(Rc,Sc,Ts)
2  % Both Rc and Sc have to be degree one polynomials
3
4  s0 = Sc(1); s1 = Sc(2);
5  r1 = Rc(2);
6  K = (s0+s1)/(1+r1);
7  N = (s1-s0*r1)/r1/(s0+s1);
8  taudbyN = -Ts*r1/(1+r1);
9  taud = taudbyN * N;
```

9.10 Problems

9.1. Obtain a 2-DOF pole placement controller for the plant

$$G(z) = \frac{z+2}{z^2+z+1}$$

so as to have a rise time of 5 samples, i.e., $N_r = 5$, and overshoot of 10%. The controller should track step changes in the reference signal, even if the plant parameters change.

9.2. The objective of this problem is to design a 2-DOF pole placement controller for a plant with open loop transfer function given by

$$G(s) = \frac{s-1}{(s-3)(s+2)}$$

(a) Design a 2-DOF pole placement controller that satisfies the following transient specifications for a unit step input:

i. Rise time should be less than or equal to 2 seconds.

ii. Overshoot should be less than or equal to 10%.

A suggested sampling time is $T_s = 0.1$.

(b) Propose an appropriate feedback configuration so that the closed loop system is stable. Simulate the closed loop system using each of Fig. A.1–A.4 on page 527.

9.3. Determine a pole placement controller for the nonminimum phase oscillator with the following transfer function:

$$y(t) = \frac{-1+2z^{-1}}{1-1.7z^{-1}+z^{-2}}z^{-2}u(t)$$

The controller should satisfy the following requirements:

(a) The closed loop characteristic polynomial $\phi_{cl}(z^{-1}) = 1 - 0.6z^{-1}$.

(b) There should be an integral term, namely $1/(1-z^{-1})$, present in the loop.

9.4. It is proposed to control a plant with the following transfer function:

$$G(z) = z^{-k}\frac{B(z)}{A(z)}b(z)$$

Here, A and B are known polynomials in z^{-1} and k is a known positive integer. Unfortunately, nothing is known about $b(z)$. In view of this, the control designer *assumes* $b = 1$ and, using our procedure, arrives at the following 2-DOF pole placement controller to achieve offset free tracking of step inputs:

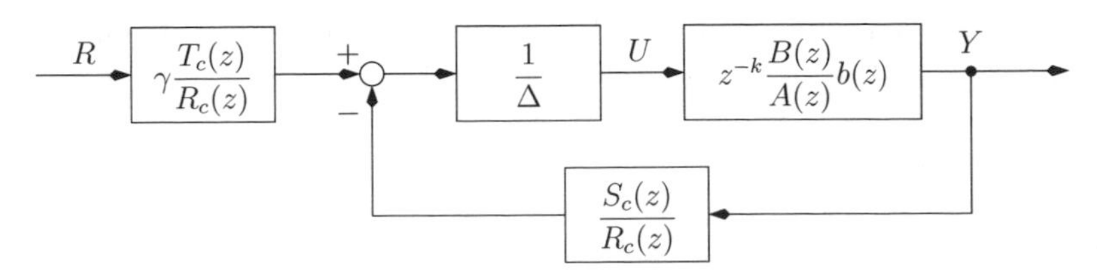

(a) For what types of $b(z^{-1})$ is the above mentioned tracking property maintained?

(b) Is this controller effective in rejecting the disturbance, denoted by V in Fig. 7.22 on page 271, for example, when b is unknown?

(c) What assumption is implicit in your answer to the above questions?

9.5. In Example 7.3 on page 257, we have presented a lead controller to meet the requirements of an antenna control system. Now design 1-DOF and 2-DOF pole placement controllers to meet the same performance specifications. Complete the following table for a unit step input:

	1-DOF		2-DOF	
	$1 - z^{-1}$	$(1 - z^{-1})^2$	$1 - z^{-1}$	$(1 - z^{-1})^2$
Initial control effort				
Actual rise time				
Overshoot				

and the following table for a ramp signal of unit slope:

	1-DOF		2-DOF	
	$1 - z^{-1}$	$(1 - z^{-1})^2$	$1 - z^{-1}$	$(1 - z^{-1})^2$
Initial control effort				
Steady state error				

For the 1-DOF controller, carry out the calculations with and without cancellation of good factors.

[Hint: The 2-DOF controller designed with the following specifications may help meet the requirements: rise time = 1.8 s, overshoot = 0.25, $T_s = 0.2$ s. Matlab code `pp_im.m`, given in M 9.16, may be useful. For the 1-DOF controller without cancellation of good factors, Matlab code `pp_pid.m`, given in M 9.20, may be useful.]

9.6. Design a pole placement controller (PPC) for the open loop transfer function

$$G(z) = \frac{z^{-1}}{1 - 0.9z^{-1}}$$

(a) Design a PPC for the system such that

i. the closed loop poles are at $-0.5 \pm 0.5j$,

ii. there is no offset to a step input.

(b) Can you implement this PPC with a PID controller? If so, what are the P, I, D settings? If not, explain why not.

Chapter 10

Special Cases of Pole Placement Control

Many controllers can be analysed from the pole placement viewpoint. In this chapter, we present the Smith predictor and internal model controller, popular design techniques in chemical engineering.

10.1 Smith Predictor

The presence of large delays reduces the achievable control performance. For example, consider our standard plant model in the mixed notation of Sec. 6.4.1:

$$y(n) = z^{-k}\frac{B(z)}{A(z)}u(n) \tag{10.1}$$

In all real life systems, k will at least be one, because all meaningful systems take a nonzero amount of time to respond to external stimuli. If there is a transport delay in implementing the control effort, k will be larger than one. Such a situation arises also when the plants are inherently sluggish and they take some time respond to control efforts. Chemical processes often have this shortcoming. In chemical engineering terminology, the time to respond to external inputs is known as the *dead time*. In all these cases, k could be a large number.

The presence of a large delay k implies that the control action will be delayed by the same extent. This can be see by writing Eq. 10.1 as

$$y(n) = \frac{B(z)}{A(z)}u(n-k)$$

It is clear that the larger the delay, the worse the control performance will be. We will illustrate this with a discrete time model of a *paper machine* system, studied by [2].

Example 10.1 The discrete time model of a paper machine is given by

$$y(n) = \frac{0.63z^{-3}}{1-0.37z^{-1}}u(n)$$

Compare the performance of a 2-DOF pole placement controller, with and without the delay.

To design a 2-DOF pole placement controller, we need to solve Aryabhatta's identity, given by Eq. 9.24 on page 339, and reproduced here for convenience:

$$A^b \Delta R_1 + z^{-k} B^b S_1 = \phi_{cl}$$

Suppose that we desire the closed loop characteristic polynomial to be $1 - 0.5z^{-1}$ and that we want an internal model of steps in the loop. Because $A^b = B^b = 1$ for this system, Aryabhatta's identity of Eq. 9.24 on page 339 becomes

$$(1 - z^{-1})R_1 + z^{-3} S_1 = 1 - 0.5z^{-1}$$

The solution is given by

$$\begin{aligned} S_1 &= 0.5 \\ R_1 &= 1 + 0.5z^{-1} + 0.5z^{-2} \end{aligned}$$

The controller parameters, obtained using Eq. 9.11b on page 330 and Eq. 9.23 on page 338, are

$$\begin{aligned} S_c &= 0.5(1 - 0.37) \\ R_c &= 0.63(1 - 0.5z^{-1} - 0.5z^{-3}) \end{aligned}$$

Because S_1 is a constant, we also obtain $\gamma T_c = S_c$. As a result, we have the property of offset free tracking of steps. M 10.1 carries out these calculations. Fig. 10.1 shows a plot, using thick lines, of the resulting y and u profiles, which have been obtained using the Simulink code given in Fig. A.1 on page 525.

Let us now examine the performance of the controller if the plant delay is reduced from the given value of three to the minimum possible value of one. We arrive at the following Aryabhatta's identity and controller parameters:

$$\begin{aligned} (1 - z^{-1})R_1 + z^{-1} S_1 &= 1 - 0.5z^{-1} \\ S_1 &= 0.5 \\ R_1 &= 1 \\ S_0 &= 0.5(1 - 0.37) \\ R_0 &= 0.63(1 - z^{-1}) \\ \gamma T_0 &= S_c \end{aligned}$$

where we have denoted R_c, S_c and T_c, respectively, by R_0, S_0 and T_0, to indicate the fact that these have been obtained when the delay is reduced to the minimum possible value.

Once again, these calculations are carried out using M 10.1, simulated using Fig. A.1 and plotted using thin lines in Fig. 10.1. From this figure, it is easy to see that the control effort $u(n)$ is identical in both cases. The plant output $y(n)$ is faster when the delay is reduced.

∎

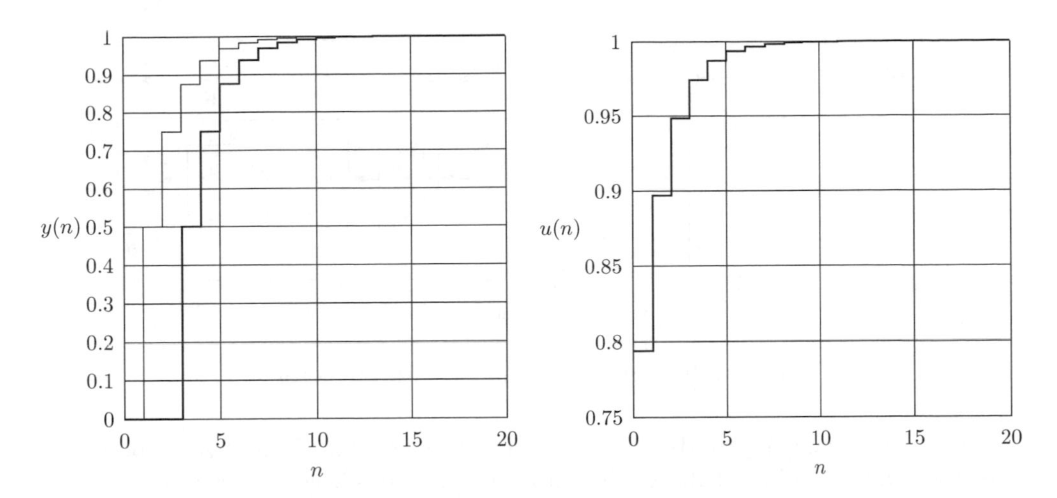

Figure 10.1: Output and input profiles for $k = 3$ (thick line) and $k = 1$ (thin line), in Example 10.1. The control effort for both cases is identical. The output is faster for smaller delay.

Because of the adverse effects of long delays in the plant, we would like to account for them. If the delay is completely compensated, however, an *algebraic loop* is created in the feedback loop, causing difficulties during simulation. This difficulty does not arise in the presence of noise or plant–model mismatch. The shortest possible delay in all real life applications is one. In this section, we propose to remove the effects of any delay larger than one, through a strategy known as the *Smith predictor*. In view of this, we will assume that in the plant model given by Eq. 10.1, $k \geq 2$. Recall that the numerator polynomial $B(z)$ has the form

$$B(z) = b_0 + b_1 z^{-1} + \cdots + b_{\mathrm{d}B} z^{-\mathrm{d}B} \tag{10.2}$$

with $b_0 \neq 0$. Defining

$$B_d(z) = z^{-1} B(z) \tag{10.3}$$

Eq. 10.1 becomes

$$G(z) = z^{-(k-1)} \frac{B_d(z)}{A(z)} \tag{10.4}$$

We have defined B_d such that it includes one delay, the minimum we expect in real applications. Now we look for ways to get rid of the adverse effects of the delay term $z^{-(k-1)}$. Towards this end, consider the following equation:

$$y_p(n) = z^{-(k-1)} \frac{B_d(z)}{A(z)} u(n) + \left[1 - z^{(k_m - 1)}\right] \frac{B_{dm}(z)}{A_m(z)} u(n) \tag{10.5}$$

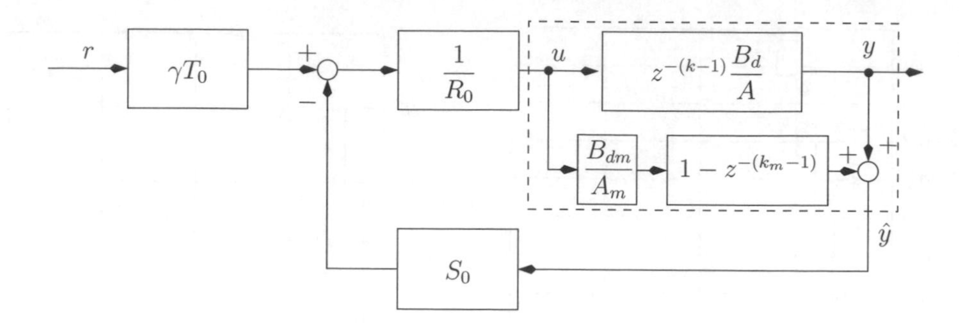

Figure 10.2: A schematic of Smith predictor

where k_m, B_{dm} and A_m can be thought of as estimates of k, B_d and A, respectively. When we have good knowledge of the plant, the estimates become exact and Eq. 10.5 becomes

$$y_p(n) = \frac{B_d(z)}{A(z)}u(n) \tag{10.6}$$

thereby getting rid of the adverse effects of $z^{-(k-1)}$ in Eq. 10.4, see Problem 10.1. We can treat this as the equivalent model of the original plant given by Eq. 10.1 and design a controller. Fig. 10.2 shows a schematic of this idea, where we have proposed a 2-DOF pole placement controller. We have used the symbols R_0, S_0 and T_0 in the place of R_c, S_c and T_c, respectively, to indicate the fact that this controller is designed for the delay free plant.

With the addition of the extra path in the control scheme, we feed back $\hat{y}$, as opposed to y. The expression for $\hat{y}$ is given by

$$\hat{y} = \frac{B_{dm}}{A_m}(1 - z^{-(k_m-1)})u + y = \frac{B_{dm}(1 - z^{-(k_m-1)})u + A_m y}{A_m} \tag{10.7}$$

The control law for this configuration is given by

$$S_0 u = \gamma T_0 r - R_0 \hat{y} \tag{10.8}$$

Substituting in the above equation the expression for $\hat{y}$ from Eq. 10.7, and simplifying, we obtain

$$u = \frac{A_m \gamma T_0 r - A_m R_0 y}{A_m S_0 + B_{dm} R_0 (1 - z^{-(k_m-1)})} \tag{10.9}$$

Using Eq. 10.4, the plant model can be written as

$$y = z^{-(k-1)}\frac{B_d}{A}u \tag{10.10}$$

Substituting into this equation the expression for u from Eq. 10.9, and simplifying, we obtain

$$\phi_{cl} = z^{-(k-1)} B_d A_m \gamma T_0 r \tag{10.11}$$

where

$$\phi_{cl} = A(A_m S_0 + B_{dm} R_0 (1 - z^{-(k_m-1)})) + z^{-(k-1)} B_d A_m R_0 \tag{10.12}$$

Suppose that we have good knowledge of the plant. As a result, the model parameters will be identical to those of the plant. That is, $A_m = A$, $B_{dm} = B_d$ and $k_m = k$. The above equation can be simplified as

$$\phi_{cl} = A(AS_0 + B_d R_0) \tag{10.13}$$

We see that the denominator of the plant transfer function, A, is a part of the closed loop characteristic polynomial. As a result, this method can be used only when the plant G is stable. Also, when the expression for ϕ_{cl} is substituted into Eq. 10.12, A cancels A_m as these are assumed equal. This leaves behind $\phi_0 = AS_0 + B_d R_0$, which is identical to the characteristic polynomial of the system in Fig. 10.2, when the blocks inside the box are replaced by their equivalent transfer function of B_d/A. The name Smith predictor refers mainly to the configuration inside the box.

We will now illustrate this procedure with an example.

Example 10.2 Apply the procedure of the Smith predictor to the paper machine problem, presented in Example 10.1.

We once again assume the presence of an internal model of a step. For comparison purposes, we take the reduced characteristic polynomial $AS_0 + B_d R_0$ to be $1 - 0.5z^{-1}$. We obtain the following expressions using the procedure outlined in this section:

$$\begin{aligned} B_d &= 0.63z^{-1} \\ S_0 &= 0.5(1 - 0.37) \\ R_0 &= 0.63(1 - z^{-1}) \end{aligned}$$

which are identical to the values obtained in Example 10.1 for $k = 1$. The control effort $u(n)$ is also the same as before. As expected, $\hat{y}$ and y values are identical to the y values of Example 10.1 corresponding to $k = 1$ and $k = 3$, respectively. M 10.2 is used to design this controller. Fig. 10.3 shows the Simulink code used to carry out the simulation.

∎

We need good knowledge of the delay for the Smith predictor to be effective. The above approach works only for stable plants. Problem 10.2 addresses unstable plants. Sec. 11.1.3 explains that the prediction error model of Sec. 11.1 plays the role of the Smith predictor when noise is present.

10.2 Internal Model Control

In this section, we will develop the internal model controller, a popular technique in the chemical engineering field. The name comes from the fact that the controller has an explicit model of the plant as its part. In this framework, if the open loop

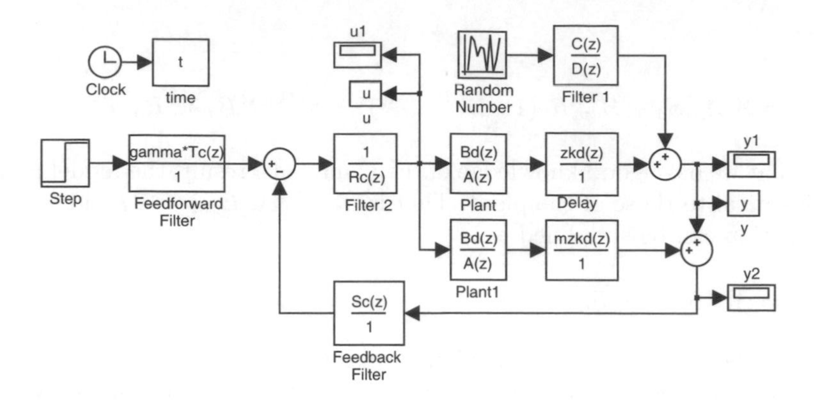

Figure 10.3: Simulink block diagram to simulate Smith predictor, used in Example 10.2. The code is available at HOME/imc/matlab/smith_disc.mdl, see Footnote 1 on page 397.

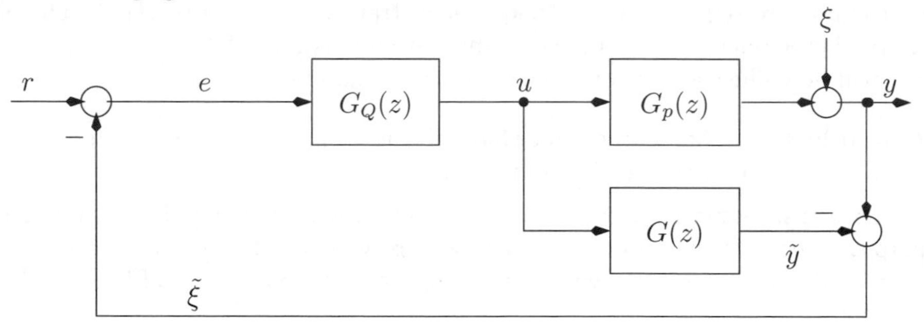

Figure 10.4: IMC feedback configuration

transfer function is stable and if the controller is stable, the closed loop system can be shown to be stable. This reduces the effort required in searching for controllers. This controller, abbreviated as IMC, can be thought of as a special case of the pole placement controller. Although it is possible to extend this idea for unstable plants, the IMC has been used mainly for stable plants. We will also focus our attention on IMC design of stable plants only.

Let the transfer function of the stable plant be denoted by $G_p(z)$. Suppose that its model is denoted by $G(z)$ or, equivalently, by

$$y(n) = G(z)u(n) + \xi(n) \tag{10.14}$$

where $y(n)$, $u(n)$ and $\xi(n)$ are plant output, input and noise, respectively. We assume that all the delays of the plant are factored in z^{-k}. Note that we use the argument z, even though the numerator and the denominator could be polynomials in powers of z^{-1}, see Footnote 5 on page 100. One popular method to control such stable plants is given in the block diagram in Fig. 10.4, which implements the well known *internal model control.* In this, G is a model of the plant and G_Q is a special type of controller. Let us consider the problem of regulation or noise rejection with $y_{sp} = 0$. We see that

$$\tilde{\xi} = y - \tilde{y} = G_p u + \xi - Gu = \xi \tag{10.15}$$

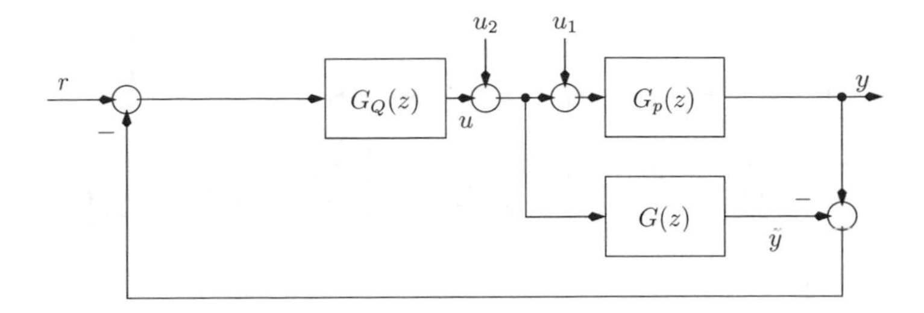

Figure 10.5: Injection of extra inputs to determine the conditions for internal stability in the IMC configuration

if $G = G_p$. If $G \neq G_p$, this mismatch gets added to the signal that is fed back. This is in agreement with our intuition that we need the feedback for stable systems mainly for noise rejection and plant–model mismatch. If in addition

$$G_Q = G_p^{-1} \tag{10.16}$$

then the plant output in the noise rejection problem, for which we can take the reference signal r to be zero, is given by

$$G_p u = -G_p G_Q \tilde{\xi} = -\xi \tag{10.17}$$

which will cancel the noise, resulting in $y = 0$. Thus, if $G_Q = G_p^{-1}$, and $G = G_p$, the noise will be rejected completely. This is one advantage of this special type of feedback connection. The additional advantage will be explained next.

A system is internally stable if and only if the transfer function between any two points in the feedback loop is stable, see Sec. 7.4.2. To evaluate internal stability, we construct the feedback diagram with extra inputs, as in Fig. 10.5. We arrive at the following transfer function matrix, assuming, $G = G_p$:

$$\begin{bmatrix} y \\ u \\ \tilde{y} \end{bmatrix} = \begin{bmatrix} GG_Q & G & (1 - GG_Q)G \\ G_Q & -GG_Q & 0 \\ GG_Q & -G^2G_Q & G \end{bmatrix} \begin{bmatrix} r \\ u_1 \\ u_2 \end{bmatrix} \tag{10.18}$$

The system is internally stable if and only if every entry in the above matrix is stable. As G is stable, internal stability is equivalent to G_Q being stable. Thus, control design is reduced to looking for any stable G_Q. This is an easier problem compared to the design of the controller in the standard configuration, G_c.

From the discussion earlier, we would want $G_Q = G^{-1}$. Combining these two requirements, we look for a stable G_Q that is an approximate inverse of G. We will refer to G_Q as the *Q form* of the IMC.

Indeed, any controller in a standard configuration for a stable plant can also be represented in this special arrangement, because of the above mentioned appealing property of the IMC. This is illustrated in Fig. 10.6, which has an extra signal through $\tilde{G}$ added and subtracted. One can verify that this is equivalent to the standard configuration. The block diagram in this figure is identical to that in Fig. 10.4 if

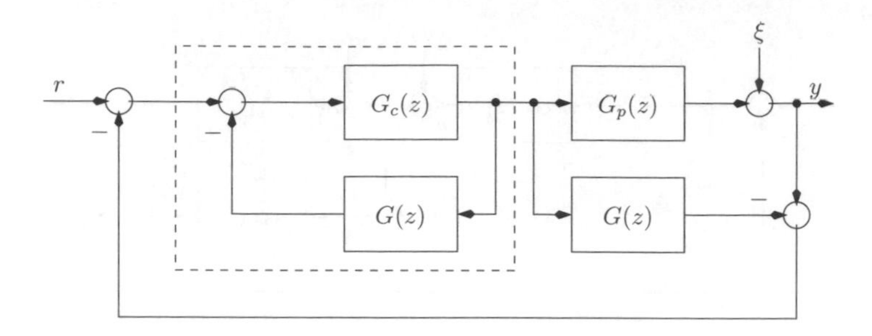

Figure 10.6: Implementation of standard controller in IMC configuration by making transfer function within dashed lines equal to G_Q

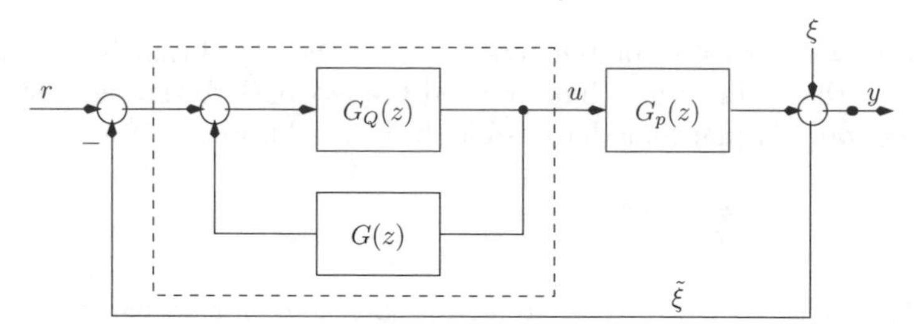

Figure 10.7: Equivalence of standard control configuration with IMC

the entries within the dashed lines are replaced with G_Q; that is, if

$$G_Q = \frac{G_c}{1 + GG_c} \tag{10.19}$$

Suppose that on the other hand we want the controller in the standard configuration, starting from Fig. 10.4. It is easy to check that this figure is identical to Fig. 10.7 if the block within the dashed lines is equal to G_c; that is,

$$G_c = \frac{G_Q}{1 - GG_Q} \tag{10.20}$$

Thus, knowing one form of the controller, we can easily get the other. As the controller has a model of the plant explicitly, it is known as the *internal model controller.*

10.2.1 IMC Design for Stable Plants

Next, we give the procedure to obtain a realizable G_Q that is stable and approximately an inverse of G.

1. Invert the delay free plant model so that G_Q is realizable. For example, if $G = z^{-k}B/A$, then $(B/A)^{-1} = A/B$ is a candidate for G_Q.

2. If the plant is of nonminimum phase, *i.e.*, if B has zeros outside the unit circle, replace these factors with reciprocal polynomials so that G_Q is stable. For example, if

$$G = z^{-k}\frac{1 - 2z^{-1}}{A} \tag{10.21}$$

then a candidate for G_Q is

$$\frac{A}{z^{-1}-2} = \frac{A}{-2+z^{-1}} \tag{10.22}$$

Notice that the zero of the plant is at 2 and that the pole of G_Q is at $1/2 = 0.5$. The reciprocal polynomial of an unstable polynomial, with its zeros strictly outside the unit circle, is guaranteed to be stable.

3. If the plant zero has negative real part, replace that factor with the steady state equivalent. For example, if

$$G = z^{-k}\frac{1+0.5z^{-1}}{A} \tag{10.23}$$

then a candidate for G_Q is

$$\frac{A}{1+0.5z^{-1}|_{z=1}} = \frac{A}{1.5} \tag{10.24}$$

Recall the discussion in Sec. 9.5 why we do not use controllers with negative poles.

4. The noise and model–mismatch have mainly high frequency components. To account for these, a low pass filter of the form

$$G_f \triangleq \frac{B_f}{A_f} = \frac{1-\alpha}{1-\alpha z^{-1}} \tag{10.25}$$

where $1 > \alpha > 0$, is added in series.

We will now summarize this procedure. Let the plant transfer function be factored as

$$G = z^{-k}\frac{B^g B^- B^{\mathrm{nm}+}}{A} \tag{10.26}$$

where B^g is the factor of B with roots inside the unit circle and with positive real parts. B^- is the factor of B with roots that have negative real parts. Note that these roots can be inside, on, or outside the unit circle. $B^{\mathrm{nm}+}$ refers to that part of B containing nonminimum zeros of B with positive real parts. Fig. 10.8 explains this diagrammatically. We will now apply this to the transfer function given in Eq. 10.21, $B^{\mathrm{nm}+} = 1 - 2z^{-1}$ and $B^g = B^- = 1$. In Eq. 10.23, $B^- = 1 + 0.5z^{-1}$ and $B^g = B^{\mathrm{nm}+} = 1$. Note that this notation is different from the conventional meaning of using + and as superscripts. Such a factorization is not considered for the denominator polynomial A because we are dealing with stable systems only and because poles with negative parts are uncommon in sampled systems.

For the stable system factored as in Eq. 10.26, the internal model controller is given by

$$G_Q = G^{\dagger}G_f \tag{10.27}$$

Here

$$G^{\dagger} = \frac{A}{B^g B_s^- B_r^{\mathrm{nm}+}} \tag{10.28}$$

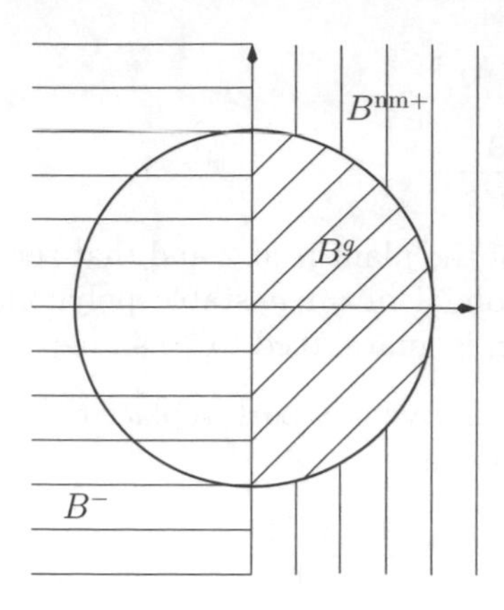

Figure 10.8: Dividing z plane region into good, bad and nonminimum phase parts

where B_s^- is the steady state equivalent of B^-,

$$B_s^- = B^-\big|_{\text{steady state}} \tag{10.29}$$

and $B_r^{\text{nm}+}$ is $B^{\text{nm}+}$ with reversed coefficients,

$$B_r^{\text{nm}+} = B^{\text{nm}+}\big|_{\text{reversed coefficients}} \tag{10.30}$$

For example, in Eq. 10.22, $B_s^- = B^-|_{z=1} = 1 + 0.5z^{-1}\big|_{z=1} = 1.5$ and in Eq. 10.24, $B_r^{\text{nm}+} = z^{-1} - 2$. The function in M 10.3 carries out the above mentioned split. Using Eq. 10.28 and Eq. 10.25 in Eq. 10.27, the IMC becomes

$$G_Q = \frac{A}{B^g B_s^- B_r^{\text{nm}+}} \frac{1-\alpha}{1-\alpha z^{-1}} \tag{10.31}$$

Note that the steady state value of G_Q is the inverse of that of the model of the plant, *i.e.*,

$$G_Q\big|_{\text{ss}} = [G_{\text{ss}}]^{-1} \tag{10.32}$$

The function in M 10.4 implements this control design. We will now illustrate this procedure with examples.

Example 10.3 Design an IMC for the viscosity control problem [34] with the open loop transfer function

$$G = z^{-1}\frac{0.51 + 1.21z^{-1}}{1 - 0.44z^{-1}}$$

Comparing with Eq. 10.26, we obtain $A = 1 - 0.44z^{-1}$, $B^g = B^{\text{nm}+} = 1$, $B^- = 0.51 + 1.21z^{-1}$. It is easy to see that $B_s^- = 1.72$ and $B_r^{\text{nm}+} = 1$. Using Eq. 10.27, Eq. 10.28 and the low pass filter as in Eq. 10.25, we obtain

$$G_Q = \frac{1 - 0.44z^{-1}}{1.72} \frac{1-\alpha}{1-\alpha z^{-1}}$$

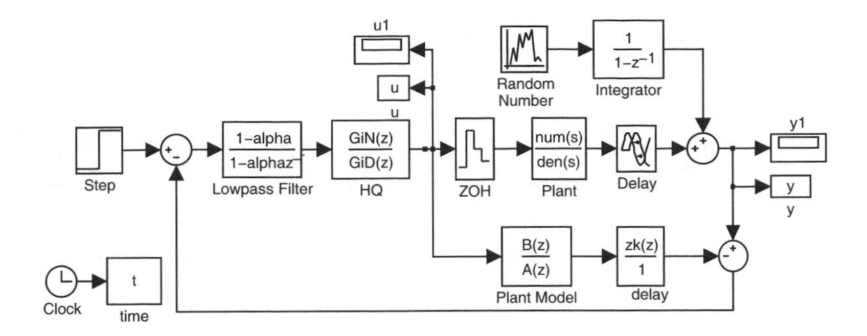

Figure 10.9: Simulink block diagram for simulating a stable system with IMC, in Q form. Code is available at `HOME/imc/matlab/imc_Q_c.mdl`, see Footnote 1 on page 397.

Matlab code M 10.6 implements this control design. The Simulink program in Fig. 10.9 shows how this system can be simulated. Note that the plant numerator B_p and denominator A_p can be made different from the assumed models, B and A, and robustness studies can be carried out. ∎

Next we present an IMC for the van de Vusse reactor, presented in Sec. 2.2.5.

Example 10.4 Design an IMC for the open loop transfer function

$$G(s) = \frac{-1.117s + 3.1472}{s^2 + 4.6429s + 5.3821}$$

where the time unit is minutes. Sampling this with $T_s = 0.1$ minute, we arrive at

$$G(z) = \frac{-0.075061z^{-1}(1 - 1.334z^{-1})}{(1 - 0.7995z^{-1})(1 - 0.7863z^{-1})}$$

Comparing with Eq. 10.26, we find that

$$\begin{aligned}
A &= (1 - 0.7995z^{-1})(1 - 0.7863z^{-1}) \\
B^{\text{nm+}} &= 1 - 1.334z^{-1} \\
B^{g} &= -0.075061 \\
B^{-} &= 1
\end{aligned}$$

We will first evaluate $B_r^{\text{nm+}}$ as

$$B_r^{\text{nm+}} = z^{-1} - 1.334$$

Using Eq. 10.27, Eq. 10.28 and the low pass filter as in Eq. 10.25, we obtain

$$\begin{aligned}
G_Q &= \frac{(1 - 0.7995z^{-1})(1 - 0.7863z^{-1})}{-0.075061(z^{-1} - 1.334)} G_f \\
&= \frac{(1 - 0.7995z^{-1})(1 - 0.7863z^{-1})}{0.1001(1 - 0.7496z^{-1})} \frac{1 - \alpha}{1 - \alpha z^{-1}}
\end{aligned}$$

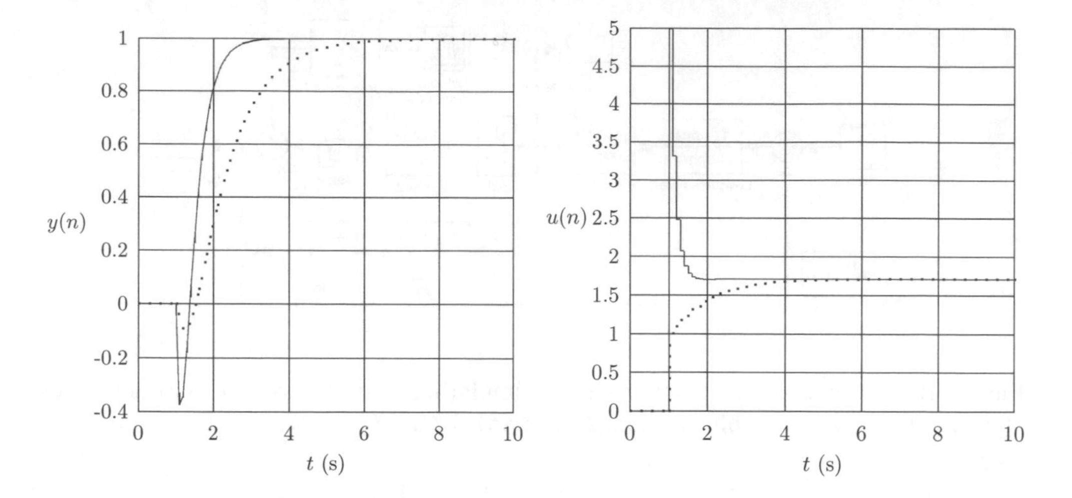

Figure 10.10: Output $y(n)$ and input $u(n)$ profiles in the van de Vusse reactor, as discussed in Example 10.4. Solid lines correspond to $\alpha = 0.5$ and dotted lines denote $\alpha = 0.9$.

This is identical to the controller obtained by [3]. The script in M 10.7 implements this control design. Fig. 10.9, with the delay block deleted, is used to simulate the efficacy of this controller. Fig. 10.10 presents the output and input profiles for α values of 0.5 and 0.9. It can be seen that the performance for $\alpha = 0.9$ is more sluggish than that for $\alpha = 0.5$. While both of them show an inverse response, it is less for $\alpha = 0.9$.

∎

We will illustrate the IMC control technique with another example.

Example 10.5 Design an IMC for the continuous transfer function

$$G(s) = \frac{1}{(10s+1)(25s+1)}$$

sampled with $T_s = 3$.

We obtain

$$G(z) = \frac{0.0157z^{-1}(1+0.8649z^{-1})}{(1-0.8869z^{-1})(1-0.7408z^{-1})}$$

Comparing with Eq. 10.26, we obtain

$$\begin{aligned} A &= (1-0.8869z^{-1})(1-0.7408z^{-1}) \\ B^g &= 0.0157 \\ B^- &= 1+0.8649z^{-1} \\ B^{\text{nm}+} &= 1 \end{aligned}$$

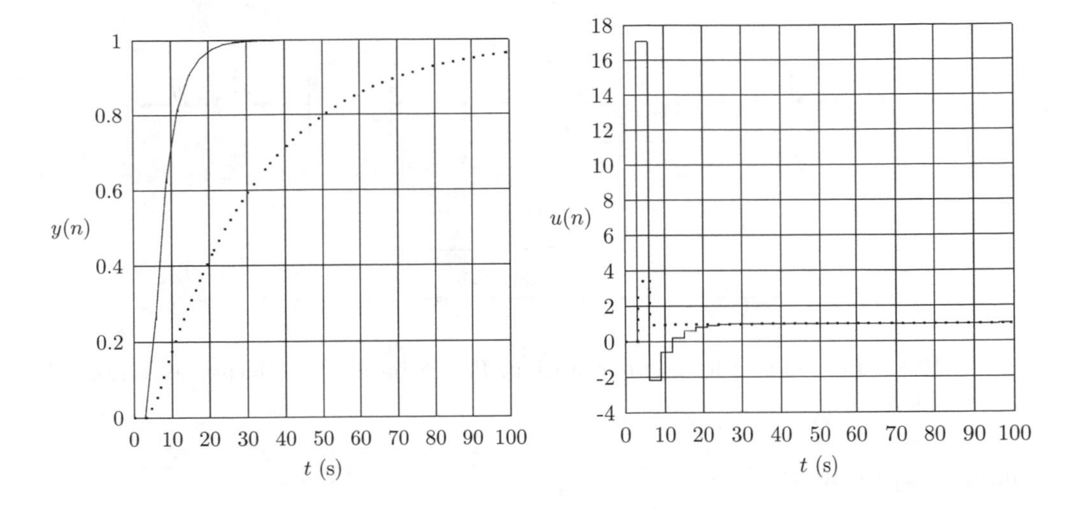

Figure 10.11: Output $y(n)$ and input $u(n)$ profiles for the problem discussed in Example 10.5. Solid lines correspond to $\alpha = 0.5$ and dotted lines denote $\alpha = 0.9$.

We will first evaluate B_s^- as

$$B_s^- = 1.8649$$

Using Eq. 10.27, Eq. 10.28 and the low pass filter as in Eq. 10.25, we obtain

$$G_Q(z) = \frac{(1 - 0.8869z^{-1})(1 - 0.7408z^{-1})}{0.0157 \times 1.8649} \frac{1 - \alpha}{1 - \alpha z^{-1}}$$

The script in M 10.8 implements this control design. Fig. 10.9, with the delay block deleted, is used to simulate the efficacy of this controller. Fig. 10.11 presents the output and input profiles for α values of 0.5 and 0.9. It can be seen that the performance for $\alpha = 0.9$ is more sluggish than that for $\alpha = 0.5$. ∎

In order to compare with other controllers, we need to express the IMC in conventional form. The next section addresses this issue.

10.2.2 IMC in Conventional Form for Stable Plants

In the last section, we have seen the benefits of the IMC structure and a way to design controllers. In this section, we will show how to obtain an equivalent controller in the conventional form. We can use either Eq. 10.20 or Fig. 10.12, which is an equivalent of Fig. 10.4 or Fig. 10.7. The IMC equivalent conventional feedback controller is given by

$$G_c = \frac{\dfrac{B_f}{A_f} \dfrac{A}{B^g B_r^{\text{nm}+} B_s^-}}{1 - \dfrac{B_f}{A_f} \dfrac{A}{B^g B_r^{\text{nm}+} B_s^-} \dfrac{B^g B^{\text{nm}+} B^-}{A} z^{-k}}$$

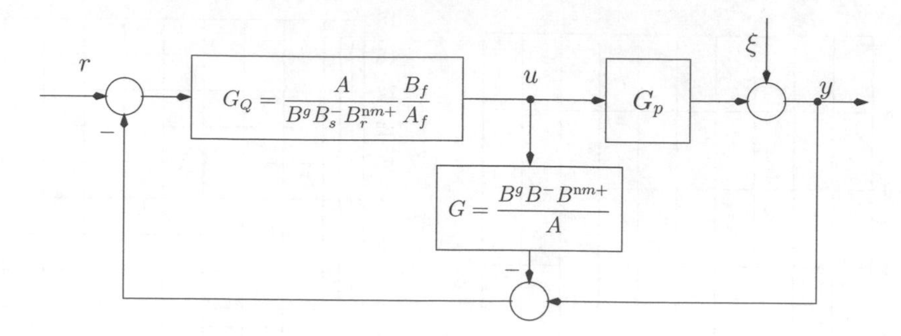

Figure 10.12: IMC closed loop configuration. It can be used to derive conventional control configuration

Simplifying, we arrive at

$$G_c = \frac{B_f A}{B^g (A_f B_r^{nm+} B_s^- - B_f B^{nm+} B^- z^{-k})} \triangleq \frac{S_c}{R_c} \tag{10.33}$$

The controller is in the form

$$R_c(z)u = T_c(z)r - S_c(z)y \tag{10.34}$$

where

$$T_c(z) = S_c(z) \tag{10.35}$$

If the filter is chosen as in Eq. 10.25, using Eq. 10.32 and Eq. 10.20, we see that

$$R_c(1) = 0 \tag{10.36}$$

implying integral control. The function in M 10.9 is an augmented version of M 10.4: it determines G_c, given G_Q. We now illustrate this approach with a few examples.

Example 10.6 Design an IMC for the first order system $G(s) = 1/(s+1)$ when sampled with $T_s = 1$ s and express it in the conventional form.

The discrete time transfer function is given by

$$G(z) = \frac{0.6321z^{-1}}{1 - 0.3679z^{-1}}$$

The Q form of the IMC is given by

$$G_Q = \frac{1 - 0.3679z^{-1}}{0.6321}$$

Suppose that the filter is chosen to be 1. Using Eq. 10.20, we obtain the controller in the conventional form as

$$\begin{aligned} G_c &= \frac{1 - 0.3679z^{-1}}{0.6321} \left[1 - \frac{0.6321z^{-1}}{1 - 0.3679z^{-1}} \frac{1 - 0.3679z^{-1}}{0.6321}\right]^{-1} \\ &= \frac{1 - 0.3679z^{-1}}{0.6321(1 - z^{-1})} \end{aligned}$$

∎

Example 10.7 Determine the IMC equivalent conventional controller for the system presented in Example 10.3.

Using Eq. 10.33, we find that

$$\begin{aligned} R_c &= 1.72(1-\alpha z^{-1}) - z^{-1}(1-\alpha)(0.51+1.21z^{-1}) \\ &= 1.72 - [1.72\alpha + (1-\alpha)0.51]z^{-1} - 1.21(1-\alpha)z^{-2} \\ &= 1.72 - (0.51+1.21\alpha)z^{-1} - 1.21(1-\alpha)z^{-2} \\ &= [1.72 + 1.21(1-\alpha)z^{-1}](1-z^{-1}) \\ S_c &= B_f A = (1-\alpha)(1-0.44z^{-1}) \end{aligned}$$

The controller is given by Eq. 10.34. It has one tuning factor, α, the filter time constant. It is easy to check that for all α, $R_c(1) = 0$. ∎

Example 10.8 Determine the IMC equivalent conventional controller for the system presented in Example 10.4.

Using Eq. 10.33, we find that

$$\begin{aligned} R_c &= -0.075061[(1-\alpha z^{-1})(-1.334+z^{-1}) - (1-\alpha)(1-1.334z^{-1})z^{-1}] \\ &= -0.075061(-1.334 + 2.334\alpha z^{-1} + (1.334 - 2.334\alpha)z^{-2}) \\ &= 0.1001(1 - 1.75\alpha z^{-1} - (1-1.75\alpha)z^{-2}) \\ &= 0.1001(1-z^{-1})[1 + (1-1.75\alpha)z^{-1}] \end{aligned}$$

It is easy to check that $R_c(1) = 0$ for all α. If we assume a filter constant $\alpha = 0.5$, this simplifies to

$$R_c = 0.1001(1-z^{-1})(1+0.875z^{-1}) = 0.1001 - 0.0876z^{-1} - 0.0125z^{-2}$$

The corresponding expression for S_c is

$$\begin{aligned} S_c &= (1-\alpha)A(1-0.7995z^{-1})(1-0.7863z^{-1}) \\ &= 0.5 - 0.7929z^{-1} + 0.3143z^{-2} \end{aligned}$$

A program to compute the coefficients is given in M 10.10. ∎

Example 10.9 Determine the IMC equivalent conventional controller for the system presented in Example 10.5.

Using Eq. 10.33, we find that

$$R_c = 0.0157[(1-\alpha z^{-1})1.8649 - (1-\alpha)(1+0.8649z^{-1})z^{-1}]$$

With $\alpha = 0.5$, this simplifies to $Rc = 0.0293 - 0.0225z^{-1} - 0.0068z^{-2}$. The corresponding expression for S_c is $S_c = 0.5 - 0.8139z^{-1} + 0.3285z^{-2}$. It is easy to check that $R_c(1) = 0$. ∎

10.2.3 PID Tuning Through IMC

In Sec. 9.8, we have presented a method to tune PID controllers using the pole placement approach. We will now use the IMC to obtain PID tuning parameters. Looking at the examples in Sec. 10.2.2, we see that the IMC generated controllers have the property of $S_c(z) = R_c(z)$ and the degree of $R_c(z) = \mathrm{d}R_c = 2$, where the controller is given by Eq. 10.34,

$$R_c(z)u = T_c(z)r - S_c(z)y \tag{10.37}$$

We see that the PID controllers presented in Sec. 8.3.5 and 8.3.6 are not suitable because, they have degree of R_c as 1. On the other hand, the PID controller presented in Sec. 8.3.7 is possibly suitable because it has $\mathrm{d}R_c = 2$, see Eq. 8.55 on page 321. This PID controller, however, has to be modified so that $S_c = T_c$, which is the required form for the IMC. This can be carried out by subjecting r also to the same treatment as y. Thus, the controller given by Eq. 8.53 becomes

$$U(z) = K\left[1 + \frac{1}{\tau_i}\frac{T_s}{2}\frac{z+1}{z-1} + \frac{\frac{z-1}{zT_s}\tau_d}{1 + \frac{z-1}{zT_s}\frac{\tau_d}{N}}\right](R(z) - Y(z)) \tag{10.38}$$

where we have used the trapezoidal rule for integration. Simplifying, we obtain

$$\begin{aligned}
&(1 - z^{-1})(1 - a_d z^{-1})U(z) \\
&\quad = K\left[(1 - z^{-1})(1 - a_d z^{-1}) + \frac{T_s}{2\tau_i}(1 + z^{-1})(1 - a_d z^{-1}) + Na_d(1 - z^{-1})^2\right] \\
&\quad \times (R(z) - Y(z))
\end{aligned} \tag{10.39}$$

This is in the standard controller form as given by Eq. 10.37 with

$$\begin{aligned}
R_c(z) &= (1 - z^{-1})(1 - a_d z^{-1}) \\
S_c(z) &= K\left[\left(1 + \frac{T_s}{2\tau_i} + Na_d\right)\right. \\
&\quad + \left(-(a_d + 1) + \frac{T_s}{2\tau_i}(1 - a_d) - 2Na_d\right)z^{-1} \\
&\quad \left. + \left(a_d - \frac{T_s}{2\tau_i}a_d + Na_d\right)z^{-2}\right] \\
T_c(z) &= S_c(z)
\end{aligned} \tag{10.40}$$

Comparing this with the IMC parameters given by Eq. 10.33, one can determine the PID parameters. We will illustrate this approach with an example.

Example 10.10 Determine the tuning parameters so that the resulting PID controller is equivalent to the IMC derived in Example 10.8.

Comparing $R_c(z)$ in Eq. 10.40 with that in the example under discussion, we find that

$$a_d = 1.75\alpha - 1$$

Equating $S_c(z)$ of Example 10.8 with the form in Eq. 10.40, we obtain three equations, which can be solved for three unknowns, K, τ_i and N. From the expression for a_d, one can determine τ_d.

∎

In this section, we have given a brief overview of tuning PID parameters through the IMC, a model based controller. We will return to this topic in Sec. 11.3.3.

10.3 Matlab Code

Matlab Code 10.1 Effect of delay in control performance. Implements the steps discussed in Example 10.1 on page 381. This code is available at HOME/imc/matlab/delay.m[1]

```
1  Ts = 1; B = 0.63; A = [1 -0.37];
2  k = input('Enter the delay as an integer: ');
3  if k<=0, k = 1; end
4  [zk,dzk] = zpowk(k);
5
6  % Desired transfer function
7  phi = [1 -0.5];
8  delta = 1; % internal model of step introduced
9
10 % Controller design
11 [Rc,Sc,Tc,gamma] = pp_im(B,A,k,phi,delta);
12
13 % simulation parameters for stb_disc.mdl
14 st = 1.0; % desired change in setpoint
15 t_init = 0; % simulation start time
16 t_final = 20; % simulation end time
17
18 % simulation parameters for stb_disc.mdl
19 N_var = 0; C = 0; D = 1; N = 1;
```

Matlab Code 10.2 Smith predictor for paper machine control, presented in Example 10.2 on page 385. This code is available at HOME/imc/matlab/smith.m

```
1  Ts = 1; B = 0.63; A = [1 -0.37]; k = 3;
2  Bd = conv(B,[0 1]);
3  kd = k - 1;
4  [zkd,dzkd] = zpowk(kd);
5  [mzkd,dmzkd] = poladd(1,0,-zkd,dzkd);
6
7  % Desired transfer function
8  phi = [1 -0.5]; delta = 1;
9
10 % Controller design
11 [Rc,Sc,Tc,gamma] = pp_im(B,A,1,phi,delta);
12
13 % simulation parameters for smith_disc.mdl
14 st = 1.0; % desired change in setpoint
```

[1] HOME stands for http://www.moudgalya.org/dc/ – first see the software installation directions, given in Appendix A.2.

```
15 t_init = 0; % simulation start time
16 t_final = 20; % simulation end time
17
18 % simulation parameters for smith_disc.mdl
19 N_var = 0; C = 0; D = 1; N = 1;
```

Matlab Code 10.3 Splitting a polynomial $B(z)$ into B^g, B^- and B^{nm+}, as discussed in Sec. 10.2.1. An example of the usage is given in M 10.4. This code is available at HOME/imc/matlab/imcsplit.m

```
1  % Splits a polynomial B into good, nonminimum with positive
2  % real & with negative real parts. All are returned in
3  % polynomial form. Gain is returned in Kp and delay in k.
4
5  function [Kp,k,Bg,Bnmp,Bm] = imcsplit(B,polynomial)
6  k = 0;
7  Kp = 1;
8  if(polynomial)
9    rts = roots(B);
10   Kp = sum(B)/sum(poly(rts));
11 else
12   rts = B;
13 end
14 Bg = 1; Bnmp = 1; Bm = 1;
15 for i = 1:length(rts),
16     rt = rts(i);
17     if rt == 0,
18        k = k+1;
19     elseif (abs(rt)<1 & real(rt)>=0)
20       Bg = conv(Bg,[1 -rt]);
21     elseif (abs(rt)>=1 & real(rt)>=0)
22       Bnmp = conv(Bnmp,[1 -rt]);
23     else
24       Bm = conv(Bm,[1 -rt]);
25     end
26 end
```

Matlab Code 10.4 Design of internal model controller, G_Q, discussed in Sec. 10.2.1. An example of the usage is given in M 10.6. This code is available at HOME/imc/matlab/imc_stable1.m

```
1  % Designs Discrete Internal Model Controller
2  % for transfer function z^{-k}B(z^{-1})/A(z^{-1})
3  % Numerator and Denominator of IMC HQ are outputs
4  % Controller is also given in R,S form
5  %
6  function [k,HiN,HiD] = imc_stable1(B,A,k,alpha)
7
8  [Kp,d,Bg,Bnmp,Bm] = imcsplit(B,logical(1));
```

```
9  Bg = Kp * Bg;
10 Bnmpr = flip(Bnmp);
11 Bms = sum(Bm);
12 HiN = A;
13 HiD = Bms * conv(Bg,Bnmpr);
14 k = k+d;
```

Matlab Code 10.5 Flipping a vector. This code is available at HOME/matlab/flip.m

```
1 function b = flip(a)
2 b = a(length(a):-1:1);
```

Matlab Code 10.6 IMC design for viscosity control problem, as discussed in Example 10.3 on page 390. This code is available at HOME/imc/matlab/visc_imc1.m

```
1 B = [0.51  1.21];
2 A = [1  -0.44];
3 k = 1;
4 alpha = 0.5;
5 [k,GiN,GiD] = imc_stable1(B,A,k,alpha);
6 [zk,dzk] = zpowk(k);
7 Bp = B; Ap = A;
8 Ts = 0.1; t0 = 0; tf = 20; Nvar = 0.01;
```

Matlab Code 10.7 IMC design for the control of van de Vusse reactor, as discussed in Example 10.4. This code is available at HOME/imc/matlab/vande_imc1.m

```
1  num = [-1.117  3.1472]; den = [1  4.6429  5.3821];
2  G = tf(num,den);
3  Ts = 0.1;
4  [B,A,k] = myc2d(G,Ts);
5  alpha = 0.9;
6  [k,GiN,GiD] = imc_stable1(B,A,k,alpha);
7  [zk,dzk] = zpowk(k);
8  Bp = B; Ap = A;
9  t0 = 0; tf = 10; st = 1; Nvar = 0;
10
11 % simulink executed with delay block deleted
```

Matlab Code 10.8 IMC design for Lewin's example, as discussed in Example 10.5 on page 392. This code is available at HOME/imc/matlab/lewin_imc1.m

```
1 num = 1; den = [250  35  1]; Ts = 3;
2 G = tf(num,den);
3 [B,A,k] = myc2d(G,Ts);
4 alpha = 0.9;
5 [k,GiN,GiD] = imc_stable1(B,A,k,alpha);
6 [zk,dzk] = zpowk(k);
```

```
7  Bp = B; Ap = A;
8  t0 = 0; tf = 100; st = 1; Nvar = 0;
9
10 % simulink executed with delay block deleted
```

Matlab Code 10.9 Design of conventional controller G_D which is an equivalent of internal model controller, G_Q, as discussed in Sec. 10.2.2. This code is available at HOME/imc/matlab/imc_stable.m

```
1  % Designs Discrete Internal Model Controller
2  % for transfer function z^{-k}B(z^{-1})/A(z^{-1})
3  % Numerator and Denominator of IMC HQ are outputs
4  % Controller is also given in R,S form
5  %
6  function [k,HiN,HiD,R,S,mu] = imc_stable(B,A,k,alpha)
7
8  [Kp,d,Bg,Bnmp,Bm] = imcsplit(B,logical(1));
9  Bg = Kp * Bg;
10 Bnmpr = flip(Bnmp);
11 Bms = sum(Bm);
12 HiN = A;
13 HiD = Bms * conv(Bg,Bnmpr);
14 k = k+d;
15 [zk,dzk] = zpowk(k);
16 Bf = (1-alpha);
17 Af = [1 -alpha];
18 S = conv(Bf,A);
19 R1 = conv(Af,conv(Bnmpr,Bms));
20 R2 = conv(zk,conv(Bf,conv(Bnmp,Bm)));
21 [R,dR] = poladd(R1,length(R1)-1,-R2,length(R2)-1);
22 R = conv(Bg,R);
```

Matlab Code 10.10 Design of conventional controller G_D for van de Vusse reactor problem, as discussed in Example 10.7 on page 395. This code is available at HOME/imc/matlab/vande_imc.m

```
1  num = [-1.117 3.1472]; den = [1 4.6429 5.3821];
2  G = tf(num,den);
3  Ts = 0.1;
4  [B,A,k] = myc2d(G,Ts);
5  alpha = 0.5;
6  [k,HiN,HiD,R,S] = imc_stable(B,A,k,alpha);
7  [zk,dzk] = zpowk(k);
8  Bp = B; Ap = A;
```

10.4 Problems

10.1. Consider the control of a process with a large inbuilt time delay (also known as dead time). Control using a delayed signal, as shown in the first figure below, is undesirable. Ideally we would like to have the configuration as in the middle figure. But we cannot poke into the process to obtain the measurement. An alternative is to go for the bottom configuration, if the time delay n is known. Show that the middle and bottom figures are equivalent. This is known as the Smith predictor. Assume that $G(z)$ includes one delay.

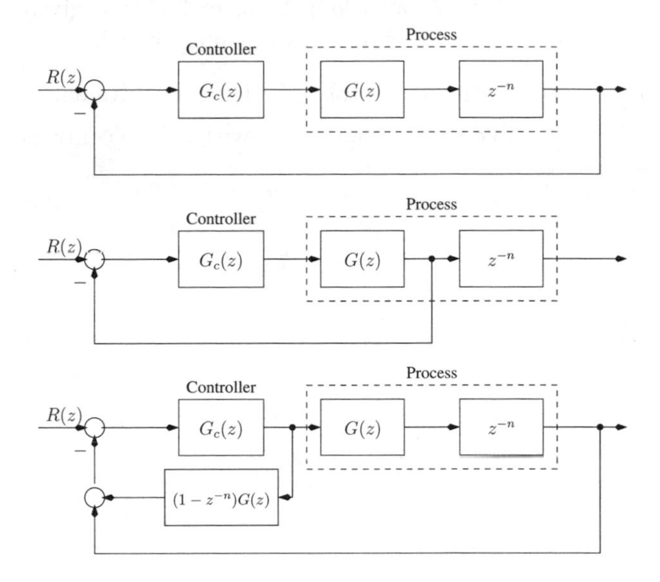

10.2. The approach presented in Sec. 10.1 works only for stable plants. In this problem, we will present a general method for Smith predictors [31]. For the system model presented in Sec. 10.1, the Smith predictor is given by the following equations:

$$R_0(z)u(n) = \gamma T_0(z)r(n) - S_0(z)\hat{y}(n+d|n) \tag{10.41}$$

$$\phi_p(z)\hat{y}(n+d|n) = F(z)y(n) + B_d(z)E(z)u(n) \tag{10.42}$$

where S_0 and R_0 are the solutions of

$$A(z)R_0(z) + B_d(z)S_0(z) = \phi_0(z) \tag{10.43}$$

and $F(z)$ and $E(z)$ are the solutions of

$$A(z)E(z) + z^{-d}F(z) = \phi_p(z) \tag{10.44}$$

with $d = k - 1$.

[Hint: Eliminate $\hat{y}$ from Eq. 10.41– 10.42 and obtain the following:

$$(\phi_p R_0 + BS_0E)u(n) = \gamma T_0\phi_p r(n) - S_0Fy(n) \tag{10.45}$$

Substitute this expression for u into $y = z^{-d}Bu/A$, simplify, and obtain

$$y(n) = \frac{z^{-d}B_d\phi_p\gamma T_0}{A(\phi_p R_0 + S_0 B_d E) + z^{-d}B_d S_0 F} \tag{10.46}$$

Using Eq. 10.43–10.44, show that the denominator of Eq. 10.46 can be written as $\phi_p\phi_0$. ϕ_p can be thought of as the characteristic polynomial of the predictor. ϕ_0 is the characteristic polynomial when $d = 0$.]

10.3. Design an IMC, G_Q, for the open loop transfer function given in Eq. 8.58 on page 322 with the value of the filter constant unspecified.

(a) What is the conventional controller G_c that is equivalent to the above G_Q?

(b) If you have to implement the above G_c with a PID controller, what are the P, I, D settings?

Chapter 11

Minimum Variance Control

Minimum variance control is concerned with the design of controllers so as to minimize the variance of the error in plant output. This controller does not worry about the control effort required to achieve the result. Minimum variance control is used as a benchmark, against which the performance of other controllers is compared. The output of minimum variance controllers could be unbounded. The generalized minimum variance control technique is one way to overcome this problem. In this chapter, we present these two families of controllers. We begin with prediction error models, which help design these controllers in the presence of noise.

11.1 j-Step Ahead Prediction Error Model

If only white noise enters the system as a disturbance, nothing can be done since white noise cannot be modelled. If, on the other hand, white noise enters through a filter, we should try to estimate it so as to take corrective action. Prediction error models are required for this purpose.

We have seen one step ahead prediction error models in Sec. 6.6. For control purposes, however, the one step ahead prediction error model is not sufficient. We look for a j step ahead prediction error model. Once we can predict how the plant will behave j steps into the future, we can design a controller that will help improve the performance of the plant over a time horizon.

In this section, we present prediction error models for systems described by noise models, such as ARMAX and ARIMAX.

11.1.1 Control Objective for ARMAX Systems

Let $u(t)$ be the input to and $y(t)$ be the output from a process that needs to be controlled. Let $\xi(t)$ be a white noise signal that affects the output. Consider an ARMAX model, described by Eq. 6.150 on page 204, reproduced here for convenience:

$$A(z)y(n) = z^{-k}B(z)u(n) + C(z)\xi(n) \tag{11.1}$$

Digital Control Kannan M. Moudgalya

Here, A, B and C are polynomials, defined in Eq. 6.130–6.132 on page 199, reproduced once again:

$$\begin{aligned} A(z) &= 1 + a_1 z^{-1} + \cdots + a_{\mathrm{d}A} z^{-\mathrm{d}A} \\ B(z) &= b_0 + b_1 z^{-1} + \cdots + b_{\mathrm{d}B} z^{-\mathrm{d}B} \\ C(z) &= 1 + c_1 z^{-1} + \cdots + c_{\mathrm{d}C} z^{-\mathrm{d}C} \end{aligned} \tag{11.2}$$

where we have used the mixed notation of Sec. 6.4.1. The effect of delay is completely factored in k so that the leading term of B is a nonzero constant. Also note that the polynomials A and C are monic, because their leading term is one. Recall that $\mathrm{d}A$, $\mathrm{d}B$ and $\mathrm{d}C$ are the corresponding polynomials. The above equation can be written as

$$y(n) = \frac{B(z)}{A(z)} u(n-k) + \frac{C(z)}{A(z)} \xi(n) \tag{11.3}$$

or

$$y(n+k) = \frac{B(z)}{A(z)} u(n) + \frac{C(z)}{A(z)} \xi(n+k) \tag{11.4}$$

As this process has a delay of k samples, any change in u will affect y only after k time instants. As a result, we cannot hope to modify the plant output at any time earlier than $n+k$. In contrast, as there is no delay between them, the noise signal $\xi(n)$ starts affecting $y(n)$ immediately. The best we can do is to predict the output at $n+j$, $j \geq k$, so as to take corrective action. In the minimum variance and generalized minimum variance controller, to be discussed in this chapter, we choose j to be equal to k. In the generalized predictive controller, to be discussed in the next chapter, we choose $j > k$.

Suppose that we want to determine the input $u(n)$ that makes $y(n+k)$ approximately zero. In Eq. 11.4, because of the presence of terms containing powers of z^{-1}, the noise element ξ could have past and future terms. If we can split these two effects, it is easy to determine the condition to minimize $y(t+k)$. We demonstrate this in the next section.

11.1.2 Prediction Error Model Through Noise Splitting

If we can split the noise term in Eq. 11.4 into past and future terms, we can zero the latter and obtain the best prediction error model. This model can then be used to determine the control law. If $C/A = 1$ in Eq. 11.4, the best prediction error model of $y(n+k)$ estimated at the current time n is

$$\hat{y}(n+k|n) = \frac{B}{A} u(n) \tag{11.5}$$

as the future noise terms are white with their best estimate being zero. The $|n$ in $\hat{y}$ says that we have used all available information until now. What do we do if C is not equal to A? We will show that we can easily handle this case by splitting C/A into E_j and F_j as follows:

$$\frac{C}{A} = E_j + z^{-j} \frac{F_j}{A} \tag{11.6}$$

where

$$E_j = e_{j,0} + e_{j,1}z^{-1} + \cdots + e_{j,j-1}z^{-(j-1)}$$
$$F_j = f_{j,0} + f_{j,1}z^{-1} + \cdots + f_{j,\mathrm{d}A-1}z^{-(\mathrm{d}A-1)} \tag{11.7}$$

We show in Sec. 11.1.4 how to arrive at Eq. 11.6. For the time being, we would only like to point out that the degree of E_j is $j-1$. Moreover, every term in F_j will be multiplied by z^{-j}, because of the presence of the latter in Eq. 11.6. We multiply Eq. 11.6 by A to obtain

$$C = E_j A + z^{-j} F_j \tag{11.8}$$

Multiplying Eq. 11.3 by $z^j E_j A$ we obtain

$$z^j E_j A y(n) = z^j E_j B u(n-k) + z^j E_j C \xi(n)$$

where the dependence on z is omitted for convenience. Substituting for $E_j A$ from Eq. 11.8 we obtain

$$z^j (C - z^{-j} F_j) y(n) = z^j E_j B u(n-k) + z^j E_j C \xi(n)$$

From this, we obtain $z^j C y(n) = F_j y(n) + z^j E_j B u(n-k) + z^j E_j C \xi(n)$. Dividing by C, we obtain

$$y(n+j) = \frac{E_j B u(n+j-k) + F_j y(n)}{C} + E_j \xi(n+j) \tag{11.9}$$

Recall that the objective of the prediction error model is to estimate the output y at a *future* time $n+j$. The last term in the above equation has the noise terms at future time instants only, as the degree of E is $j-1$, see Eq. 11.7. As we cannot estimate the future noise, which is supposed to be random, the best estimate of the output for the model in Eq. 11.9 is an identical one without the noise term. That is, the estimate of $y(n+j)$ is $\hat{y}(n+j|n)$, given by

$$\hat{y}(n+j|n) = \frac{E_j B u(n+j-k) + F_j y(n)}{C} \tag{11.10}$$

We define a new variable G_j as follows:

$$G_j = E_j(z) B(z) \tag{11.11}$$

Eq. 11.10 becomes

$$C\hat{y}(n+j|n) = G_j u(n+j-k) + F_j y(n) \tag{11.12}$$

As this model can be used to predict the relationship between $u(n)$ and $y(n+j)$ so as to reduce the effect of noise, it is known as the prediction error model. Comparing Eq. 11.9 and Eq. 11.10, we obtain

$$y(n+j) = \hat{y}(n+j|n) + \tilde{y}(n+j|n) \tag{11.13}$$

where $\tilde{y}(n+j|n)$ is the error in this prediction and it is given by

$$\tilde{y}(n+j|n) = E_j \xi(n+j) \tag{11.14}$$

When the context is clear, we will write $\hat{y}(n+j|n)$ simply as $\hat{y}(n+j)$.

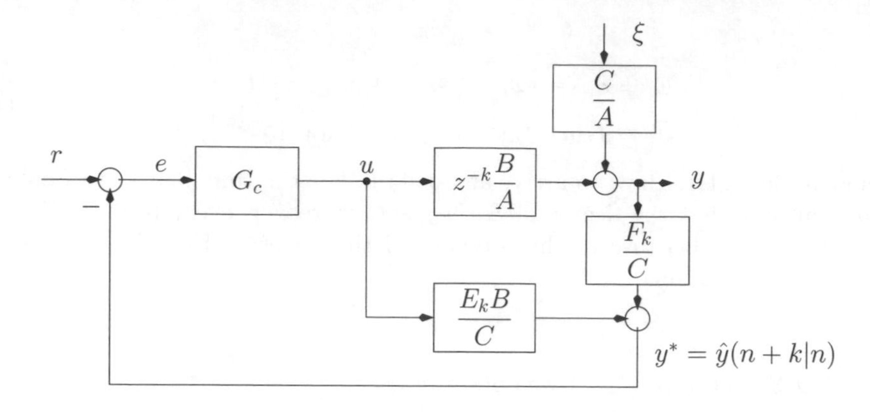

Figure 11.1: An interpretation of the prediction error model

11.1.3 Interpretation of the Prediction Error Model

In Sec. 10.1, we have shown that the Smith predictor can help overcome the adverse effects of time delays in the plant. Unfortunately, however, the Smith predictor does not work when noise is present. We will demonstrate in this section that the prediction error model plays the role of Smith predictor when noise is present [18]. For this purpose, we make use of the k step ahead prediction error model, with which we will derive two popular control strategies in the latter part of this chapter.

We will study the prediction error model of the ARMAX model in detail in this section. First, let us recall the ARMAX model of Eq. 11.1 and reproduce it:

$$y = z^{-k}\frac{B}{A}u + \frac{C}{A}\xi \tag{11.15}$$

If we substitute k in place of j in Eq. 11.10, we obtain the following k step ahead prediction error model:

$$\hat{y}(n+k|n) = \frac{E_kB}{C}u(n) + \frac{F_k}{C}y(n)$$

A schematic of this model is given in Fig. 11.1. We denote the predicted output as y^* in this figure. We have

$$y^* = \frac{E_kB}{C}u + \frac{F_k}{C}y$$

We now substitute for y using Eq. 11.15 and simplify. We obtain

$$\begin{aligned} y^* &= \frac{E_kB}{C} + \frac{F_k}{C}\left[z^{-k}\frac{B}{A}u + \frac{C}{A}\xi\right] \\ &= \frac{E_kB}{C}u + z^{-k}\frac{BF_k}{AC}u + \frac{F_k}{A}\xi = \frac{B}{C}\left(E_k + z^{-k}\frac{F_k}{A}\right)u + \frac{F_k}{A}\xi \end{aligned}$$

We now make use of Eq. 11.8, with k in place of j, to obtain

$$y^* = \frac{B}{C}\frac{C}{A}u + \frac{F_k}{A}\xi = \frac{B}{A}u + \frac{F_k}{A}\xi$$

Comparing this with Eq. 11.15, we see that the first term appears without k, while the C/A term has been replaced by F_k/A. Let us now try to understand what this means. Substituting for C/A using Eq. 11.6 in Eq. 11.15, we obtain

$$y = z^{-k}\frac{B}{A}u + E_k\xi + \frac{F_k}{A}z^{-k}\xi$$

The prediction error model is obtained by zeroing future noise terms. We obtain

$$\hat{y} = z^{-k}\frac{B}{A}u + \frac{F_k}{A}z^{-k}\xi = \left(\frac{B}{A}u + \frac{F_k}{A}\xi\right)z^{-k}$$

Thus, y^* is the k step ahead predictor of $\hat{y}$. In other words, the predictor $y^* = \hat{y}(n+k|n)$ takes care of the task that cannot be handled by the Smith predictor.

An important technique that we have used to arrive at the prediction error model is the splitting of C/A into past and future contributions, using Eq. 11.6. In the next section, we present a few methods to carry out this calculation.

11.1.4 Splitting Noise into Past and Future Terms

We have arrived at the prediction error model using the relation given in Eq. 11.6. In this section, we show how E_j and F_j can be calculated, given j. We begin with an example.

Example 11.1 Divide the noise term arising out of

$$C(z) = 1 + 0.5z^{-1}$$
$$A(z) = (1 + 0.2z^{-1})(1 - 0.8z^{-1}) = 1 - 0.6z^{-1} - 0.16z^{-2}$$

into E_j and F_j, as in Eq. 11.6, for $j = 2$.

We carry out the following long division:

$$\begin{array}{r|llll}
 & 1 & +1.1z^{-1} & & \\
\hline
1 - 0.6z^{-1} - 0.16z^{-2} & 1 & +0.5z^{-1} & & \\
 & 1 & -0.6z^{-1} & -0.16z^{-2} & \\
\hline
 & & +1.1z^{-1} & +0.16z^{-2} & \\
 & & +1.1z^{-1} & -0.66z^{-2} & -0.176z^{-3} \\
\hline
 & & & +0.82z^{-2} & +0.176z^{-3} \\
\hline
\end{array}$$

Thus we obtain

$$\begin{aligned} E_2 &= 1 + 1.1z^{-1} \\ F_2 &= 0.82 + 0.176z^{-1} \end{aligned} \tag{11.16}$$

In other words

$$\frac{1 + 0.5z^{-1}}{1 - 0.6z^{-1} - 0.16z^{-2}} = (1 + 1.1z^{-1}) + z^{-2}\frac{0.82 + 0.176z^{-1}}{1 - 0.6z^{-1} - 0.16z^{-2}}$$

∎

From the above example, it is easy to see the following relations for the degrees of E_j and F_j:

$$\begin{aligned} \mathrm{d}E_j &= j - 1 \\ \mathrm{d}F_j &= \max(\mathrm{d}C - j, \mathrm{d}A - 1) \end{aligned} \tag{11.17}$$

We can also compute E_j and F_j recursively. First we illustrate this concept with the problem considered in the above example.

Example 11.2 Recursively compute E_j and F_j for the system discussed in Example 11.1.

Rewrite the division results obtained in Example 11.1 as follows:

$$\begin{aligned} &\frac{1 + 0.5z^{-1}}{1 - 0.6z^{-1} - 0.16z^{-2}} \\ &\quad = 1 + \frac{0.5z^{-1} - (-0.6z^{-1} - 0.16z^{-2})1}{1 - 0.6z^{-1} - 0.16z^{-2}} \\ &\quad = 1 + \frac{1.1z^{-1} + 0.16z^{-2}}{1 - 0.6z^{-1} - 0.16z^{-2}} \\ &\quad = 1 + 1.1z^{-1} + \frac{0.16z^{-2} - (-0.6z^{-1} - 0.16z^{-2})1.1z^{-1}}{1 - 0.6z^{-1} - 0.16z^{-2}} \\ &\quad = 1 + 1.1z^{-1} + \frac{0.82z^{-2} + 0.176z^{-3}}{1 - 0.6z^{-1} - 0.16z^{-2}} \\ &\quad = 1 + 1.1z^{-1} + 0.82z^{-2} + \frac{0.176z^{-3} - (-0.6z^{-1} - 0.16z^{-2})0.82z^{-2}}{1 - 0.6z^{-1} - 0.16z^{-2}} \\ &\quad = 1 + 1.1z^{-1} + 0.82z^{-2} + \frac{0.668z^{-3} + 0.1312z^{-4}}{1 - 0.6z^{-1} - 0.16z^{-2}} \end{aligned}$$

We observe the following:

$$\begin{aligned} F_0 &= C \\ E_1 &= 1 \end{aligned}$$

The leading coefficients of both C and A are both equal to one:

$$\begin{aligned} F_1 z^{-1} &= 0.5z^{-1} - (-0.6z^{-1} - 0.16z^{-2}) - (1.1 + 0.16z^{-1})z^{-1} \\ &= \overline{F}_0 - \overline{A}F_{0,0} \end{aligned}$$

$\overline{F}_0$ is equal to all of F_0, except its leading term. In the above calculations, for $F_0 = C = 1 + 0.5z^{-1}$, we obtain $\overline{F}_0 = 0.5z^{-1}$. Similarly, $\overline{A}$ is all of A, except its leading term. With $A = 1 - 0.6z^{-1} - 0.16z^{-2}$, we obtain $\overline{A} = -0.16z^{-1} - 0.16z^{-2}$. $F_{0,0}$ denotes the leading term of F_0, which is one. We proceed with the above type of calculations:

$$E_2 = 1 + 1.1z^{-1} = E_1 + F_{1,0}z^{-1}$$

Notice that $F_{1,0}$ is the leading term of F_1, which is 1.1. We obtain

$$\begin{aligned}
F_2 z^{-2} &= 0.16z^{-2} - (-0.6z^{-1} - 0.16z^{-2})1.1z^{-1} = (0.82 + 0.176z^{-1})z^{-2} \\
&= \overline{F}_1 - \overline{A}F_{1,0} \\
E_3 &= 1 + 1.1z^{-1} + 0.82z^{-2} = E_2 + F_{2,0}z^{-2} \\
F_3 z^{-3} &= 0.176z^{-3} - (-0.6z^{-1} - 0.16z^{-2})0.82 \\
&= (0.668 + 0.1312z^{-1})z^{-3} \\
&= \overline{F}_2 - \overline{A}F_{2,0}
\end{aligned}$$

M 11.2 demonstrates how to implement these calculations in Matlab. ∎

The above procedure to compute E_j and F_j through recursive means can be summarized by the following algorithm:

$F_0 = C$
$E_1 = 1$
for all $j < k$ **do**
  $F_j = \overline{F}_{j-1} - \overline{A}F_{j-1,0}z^{-(j-1)}$
  $E_{j+1} = E_j + F_{j,0}z^{-j}$
end for
$F_k = \overline{F}_{k-1} - \overline{A}F_{k-1,0}z^{-(k-1)}$

where the bar denotes the variable with the constant term removed and ,0 denotes the constant term of the variable.

We can also find E_j and F_j by solving Aryabhatta's identity, given in Eq. 11.8. In the next example, we demonstrate how to obtain E_j and F_j for the problem discussed in Eq. 11.1 with the use of the code `xdync.m` described by Moudgalya [39], which is a Matlab implementation of the algorithm described by Chang and Pearson [8].

Example 11.3 Solve the problem discussed in Example 11.1 with the use of the Matlab code `xdync.m`.

$$\begin{aligned}
C &= 1 + 0.5z^{-1}, \quad dC = 1 \\
A &= 1 - 0.6z^{-1} - 1.6z^{-2}, \quad dA = 2
\end{aligned}$$

We invoke the commands in M 11.3 to arrive at, as before,

$$\begin{aligned}
E_2 &= 1 + 1.1z^{-1} \\
F_2 &= 0.82 + 0.176z^{-1}
\end{aligned}$$

∎

In the previous sections, we have seen how to obtain the j-step ahead prediction error model for the system presented in Eq. 11.1. In some industries, noise enters the system through an integral term. We consider such models in the next two sections.

11.1.5 ARIX Prediction Error Model

To simplify the controller design procedure, we often use simple noise models. It turns out that models of the following form

$$A(z)y(n) = z^{-k}B(z)u(n) + \frac{1}{\Delta}\xi(n) \tag{11.18}$$

are often adequate to describe chemical processes, where the noise occasionally comes in the form of random steps. Here, Δ is the backward shift operator

$$\Delta(z) = 1 - z^{-1} \tag{11.19}$$

and A and B are as defined in Eq. 11.2. Comparing this with Sec. 6.6.5, we refer to it as *autoregressive integrated exogeneous* (*ARIX*) input model. As $\xi(n)$ stands for a random signal, ξ/Δ denotes random steps. As before, we would like to obtain an estimate of the plant output at $t+j$, $j \geq k$. We multiply Eq. 11.18 by Δ to arrive at

$$\Delta A(z)y(n) = z^{-k}B(z)\Delta u(n) + \xi(n) \tag{11.20}$$

In the previous section, we separated the coefficient of ξ (namely, C) divided by that of y (namely, A) into E_j and F_j. We do the same in this section, *i.e.*, divide 1 by ΔA. Let

$$\frac{1}{\Delta A(z)} = E_j(z) + z^{-j}\frac{F_j(z)}{\Delta A(z)} \tag{11.21}$$

As before, E_j is of degree $j-1$ while the degree of F_j is one less than that of ΔA. That is,

$$\mathrm{d}F_j = \mathrm{d}A \tag{11.22}$$

Because $C = 1$, the degree condition of F_j, given in Eq. 11.17, is simplified to the above. Cross multiplying Eq. 11.21, we obtain

$$1 = E_j\Delta A + z^{-j}F_j \tag{11.23}$$

We use this relation to obtain the prediction error model. Multiplying Eq. 11.20 by z^jE_j, we obtain

$$z^jE_j\Delta Ay(n) = z^jE_jB\Delta u(n-k) + z^jE_j\xi(n) \tag{11.24}$$

Substituting for $E_j\Delta A$ from Eq. 11.23,

$$z^j(1 - z^{-j}F_j)y(n) = E_jB\Delta u(n+j-k) + E_j\xi(n+j) \tag{11.25}$$

Simplifying this we arrive at

$$y(n+j) = G_j\Delta u(n+j-k) + F_jy(n) + E_j\xi(n+j) \tag{11.26}$$

where G_j is as defined in Eq. 11.11. As the degree of E_j is $j-1$, $E_j\xi(n+j)$ has only terms of the form $\xi(n+i)$, $i>0$, *i.e.*, only the future noise values. As a result, we arrive at the following optimal prediction error model:

$$\hat{y}(n+j) = G_j\Delta u(n+j-k) + F_j y(n) \tag{11.27}$$

The prediction error is once again given by Eq. 11.14. We will now generalize the model studied in this section.

11.1.6 ARIMAX Prediction Error Model

The noise model of the previous section included only random steps. We generalize the treatment in this section. Consider a system with the model

$$A(z)y(n) = z^{-k}B(z)u(n) + \frac{C(z)}{\Delta}\xi(n) \tag{11.28}$$

where as before, u is the input, y is the output, ξ is white noise and Δ is the backward difference operator $1-z^{-1}$. If we follow the approach of the previous section, because of the presence of C, the last term in Eq. 11.26 will have past and future terms, making the prediction error model not so obvious as in Eq. 11.27. As a result, we proceed as follows. First we solve the following Aryabhatta's identity for E_j and F_j:

$$C = E_j\Delta A + z^{-j}F_j \tag{11.29}$$

The degrees of E_j and F_j are

$$\begin{aligned} \mathrm{d}E_j &= j-1 \\ \mathrm{d}F_j &= \mathrm{d}\Delta A - 1 = \mathrm{d}A \end{aligned} \tag{11.30}$$

which is the same as in Eq. 11.22. Substituting for C from Eq. 11.29 in Eq. 11.28 and multiplying by z^j, we obtain

$$Ay(n+j) = Bu(n+j-k) + E_jA\xi(n+j) + \frac{F_j}{\Delta}\xi(n) \tag{11.31}$$

From Eq. 11.28, we see that

$$\frac{1}{\Delta}\xi(n) = (Ay(n) - Bu(n-k))\frac{1}{C} \tag{11.32}$$

Substituting this in Eq. 11.31, we obtain

$$Ay(n+j) = Bu(n+j-k) - \frac{F_jB}{C}u(n-k) + \frac{F_jA}{C}y(n) + E_jA\xi(n+j)$$

which can be simplified as

$$Ay(n+j) = \left[1 - z^{-j}\frac{F_j}{C}\right]Bu(n+j-k) + \frac{F_jA}{C}y(n) + E_jA\xi(n+j) \tag{11.33}$$

In view of Eq. 11.29, this becomes

$$y(n+j) = \frac{E_j B}{C}\Delta u(n+j-k) + \frac{F_j}{C}y(n) + E_j\xi(n+j) \tag{11.34}$$

As $dE_j = j - 1$, the noise term has only future values. As a result, the prediction error model of $y(n+j)$ is the above equation without the noise term:

$$C\hat{y}(n+j) = E_j B\Delta u(n+j-k) + F_j y(n) \tag{11.35}$$

Recall that in Sec. 11.1.1, we motivated the need to study prediction error models so as to design controllers. We now return to the task of designing controllers.

11.2 Minimum Variance Controller

As mentioned earlier, the minimum variance controller tries to minimize the variance of the error in plant output. This controller is used as a benchmark, against which other controllers are compared. We will design a minimum variance control law for different types of disturbance.

11.2.1 Minimum Variance Controller for ARMAX Systems

The objective of the minimum variance controller is to minimize the variance of the output of the system presented in Eq. 11.1. As discussed in the previous section, $n+k$ is the earliest time at which the current control action will have any effect on y. So we would like to minimize the following performance index by suitably choosing $u(n)$:

$$J = \mathscr{E}\left[y^2(n+k)\right] \tag{11.36}$$

Substituting for $y(n+k)$ from Eq. 11.9, noting that $\xi(n+k)$ is independent of $u(n)$ and $y(n)$, we obtain

$$\mathscr{E}\left[y^2(n+k)\right] = \mathscr{E}\left[\left(\frac{G_k u(n) + F_k y(n)}{C}\right)^2\right] + \mathscr{E}\left[\left(E_k\xi(n+k)\right)^2\right] \tag{11.37}$$

The second term on the right-hand side does not depend on $u(n)$ and hence its value cannot be changed. The first term, on the other hand, can be modified by $u(n)$. Indeed, by a proper choice of $u(n)$, it can be made zero as our objective is to minimize the variance:

$$G_k u(n) + F_k y(n) = 0 \tag{11.38}$$

which is equivalent to $\hat{y}(n+k|n) = 0$ from Eq. 11.10. That is, the smallest possible value for the variance of the output is achieved by making the output of the prediction model at $n+k$ to be zero; we cannot minimize the index any further. Solving this equation for u, we obtain

$$u(n) = -\frac{F_k}{G_k}y(n) \tag{11.39}$$

or, using Eq. 11.11,

$$u(n) = -\frac{F_k(z)}{E_k(z)B(z)}y(n) \tag{11.40}$$

where F_k and E_k are the solutions of Aryabhatta's identity,

$$C = E_k A + z^{-k} F_k \tag{11.41}$$

obtained by substituting $j = k$ in Eq. 11.8. The law in Eq. 11.40 is known as the *minimum variance control law.* As $\hat{y}(n+k|t) = 0$, from Eq. 11.9, we obtain the following:

$$y(n+k) = E_k \xi(n+k) \tag{11.42}$$

$$\mathscr{E}\left[y^2(n+k)\right] = \mathscr{E}\left[(E_k \xi(n+k))^2\right] \tag{11.43}$$

Using the form of E_k defined in Eq. 11.7, we obtain

$$\mathscr{E}\left[y^2(n+k)\right] = \mathscr{E}\left[(e_{k,0}\xi(n+k) + \cdots + e_{k,k-1}\xi(n+1))^2\right]$$

As ξ is white, recalling its properties from Sec. 6.3.2, we obtain

$$\mathscr{E}\left[y^2(n+k)\right] = \mathscr{E}\left[(e_{k,0}\xi(n+k))^2 + \cdots + (e_{k,k-1}\xi(n+1))^2\right]$$

As the variance of ξ is σ_ξ^2, we obtain

$$\mathscr{E}\left[y^2(n+k)\right] = \sigma_\xi^2\left[e_{k,0}^2 + \cdots + e_{k,k-1}^2\right] \tag{11.44}$$

The expression given by Eq. 11.44 is the smallest variance of y and it cannot be reduced any further. As a result, the control law that helps achieve this, namely Eq. 11.40, is known as the minimum variance control law.

Example 11.4 We now discuss an example presented by MacGregor [34]. We will refer to this as *MacGregor's first control* problem. Find the minimum variance control for the system

$$y(n) = \frac{1-a}{1-az^{-1}}u(n-1) + \frac{1}{1-cz^{-1}}\xi(n)$$

where $a = 0.5$, $c = 0.9$ and $\xi(n)$ is white noise.

Substituting these values, the model becomes

$$y(n) = \frac{0.5}{1-0.5z^{-1}}u(n-1) + \frac{1}{1-0.9z^{-1}}\xi(n) \tag{11.45}$$

If this plant is operated in open loop, it is equivalent to keeping u at a constant value. The variation in y is due to ξ only. In Example 11.5, we will explain how to carry out these calculations. The same example shows how the Matlab function covar can be used for this purpose. The output variance σ_y^2 is found to be 5.26, for $\xi(n)$ of unit variance.

We may use M 11.4 to design the minimum variance controller for this problem. The above equation for y is in the form of Eq. 11.3 on page 404 with

$$
\begin{aligned}
A &= (1 - 0.5z^{-1})(1 - 0.9z^{-1}) = 1 - 1.4z^{-1} + 0.45z^{-2} \\
B &= 0.5(1 - 0.9z^{-1}) \\
C &= (1 - 0.5z^{-1}) \\
k &= 1
\end{aligned}
$$

Recall that in minimum variance control we zero the predicted output at $n + k$. That is, we choose j as k. Thus Eq. 11.8 becomes for $j = k = 1$,

$$1 - 0.5z^{-1} = E_1(1 - 1.4z^{-1} + 0.45z^{-2}) + z^{-1}F_1$$

Solving this equation, we obtain the following result:

$$
\begin{aligned}
E_1 &= 1 \\
F_1 &= 0.9 - 0.45z^{-1} \\
G_1 &= E_1 B = 0.5(1 - 0.9z^{-1})
\end{aligned}
$$

The control law is given by

$$u(n) = -\frac{0.9 - 0.45z^{-1}}{0.5(1 - 0.9z^{-1})}y(n) = -0.9\frac{2 - z^{-1}}{1 - 0.9z^{-1}}y(n) \tag{11.46}$$

As $E_k = 1$, using Eq. 11.44, we find the variance of the error to be σ^2. Closing of the loop is equivalent to substituting the above expression for u, in terms of y, into Eq. 11.45 and getting an expression for y in terms of ξ. It is easy to see that we arrive at the expression $y(n) = \xi(n)$. As a result, we find the variance of y also to be one in the closed loop.

If we substitute the expression for $y(n)$ in terms of $\xi(n)$ into the control law, we will obtain $u(n)$ also as a function $\xi(n)$. In this example, because $y(n) = \xi(n)$, Eq. 11.46 becomes

$$u(n) = -0.9\frac{2 - z^{-1}}{1 - 0.9z^{-1}}\xi(n)$$

Observe that this is the relation between the control input and white noise in the closed loop. Using the procedure to be explained in Sec. 11.2.2, the variance of u can also be calculated, see Problem 11.2 to obtain a value of 5.97. In other words, in order to bring the variance of y from 5.26 to 1, the variance of the input has to go up from 0 to 5.97.

M 11.4 implements the calculations indicated above. It calls M 11.5, which designs the controller, and M 11.6 to determine the closed loop transfer functions. The last code calls M 11.7 to cancel common terms and to determine the covariances.

∎

Comparison of the variance is one way to compare different control design techniques. In view of this, it is important to discuss how to calculate the variance. We take up this topic in the next section.

11.2.2 Expression for Sum of Squares

In this section, we will present a method to calculate the variance of the filtered white noise process, of the form

$$v(n) = H(z)\xi(n)$$

where $\xi(n)$ is white noise. Recall from the mixed notation of Sec. 6.4.1 that this stands for

$$v(n) = h(n) * \xi(n)$$

where $h(n)$ is the inverse Z-transform of $H(z)$. Using Eq. 6.86 and Eq. 6.87 on page 183, we obtain

$$\Gamma_{vv}(z) = H(z)H(z^{-1})$$

for a white noise process of variance 1. Let us invert this expression. Using Eq. 4.37 on page 91, we obtain

$$\gamma_{vv}(n) = \frac{1}{2\pi j}\oint_C H(z)H(z^{-1})z^{n-1}dz$$

When we let $n = 0$ in this equation, we obtain the expression for variance that we are looking for. For $n = 0$, we obtain the expression for $\sigma_v^2 = \gamma_{vv}(0)$ as

$$\sigma_v^2 = \frac{1}{2\pi j}\oint_C H(z)H(z^{-1})\frac{dz}{z} \tag{11.47}$$

Example 11.5 Find the variance of $v(n)$, where v is given as the filtered white noise:

$$v(n) = \frac{1}{1 - 0.9z^{-1}}\xi(n)$$

The transfer function $H(z)$ is given by

$$H(z) = \frac{1}{1 - 0.9z^{-1}} = \frac{z}{z - 0.9}$$

Hence, by replacing all the occurrences of z in $H(z)$ with z^{-1}, we obtain $H(z^{-1})$:

$$H(z^{-1}) = \frac{z^{-1}}{z^{-1} - 0.9}$$

We are now ready to make use of Eq. 11.47. We obtain

$$\begin{aligned}\sigma_v^2 &= \frac{1}{2\pi j}\oint_C \left(\frac{z}{z - 0.9}\right)\left(\frac{z^{-1}}{z^{-1} - 0.9}\right)\frac{1}{z}dz \\ &= \frac{1}{2\pi j}\oint_C \frac{1}{(z - 0.9)(1 - 0.9z)}dz = \frac{1}{2\pi j}\oint_C \frac{f(z)}{z - 0.9}dz\end{aligned}$$

where $f(z) = 1/(1-0.9z)$, which does not have a pole inside the unit circle. Now using the Cauchy residue theorem, given by Eq. 4.38 on page 91, we obtain

$$\sigma_v^2 = \left.\frac{1}{1-0.9z}\right|_{z=0.9} = \frac{1}{1-0.81} = 5.2632$$

M 11.8 shows how to carry out this calculation in Matlab. ∎

If the transfer function is a complicated one, evaluation of the integral in Eq. 11.47 could be cumbersome. Matlab function `covar`, used in the above example, implements it using efficient techniques.

11.2.3 Control Law for Nonminimum Phase Systems

The control law given by Eq. 11.40 will produce unbounded values in u when the plant is nonminimum phase, *i.e.*, when $B(z)$ has zeros outside the unit circle. This will result in the saturation of control signals and damage to equipment. The solution to this problem is involved and will be presented in Sec. 13.3. Here we just state the solution. Let the numerator polynomial of B be decomposed into good and bad factors,

$$B(z) = B^g(z)B^b(z) \tag{11.48}$$

and let $B_r^b(z)$ be the reciprocal polynomial, as discussed in Sec. 10.2.1. Let R_c and S_c be the solutions of Aryabhatta's identity:

$$A(z)R_c(z) + z^{-k}B(z)S_c(z) = C(z)B^g(z)B_r^b(z). \tag{11.49}$$

The control law is given by

$$u(n) = -\frac{S_c(z)}{R_c(z)}y(n) \tag{11.50}$$

M 11.9 implements this controller. We illustrate this with an example, taken from [2].

Example 11.6 Design the minimum variance control law for a system described by an ARMAX model with

$$\begin{aligned} A(z) &= (1-z^{-1})(1-0.7z^{-1}) \\ B(z) &= 0.9 + z^{-1} \\ C(z) &= 1 - 0.7z^{-1} \\ k &= 1 \end{aligned}$$

Splitting B into good and bad factors, we obtain

$$\begin{aligned} B^g &= 1 \\ B^b &= 0.9 + z^{-1} \end{aligned}$$

We obtain the reciprocal polynomial as

$$B_r^b = 1 + 0.9z^{-1}$$

Substituting in Eq. 11.49, we obtain the polynomial equation

$$(1 - z^{-1})(1 - 0.7z^{-1})R_c + z^{-1}(0.9 + z^{-1})S_c = (1 + 0.9z^{-1})(1 - 0.7z^{-1})$$

Solving this Aryabhatta's identity, we obtain

$$R_c = 1 + z^{-1}$$
$$S_c = 1 - 0.7z^{-1}$$

The control law is given by

$$u(n) = -\frac{1 - 0.7z^{-1}}{1 + z^{-1}} y(n)$$

M 11.10 implements this example. The variances σ_y^2 and σ_u^2 are given, respectively, by `yvar` and `uvar` in this code. We obtain $\sigma_y^2 = 1.0526$ and $\sigma_u^2 = 14.4737$. The minimum variance control law of Eq. 11.39, instead, would have given $\sigma_y^2 = 1$ with an infinite σ_u^2. Through the new procedure, the control variance has been made finite, at the expense of only 5% increase in the output variance. ∎

It is easy to see that the control law given in Eq. 11.49–11.50 reduces to that given in Eq. 11.40–11.41 when the plant is minimum phase. When the plant is minimum phase, we obtain $B^g = B$ and $B^b = B_r^b - 1$. As a result, Eq. 11.49 becomes

$$AR_c + z^{-k}BS_c = CB \tag{11.51}$$

where we have dropped the dependence on z for convenience. As the second and the third terms have B, the first term should also be divisible by it. Since A and B do not have a common factor, R_c should be divisible by B. In view of this, we obtain

$$R_c = R_1 B \tag{11.52}$$

Substituting this in Eq. 11.51, and cancelling the common factor B, we arrive at

$$AR_1 + z^{-k}S_c = C \tag{11.53}$$

Comparing this with Eq. 11.41, we identify R_1 with E_k and S_c with F_k. It is now straightforward to see that the control law given in Eq. 11.50 reduces to that in Eq. 11.40, because $R_c = R_1 B$.

The reader is warned, however, not to use Eq. 11.51 when the plant is minimum phase. The reason for this is illustrated in the next example.

Example 11.7 Examine the feasibility of using the polynomial equation given in Eq. 11.49 to solve the problem of Example 11.4.

As explained above, because this plant is minimum phase, we explore solving Eq. 11.51. We obtain

$$(1 - 1.4z^{-1} + 0.45z^{-2})R_c + z^{-1}(0.5 - 0.45z^{-1})S_c =$$
$$(1 - 0.5z^{-1})(0.5 - 0.45z^{-1})$$

This problem is solved through M 11.4, with a call to `mv_nm` of M 11.9, as opposed to calling `mv`. We obtain the solution as $R_c = 0.5$ and $S_c = 0$. Of course, in this simple problem, the solution is obvious. Because this is equivalent to running the system in open loop, this solution is unacceptable and it is different from the one obtained in Example 11.4.

∎

When does the difficulty experienced in the above example occur? Is this a chance occurrence? To answer this, consider the output error model, a schematic of which is given in Fig. 6.15 on page 208. Recall that because white noise $\xi(n)$ directly adds to the output w, it is called the output error model. Suppose that we represent this model by the following equation:

$$y(n) = z^{-k}\frac{B}{A}u(n) + \xi(n)$$

Multiplying throughout by A, we obtain

$$Ay(n) = z^{-k}Bu(n) + A\xi(n)$$

Comparing this equation with the standard ARMAX model used in this section, namely with Eq. 11.1 on page 403, we see that

$$C(z) = A(z)$$

In view of this fact, Eq. 11.53 becomes

$$AR_1 + z^{-k}S_c = A$$

This equation always has the trivial solution $R_1 = 1$, $S_c = 0$, which says that no control action is required, as obtained in the above example. Thus, the difficulty explained in the above example always occurs in output error models.

11.2.4 Minimum Variance Controller for ARIMAX Systems

We next derive the minimum variance control for the ARIX model given in Eq. 11.18. By requiring that the prediction model output vanish at k, we obtain from Eq. 11.27

$$G_k \Delta u(n) = -F_k y(n) \tag{11.54}$$

or the required control effort as

$$\Delta u(n) = -\frac{F_k}{G_k}y(n) \tag{11.55}$$

where $G_k = E_kB$, as given by Eq. 11.11. Because $\Delta = 1 - z^{-1}$, we obtain

$$u(n) = u(n-1) - \frac{F_k}{G_k}y(n) \tag{11.56}$$

As the predicted model output given by Eq. 11.27 is made zero, the plant output is

$$y(n+k) = E_k\xi(n+k) \tag{11.57}$$

from Eq. 11.26. This is identical to Eq. 11.42 and hence we obtain once again the variance expression as in Eq. 11.44.

For ARMAX models, using Eq. 11.35 and using the condition $\hat{y}(n+k|n)$, the control law becomes

$$\Delta u(n) = -\frac{F_k}{E_kB}y(n) \tag{11.58}$$

which is identical to Eq. 11.55 as $G_k = E_kB$. So we obtain the same control expression as in Eq. 11.56. From Eq. 11.34, we see once again that the output satisfies Eq. 11.57 and hence the variance is given by Eq. 11.44.

If the system is nonminimum phase, *i.e.*, $B(z)$ has its zeros outside the unit circle, we have a situation similar to the one in Sec. 11.2.3. The control law is now given by

$$\Delta u(n) = -\frac{S_c(z)}{R_1(z)}y(n) \tag{11.59}$$

where R_1 and S_c are the solutions of Aryabhatta's identity,

$$A(z)\Delta R_1(z) + z^{-k}B(z)S_c(z) = C(z)B^g(z)B_r^b(z) \tag{11.60}$$

where $\Delta = 1 - z^{-1}$. Note that we now have an extra Δ in the above equations, as compared to Eq. 11.49–11.50. The derivation of this control law is presented in Sec. 13.3. M 11.9 shows how to calculate this controller. We illustrate this approach with an example.

Example 11.8 Design the minimum variance control law for the *viscosity control* problem, presented by [34]:

$$y(n) = \frac{0.51 + 1.21z^{-1}}{1 - 0.44z^{-1}}u(n-1) + \frac{1}{1 - z^{-1}}\xi(n)$$

We see that

$$\begin{aligned} A &= 1 - 0.44z^{-1} \\ B &= 0.51 + 1.21z^{-1} \\ C &= 1 - 0.44z^{-1} \\ k &= 1 \end{aligned}$$

We obtain

$$\begin{aligned} B^g &= 0.51 \\ B^b &= 1 + 2.3725z^{-1} \\ B_r^b &= 2.3725 + z^{-1} \end{aligned}$$

Because this system is nonminimum phase, we solve Aryabhatta's identity, given by Eq. 11.60:

$$(1 - 0.44z^{-1})(1 - z^{-1})R_1 + z^{-1}(0.51 + 1.21z^{-1})S_c =$$
$$0.51(1 - 0.44z^{-1})(2.3725 + z^{-1})$$

The solution is obtained as

$$R_1 = 1.21 + 1.21z^{-1}$$
$$S_c = 1 - 0.44z^{-1}$$

with the controller given by Eq. 11.59. M 11.11 carries out the above mentioned calculations. It is easy to check that this controller stabilizes the closed loop. The transfer function between y and ξ, stored in Ny/Dy of this code, and that between u and ξ, stored in Nu/Du, are obtained as

$$y(n) = \frac{1 + z^{-1}}{1 + 0.4215z^{-1}}\xi(n)$$
$$u(n) = \frac{0.8264 - 0.3636z^{-1}}{1 + 0.4215z^{-1}}\xi(n)$$

We obtain the variance, stored in yvar and uvar of this code, as $\sigma_y^2 = 1.4070$ and $\sigma_u^2 = 1.2994$, respectively. If we had used the control law of Eq. 11.40 instead, we would have ended up with $\sigma_y^2 = 1$, with infinite variance in u.

∎

11.3 Generalized Minimum Variance Controller

Unbounded control effort can result from minimum variance control. It is possible to solve this problem by minimizing a weighted sum of the setpoint error and the control effort. A rigorous solution to this problem yields LQG controller, to be discussed in detail Chapter 13, and briefly in Chapter 14. In this section, we present an easier, but approximate solution method to this problem and derive the *generalized minimum variance controller* (*GMVC*).

11.3.1 GMVC for ARMAX Model

When the system is nonminimum, the control law of Eq. 11.40 or Eq. 11.58 gives rise to unbounded signals leading to saturation and damage to equipment. One way to solve this problem has been outlined in Sec. 11.2.3. Nevertheless, owing to plant uncertainties, it is not always possible to do this. In any case, keeping the control effort minimum is not the objective of the minimum variance control law. A practical approach is to constrain u by minimizing the following objective function:

$$J = \mathscr{E}\left[y^2(n+k) + \rho u^2(n)\right] \tag{11.61}$$

with $\rho > 0$, instead of the performance index in Eq. 11.36. To simplify the discussion, we present only the regulation or disturbance rejection problem in this section. A way to accommodate the tracking problems is presented in Sec. 11.3.3. Substituting from Eq. 11.9 for $j = k$, we obtain

$$J = \mathscr{E}\left[\left(\frac{G_k u(n) + F_k y(n)}{C} + E_k \xi(n+k)\right)^2 + \rho u^2(n)\right] \tag{11.62}$$

As $\xi(n+k)$ is not correlated with $u(n)$ and $y(n)$, we obtain

$$J = \mathscr{E}\left[\left(\frac{G_k u(n) + F_k y(n)}{C}\right)^2 + (E_k \xi(n+k))^2 + \rho u^2(n)\right] \tag{11.63}$$

An approximate method to solve this problem is to differentiate the argument of the expectation operator with respect to $u(n)$ and to equate it to zero. This approach is shown to be an approximation by MacGregor [34]. Nevertheless, as this method has given rise to several useful controllers, we present this approximate method here. Differentiating the argument of $\mathscr{E}$ by $u(n)$ and equating to zero, we obtain

$$2\frac{G_k u(n) + F_k y(n)}{C}\alpha_0 + 2\rho u(n) = 0 \tag{11.64}$$

where

$$\alpha_0 = \text{constant term of } \frac{G_k(z)}{C(z)} \tag{11.65}$$

Simplifying, we obtain

$$u(n) = -\frac{\alpha_0 F_k(z)}{\alpha_0 G_k(z) + \rho C(z)} y(n) \tag{11.66}$$

where $G(z) = E_k(z)B(z)$. Notice that by choosing ρ to be large, one can move the poles away from the zeros of G_k. We can see that when $\rho = 0$, the above equation reduces to the minimum variance control law given by Eq. 11.39. See M 11.12 for the implementation of GMVC design.

Example 11.9 We now continue with MacGregor's problem presented in Example 11.4. We would like to find the generalized minimum variance control law with $\rho = 1$. We have

$$\begin{aligned}
C &= 1 - 0.5z^{-1} \\
F_1 &= 0.9 - 0.45z^{-1} \\
G_1 &= 0.5(1 - 0.9z^{-1}) \\
k &= 1
\end{aligned}$$

From Eq. 11.65,

$$\alpha_0 = \text{constant value of } \frac{0.5(1 - 0.9z^{-1})}{1 - 0.5z^{-1}} = 0.5$$

From Eq. 11.66, we obtain

$$u(n) = -\frac{0.5(0.9 - 0.45z^{-1})}{0.5 \times 0.5(1 - 0.9z^{-1}) + (1 - 0.5z^{-1})} y(n)$$

Simplifying, we obtain the expression

$$u(n) = -\frac{0.45 - 0.225z^{-1}}{1.25 - 0.725z^{-1}} y(n) = -\frac{0.36(1 - 0.5z^{-1})}{1 - 0.58z^{-1}} y(n)$$

Refer to M 11.13 to find out how to do these calculations in Matlab.

∎

11.3.2 GMVC for ARIMAX Model

As mentioned earlier, integrated noise models are common in process industries. We will derive the GMVC for ARIMAX model, described by Eq. 11.28, also known as the integrated coloured noise model. Then by substituting $C = 1$, we can obtain the expression for the ARIX model.

For the plant model in Eq. 11.28, we would like to minimize a combination of the input and output variances. As the noise is assumed to be in the form of random *steps*, we cannot hope to control the absolute value of the control effort, but control only *changes* in it. That is, the objective function that we want to minimize becomes

$$J = \mathscr{E}\left[y^2(n+k) + \rho(\Delta u(n))^2\right] \tag{11.67}$$

We proceed as before. Substituting for $y(n+k)$ from Eq. 11.34 on page 412, and because of the fact that the future noise is not correlated with the current input and output of the system, we obtain

$$J = \mathscr{E}\left[\left(\frac{G_k \Delta u(n) + F_k y(n)}{C}\right)^2 + (E_k \xi(n+k))^2 + \rho \Delta u^2(n)\right] \tag{11.68}$$

We now differentiate this expression with respect to Δu and equate it with zero to arrive at

$$2\frac{G_k \Delta u(n) + F_k y(n)}{C} \alpha_0 + 2\rho \Delta u(n) = 0 \tag{11.69}$$

Simplifying, we obtain the following expression for control effort:

$$\Delta u(n) = -\frac{\alpha_0 F_k}{\alpha_0 G_k + \rho C} y(n) \tag{11.70}$$

It is interesting to note that the right-hand side of the above equation and that of Eq. 11.66 are identical. We demonstrate this approach with the following example.

Example 11.10 Design the generalized minimum variance control law for the viscosity control problem presented in Example 11.8,

$$y(n) = \frac{0.51 + 1.21z^{-1}}{1 - 0.44z^{-1}} u(n-1) + \frac{1}{1 - z^{-1}} \xi(n)$$

by minimizing the objective function given in Eq. 11.67 with $\rho = 1$.

We have

$$\begin{aligned} A &= 1 - 0.44z^{-1} \\ B &= 0.51 + 1.21z^{-1} \\ C &= 1 - 0.44z^{-1} \end{aligned}$$

Eq. 11.29 is next solved for E_1 and F_1:

$$\begin{aligned} E_1 &= 1 \\ F_1 &= 1 - 0.44z^{-1} \end{aligned}$$

Because G_j has been defined as $E_j B$ in Eq. 11.11 on page 405, $G_1 = B$. Using Eq. 11.65, we see that

$$\alpha_0 = \text{constant value of } \frac{G}{C} = 0.51$$

Using Eq. 11.70, we obtain

$$\Delta u(n) = -\frac{0.51(1 - 0.44z^{-1})}{0.51(0.51 + 1.21z^{-1}) + (1 - 0.44z^{-1})} y(n)$$

which is simplified as

$$\Delta u(n) = -\frac{0.51 - 0.2244z^{-1}}{1.2601 + 0.1771z^{-1}} y(n) = -\frac{0.4047(1 - 0.44z^{-1})}{1 + 0.1405z^{-1}} y(n)$$

see M 11.14. Note that even though the noise term is assumed to have random steps, the output will have finite variance so long as the closed loop is stable. This can be seen by substituting the expression for $\Delta u(n)$ in Eq. 11.34. In the code discussed above, `Ny/Dy` gives the transfer function between y and ξ, while `Nu/Du` gives the transfer function between u and ξ, in the closed loop. We obtain

$$\begin{aligned} y(n) &= \frac{1 + 0.1405z^{-1}}{1 - 0.6530z^{-1} + 0.3492z^{-2}} \xi(n) \\ u(n) &= \frac{0.4047 - 0.1781z^{-1}}{1 - 0.6530z^{-1} + 0.3492z^{-2}} \xi(n) \end{aligned}$$

Once the closed loop transfer functions are obtained, we can determine the variances. The variances of y and u, as given by `yvar` and `uvar`, respectively, are given by $\sigma_y^2 = 1.719$ and $\sigma_u^2 = 0.187$. Comparing these values with those of Example 11.8, we see that the variance of u is now made smaller, but at the expense of the variance of y.

∎

11.3.3 PID Tuning Through GMVC

In Sec. 9.8, we have seen the procedure to tune PID controllers using pole placement controllers. In Sec. 10.2.3, we have used the IMC for the same purpose. In this section, we will use the GMVC for the same purpose. This method was proposed by [60]. We will assume that the plant to be controlled can be described by the usual ARIX model, namely

$$Ay(n) = Bu(n-k) + \frac{1}{\Delta}e(n) \tag{11.71}$$

For such models, an index of the form given in Eq. 11.67 has to be minimized. We propose to minimize the index

$$J = \mathscr{E}\left[\{P(z)y(n+k) + Q(z)\Delta u(n) - P(1)r(n)\}\right] \tag{11.72}$$

The method to choose P and Q will be explained shortly. We require P and Q to be of the following form:

$$P(z) = p_0 + p_1 z^{-1} + p_2 z^{-2} \tag{11.73}$$

$$Q(z) = q_0 + q_1 z^{-1} + \cdots + q_{m-1} z^{-m+1} \tag{11.74}$$

It is customary in the GMVC to minimize the instantaneous value of J given in Eq. 11.72. This is obtained by letting the argument, after the use of the prediction model, go to zero, *i.e.*,

$$P(z)y(n+k) + Q(z)\Delta u(n) - P(1)r(n) = 0 \tag{11.75}$$

Using the relation

$$P(z) = p_0 \Delta A(z)E(z) + z^{-k}F(z) \tag{11.76}$$

where

$$E(z) = 1 + e_1 z^{-1} + \cdots + e_{k-1} z^{-k+1} \tag{11.77}$$

$$F(z) = f_0 + f_1 z^{-1} + f_2 z^{-2} \tag{11.78}$$

Eq. 11.75 becomes

$$F(z)y(n) + p_0 \Delta A(z)E(z)y(n+k) + Q(z)\Delta u(n) - P(1)r(n) = 0 \tag{11.79}$$

The prediction model for the ARIMAX model is

$$\Delta A\hat{y}(n+k) = \Delta Bu(n) \tag{11.80}$$

Using this, Eq. 11.79 becomes

$$F(z)y(n) + (p_0 E(z)B(z) + Q(z))\Delta u(n) - P(1)r(n) = 0 \tag{11.81}$$

We obtain the control law as

$$\Delta u(n) = \frac{P(1)r(n) - Fy(n)}{p_0 EB + Q} \tag{11.82}$$

Substituting this in Eq. 11.71, we obtain

$$A\Delta y(n) = z^{-k}B\frac{P(1)r(n) - Fy(n)}{p_0EB + Q} + e(n)$$

$$\frac{p_0EB + Q}{z^{-k}B}A\Delta y(n) = P(1)r(n) - Fy(n) + \frac{p_0EB + Q}{z^{-k}B}e(n)$$

Simple manipulation gives

$$\frac{(p_0\Delta AE + z^{-k}F)B + QA\Delta}{z^{-k}B}y(n) = P(1)r(n) + \frac{p_0EB + Q}{z^{-k}B}e(n)$$

Now using Eq. 11.76 and with the desired closed loop transfer function ϕ_{cl} split as

$$\phi_{cl} = PB + QA\Delta \tag{11.83}$$

we obtain the following closed loop expression:

$$y(n) = \frac{z^{-k}BP(1)}{\phi_{cl}}r(n) + \frac{p_0EB + Q}{\phi_{cl}}e(n) \tag{11.84}$$

If the coefficient of $\Delta u(n)$ were a constant in Eq. 11.81, we could divide the entire equation by this coefficient and arrive at an equation similar to Eq. 8.51 on page 319. In general, however, this will not be the case. If we approximate this coefficient by its steady state value, then the GMVC relation is satisfied at least in the steady state. In view of this, we let

$$\nu = p_0E(1)B(1) + Q(1) \tag{11.85}$$

so that Eq. 11.81 can be written approximately as

$$\frac{F(z)}{\nu}y(n) + \Delta u(n) - \frac{P(1)}{\nu}r(n) = 0 \tag{11.86}$$

Comparing this equation with the PID control law given by Eq. 8.51 on page 319, we arrive at the relation

$$L(z) = \frac{F(z)}{\nu} \tag{11.87a}$$

From the definition of L given in Eq. 8.50 and that of F in Eq. 11.78, we obtain

$$K\left(1 + \frac{T_s}{\tau_i} + \frac{\tau_d}{T_s}\right) = \frac{f_0}{\nu} \tag{11.87b}$$

$$K\left(1 + \frac{2\tau_d}{T_s}\right) = -\frac{f_1}{\nu} \tag{11.87c}$$

$$K\frac{\tau_d}{T_s} = \frac{f_2}{\nu} \tag{11.87d}$$

Solving these equations, we obtain the PID parameters,

$$K = -\frac{1}{\nu}(f_1 + 2f_2) \tag{11.88a}$$

$$\tau_i = -\frac{f_1 + 2f_2}{f_0 + f_1 + f_2} T_s \tag{11.88b}$$

$$\tau_d = -\frac{f_2}{f_1 + 2f_2} T_s \tag{11.88c}$$

We now summarize these steps:

1. Determine the desired closed loop characteristic polynomial, perhaps through the approach presented in Sec. 7.7.
2. Solve Aryabhatta's identity given in Eq. 11.83 for P and Q.
3. Solve Aryabhatta's identity given in Eq. 11.76 for E and F, with F as defined in Eq. 11.78.
4. Define ν as in Eq. 11.85.
5. The PID parameters are given by Eq. 11.88.
6. Implement the controller as in Eq. 8.48 on page 319.

We will illustrate this approach with an example, considered by [37].

Example 11.11 Consider the unstable system given by Eq. 11.71, with

$$A = 1 - 1.95z^{-1} + 0.935z^{-2} \tag{11.89}$$

$$B = -0.015 \tag{11.90}$$

$$k = 1 \tag{11.91}$$

Assume the sampling time to be $T_s = 1$ s. Design a PID controller with its parameters tuned by the GMVC, such that the rise time to a step input is 15 s and overshoot is $\varepsilon = 0.1$ [21].

The solution to this problem is implemented in M 11.15. Using the procedure outlined in Sec. 7.7, we find the characteristic polynomial to be

$$\phi_{cl} = 1 - 1.8421z^{-1} + 0.8577z^{-2}$$

M 11.17 calculates the GMVC equivalent PID parameters. We obtain

$$K_c = -10.4869$$
$$\tau_i = 10.0802$$
$$\tau_d = 5.9439$$
$$L = -73.8606 + 135.1536z^{-1} - 62.3333z^{-2}$$

Simulation is carried out using the Simulink program in Fig. 11.2. A unit step change in the setpoint begins at 50 s and ends at 150 s. A disturbance in the form of a step of height 0.1 appears at 300 s and ends at 400 s. The resulting output (y) and the control effort (u) are plotted in Fig. 11.3. It is clear that the controller helps track the setpoint change and to reject the disturbance. ∎

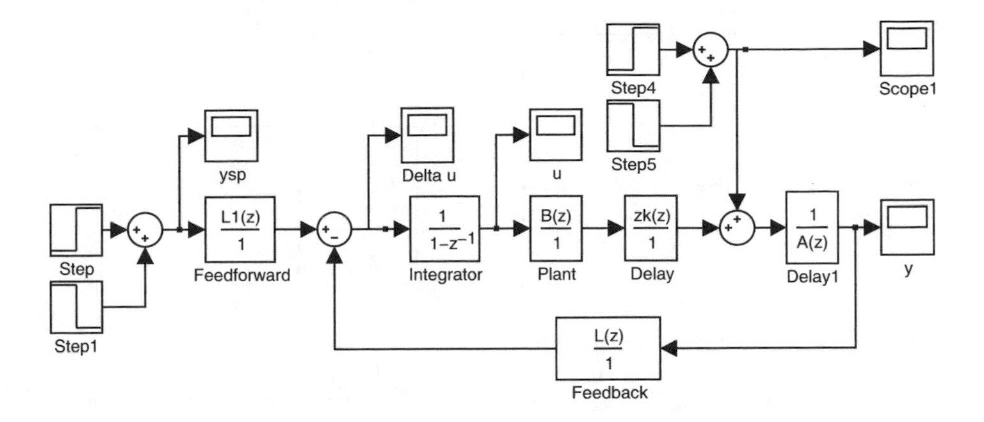

Figure 11.2: Simulink block diagram to demonstrate the efficacy of GMVC tuned PID control law on example presented by [37]. The code is available at HOME/pid/matlab/GMVC_pid.mdl, see Footnote 1 on the next page. Inputs for the simulation are established by first executing the script in M 11.15.

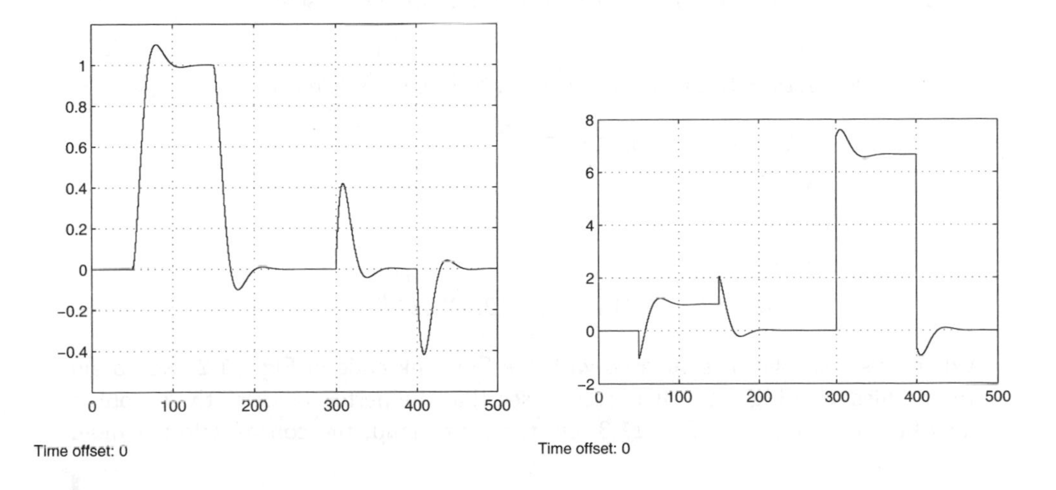

Figure 11.3: Plant output (left) and control effort (right) with GMVC tuned PID controller, as discussed in Example 11.11. The controller is designed for rise time = 15 s and overshoot = 0.1. It helps track the unit step change in the setpoint from 50 to 150 s and reject a step disturbance of magnitude 0.1 from 300 s to 400 s.

We observe in the above example that the disturbance produces a large change of 0.4 units in the output. This is essentially because the control does not act fast enough. We can achieve this by requiring the rise time and overshoot to be less. This is discussed in the next example.

Example 11.12 Control the system discussed in Example 11.11 such that the rise time to a step input is 5 s and overshoot is $\varepsilon = 0.05$.

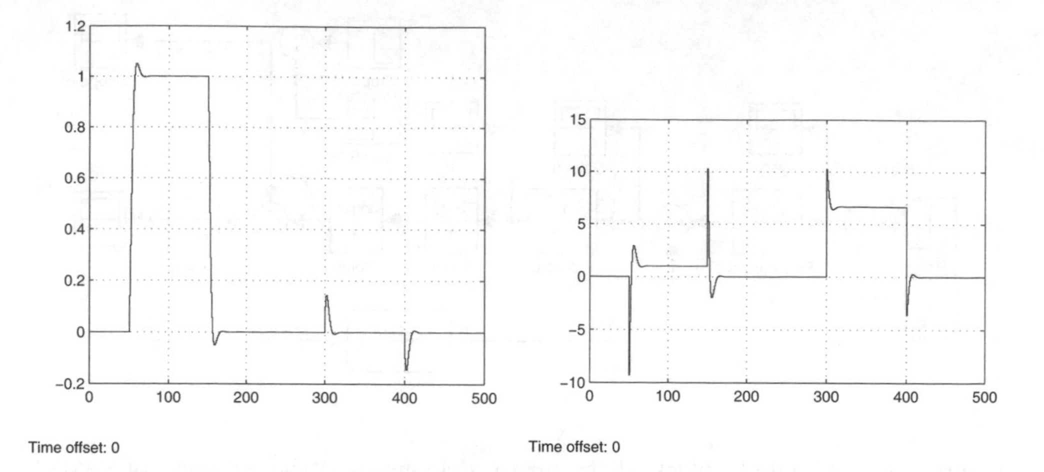

Figure 11.4: Plant output (left) and control effort (right) for a more aggressive controller than that in Fig. 11.3. Now rise time = 5 s and overshoot = 0.05. While the plant output is better, the control action is more aggressive.

We rerun M 11.15 with `N=5, epsilon=0.05` in Line 6. We obtain

$$\begin{aligned} T &= 1 - 1.4097z^{-1} + 0.5493z^{-2} \\ K_c &= -31.048 \\ \tau_i &= 3.3371 \\ \tau_d &= 2.0076 \\ L &= -102.6852 + 155.7146z^{-1} - 62.3333z^{-2} \end{aligned}$$

When this controller is simulated with the Simulink code in Fig. 11.2, we obtain the profiles as in Fig. 11.4. It is easy to see that the performance of the output is now better than that in Fig. 11.3. On the other hand, the control effort is more aggressive.

∎

11.4 Matlab Code

Matlab Code 11.1 Recursive computation of E_j and F_j, as discussed in Sec. 11.1.4. An example of usage is in M 11.2. This code is available at
`HOME/minv/matlab/recursion.m` [1]

```
1 % function [Fj,dFj,Ej,dEj] = recursion(A,dA,C,dC,j)
2 %
3 function [Fj,dFj,Ej,dEj] = recursion(A,dA,C,dC,j)
```

[1] `HOME` stands for `http://www.moudgalya.org/dc/` – first see the software installation directions, given in Appendix A.2.

```
4  Fo = C; dFo = dC;
5  Eo = 1; dEo = 0;
6  A_z = A(2:dA+1); dA_z = dA-1;
7  zi = 1; dzi = 0;
8  for i = 1:j-1
9      if (dFo == 0)
10         Fn1 = 0;
11     else
12         Fn1 = Fo(2:(dFo+1));
13     end
14     dFn1 = max(dFo-1,0);
15     Fn2 = -Fo(1)*A_z;  dFn2 = dA-1;
16     [Fn,dFn] = poladd(Fn1,dFn1,Fn2,dFn2);
17     zi = conv(zi,[0,1]); dzi = dzi + 1;
18     En2 = Fn(1)*zi; dEn2 = dzi;
19     [En,dEn] = poladd(Eo,dEo,En2,dEn2);
20     Eo = En; Fo = Fn;
21     dEo = dEn; dFo = dFn;
22 end
23 if (dFo == 0)
24     Fn1 = 0;
25 else
26 Fn1 = Fo(2:(dFo+1));
27 end
28 dFn1 = max(dFo-1,0);
29 Fn2 = -Fo(1)*A_z;  dFn2 = dA-1;
30 [Fn,dFn] = poladd(Fn1,dFn1,Fn2,dFn2);
31 Fj = Fn; dFj = dFn;
32 Ej = Eo; dEj = dEo;
```

Matlab Code 11.2 Recursive computation of E_j and F_j for the system presented in Example 11.2 on page 408. This code is available at HOME/minv/matlab/recursion_ex1.m

```
1  C = [1  0.5]; dC = 1;
2  A = [1  -0.6  -0.16]; dA = 2;
3  j = 2;
4  [Fj,dFj,Ej,dEj] = recursion(A,dA,C,dC,j)
```

Matlab Code 11.3 Solution of Aryabhatta's identity Eq. 11.8, as discussed in Example 11.3 on page 409. This code is available at HOME/minv/matlab/pm_10.m

```
1  C = [1  0.5]; dC = 1;
2  A = [1  -0.6  -0.16]; dA = 2; j = 2;
3  zj = zeros(1,j+1); zj(j+1) = 1;
4  [Fj,dFj,Ej,dEj] = xdync(zj,j,A,dA,C,dC)
```

Matlab Code 11.4 MacGregor's first control problem, discussed in Example 11.4 on page 413. This code is available at HOME/minv/matlab/mv_mac1.m

```
1  % MacGregor's first control problem
2  %
3  A = [1 -1.4 0.45]; dA = 2; C = [1 -0.5]; dC = 1;
4  B = 0.5*[1 -0.9]; dB = 1; k = 1; int = 0;
5  [Sc,dSc,Rc,dRc] = mv(A,dA,B,dB,C,dC,k,int)
6  [Nu,dNu,Du,dDu,Ny,dNy,Dy,dDy,yvar,uvar] = ...
7  cl(A,dA,B,dB,C,dC,k,Sc,dSc,Rc,dRc,int);
8
9  % Simulation parameters for stb_disc.mdl
10 Tc = Sc; gamma = 1; [zk,dzk] = zpowk(k);
11 D = 1; N_var = 0.1; Ts = 1; st = 0;
12 t_init = 0; t_final = 1000;
13
14 open_system('stb_disc.mdl')
```

Matlab Code 11.5 Minimum variance control law design, given by Eq. 11.40 on page 413. This code is available at `HOME/minv/matlab/mv.m`

```
1  % function [Sc,dSc,Rc,dRc] = mv(A,dA,B,dB,C,dC,k,int)
2  % implements the minimum variance controller
3  % if int>=1, integrated noise is assumed; otherwise,
4  % it is not integrated noise
5  %
6  function [Sc,dSc,Rc,dRc] = mv(A,dA,B,dB,C,dC,k,int)
7  zk = zeros(1,k+1); zk(k+1) = 1;
8  if int>=1, [A,dA] = polmul([1 -1],1,A,dA); end
9  [Fk,dFk,Ek,dEk] = xdync(zk,k,A,dA,C,dC);
10 [Gk,dGk] = polmul(Ek,dEk,B,dB);
11 Sc = Fk; dSc = dFk; Rc = Gk; dRc = dGk;
```

Matlab Code 11.6 Calculation of closed loop transfer functions. For an example of the usage, see M 11.4. This code is available at `HOME/matlab/cl.m`

```
1  % function [Nu,dNu,Du,dDu,Ny,dNy,Dy,dDy,yvar,uvar] = ...
2  %     cl(A,dA,B,dB,C,dC,k,Sc,dSc,Rc,dRc,int)
3  % int>=1 means integrated noise and control law:
4  % delta u = - (Sc/Rc)y
5  % Evaluates the closed loop transfer function and
6  % variances of input and output
7  %
8  function [Nu,dNu,Du,dDu,Ny,dNy,Dy,dDy,yvar,uvar] = ...
9      cl(A,dA,B,dB,C,dC,k,Sc,dSc,Rc,dRc,int)
10 [zk,dzk] = zpowk(k);
11 [BSc,dBSc] = polmul(B,dB,Sc,dSc);
12 [zBSc,dzBSc] = polmul(zk,dzk,BSc,dBSc);
13 [RcA,dRcA] = polmul(Rc,dRc,A,dA);
14 if int>=1, [RcA,dRcA] = polmul(RcA,dRcA,[1 -1],1); end
15 [D,dD] = poladd(RcA,dRcA,zBSc,dzBSc);
16 [Ny,dNy] = polmul(C,dC,Rc,dRc);
```

```
17 [Nu,dNu] = polmul(C,dC,Sc,dSc);
18 [Nu,dNu,Du,dDu,uvar] = tfvar(Nu,dNu,D,dD);
19 [Ny,dNy,Dy,dDy,yvar] = tfvar(Ny,dNy,D,dD);
```

Matlab Code 11.7 Cancellation of common factors and determination of covariance. For an example of the usage, see M 11.6. This code is available at `HOME/matlab/tfvar.m`

```
1 % function [N,dN,D,dD,yvar] = tfvar(N,dN,D,dD)
2 % N and D polynomials in z^{-1} form; discrete case
3 %
4 function [N,dN,D,dD,yvar] = tfvar(N,dN,D,dD)
5 [N,dN,D,dD] = l2r(N,dN,D,dD);
6 N = N/D(1); D = D/D(1);
7 LN = length(N); LD = length(D);
8 D1 = D;
9 if LD<LN, D1 = [D zeros(1,LN-LD)]; dD1 = dD+LN-LD; end
10 H = tf(N,D1,1);
11 yvar = covar(H,1);
```

Matlab Code 11.8 Computing sum of squares, as presented in Example 11.5 on page 415. This code is available at `HOME/Z-trans/matlab/sumsq.m`

```
1 Y = tf([1 0],[1 -0.9],-1);
2 covar(Y,1)
```

Matlab Code 11.9 Minimum variance control for nonminimum phase systems, given by Eq. 11.50 on page 416 and Eq. 11.59 on page 419. This code is available at `HOME/minv/matlab/mv_nm.m`

```
1 % function [Sc,dSc,Rc,dRc] = mv_nm(A,dA,B,dB,C,dC,k,int)
2 % implements the minimum variance controller
3 % if int>=1, integrated noise is assumed; otherwise,
4 % it is not integrated noise
5 %
6 function [Sc,dSc,Rc,dRc] = mv_nm(A,dA,B,dB,C,dC,k,int)
7 if int>=1, [A,dA] = polmul([1 -1],1,A,dA); end
8 [zk,dzk] = zpowk(k);
9 [Bzk,dBzk] = polmul(B,dB,zk,dzk);
10 [Bg,Bb] = polsplit3(B); Bbr = flip(Bb);
11 RHS = conv(C,conv(Bg,Bbr)); dRHS = length(RHS)-1;
12 [Sc,dSc,Rc,dRc] = xdync(Bzk,dBzk,A,dA,RHS,dRHS);
```

Matlab Code 11.10 Minimum variance control for nonminimum phase example of Example 11.6 on page 416. This code is available at `HOME/minv/matlab/ast_12p9.m`

```
1 A = conv([1 -1],[1 -0.7]); dA = 2;
2 B = [0.9 1]; dB = 1; k = 1;
3 C = [1 -0.7]; dC = 1; int = 0;
```

```
4 [Sc,dSc,Rc,dRc] = mv_nm(A,dA,B,dB,C,dC,k,int)
5 [Nu,dNu,Du,dDu,Ny,dNy,Dy,dDy,yvar,uvar] = ...
6 cl(A,dA,B,dB,C,dC,k,Sc,dSc,Rc,dRc,int);
```

Matlab Code 11.11 Minimum variance control of viscosity control problem, as presented in Example 11.8 on page 419. This code is available at `HOME/minv/matlab/mv_visc.m`

```
1 % Viscosity control problem of MacGregor
2 %
3 A = [1 -0.44]; dA = 1; B = [0.51 1.21]; dB = 1;
4 C = [1 -0.44]; dC = 1; k = 1; int = 1;
5 [Sc,dSc,R1,dR1] = mv_nm(A,dA,B,dB,C,dC,k,int)
6 [Nu,dNu,Du,dDu,Ny,dNy,Dy,dDy,yvar,uvar] = ...
7 cl(A,dA,B,dB,C,dC,k,Sc,dSc,R1,dR1,int);
```

Matlab Code 11.12 Generalized minimum variance controller design, as given by Eq. 11.66 on page 421 and Eq. 11.70 on page 422. This code is available at `HOME/minv/matlab/gmv.m`

```
1 % function [Sc,dSc,R,dR] = gmv(A,dA,B,dB,C,dC,k,rho,int)
2 % implements the generalized minimum variance controller
3 % if int>=1, integrated noise is assumed; otherwise,
4 % it is not integrated noise
5 %
6 function [Sc,dSc,R,dR] = gmv(A,dA,B,dB,C,dC,k,rho,int)
7 zk = zeros(1,k+1); zk(k+1) = 1;
8 if int>=1, [A,dA] = polmul([1 -1],1,A,dA); end
9 [Fk,dFk,Ek,dEk] = xdync(zk,k,A,dA,C,dC);
10 [Gk,dGk] = polmul(Ek,dEk,B,dB);
11 alpha0 = Gk(1)/C(1);
12 Sc = alpha0 * Fk; dSc = dFk;
13 [R,dR] = poladd(alpha0*Gk,dGk,rho*C,dC);
```

Matlab Code 11.13 GMVC design of MacGregor's first example, as discussed in Example 11.9 on page 421. This code is available at `HOME/minv/matlab/gmv_mac1.m`

```
1 % MacGregor's first control problem by gmv
2 %
3 A = [1 -1.4 0.45]; dA = 2; C = [1 -0.5]; dC = 1;
4 B = 0.5*[1 -0.9]; dB = 1; k = 1; int = 0;
5 rho = 1;
6 [Sc,dSc,Rc,dRc] = gmv(A,dA,B,dB,C,dC,k,rho,int)
7 [Nu,dNu,Du,dDu,Ny,dNy,Dy,dDy,yvar,uvar] = ...
8         cl(A,dA,B,dB,C,dC,k,Sc,dSc,Rc,dRc,int);
```

Matlab Code 11.14 GMVC design of viscosity problem, as discussed in Example 11.10 on page 423. This code is available at `HOME/minv/matlab/gmv_visc.m`

```
1  % MacGregor's Viscosity control problem by gmv
2  %
3  A = [1 -0.44]; dA = 1; B = [0.51 1.21]; dB = 1;
4  C = [1 -0.44]; dC = 1; k = 1; int = 1;
5  rho = 1;
6  [Sc,dSc,R1,dR1] = gmv(A,dA,B,dB,C,dC,k,rho,int)
7  [Nu,dNu,Du,dDu,Ny,dNy,Dy,dDy,yvar,uvar] = ...
8          cl(A,dA,B,dB,C,dC,k,Sc,dSc,R1,dR1,int);
```

Matlab Code 11.15 PID tuning through GMVC law, as discussed in Example 11.11. This code is available at `HOME/pid/matlab/miller.m`

```
1  % GMVC PID tuning of example given by Miller et al.
2  % Model
3  A = [1 -1.95 0.935]; B = -0.015; k = 1; Ts = 1;
4  %
5  % Transient specifications
6  N = 15; epsilon = 0.1; Ts = 1;
7  T = desired(Ts,N,epsilon);
8  %
9  % Controller Design
10 [Kc,tau_i,tau_d,L] = gmvc_pid(A,B,k,T,Ts);
11 L1 = filtval(L,1);
12 zk - zpowk(k);
```

Matlab Code 11.16 Value of polynomial $p(x)$, evaluated at x. This code is available at `HOME/matlab/filtval.m`

```
1  % finds the value of a polynomial in powers of z^{-1}
2  % function Y = filtval(P,z)
3
4  function Y = filtval(P,z)
5
6  N = length(P)-1;
7  Y = polyval(P,z)/z^N;
```

Matlab Code 11.17 PID tuning through GMVC law, as discussed in Sec. 11.3.3. This code is available at `HOME/pid/matlab/gmvc_pid.m`

```
1  % function [Kc,tau_i,tau_d,L] = gmvc_pid(A,B,k,T,Ts)
2  % Determines p,i,d tuning parameters using GMVC
3  % Plant model: Integrated white noise
4  % A, B in discrete time form
5
6  function [Kc,tau_i,tau_d,L] = gmvc_pid(A,B,k,T,Ts)
7  dA = length(A)-1; dB = length(B)-1;
8  dT = length(T)-1;
9  if dA > 2,
10    echo('degree of A cannot be more than 2')
```

```
11      exit
12   elseif dB > 1,
13      echo('degree_of_B_cannot_be_more_than_1')
14      exit
15   elseif dT > 2,
16      echo('degree_of_T_cannot_be_more_than_2')
17      exit
18   end
19   delta = [1 -1]; ddelta = 1;
20   [Adelta,dAdelta] = polmul(A,dA,delta,ddelta);
21   [Q,dQ,P,dP] = ...
22   xdync(Adelta,dAdelta,B,dB,T,dT);
23   PAdelta = P(1)*Adelta;
24   [zk,dzk] = zpowk(k);
25   [E,degE,F,degF] = ...
26   xdync(PAdelta,dAdelta,zk,dzk,P,dP);
27   nu = P(1)*E(1)*B(1);
28   Kc = -1/nu*(F(2)+2*F(3));
29   tau_i = -(F(2)+2*F(3))/(F(1)+F(2)+F(3))*Ts;
30   tau_d = -F(3)/(F(2)+2*F(3))*Ts;
31   L(1) = 1+Ts/tau_i+tau_d/Ts;
32   L(2) = -(1+2*tau_d/Ts);
33   L(3) = tau_d/Ts;
34   L = Kc * L;
```

11.5 Problems

11.1. This problem is concerned with the design of the minimum variance controller (MVC).

(a) For the open loop transfer function in Eq. 8.58 with integrated white noise, *i.e.*, for the following transfer function,

$$(1 - 0.9z^{-1})y(t) = z^{-1}u(t) + \frac{1}{\Delta}\xi(t) \tag{11.92}$$

design an MVC that will express $\Delta u(t)$ as a function of $y(t)$.

(b) If you have to implement this MVC with a PID controller, what are the P, I, D settings?

11.2. Determine the variance of $u(n)$, if it is related to white noise $\xi(n)$ through the following transfer function:

$$u(n) = \frac{1.8 - 0.9z^{-1}}{1 - 0.9z^{-1}}\xi(n)$$

[Hint: Using the method of Sec. 11.2.2, obtain the variance as

$$\frac{1}{2\pi j}\oint_C \left(\frac{1.8z - 0.9}{z - 0.9}\frac{1.8 - 0.9z}{1 - 0.9z}\right)\frac{dz}{z}$$

which can be shown to be equal to the sum of the residue at $z = 0.9$, which is 4.17, and the residue at $z = 0$, which can also be shown to be equal to 1.8.]

11.3. Determine the variance of $u(n)$, if it is related to white noise $\xi(n)$ through the following transfer function:

$$u(n) = \frac{1}{(1 - 0.9z^{-1})(1 - 0.8z^{-1})}\xi(n)$$

Chapter 12

Model Predictive Control

Model predictive control (*MPC*) refers to a class of control techniques that are popular in the chemical, petroleum and similar industries that are characterized by slow and large dimensional systems. This approach involves model identification from plant data, designing a sequence of control moves by minimizing a sum of squares between the desired and estimated trajectories of the plant output and implementing the first control move. The MPC approach allows one explicitly state the constraints on the control effort.

Depending on the modelling strategy used to describe the plants under question, one arrives at different MPC strategies. The literature on MPC is vast and hence we can only present a glimpse of it in this book. In this chapter, we give a brief introduction to the design of generalized predictive control and dynamic matrix control. We will let *GPC* and *DMC* denote the generalized predictive control and dynamic matrix control strategies, respectively, as well as the resulting controllers. While the GPC makes use of the j-step ahead prediction error model described in Sec. 11.1, the DMC makes use of step response models. We will begin the discussion with GPC.

12.1 Generalized Predictive Control

Generalized predictive control was first proposed by Clarke *et al.* [11] as an effective methodology to control industrial plants. This is based on minimizing a weighted sum of the setpoint error and the control effort. It also allows plant models to be updated frequently. We will describe the GPC design procedure in this section.

12.1.1 GPC for ARIX Model

In this section, we will develop a controller for plants with models in the form of Eq. 11.18 on page 410, which is reproduced here for convenience:

$$A(z)y(n) = z^{-k}B(z)u(n) + \frac{1}{\Delta}\xi(n) \tag{12.1}$$

We would like the plant output to follow a desired trajectory w. As the plant has a delay of k, the earliest time when the current input $u(n)$ can influence the output is $t+k$. As a result, we would like the plant output to follow a reference trajectory w from $t+k$ onwards. Thus we are interested in minimizing the index

$$[\hat{y}(n+k) - r(n+k)]^2 + [\hat{y}(n+k+1) - r(n+k+1)]^2 + \cdots \tag{12.2}$$

where $\hat{y}$ refers to an estimate of y. But as this should not result in large control, we would like to constrain u also. As the noise is assumed to have steps, we may not be able to constrain the absolute value of $u(n)$, but only *changes* in it. Thus, we would like to add to the above index the terms

$$\rho(\Delta u(n))^2 + \rho(\Delta u(n+1))^2 + \cdots \tag{12.3}$$

where $\rho > 0$. As we would expect the plant output to become constant and close to the setpoint after N intervals, we would like to have terms up to $t+k+N$ only in Eq. 12.2. As the output will be close to the setpoint, the input will become constant or Δu will become zero. Thus we need to have terms of Δu only up to $t+k+N$ in Eq. 12.3. It is possible to have a different number of terms in these indexes, but here we follow the simplified approach of [7]. The performance index that we wish to minimize becomes

$$\begin{aligned} J_{\mathrm{GPC}} = {} & [\hat{y}(n+k) - r(n+k)]^2 + \cdots \\ & + [\hat{y}(n+k+N) - r(n+k+N)]^2 \\ & + \rho[\Delta u(n)]^2 + \cdots + \rho[\Delta u(n+N)]^2 \end{aligned} \tag{12.4}$$

We now derive expressions for each of the terms in the above equation. The predictive model for this plant is given by Eq. 11.27, which is reproduced here convenicnce:

$$\hat{y}(n+j) = G_j \Delta u(n+j-k) + F_j y(n) \tag{12.5}$$

with G_j given by Eq. 11.11, which is reproduced below:

$$G_j = E_j(z)B(z) \tag{12.6}$$

E_j and F_j are obtained by solving Aryabhatta's identity, given by Eq. 11.23 on page 410, reproduced next:

$$1 = E_j \Delta A + z^{-j} F_j \tag{12.7}$$

We next derive expressions for the terms in Eq. 12.4 so as to express the performance index in terms of past values of y and u and future values of u. The first term in Eq. 12.5 can be written as

$$\begin{aligned} G_j \Delta u(n+j-k) = {} & g_{j,0} \Delta u(n+j-k) + \cdots \\ & + g_{j,\mathrm{d}G_j} \Delta u(n+j-k-\mathrm{d}G_j) \\ = {} & g_{j,0} \Delta u(n+j-k) + \cdots \\ & + g_{j,\mathrm{d}G_j} \Delta u(n-k+1-\mathrm{d}B) \end{aligned} \tag{12.8}$$

since

$$dG_j = dE_j + dB = j - 1 + dB \tag{12.9}$$

In Eq. 12.8, $g_{j,0}$, ... are coefficients of the polynomial $G_j(z)$. Substituting for $j = k$ to $k + N$ in Eq. 12.8, and stacking them one below the other, we obtain the first term of Eq. 12.5 as

$$\begin{bmatrix} g_{k,0} & 0 & \cdots & 0 \\ g_{k+1,1} & g_{k+1,0} & \cdots & 0 \\ \vdots & & & \\ g_{k+N,N} & g_{k+N,N-1} & \cdots & g_{k+N,0} \end{bmatrix} \begin{bmatrix} \Delta u(n) \\ \Delta u(n+1) \\ \vdots \\ \Delta u(n+N) \end{bmatrix} + \begin{bmatrix} g_{k,1} & \cdots & g_{k,dG_k} \\ g_{k+1,2} & \cdots & g_{k+1,dG_{k+1}} \\ \vdots & & \\ g_{k+N,N+1} & \cdots & g_{k+N,dG_{k+N}} \end{bmatrix} \begin{bmatrix} \Delta u(n-1) \\ \Delta u(n-2) \\ \vdots \\ \Delta u(n-k+1-dB) \end{bmatrix}$$

It is easy to see that the second term of Eq. 12.5 gives rise to

$$\begin{bmatrix} f_{k,0} & \cdots & f_{k,dA} \\ f_{k+1,0} & \cdots & f_{k+1,dA} \\ \vdots & & \\ f_{k+N,0} & \cdots & f_{k+N,dA} \end{bmatrix} \begin{bmatrix} y(n) \\ y(n-1) \\ \vdots \\ y(n-dA) \end{bmatrix}$$

Combining these two terms, Eq. 12.5 becomes

$$\underline{\hat{y}} = G\underline{u} + H_1\underline{u}_{\text{old}} + H_2\underline{y}_{\text{old}} \tag{12.10}$$

where

$$\underline{\hat{y}} = \begin{bmatrix} \hat{y}(n+k) \\ \vdots \\ \hat{y}(n+k+N) \end{bmatrix}, \quad \underline{u} = \begin{bmatrix} \Delta u(n) \\ \vdots \\ \Delta u(n+N) \end{bmatrix},$$
$$\underline{u}_{\text{old}} = \begin{bmatrix} \Delta u(n-1) \\ \vdots \\ \Delta u(n-k+1-dB) \end{bmatrix}, \quad \underline{y}_{\text{old}} = \begin{bmatrix} y(n) \\ \vdots \\ y(n-dA) \end{bmatrix} \tag{12.11}$$

The definition of G, F_1 and F_2 should be clear from the above derivation. Repeated solutions of Aryabhatta's identity given by Eq. 12.7, for different values of j, are implied in the above derivation. F_j and G_j can also be computed recursively, as explained in Sec. 11.1.4. We will refer to this as the *GPC model*.

Example 12.1 Derive the GPC model for the system described by [7]

$$(1 - 0.8z^{-1})y(n) = (0.4 + 0.6z^{-1})z^{-1}u(n) + \frac{1}{\Delta}\xi(n)$$

for $N = 3$.

The model has unit delay, *i.e.*, $k = 1$. In Eq. 12.4, we have to obtain expressions for $\hat{y}(n+1)$ to $\hat{y}(n+4)$. M 12.1 solves Aryabhatta's identity given by Eq. 12.7 for E_j and F_j; computes G_j through Eq. 12.6 and stacks them all up to compute G, H_1 and H_2 matrices of Eq. 12.10:
We obtain the following:

$$F_1 = 1.8000 - 0.8000z^{-1}$$
$$F_2 = 2.4400 - 1.4400z^{-1}$$
$$F_3 = 2.9520 - 1.9520z^{-1}$$
$$F_4 = 3.3616 - 2.3616z^{-1}$$
$$E_1 = 1$$
$$E_2 = 1 + 1.8z^{-1}$$
$$E_3 = 1 + 1.8z^{-1} + 2.44z^{-2}$$
$$E_4 = 1 + 1.8z^{-1} + 2.44z^{-2} + 2.9520z^{-3}$$
$$G_1 = 0.4 + 0.60z^{-1}$$
$$G_2 = 0.4 + 1.32z^{-1} + 1.0800z^{-2}$$
$$G_3 = 0.4 + 1.32z^{-1} + 2.0560z^{-2} + 1.4640z^{-3}$$
$$G_4 = 0.4 + 1.32z^{-1} + 2.0560z^{-2} + 2.6448z^{-3} + 1.7712z^{-4}$$

Stacking these up, we arrive at the following G, H_1 and H_2 matrices:

$$G = \begin{bmatrix} 0.4000 & 0 & 0 & 0 \\ 1.3200 & 0.4000 & 0 & 0 \\ 2.0560 & 1.3200 & 0.4000 & 0 \\ 2.6448 & 2.0560 & 1.3200 & 0.4000 \end{bmatrix}$$

$$H_2 = \begin{bmatrix} 1.8000 & -0.8000 \\ 2.4400 & -1.4400 \\ 2.9520 & -1.9520 \\ 3.3616 & -2.3616 \end{bmatrix}, \quad H_1 = \begin{bmatrix} 0.6000 \\ 1.0800 \\ 1.4640 \\ 1.7712 \end{bmatrix}$$

It is easy to verify these results using M 12.1. ∎

Note that we are interested in minimizing errors of the form Eq. 12.2. Subtracting $r(n+j)$, $k \le j \le k+N$, from both sides of every equation of Eq. 12.10, we obtain terms of the form $\hat{y}(n+k) - r(n+k)$, $\hat{y}(n+k+1) - r(n+k+1)$, ... Stacking these as before, Eq. 12.10 becomes

$$\underline{\hat{y}} - \underline{r} = G\underline{u} + H_1\underline{u}_{\text{old}} + H_2\underline{y}_{\text{old}} - \underline{r} \tag{12.12}$$

where $\underline{r}$ is a trajectory of reference signals:

$$\underline{r} = \begin{bmatrix} r(n+k) & \cdots & r(n+k+N) \end{bmatrix}^T \tag{12.13}$$

To minimize sums of squares of components of $\underline{\hat{y}} - \underline{r}$, we could solve $\underline{\hat{y}} - \underline{r} = 0$ or

$$G\underline{u} = \underline{r} - H_2\underline{y}_{\text{old}} - H_1\underline{u}_{\text{old}} \tag{12.14}$$

in a least squares sense. The solution to this problem is

$$\underline{u} = K_1\underline{r} - K_1H_2\underline{y}_{\text{old}} - K_1H_1\underline{u}_{\text{old}} \tag{12.15}$$

where

$$K_1 = (G^TG)^{-1}G^T \tag{12.16}$$

To minimize the control effort as well, we could augment Eq. 12.14 with $\rho\underline{u} = 0$ to arrive at

$$\begin{bmatrix} G \\ \rho I \end{bmatrix} \underline{u} = \begin{bmatrix} \underline{r} - H_2\underline{y}_{\text{old}} - H_1\underline{u}_{\text{old}} \\ 0 \end{bmatrix} \tag{12.17}$$

The least squares solution to this problem is

$$\underline{u} = K\underline{r} - KH_2\underline{y}_{\text{old}} - KH_1\underline{u}_{\text{old}} \tag{12.18}$$

where

$$K = (G^TG + \rho^2 I)^{-1}G^T \tag{12.19}$$

M 12.2 carries out these calculations. We only implement the first row of the above control law. From the definition of $\underline{u}_{\text{old}}$ in Eq. 12.11, we can see that this corresponds to $\Delta u(n)$.

Example 12.2 Derive the GPC law for Example 12.1.

After executing the Matlab commands given in Example 12.1, calculations indicated by Eq. 12.18 are carried out. M 12.3 has a complete listing of the basic GPC algorithm:

We obtain

$$K = \begin{bmatrix} 0.1334 & 0.2864 & 0.1496 & -0.0022 \\ -0.1538 & -0.1968 & 0.1134 & 0.1496 \\ -0.0285 & -0.2155 & -0.1968 & 0.2864 \\ 0.0004 & -0.0285 & -0.1538 & 0.1334 \end{bmatrix}$$

$$KH_2 = \begin{bmatrix} 1.3732 & -0.8060 \\ 0.0807 & -0.1683 \\ -0.1953 & 0.0409 \\ -0.0744 & 0.0259 \end{bmatrix}, \quad KH_1 = \begin{bmatrix} 0.6045 \\ 0.1262 \\ -0.0307 \\ -0.0194 \end{bmatrix}$$

When these are substituted in Eq. 12.18, the first row is obtained as

$$\begin{aligned} \Delta(z)u(n) = {} & 0.1334r(n+1) + 0.2864r(n+2) + 0.1496r(n+3) \\ & - 0.0022r(n+4) \\ & - 1.3732y(n) + 0.8060y(n-1) - 0.6045\Delta u(n-1) \end{aligned}$$

which can be expressed in the standard 2-DOF control configuration:

$$R_1(z)\Delta(z)u(n) = T_c(z)r(n) - S_c(z)y(n)$$

where

$$T_c(z) = 0.1334 + 0.2864z^{-1} + 0.1496r(n+3)z^{-2} - 0.0022z^{-3}$$
$$S_c(z) = 1.3732 - 0.8060z^{-1}, \quad R_1(z) = 1 + 0.6045z^{-1}$$

It is easy to see that the entries of T_c and S_c are identical to those of the first rows of K and KH_2, respectively. But for a leading term of 1, the entry of R_1 is identical to that of KH_1. By executing M 12.3, one can verify these results. ∎

Until now, we have assumed that the error signal and the control effort are weighted over the same length of time, see Eq. 12.4, in which both terms are weighted over $N+1$ intervals. In general, however, the control effort is usually weighted over a shorter interval. This implies u becoming a constant sooner than $k+N$ intervals in Eq. 12.4. This is equivalent to Δu becoming zero sooner than $k+N$ intervals. We generalize this situation by requiring that we minimize the error from $n+k+N_1$ to $n+k+N_2$, $N_2 \geq N_1$, and the control effort from n to $n+N_u$. That is, we now wish to minimize

$$\begin{aligned} J_{\text{GPC}} = & [\hat{y}(n+k+N_1) - r(n+k+N_1)]^2 + \cdots \\ & + [\hat{y}(n+k+N_2) - r(n+k+N_2)]^2 \\ & + \rho[\Delta u(n)]^2 + \cdots + \rho[\Delta u(n+N_u)]^2 \end{aligned} \tag{12.20}$$

As a result of this, $\underline{\hat{y}}$ and $\underline{u}$ of Eq. 12.11 become

$$\underline{\hat{y}} = \begin{bmatrix} \hat{y}(n+k+N_1) \\ \hat{y}(n+k+N_1+1) \\ \vdots \\ \hat{y}(n+k+N_2) \end{bmatrix}, \quad \underline{u} = \begin{bmatrix} \Delta u(n) \\ \Delta u(n+1) \\ \vdots \\ \Delta u(n+N_u) \end{bmatrix} \tag{12.21}$$

In general, because $N_u < N_2 - N_1$, the G matrix of Eq. 12.10 will not be square, but tall. Instead of repeating the derivation, we will now illustrate this approach by controlling the system discussed in Example 12.1 and Example 12.2 with different parameters.

Example 12.3 Design the GPC law for the system

$$(1 - 0.8z^{-1})y(n) = (0.4 + 0.6z^{-1})z^{-1}u(n) + \frac{1}{\Delta}\xi(n)$$

with $N_1 = 0$, $N_2 = 3$, $N_u = 2$ and $\rho = 0.8$.

M 12.4 and M 12.5 are used for this purpose. We obtain

$$G = \begin{bmatrix} 0.4000 & 0 & 0 \\ 1.3200 & 0.4000 & 0 \\ 2.0560 & 1.3200 & 0.4000 \\ 2.6448 & 2.0560 & 1.3200 \end{bmatrix}$$

$$H_2 = \begin{bmatrix} 1.8000 & -0.8000 \\ 2.4400 & -1.4400 \\ 2.9520 & -1.9520 \\ 3.3616 & -2.3616 \end{bmatrix}, \quad H_1 = \begin{bmatrix} 0.6000 \\ 1.0800 \\ 1.4640 \\ 1.7712 \end{bmatrix}$$

Compare this with the results of Example 12.1. Because u is weighted over one less interval, the vector $\underline{u}$ is of length one less than that in Examples 12.1 and 12.2. Correspondingly, the number of columns of G is now one less. Using the procedure of Example 12.2, we obtain the following results:

$$K = \begin{bmatrix} 0.1334 & 0.2864 & 0.1497 & -0.0023 \\ -0.1538 & 0.1986 & 0.1037 & 0.1580 \\ -0.0284 & -0.2189 & -0.2154 & 0.3025 \end{bmatrix}$$

$$KH_2 = \begin{bmatrix} 1.3733 & -0.8060 \\ 0.0760 & -0.1667 \\ -0.2043 & 0.0440 \end{bmatrix}, \quad KH_1 = \begin{bmatrix} 0.6045 \\ 0.1250 \\ -0.0330 \end{bmatrix}$$

$$T_c = 0.1334 + 0.2864z^{-1} + 0.1497z^{-2} - 0.0023z^{-3}$$

$$S_c = 1.3733 - 0.8060z^{-1}, \quad R_1 = 1 + 0.6045z^{-1}$$

∎

How does one tune this controller? The tuning parameters are N_1, N_2, N_u and ρ.

- One typically chooses N_1 to be zero. N_2 is chosen approximately as the settling time, divided by the sampling time. Large values of N_2, of the order of about 100 are quite common in the chemical industry. The control action becomes aggressive as N_2 is made smaller, the reason being that the actual output should reach the reference trajectory in a short time.

- N_u is generally taken to be about one-half to one-third of $N_2 - N_1$ for plants that have large time constants, a common example being chemical processes. The control action tends to become aggressive for large N_u, the reason being that the effect of aggressive control action in one move is compensated by subsequent moves. With a large number of control moves, it is possible to make large control moves and to compensate their effects subsequently. Thus, if the control action has to be less aggressive, we have to choose a smaller N_u.

- The control weighting parameter ρ is to be chosen large or small, depending on whether we want less aggressive or more aggressive control action. This parameter becomes especially useful in multi-input systems, where we may want to apply different weightings to different control efforts.

The performance of the GPC is improved by the introduction of a better noise model C, which is often tuned. In the next section, we show how to design the GPC for this case.

We conclude this section with a brief discussion on how to handle the constraints in the control effort u. In this case, we could minimize the sum of squares of residuals indicated by Eq. 12.12, subject to the constraints. These problems do not have explicit solutions and hence one may have to invoke methods such as sequential quadratic programming.

12.1.2 ARIMAX Model

The noise model of the previous section included only random steps. We generalize this in this section. Consider a system with the model

$$A(z)y(n) = z^{-k}B(z)u(n) + \frac{C(z)}{\Delta}\xi(n) \tag{12.22}$$

where as before, u is the input, y is the output, ξ is white noise and Δ is the backward difference operator $1 - z^{-1}$. As mentioned earlier, $C(z)$ is often used as a tuning parameter. As before, we would like this plant output to follow a reference trajectory w. We once again propose to achieve this by minimizing the performance index Eq. 12.4. The predictive model for this case is given by Eq. 11.35 and it is reproduced here for convenience:

$$C\hat{y}(n+j) = E_j B\Delta u(n+j-k) + F_j y(n) \tag{12.23}$$

with E_j and F_j obtained by solving Aryabhatta's identity of Eq. 11.29, which is also reproduced below:

$$C = E_j \Delta A + z^{-j} F_j \tag{12.24}$$

This predictive model has an expression for $C\hat{y}$, although we would like $\hat{y}$. In addition, unlike before, now $C\hat{y}(n+j)$ could contain past values of output as well. As a result, we split these by solving the following Aryabhatta's identity for M_j and N_j:

$$1 = CM_j + z^{-j}N_j \tag{12.25}$$

We see that

$$\begin{aligned} \mathrm{d}M_j &= j - 1 \\ \mathrm{d}N_j &= \mathrm{d}C - 1 \end{aligned} \tag{12.26}$$

Multiplying Eq. 12.23 by M_j and substituting for CM_j from Eq. 12.25, we obtain

$$(1 - z^{-j}N_j)\hat{y}(n+j) = M_j E_j B\Delta u(n+j-k) + M_j F_j y(n) \tag{12.27}$$

Simplifying, we obtain

$$\hat{y}(n+j) = M_j E_j B\Delta u(n+j-k) + M_j F_j y(n) + N_j \hat{y}(n) \tag{12.28}$$

The last term does not contain future values, and as a result, we can use $y(n)$ in the place of $\hat{y}(n)$ to arrive at

$$\hat{y}(n+j) = M_j E_j B\Delta u(n+j-k) + (M_j F_j + N_j)y(n) \tag{12.29}$$

We define

$$\begin{aligned} G_j(z) &= M_j(z)E_j(z)B(z) \\ P_j(z) &= M_j(z)F_j(z) + N_j(z) \end{aligned} \tag{12.30}$$

to arrive at

$$\hat{y}(n+j) = G_j \Delta u(n+j-k) + P_j y(n) \tag{12.31}$$

which is in the same form as Eq. 12.5. From Eq. 12.30, Eq. 12.26 and Eq. 11.30, we see that

$$\begin{aligned} \mathrm{d}G_j &= j-1+j-1+\mathrm{d}B = 2j-2+\mathrm{d}B \\ \mathrm{d}P_j &= \max\left(j-1+\mathrm{d}A, \mathrm{d}C-1\right) \end{aligned} \tag{12.32}$$

As in the previous section, substituting for $j = k$ to $k+N$ and stacking them one below the other, the first term of Eq. 12.31 becomes

$$\begin{bmatrix} G_{k,0} & & & \\ G_{k+1,1} & G_{k+1,0} & & \\ \vdots & & & \\ G_{k+N,N} & G_{k+N,N-1} & \cdots & G_{k+N,0} \end{bmatrix} \begin{bmatrix} \Delta u(n) \\ \Delta u(n+1) \\ \vdots \\ \Delta u(n+N) \end{bmatrix} + \begin{bmatrix} G_{k,1} & \cdots & G_{k,\mathrm{d}G_k} & & \\ G_{k+1,2} & & \cdots & G_{k+1,\mathrm{d}G_{k+1}} & \\ \vdots & & & & \\ G_{k+N,N+1} & & & \cdots & G_{k+N,\mathrm{d}G_{k+N}} \end{bmatrix} \begin{bmatrix} \Delta u(n-1) \\ \Delta u(n-2) \\ \vdots \\ \Delta u(n-M) \end{bmatrix}$$

where

$$M = 2k + N - 2 + \mathrm{d}B \tag{12.33}$$

The second term of Eq. 12.31 becomes

$$\begin{bmatrix} P_{k,0} & \cdots & P_{k,\mathrm{d}P_k} & & \\ P_{k+1,0} & & \cdots & P_{k+1,\mathrm{d}P_{k+1}} & \\ \vdots & & & & \\ P_{k+N,0} & & & \cdots & P_{k+N,\mathrm{d}P_{k+N}} \end{bmatrix} \begin{bmatrix} y(n) \\ y(n-1) \\ \vdots \\ y(n-\mathrm{d}P_{k+N}) \end{bmatrix}$$

We arrive at Eq. 12.10, reproduced here for convenience:

$$\hat{y} = G\underline{u} + H_1\underline{u}_{\mathrm{old}} + H_2\underline{y}_{\mathrm{old}} \tag{12.34}$$

where the definitions of $\hat{y}$ and $\underline{u}$ are the same as in Eq. 12.11, but now

$$\underline{u}_{\mathrm{old}} = \begin{bmatrix} \Delta u(n-1) \\ \vdots \\ \Delta u(n-M) \end{bmatrix}, \quad \underline{y}_{\mathrm{old}} = \begin{bmatrix} y(n) \\ \vdots \\ y(n-\mathrm{d}P_{k+N}) \end{bmatrix} \tag{12.35}$$

where M is given by Eq. 12.33. The definition of the G, H_1 and H_2 matrices should be clear from the derivation. As before, the control law minimizing the performance index Eq. 12.4 is obtained by minimizing the residuals defined by Eq. 12.12. The solution is given by Eq. 12.18 and Eq. 12.19, reproduced here for convenience:

$$\begin{aligned} \underline{u} &= K\underline{r} - KH_2\underline{y}_{\text{old}} - KH_1\underline{u}_{\text{old}} \\ K &= (G^TG + \rho^2 I)^{-1}G^T \end{aligned} \tag{12.36}$$

M 12.6 helps carry out these calculations.

Example 12.4 Solve the viscosity control problem, discussed in Example 11.10 on page 423, by the GPC method, with $N = 2$ and $\rho = 1$.

The listing in M 12.7 solves this problem. We obtain

$$\begin{aligned}
F_1 &= 1.0000 - 0.4400z^{-1} \\
E_1 &= 1 \\
M_1 &= 1 \\
N_1 &= 0.4400 \\
G_1 &= 0.5100 + 1.2100z^{-1} \\
P_1 &= 1.4400 - 0.4400z^{-1} \\
F_2 &= 1.0000 - 0.4400z^{-1} \\
E_2 &= 1 + z^{-1} \\
M_2 &= 1.0000 + 0.4400z^{-1} \\
N_2 &= 0.1936 \\
G_2 &= 0.5100 + 1.9444z^{-1} + 1.9668z^{-2} + 0.5324z^{-3} \\
P_2 &= 1.1936 - 0.1936z^{-2} \\
F_3 &= 1 - 0.4400z^{-1} \\
E_3 &= 1 + z^{-1} + z^{-2} \\
M_3 &= 1 + 0.4400z^{-1} + 0.1936z^{-2} \\
N_3 &= 0.0852 \\
G_3 &= 0.5100 + 1.9444z^{-1} + 2.5755z^{-2} + 2.2998z^{-3} \\
&\quad + 0.8654z^{-4} + 0.2343z^{-5} \\
P_4 &= 1.0852 - 0.0852z^{-3}
\end{aligned}$$

The stacked variables are

$$G = \begin{bmatrix} 0.5100 & 0 & 0 \\ 1.9444 & 0.5100 & 0 \\ 2.5755 & 1.9444 & 0.5100 \end{bmatrix}, \quad H_1 = \begin{bmatrix} 1.2100 & 0 & 0 \\ 1.9668 & 0.5324 & 0 \\ 2.2998 & 0.8654 & 0.2343 \end{bmatrix}$$

$$H_2 = \begin{bmatrix} 1.4400 & -0.4400 & 0 & 0 \\ 1.1936 & 0 & -0.1936 & 0 \\ 1.0852 & 0 & 0 & -0.0852 \end{bmatrix}$$

The variables that define the control law of Eq. 12.36 are given by

$$K = \begin{bmatrix} 0.1129 & 0.2989 & 0.0543 \\ -0.1316 & -0.2286 & 0.2989 \\ -0.0141 & -0.1316 & 0.1129 \end{bmatrix}$$

$$KH_2 = \begin{bmatrix} 0.5783 & -0.0497 & -0.0579 & -0.0046 \\ -0.1381 & 0.0579 & 0.0443 & -0.0255 \\ -0.0549 & 0.0062 & 0.0255 & -0.0096 \end{bmatrix}$$

$$KH_1 = \begin{bmatrix} 0.8494 & 0.2061 & 0.0127 \\ 0.0784 & 0.1369 & 0.0700 \\ -0.0163 & 0.0276 & 0.0264 \end{bmatrix}$$

When these are substituted in Eq. 12.36, the first row is obtained as

$$\begin{aligned} \Delta u(n) &= 0.1129r(n) + 0.2989r(n+1) + 0.0543r(n+2) \\ &\quad - 0.5783y(n) + 0.0497y(n-1) + 0.0579y(n-2) + 0.0046y(n-3) \\ &\quad - 0.8494\Delta u(n-1) - 0.2061\Delta u(n-2) - 0.0127\Delta u(n-3) \end{aligned}$$

which can be expressed in the standard 2-DOF control configuration:

$$R_1(z)\Delta(z)u(n) = T_c(z)r(n) - S_c(z)y(n)$$

where

$$\begin{aligned} T_c(z) &= 0.1129 + 0.2989z^{-1} + 0.0543z^{-2} \\ S_c(z) &= 0.5783 - 0.0497z^{-1} - 0.0579z^{-2} - 0.0046z^{-3} \\ R_1(z) &= 1 + 0.8494z^{-1} + 0.2061z^{-2} + 0.0127z^{-3} \end{aligned}$$

It is easy to see that the entries of T_c and S_c are identical to those of the first rows of K and KH_2, respectively. But for a leading term of 1, the entry of R_1 is identical to that of KH_1. By executing M 12.7, one can verify these results.

As explained in Example 12.3, it is easy to accommodate the case of weighting u and y over different time intervals, in order to minimize the index of Eq. 12.20. For example, if N_u is made smaller by 2, the number of columns of G will also come down by 2. M 12.8 and M 12.9 may be used for this purpose. For $N_1 = 0$, $N_2 = 0$, $N_u = 0$ and $\rho = 1$, we obtain G as a column vector, with H_1 and H_2 unchanged. That is:

$$\begin{aligned} G &= \begin{bmatrix} 0.51 & 1.9444 & 2.5755 \end{bmatrix}^T \\ T_c &= 0.0437 + 0.1666z^{-1} + 0.2206z^{-2} \\ S_c &= 0.5011 - 0.0192z^{-1} - 0.0322z^{-2} - 0.0188z^{-3} \\ R_1 &= 1 + 0.8878z^{-1} + 0.2796z^{-2} + 0.0517z^{-3} \end{aligned}$$

∎

Because the method described in this section generalizes the result of Sec. 12.1.1, it is possible to use the algorithm of the current section for both. In other words, it is possible to solve the problem in Example 12.2 using M 12.6. All that one has to do is to comment Lines 4–5 and uncomment Lines 6–7 in M 12.3.

GPC has been very successful in a lot of applications in industry. Naturally, there have been several modifications and enhancements to it. We will look at one such modification in the next section.

12.2 Steady State Weighted Generalized Predictive Control (γ-GPC)

Practical experience suggests that it takes a large number of terms for the GPC to work well. This results in a lot of computations. It has been found [37] that from the inclusion of steady state weighting, the number of terms required in the controller can be reduced. We will present one such approach in this section.

12.2.1 Model Derivation

Consider the model

$$A(z)y(n) = B(z)u(n-1) + \frac{C}{\Delta}\xi(n) \tag{12.37}$$

We divide C by $A\Delta$ and obtain a quotient E_j and the reminder F_j as per the following relation:

$$C = E_j A\Delta + z^{-j}F_j \tag{12.38}$$

The degrees of E_j and F_j are $j-1$ and dA, respectively. Multiplying Eq. 12.37 by $E_j z^j \Delta$ and substituting for $E_j A\Delta$ from Eq. 12.38, we obtain

$$(C - z^{-j}F_j)y(n+j) = E_j B\Delta u(n+j-1) + F_j y(n) + E_j C\xi(n+j) \tag{12.39}$$

Dividing $E_j B$ by C we obtain a quotient G_j and the reminder H_j as per the following relation:

$$E_j B = G_j C + z^{-j}H_j \tag{12.40}$$

Using this, Eq. 12.39 becomes

$$y(n+j) = \left(G_j + z^{-j}\frac{H_j}{C}\right)\Delta u(n+j-1) + \frac{F_j}{C}y(n) + E_j\xi(n+j) \tag{12.41}$$

Defining filtered variables

$$u_f = \frac{u}{C}, \quad y_f = \frac{y}{C} \tag{12.42}$$

we arrive at the following relation:

$$y(n+j) = G_j\Delta u(n+j-1) + H_j\Delta u_f(n-1)F_j y_f(n) + E_j\xi(n+j) \tag{12.43}$$

As the last term has only future noise, we obtain the following prediction model:

$$\hat{y}(n+j) = G_j \Delta u(n+j-1) + H_j \Delta u_f(n-1) F_j y_f(n) \tag{12.44}$$

Since G_j is a polynomial of degree $j-1$, it can be written as

$$G_j = g_0 + g_1 z^{-1} + \cdots + g_{j-1} z^{-j+1}, \quad j \geq \mathrm{nu} \tag{12.45}$$

where nu is the number of control moves to be used. Substituting the above expression for G_j in Eq. 12.44, we obtain

$$\begin{aligned}\hat{y}(n+j) = {} & g_{j-1}\Delta u(n) + g_{j-2}\Delta u(n+1) + \cdots + g_{j-\mathrm{nu}}\Delta u(n+\mathrm{nu}-1) \\ & + g_{j-\mathrm{nu}-1}\Delta u(n+\mathrm{nu}) + \cdots + g_0 \Delta u(n+j-1) \\ & + H_j \Delta u_f(n-1) + F_j y_f(n)\end{aligned} \tag{12.46}$$

As only nu control moves are used, u becomes constant from $(n+\mathrm{nu})$. In view of this, the second line in the above equation becomes zero. Allowing j to vary from N_1 to N_2 and stacking the resulting equations one below another, we obtain the following vector equation:

$$\begin{bmatrix} \hat{y}(n+N_1) \\ \hat{y}(n+N_1+1) \\ \vdots \\ \hat{y}(n+\mathrm{nu}) \\ \hat{y}(n+\mathrm{nu}+1) \\ \vdots \\ \hat{y}(n+N_2) \end{bmatrix} = \begin{bmatrix} g_{N_1-1} & g_{N_1-2} & \cdots & g_0 & 0 & 0 & \cdots & 0 \\ g_{N_1} & g_{N_1-1} & \cdots & g_1 & g_0 & 0 & \cdots & 0 \\ \vdots & & & & & & & \\ g_{\mathrm{nu}-1} & g_{\mathrm{nu}-2} & \cdots & & & & & g_0 \\ g_{\mathrm{nu}} & g_{\mathrm{nu}-1} & \cdots & & & & & g_1 \\ \vdots & & & & & & & \\ g_{N_2-1} & g_{N_2-2} & \cdots & & & & & g_{N_2-\mathrm{nu}} \end{bmatrix} \underline{u}$$
$$+ \begin{bmatrix} H_{N_1} \\ H_{N_1+1} \\ \vdots \\ H_{N_2} \end{bmatrix} \Delta u_f(n-1) + \begin{bmatrix} F_{N_1} \\ F_{N_1+1} \\ \vdots \\ F_{N_2} \end{bmatrix} y_f(n) \tag{12.47}$$

where

$$\underline{u} = \begin{bmatrix} \Delta u(n) \\ \Delta u(n+1) \\ \vdots \\ \Delta u(n+\mathrm{nu}-1) \end{bmatrix} \tag{12.48}$$

We arrive at the following vector relation:

$$\underline{y} = G\underline{u} + \underline{H_1}\Delta u_f(n-1) + \underline{H_2} y_f(n) \triangleq G\underline{u} + \underline{f} \tag{12.49}$$

By taking limits as $t \to \infty$, we obtain

$$\begin{bmatrix} \hat{y}(s|n) \\ \hat{y}(s|n+1) \\ \vdots \\ \hat{y}(s|n+\mathrm{nu}-1) \end{bmatrix} = \begin{bmatrix} g_s & & & \\ g_s & g_s & & \\ \vdots & \vdots & \ddots & \\ g_s & \cdots & \cdots & g_s \end{bmatrix} \underline{u} + \begin{bmatrix} H_s \\ H_s \\ \vdots \\ H_s \end{bmatrix} \Delta u_f(n-1) + \begin{bmatrix} F_s \\ F_s \\ \vdots \\ F_s \end{bmatrix} y_f(n) \tag{12.50}$$

Thus, we arrive at the following vector equation:

$$\underline{r_s} = G_s\underline{u} + \underline{H_{1s}}\Delta u_f(n-1) + \underline{H_{2s}}y_f(n) \triangleq G_s\underline{u} + \underline{f_s} \tag{12.51}$$

12.2.2 Optimization of Objective Function

As in GPC, we would like to arrive at the control action by minimizing an objective function. We now have an additional term that takes into account steady state weighting:

$$\begin{aligned} J = & \sum_{j=N_1}^{N_2} \gamma_y(j)[\hat{y}(n+j) - w(n+j)]^2 \\ & + \sum_{j=1}^{nu} \lambda(j)[\Delta u(n+j-1)]^2 + \sum_{j=1}^{nu} \gamma(j)[\hat{y}(s|t+j-1) - w(s)]^2 \end{aligned} \tag{12.52}$$

This can be written as

$$J = [\underline{y} - \underline{r}]^T\Gamma_y[\underline{y} - \underline{r}] + \underline{u}^T\Lambda\underline{u} + [\underline{y_s} - \underline{r_s}]^T\Gamma[\underline{y_s} - \underline{r_s}] \tag{12.53}$$

Substituting for $\underline{y}$ and $\underline{y}_s$, respectively, from Eq. 12.49 and Eq. 12.51,

$$\begin{aligned} J = & [G\underline{u} + (\underline{f} - \underline{r})]^T\Gamma_y[G\underline{u} + (\underline{f} - \underline{r})] + \underline{u}^T\Lambda\underline{u} \\ & + [G_s\underline{u} + (\underline{f}_s - \underline{r_s})]^T\Gamma[G_s\underline{u} + (\underline{f}_s - \underline{r_s})] \end{aligned}$$

Expanding and including only terms with $\underline{u}$,

$$\begin{aligned} J = & 2\underline{u}^TG^T\Gamma_y(\underline{f} - \underline{r}) + \underline{u}^TG^T\Gamma_yG\underline{u} + \underline{u}^T\Lambda\underline{u} \\ & + 2\underline{u}^TG_s^T\Gamma(\underline{f}_s - \underline{r_s}) + \underline{u}^TG_s^T\Gamma G_s\underline{u} \end{aligned} \tag{12.54}$$

We would like to choose an optimal $\underline{u}$ that will minimize J. In view of this, we differentiate J with respect to $\underline{u}$ and equate to zero:

$$\frac{\partial J}{\partial \underline{u}} = 2G^T\Gamma_y(\underline{f} - \underline{r}) + 2G^T\Gamma_yG\underline{u} + 2\Lambda\underline{u} + 2G_s^T\Gamma(\underline{f}_s - \underline{r_s}) + 2G_s^T\Gamma G_s\underline{u} = \underline{0}$$

Solving this for $\underline{u}$, we obtain

$$\underline{u} = [G^T\Gamma_yG + \Lambda + G_s^T\Gamma G_s]^{-1}[G^T\Gamma_y(\underline{r} - \underline{f}) + G_s^T\Gamma(\underline{r_s} - \underline{f}_s)] \tag{12.55}$$

Here, $G^T\Gamma_yG$ and $G_s^T\Gamma G_s$ are both matrices of dimension $nu \times nu$. The dynamic matrix G contains all the step response coefficients arranged in a lower triangular structure. The term $G_s^T\Gamma G_s$ is of full rank and helps ensure the existence of the inverse in the above equation, even if $G^T\Gamma_yG$ is ill conditioned due to large time delays or too short an output prediction horizon.

When Γ is set to zero, Eq. 12.55 reduces to the basic GPC law. As a result, this control law is referred to as *GPC with γ weighting*. As in the GPC law, we write the first equation of Eq. 12.55:

$$\Delta u(n) = \underline{h}(\underline{r} - \underline{f}) + h_s(r_s - f_s) \tag{12.56}$$

where

$$
\begin{aligned}
\underline{h} &= \text{first row of } [G^T \Gamma_y G + \Lambda + G_s^T \Gamma G_s]^{-1} G^T \Gamma_y \\
h_s &= \sum \text{first row of } [G^T \Gamma_y G + \Lambda + G_s^T \Gamma G_s]^{-1} G_s^T \Gamma
\end{aligned}
\tag{12.57}
$$

where h_s is the sum of elements of the first row of the indicated matrix. Note that $\underline{r} - \underline{f}$ is a vector, while $r_s - f_s$ is a scalar. Substituting the expressions for $\underline{h}$ and h_s from Eq. 12.57 and for $\underline{r}$ and $\underline{f}$ in Eq. 12.49 into Eq. 12.56, we obtain

$$
\Delta u(n) = \sum h_j \left(r_j - \frac{F_j y}{C} - z^{-1} \frac{H_j \Delta u}{C} \right) + h_s (r_s - f_s) \tag{12.58}
$$

Simplifying, we obtain

$$
\left(1 + z^{-1} \frac{\sum h_j H_j}{C} \right) \Delta u(n) = \sum h_j r_j + h_s (r_s - f_s) - \sum h_j \frac{F_j y}{C} \tag{12.59}
$$

Substituting for f_s now,

$$
\begin{aligned}
\left(1 + z^{-1} \frac{\sum h_j H_j}{C} \right) \Delta u(n) &= \sum h_j r_j + h_s r_s \\
&\quad - h_s \left(z^{-1} H_s \frac{\Delta u}{C} + \frac{F_s}{C} y \right) - \sum h_j \frac{F_j y}{C}
\end{aligned}
\tag{12.60}
$$

Simplifying this further, we obtain

$$
\begin{aligned}
\left(C + z^{-1} \left(\sum h_j H_j + h_s H_s \right) \right) \Delta u(n) &= C \left(\sum h_j r_j + h_s r_s \right) \\
&\quad - \left(\sum F_j h_j + F_s h_s \right) y
\end{aligned}
\tag{12.61}
$$

If r is a step, $r_j = r_s = r$, and the above equation simplifies to

$$
\begin{aligned}
\left(C + z^{-1} \left(\sum h_j H_j + h_s H_s \right) \right) \Delta u(n) &= C \left(\sum h_j + h_s \right) r(n) \\
&\quad - \left(\sum F_j h_j + F_s h_s \right) y(n)
\end{aligned}
\tag{12.62}
$$

which is in the standard controller form, given by Eq. 9.5 on page 329, reproduced here for convenience:

$$
R_c(z) u(n) = T_c(z) r(n) - S_c(z) y(n) \tag{12.63}
$$

where we have chosen the coefficient of T_c to be 1. Comparing coefficients, we obtain

$$
\begin{aligned}
R_c(z) &= \left(C + z^{-1} \left(\sum h_j H_j + h_s H_s \right) \right) \Delta \\
T_c(z) &= C \left(\sum h_j + h_s \right) \\
S_c(z) &= \left(\sum F_j h_j + F_s h_s \right)
\end{aligned}
\tag{12.64}
$$

with the order of the polynomials being

$$\begin{aligned} \mathrm{d}R_c &= \max(\mathrm{d}B, \mathrm{d}C) \\ \mathrm{d}T_c &= \mathrm{d}C \\ \mathrm{d}S_c &= \mathrm{d}A \end{aligned} \tag{12.65}$$

We will use this approach to tune PID controllers in the next section.

12.2.3 Predictive PID, Tuned with γ-GPC

In this section, we will present a method [37] to tune the PID controller using the γ-GPC for a special kind of plant. Defining

$$\begin{aligned} K_I &= \frac{KT_s}{\tau_i} \\ K_D &= \frac{K\tau_d}{T_s} \end{aligned} \tag{12.66}$$

Eq. 8.49 becomes

$$\Delta u(n) = K_I r(n) - [(K + K_I + K_D) - (K + 2K_D)z^{-1} + K_D z^{-2}]y(n) \tag{12.67}$$

When $C = 1$, A is a second degree polynomial and B is of zero degree. From Eq. 12.65, we arrive at the following degree relations:

$$\begin{aligned} \mathrm{d}R &= 0 \\ \mathrm{d}T &= 0 \\ \mathrm{d}S &= 2 \end{aligned} \tag{12.68}$$

With these, Eq. 12.63 becomes

$$\Delta u(n) = r_0 r(n) - [s_0 - s_1 z^{-1} + s_2 z^{-2}]y(n) \tag{12.69}$$

Comparing the coefficients of the polynomial on the right-hand side with those in Eq. 12.67, we arrive at the following relations:

$$\begin{aligned} K_I &= r_0 \\ K_D &= s_2 \\ K &= -s_1 - 2s_2 = s_0 - r_0 - s_2 \end{aligned} \tag{12.70}$$

where the last equation follows from the condition of no steady state offset, namely $S(1) = T(1)$, see Eq. 8.23 on page 311. This controller is known as the *predictive PID* controller. We will illustrate this approach with the example presented by [37].

Example 12.5 Control the system discussed in Example 11.11 with $N_1 = 1$, $N_2 = 5$, $\mathrm{nu} = 2$, $\lambda = 0.02$, $\gamma = 0.05$, $\gamma_y = 1$ [21].

We have $A = 1 - 1.95z^{-1} + 0.935z^{-2}$, $B = -0.015$. As this procedure assumes a delay of one sample time, the condition of $k = 1$ is already taken care of. Solving Eq. 12.38 for $j = 1$ to 5, we obtain

$$\begin{aligned}
E_1 &= 1 \\
F_1 &= 2.95 - 2.885z^{-1} + 0.935z^{-2} \\
E_2 &= 1 + 2.95z^{-1} \\
F_2 &= 5.8175 - 7.5757z^{-1} + 2.7582z^{-2} \\
E_3 &= 1 + 2.95z^{-1} + 5.8175z^{-2} \\
F_3 &= 9.5859 - 14.0252z^{-1} + 5.4394z^{-2} \\
E_4 &= 1 + 2.95z^{-1} - 2 + 5.8175z^{-2} + 9.5859z^{-3} \\
F_4 &= 14.2531 - 22.2159z^{-1} + 8.9628z^{-2} \\
E_5 &= 1 + 2.95z^{-1} - 2 + 5.8175z^{-2} + 9.5859z^{-3} + 14.2531z^{-4} \\
F_5 &= 19.8307 - 32.1574z^{-1} + 13.3266z^{-2}
\end{aligned}$$

Solving Eq. 12.40 for $j = 1$ to 5, we obtain

$$\begin{aligned}
G_1 &= -0.015 \\
G_2 &= -0.015 - 0.0442z^{-1} \\
G_3 &= -0.015 - 0.0442z^{-1} - 0.0873z^{-2} \\
G_4 &= -0.015 - 0.0442z^{-1} - 0.0873z^{-2} - 0.1438z^{-3} \\
G_5 &= -0.015 - 0.0442z^{-1} - 0.0873z^{-2} - 0.1438z^{-3} - 0.2138z^{-4}
\end{aligned}$$

We also have $H_j = 0, \ \forall j$. Thus, we obtain the matrix G of Eq. 12.49 as

$$G = \begin{bmatrix} -0.015 & 0 \\ -0.0442 & -0.015 \\ -0.0873 & -0.0442 \\ -0.1438 & -0.0873 \\ -0.2138 & -0.1438 \end{bmatrix}$$

We also obtain

$$\begin{aligned}
e_s &= \frac{C(1)}{A(1)} = -66.6667 \\
g_s &= \frac{B(1)}{A(1)} = 1 \\
F_s &= e_s A = -66.6667 + 130z^{-1} - 62.3333z^{-2} \\
H_s\Delta &= g_s C - e_s B = 0
\end{aligned}$$

As a result, $H_s = 0$. Thus, G_s of Eq. 12.51 becomes

$$G_s = \begin{bmatrix} 1 & 0 \\ 1 & 1 \end{bmatrix}$$

We also obtain $\underline{h}$ and h_s of Eq. 12.56 as

$$\underline{h} = \begin{bmatrix} -0.1486 & -0.2937 & -0.4376 & -0.5828 & -0.7311 \end{bmatrix}^T$$
$$h_s = 0.5084$$

Using Eq. 12.64, we obtain $R_c = 1$, $T_c = -1.6854$, $S_c = -63.0387 + 111.338z^{-1} - 49.9848z^{-2}$. Using Eq. 12.70, we obtain $K = -11.3685$, $K_I = -1.6854$, $K_D = -49.9842$. These are in agreement with the findings of [37]. M 12.10 sets up this problem, while M 12.11 solves it. ∎

12.3 Dynamic Matrix Control

In this section, we will present dynamic matrix control (*DMC*), another technique of MPC. model predictive control. While in the previous section we have required parametric models, in this section, we will work with a step response model, which is a nonparametric model. Although DMC does not have extensive noise models, it is simple and easily extendible to multivariable systems. In view of these reasons, DMC is popular in the chemical industry. In this section, we will present a brief introduction to DMC.

We start with the step response model given by Eq. 3.38 on page 53. We further add a bias term, b, to account for the difference between the model prediction and the actual output. Suppose that we start applying control effort u from time instant k onwards. At the time instant $k+1$, we obtain

$$\hat{y}(k+1) = y_x(k+1) + s(1)\Delta u(k) + b(k)$$

where we have used the superposition principle, given by Eq. 3.10 on page 39. Note that we have taken $s(0) = 1$. Similarly, at future time instants, we obtain

$$\begin{aligned} \hat{y}(k+2) &= y_x(k+2) + s(2)\Delta u(k) + s(1)\Delta u(k+1) + b(k+2) \\ &\vdots \\ \hat{y}(k+N_u+1) &= y_x(k+N_u+1) + s(N_u+1)\Delta u(k) + \cdots \\ &\quad + s(1)\Delta u(k+N_u) + b(k+N_u+1) \end{aligned}$$

where N_u determines the control horizon, *i.e.*, we apply the control effort up to $k+N_u$ and keep it afterwards. In other words, $\Delta u(k+m) = 0$, for all $m > N_u$. Thus, we obtain

$$\begin{aligned} \hat{y}(k+N_u+2) &= y_x(k+N_u+2) + s(N_u+2)\Delta u(k) + \cdots \\ &\quad + s(2)\Delta u(k+N_u) + b(k+N_u+2) \\ &\vdots \\ \hat{y}(k+N) &= y_x(k+N) + s(N)\Delta u(k) + s(N-1)\Delta u(k+1) + \cdots \\ &\quad + s(N-N_u)\Delta u(k+N_u) + b(k+N) \end{aligned}$$

where N is the prediction horizon. As mentioned earlier, the prediction horizon is as large as at least one settling time. Usually, $N > N_u$ by a factor of two to three. Stacking all of the above equations one below another, we obtain

$$\underline{\hat{y}}(k+1) = \underline{y}_x(k+1) + \underline{s}\,\underline{u}(k) + \underline{b}(k+1)$$

where

$$\underline{\hat{y}}(k+1) = \begin{bmatrix} \hat{y}(k+1) \\ \hat{y}(k+2) \\ \vdots \\ \hat{y}(k+N) \end{bmatrix}, \quad \underline{y}_x(k+1) = \begin{bmatrix} y_x(k+1) \\ y_x(k+2) \\ \vdots \\ y_x(k+N) \end{bmatrix}$$

$$\underline{s} = \begin{bmatrix} s(1) & & & \cdots & 0 \\ s(2) & s(1) & & \cdots & 0 \\ \vdots & & & & \\ s(N_u) & s(N_u-1) & s(N_u-2) & \cdots & 0 \\ s(N_u+1) & s(N_u) & s(N_u-1) & \cdots & s(2) \\ \vdots & & & & \\ s(N) & s(N-1) & s(N-2) & \cdots & s(N-N_u) \end{bmatrix},$$

$$\underline{u}(k) = \begin{bmatrix} \Delta u(k) \\ \Delta u(k+1) \\ \vdots \\ \Delta u(k+N_u) \end{bmatrix}, \quad \underline{b}(k+1) = \begin{bmatrix} b(k+1) \\ b(k+2) \\ \vdots \\ b(k+N) \end{bmatrix}$$

The variable $\underline{s}$ has N rows and $N_u + 1$ columns, $N > N_u$, and $\underline{u}$ has $N_u + 1$ rows. It is called the system's *dynamic matrix*, consisting of the collective effect of unmodelled disturbances. The argument k indicates the origin. The subscript x indicates prediction in the *absence* of further control action. The objective is to determine $\underline{u}(k)$, which consists of $N_u + 1$ control moves over time, $k, k+1, \ldots, k+N_u$, to move the system to the desired trajectory, given by

$$\underline{r}(k+1) = \begin{bmatrix} r(k+1) & r(k+2) & \cdots & r(k+N) \end{bmatrix}^T$$

Thus we get

$$\underline{y}_x(k+1) + \underline{s}\,\underline{u}(k) + \underline{b}(k+1) = \underline{r}(k+1)$$

By rearranging the terms, we require

$$\underline{s}\,\underline{u}(k) - \left[\underline{r}(k+1) - \underline{y}_x(k+1) - \underline{b}(k+1)\right] = 0$$

Defining the terms within the square brackets as $\underline{e}(k+1)$, we want

$$\underline{s}\,\underline{u}(k) - \underline{e}(k+1) = 0$$

The second term is the predicted deviation of plant output from the desired setpoint *in the absence of further control action.* We calculate $\underline{u}$ using the above equation, which has N equations in $N_u + 1$ unknowns, with N being two to three times N_u. As a result, we cannot get an exact solution. The least squares solution to this problem is given by

$$\underline{u}(k) = (\underline{s}^T \underline{s})^{-1} \underline{s}^T \underline{e}(k+1)$$

In order that excessive control action is not applied, the following method of control action calculation is done:

$$\underline{u}(k) = \left[\underline{s}^T \underline{s} + \rho^2 I\right]^{-1} \underline{s}^T \underline{e}(k+1)$$

But for the bias term, the tuning procedure is similar to that of GPC, see the discussion after Example 12.3 on page 442. We determine the current bias and assume that it is constant for the rest of the control moves. In other words, we let $b(k+1) = y(k) - \hat{y}(k)$, $i = 1, 2, \ldots, N$, where y is the measured value of output.

12.4 Matlab Code

Matlab Code 12.1 Model derivation for GPC design in Example 12.1 on page 439. This code is available at `HOME/mpc/matlab/gpc_ex11.m`[1]

```
1  % Camacho and Bordon's GPC example; model formation
2  %
3  A=[1 -0.8]; dA=1; B=[0.4 0.6]; dB=1; N=3; k=1;
4  D=[1 -1]; dD=1; AD=conv(A,D); dAD=dA+1; Nu=N+1;
5  zj = 1; dzj = 0; G = zeros(Nu);
6  H1 = zeros(Nu,k-1+dB); H2 = zeros(Nu,dA+1);
7  for j = 1:Nu,
8      zj = conv(zj,[0,1]); dzj = dzj + 1;
9      [Fj,dFj,Ej,dEj] = xdync(zj,dzj,AD,dAD,1,0);
10     [Gj,dGj] = polmul(B,dB,Ej,dEj);
11     G(j,1:dGj) = flip(Gj(1:dGj));
12     H1(j,1:k-1+dB) = Gj(dGj+1:dGj+k-1+dB);
13     H2(j,1:dA+1) = Fj;
14 end
15 G, H1, H2
```

Matlab Code 12.2 Calculates the GPC law given by Eq. 12.19 on page 441. A sample usage is given in M 12.3. This code is available at `HOME/mpc/matlab/gpc_bas.m`

```
1  function [K,KH1,KH2,Tc,dTc,Sc,dSc,R1,dR1] = ...
2  gpc_bas(A,dA,B,dB,N,k,rho)
3  D=[1 -1]; dD=1; AD=conv(A,D); dAD=dA+1; Nu=N+1;
4  zj = 1; dzj = 0; G = zeros(Nu);
5  H1 = zeros(Nu,k-1+dB); H2 = zeros(Nu,dA+1);
6  for j = 1:Nu,
```

[1] HOME stands for `http://www.moudgalya.org/dc/` – first see the software installation directions, given in Appendix A.2.

```
7        zj = conv(zj,[0,1]); dzj = dzj + 1;
8        [Fj,dFj,Ej,dEj] = xdync(zj,dzj,AD,dAD,1,0);
9        [Gj,dGj] = polmul(B,dB,Ej,dEj);
10       G(j,1:dGj) = flip(Gj(1:dGj));
11       H1(j,1:k-1+dB) = Gj(dGj+1:dGj+k-1+dB);
12       H2(j,1:dA+1) = Fj;
13   end
14   K = inv(G'*G+rho*eye(Nu))*G';
15   % Note: inverse need not be calculated
16   KH1 = K * H1; KH2 = K * H2;
17   R1 = [1 KH1(1,:)]; dR1 = length(R1)-1;
18   Sc = KH2(1,:); dSc = length(Sc)-1;
19   Tc = K(1,:); dTc = length(Tc)-1;
```

Matlab Code 12.3 GPC design for the problem discussed in Example 12.2 on page 441. This code is available at HOME/mpc/matlab/gpc_ex12.m

```
1   % Camacho and Bordon's GPC example; Control law
2   %
3   A=[1 -0.8]; dA=1; B=[0.4 0.6]; dB=1; N=3; k=1; rho=0.8;
4   [K,KH1,KH2,Tc,dTc,Sc,dSc,R1,dR1] = ...
5   gpc_bas(A,dA,B,dB,N,k,rho)
6   % C=1; dC=0; [K,KH1,KH2,Tc,dTc,Sc,dSc,R1,dR1] = ...
7   % gpc_col(A,dA,B,dB,C,dC,N,k,rho)
```

Matlab Code 12.4 GPC design for the problem discussed in Example 12.3. This code is available at HOME/mpc/matlab/gpc_wt.m

```
1   A=[1 -0.8]; dA=1; B=[0.4 0.6]; dB=1;
2   rho = 0.8; k = 1;
3   N1 = 0; N2 = 3; Nu = 2;
4   [K,KH1,KH2,Tc,dTc,Sc,dSc,R1,dR1] = ...
5   gpc_N(A,dA,B,dB,k,N1,N2,Nu,rho)
```

Matlab Code 12.5 Calculates the GPC law given by Eq. 12.36 on page 446. A sample usage is given in M 12.4. This code is available at HOME/mpc/matlab/gpc_N.m

```
1   function [K,KH1,KH2,Tc,dTc,Sc,dSc,R1,dR1] = ...
2   gpc_N(A,dA,B,dB,k,N1,N2,Nu,rho)
3   D=[1 -1]; dD=1; AD=conv(A,D); dAD=dA+1;
4   zj = 1; dzj = 0;
5   for i = 1:N1+k-1
6        zj = conv(zj,[0,1]); dzj = dzj + 1;
7   end
8   G = zeros(N2-N1+1,Nu+1);
9   H1 = zeros(N2-N1+1,k-1+dB); H2 = zeros(N2-N1+1,dA+1);
10  for j = k+N1:k+N2
11       zj = conv(zj,[0,1]); dzj = dzj + 1;
```

```
12      [Fj,dFj,Ej,dEj] = xdync(zj,dzj,AD,dAD,1,0);
13      [Gj,dGj] = polmul(B,dB,Ej,dEj);
14      if (j-k >= Nu)
15      G(j-(k+N1-1),1:Nu+1) = flip(Gj(j-k-Nu+1:j-k+1));
16  else
17      G(j-(k+N1-1),1:j-k+1) = flip(Gj(1:j-k+1));
18  end
19      H1(j-(k+N1-1),1:k-1+dB) = Gj(j-k+2:j+dB);
20      H2(j-(k+N1-1),1:dA+1) = Fj;
21  end
22  K = inv(G'*G+rho*eye(Nu+1))*G';
23  % Note: inverse need not be calculated
24  KH1 = K * H1; KH2 = K * H2;
25  R1 = [1 KH1(1,:)]; dR1 = length(R1)-1;
26  Sc = KH2(1,:); dSc = length(Sc)-1;
27  Tc = K(1,:); dTc = length(Tc)-1;
```

Matlab Code 12.6 Calculates the GPC law given by Eq. 12.36 on page 446. A sample usage is given in M 12.7. This code is available at HOME/mpc/matlab/gpc_col.m

```
1   function [K,KH1,KH2,Tc,dTc,Sc,dSc,R1,dR1] = ...
2   gpc_col(A,dA,B,dB,C,dC,N,k,rho)
3   D=[1 -1]; dD = 0; AD=conv(A,D); dAD=dA+1; zj=1; dzj=0;
4   Nu = N+1; G=zeros(Nu); H1=zeros(Nu,2*k+N-2+dB);
5   H2 = zeros(Nu,k+N+dA);
6   for j = 1:Nu,
7       zj = conv(zj,[0,1]); dzj = dzj + 1;
8       [Fj,dFj,Ej,dEj] = ...
9           xdync(zj,dzj,AD,dAD,C,dC);
10      [Nj,dNj,Mj,dMj] = ...
11          xdync(zj,dzj,C,dC,1,0);
12      [Gj,dGj] = polmul(Mj,dMj,Ej,dEj);
13      [Gj,dGj] = polmul(Gj,dGj,B,dB);
14      [Pj,dPj] = polmul(Mj,dMj,Fj,dFj);
15      [Pj,dPj] = poladd(Nj,dNj,Pj,dPj);
16      j,Fj,Ej,Mj,Nj,Gj,Pj
17      G(j,1:j) = flip(Gj(1:j));
18      H1(j,1:dGj-j+1) = Gj(j+1:dGj+1);
19      H2(j,1:dPj+1) = Pj;
20  end
21  K = inv(G'*G+rho*eye(Nu))*G'
22  % Note: inverse need not be calculated
23  KH1 = K * H1; KH2 = K * H2;
24  R1 = [1 KH1(1,:)]; dR1 = length(R1)-1;
25  Sc = KH2(1,:); dSc = length(Sc)-1;
26  Tc = K(1,:); dTc = length(Tc)-1;
```

Matlab Code 12.7 GPC design for viscosity control in Example 12.4 on page 446. This code is available at `HOME/mpc/matlab/gpc_ex2.m`

```
% GPC control of viscosity problem
%
A=[1 -0.44]; dA=1; B=[0.51 1.21]; dB=1; N=2; k=1;
C = [1 -0.44]; dC = 1; rho = 1;

[K,KH1,KH2,Tc,dTc,Sc,dSc,R1,dR1] = ...
gpc_col(A,dA,B,dB,C,dC,N,k,rho)
```

Matlab Code 12.8 GPC design for the problem discussed in Example 12.3. This code is available at `HOME/mpc/matlab/gpc_wtc.m`

```
A=[1 -0.8]; dA=1; B=[0.4 0.6]; dB=1;
rho = 0.8; k = 1;
N1 = 0; N2 = 3; Nu = 2;
[K,KH1,KH2,Tc,dTc,Sc,dSc,R1,dR1] = ...
gpc_N(A,dA,B,dB,k,N1,N2,Nu,rho)
```

Matlab Code 12.9 Calculates the GPC law for different prediction and control horizons. A sample usage is given in M 12.8. This code is available at
`HOME/mpc/matlab/gpc_Nc.m`

```
function [K,KH1,KH2,Tc,dTc,Sc,dSc,R1,dR1] = ...
gpc_Nc(A,dA,B,dB,C,dC,k,N1,N2,Nu,rho)
D=[1 -1]; dD=1; AD=conv(A,D); dAD=dA+1;
zj = 1; dzj = 0;
for i = 1:N1+k-1
    zj = conv(zj,[0,1]); dzj = dzj + 1;
end
M = 2*k+N2-2+dB;  P = max(k+N2+dA-1,dC-1)
G = zeros(N2-N1+1,Nu+1); H1 = zeros(N2-N1+1,M);
H2 = zeros(N2-N1+1,P+1);
for j = k+N1:k+N2
    zj = conv(zj,[0,1]); dzj = dzj + 1;
    [Fj,dFj,Ej,dEj] = xdync(zj,dzj,AD,dAD,C,dC);
    [Nj,dNj,Mj,dMj] = xdync(zj,dzj,C,dC,1,0);
    [Gj,dGj] = polmul(Mj,dMj,Ej,dEj);
    [Gj,dGj] = polmul(Gj,dGj,B,dB);
    [Pj,dPj] = polmul(Mj,dMj,Fj,dFj);
    [Pj,dPj] = poladd(Nj,dNj,Pj,dPj);
    if (j-k >= Nu)
   G(j-(k+N1-1),1:Nu+1) = flip(Gj(j-k-Nu+1:j-k+1));
else
   G(j-(k+N1-1),1:j-k+1) = flip(Gj(1:j-k+1));
end
    H1(j-(k+N1-1),1:j+k-2+dB) = Gj(j-k+2:2*j+dB-1);
    dPj = max(j-1+dA,dC-1);
```

```
26      H2(j-(k+N1-1),1:dPj+1) = Pj;
27    end
28    K = inv(G'*G+rho*eye(Nu+1))*G';
29    % Note: inverse need not be calculated
30    KH1 = K * H1; KH2 = K * H2;
31    R1 = [1 KH1(1,:)]; dR1 = length(R1)-1;
32    Sc = KH2(1,:); dSc = length(Sc)-1;
33    Tc = K(1,:); dTc = length(Tc)-1;
```

Matlab Code 12.10 PID controller, tuned with GPC, as discussed in Example 12.5 on page 452. This code is available at HOME/pid/matlab/gpc_pid_test.m

```
1   clear
2   A = [1 -1.95 0.935];
3   B=-0.015;
4   C=1;
5   degA=2;
6   degB=0;
7   degC=0;
8   N1=1;
9   N2=5;
10  Nu=2;
11  gamma=0.05;
12  gamma_y=1;
13  lambda=0.02;
14  [Kp,Ki,Kd] = ...
15  gpc_pid(A,degA,B,degB,C,degC,N1,N2,Nu,lambda,gamma,gamma_y)
```

Matlab Code 12.11 Predictive PID, tuned with GPC, as explained in Sec. 12.2.3. This code is available at HOME/pid/matlab/gpc_pid.m

```
1   function [Kp,Ki,Kd] = ...
2   gpc_pid(A,dA,B,dB,C,dC,N1,N2,Nu,lambda,gamma,gamma_y)
3   Adelta=conv(A,[1 -1]); G=[];
4   for i=N1:N2
5       zi=zpowk(i);
6       [E,dE,F,dF]=xdync(Adelta,dA+1,zi,i,C,dC);
7       [Gtilda,dGtilda,Gbar,dGbar] = ...
8          xdync(C,dC,zi,i,E*B,dE+dB);
9       for j = 1:i, Gtilda1(j)=Gtilda(i+1-j); end
10      if i<=Nu-1
11         G=[G;[Gtilda1,zeros(1,Nu-i)]];
12      else
13         G=[G;Gtilda1(1:Nu)];
14      end
15  end
16  es=sum(C)/sum(A); gs=sum(B)/sum(A); F_s=es*A; G_s=[];
17  for i=1:Nu
18      row=gs*ones(1,i); row=[row,zeros(Nu-i)];
```

```
19        G_s=[G_s;row];
20  end
21  lambda_mat=lambda*(diag(ones(1,Nu)));
22  gamma_mat=gamma*(diag(ones(1,Nu)));
23  gamma_y_mat=gamma_y*(diag(ones(1,N2-N1+1)));
24  mat1=inv(G'*gamma_y_mat*G+lambda_mat+G_s'*gamma_mat*G_s);
25  % Note: inverse need not be calculated
26  mat2=mat1*(G'*gamma_y_mat);
27  mat2_s=mat1*(G_s'*gamma_mat);
28  h_s=sum(mat2_s(1,:)); h=mat2(1,:);
29  T=C; R=C*(sum(h(:))+h_s); S=0;
30  for i=N1:N2
31        zi=zpowk(i);
32        [E,dE,F,dF]=xdync(Adelta,dA+1,zi,i,C,dC);
33        [Gtilda,dGtilda,Gbar,dGbar]=...
34          xdync(C,dC,zi,i,E*B,dE+dB);
35        S=S+F*h(i);
36  end
37  S=S+F_s*h_s;
38  if length(A)==3
39      Kp=S(1)-R-S(3); Ki=R; Kd=S(3);
40  else
41      Kp=S(1)-R; Ki=R; Kd=0;
42  end
```

12.5 Problems

12.1. Derive the GPC model (*i.e.*, Eq. 12.10 on page 439) for the system described by

$$(1 - 0.9z^{-1})y(t) = z^{-1}u(t) + \frac{1}{\Delta}\xi(t)$$

for the two sets of conditions given below. (Here, $y(t)$, $u(t)$ and $\xi(t)$ are, respectively, output, input and white noise, and $\Delta = 1 - z^{-1}$.)

(a) Using predictive models for $y(t+1)$ and $y(t+2)$ (*i.e.*, $N_y = 1$) and varying $\Delta u(t)$ and $\Delta u(t+1)$ (*i.e.*, $N_u = 1$).

(b) Same as above, but now $N_u = 0$, *i.e.*, only $\Delta u(t)$ is varied.

[Hint: Long division may be useful.]

12.2. For the system defined by Eq. 11.92, design a GPC using predictive models for $\hat{y}(t+1)$ and $\hat{y}(t+2)$ (*i.e.*, $N_y = 1$) by varying only $\Delta u(t)$ and making $u(t+j) =$ constant for $j \geq 1$ (*i.e.*, $N_u = 0$). Assume $\rho = 1$ for the control weighting parameter.

(a) Comment on implementing this controller with a PID controller.

(b) If $N_y = 0$ in part (a), what is the resulting controller known as? Note – you don't have to design this controller.

(c) If Δ is replaced by 1 in the system model, how would you proceed to design the controller required in part (a)? You don't have to *design* the controller, it is enough if you point out the differences. Is it possible implement this controller through PID knobs?

Chapter 13

Linear Quadratic Gaussian Control[1]

In this chapter, we present the important design technique of linear quadratic Gaussian (LQG) control. In this method, one minimizes the expectation of the weighted sum of regulation error and control effort. This is the correct method, compared to the approximate approach used in GMVC design in Sec. 11.3 [34].

It is through the LQG technique that one can design the important minimum variance controller for nonminimum phase systems, see Sec. 13.3. The LQG controller may be used as a standard, against which other controllers can be compared, using the performance curve, to be presented in Sec. 13.4. The reason is that as the control weighting is reduced to zero, the performance of an LQG controller becomes that of the minimum variance controller, while satisfying closed loop stability. In contrast, closed loop systems with GMVC and GPC of Sec. 12.1 become unstable for nonminimum phase systems when the control effort goes to zero.

Unlike the notation referred to in Footnote 5 on page 100, in the whole of this chapter we will use the argument of z^{-1} to indicate polynomials in powers of z^{-1} and the argument z to indicate polynomials in powers of z. The reason is that we need to use both of these polynomials simultaneously.

We begin this chapter with a study of spectral factorization, one use of which has been identified in Sec. 6.4.4. Using this approach, we will design the LQG controller.

13.1 Spectral Factorization

In this section, we will briefly discuss the topic of spectral factorization, by which we can split a given polynomial into two factors, one with all zeros inside the unit circle and the other with all zeros outside. Although good numerical methods are required for spectral factorization, we will not discuss them. We will restrict our attention to a qualitative discussion of spectral factorization only.

[1]This chapter may be skipped in a first reading.

Digital Control Kannan M. Moudgalya

We start our discussion by explaining the concept of self-reciprocal polynomials. To understand this concept, let us consider a polynomial in z^{-1} of the following form:

$$A(z^{-1}) = a_0 + a_1 z^{-1} + \cdots + a_n z^{-n} \tag{13.1}$$

and the polynomial with reversed coefficients:

$$A_r(z^{-1}) = a_0 z^{-n} + a_1 z^{-n+1} + \cdots + a_n \tag{13.2}$$

We can derive a relation between $A_r(z^{-1})$ and $A(z)$ in powers of z. We can write Eq. 13.2 as $A_r(z^{-1}) = z^{-n}(a_0 + a_1 z^1 + \cdots + a_n z^n)$. Thus we obtain the useful property

$$A_r(z^{-1}) = z^{-n} A(z) \tag{13.3}$$

A polynomial $\varsigma(z^{-1})$ is self-reciprocal if

$$\varsigma(z^{-1}) = \varsigma_r(z^{-1}) \tag{13.4}$$

It is easy to see that $A(z^{-1})A_r(z^{-1})$ is self-reciprocal. From

$$A(z^{-1})A_r(z^{-1}) = (a_0 + \cdots + a_n z^{-n})(a_0 z^{-n} + \cdots + a_n) \tag{13.5}$$

we obtain the coefficients of powers of z^{-1} as

$$\begin{aligned} z^0 &: a_0 a_n \\ z^{-1} &: a_0 a_{n-1} + a_1 a_n \\ &\vdots \end{aligned}$$

Now we change the order of multiplication:

$$A_r(z^{-1})A(z^{-1}) = (a_0 z^{-n} + \cdots + a_n)(a_0 + \cdots + a_n z^{-n})$$

and obtain the coefficients of powers of z^{-2n}, z^{-2n+1}, ... as

$$\begin{aligned} z^{-2n} &: a_0 a_1 \\ z^{-2n+1} &: a_0 a_{n-1} + a_1 a_n \\ &\vdots \end{aligned}$$

One can see that the coefficient of z^{-i} equals the coefficient of z^{-2n+i}. Thus, $A(z^{-1}) \times A_r(z^{-1})$ is of the form

$$\alpha_0 z^{-2n} + \alpha_1 z^{-2n+1} + \cdots + \alpha_{n-1} z^{-n-1} + \alpha_n z^{-n} + \alpha_{n-1} z^{-n+1} + \cdots + \alpha_1 z^{-1} + \alpha_0$$

which is a self-reciprocal polynomial. Note that the coefficients are symmetric about z^{-n}, where n is the degree of $A(z^{-1})$.

Now suppose that $B(z^{-1})$ is another polynomial with degree $\mathrm{d}B$ given by

$$\mathrm{d}B = \mathrm{d}A - k = n - k, \quad n \geq 0 \tag{13.6}$$

By the above logic, $B(z^{-1})B_r(z^{-1})$ will be a self-reciprocal polynomial with coefficients symmetric about z^{-n+k}. It follows that $z^{-k}B(z^{-1})B_r(z^{-1})$ will be a self-reciprocal polynomial with coefficients symmetric about z^{-n}. It is easy to see that $\rho A(z^{-1})A_r(z^{-1}) + z^{-k}B(z^{-1})B_r(z^{-1})$ is self-reciprocal, where $\rho \geq 0$ is an arbitrary scalar.

Example 13.1 Carry out the above indicated calculations for the polynomials

$$\begin{aligned} A(z^{-1}) &= a_0 + a_1 z^{-1} + a_2 z^{-2}, \quad n = 2 \\ B(z^{-1}) &= b_0 + b_1 z^{-1}, \quad k = 2 - 1 = 1 \end{aligned}$$

It is easy to see that in

$$\begin{aligned} A(z^{-1})A_r(z^{-1}) = a_0 a_2 &+ (a_0 a_1 + a_1 a_2)z^{-1} + (a_0^2 + a_1^2 + a_2^2)z^{-2} \\ &+ (a_0 a_1 + a_1 a_2)z^{-3} + a_0 a_2 z^{-4} \end{aligned}$$

the coefficients are symmetric about z^{-2}. For B,

$$B(z^{-1})B_r(z^{-1}) = b_0 b_1 + (b_0^2 + b_1^2)z^{-1} + b_0 b_1 z^{-2}$$

the coefficients are symmetric about z^{-1}. In

$$z^{-1}B(z^{-1})B_r(z^{-1}) = b_0 b_1 z^{-1} + (b_0^2 + b_1^2)z^{-2} + b_0 b_1 z^{-3}$$

the coefficients are symmetric about z^{-2}. As the coefficients are symmetric, $\rho A(z^{-1})A_r(z^{-1}) + z^{-1}B(z^{-1})B_r(z^{-1})$ is self-reciprocal.

∎

The zeros of a self-reciprocal polynomial have an interesting property. Let $\varsigma(z^{-1})$ be a self-reciprocal polynomial of degree $2n$ in z^{-1}. Using Eq. 13.3, we obtain

$$\varsigma_r(z^{-1}) = z^{-2n}\varsigma(z) \tag{13.7}$$

As $\varsigma(z^{-1})$ is self-reciprocal, using Eq. 13.4,

$$\varsigma(z^{-1}) = z^{-2n}\varsigma(z) \tag{13.8}$$

Thus if $\varsigma(z^{-1})$ can be factored as

$$\varsigma(z^{-1}) = (z^{-1} - \varsigma_1)\cdots(z^{-1} - \varsigma_{2n}) \tag{13.9}$$

$\varsigma(z)$ can be factored as

$$\varsigma(z) = (z - \varsigma_1)\cdots(z - \varsigma_{2n}) \tag{13.10}$$

Using Eq. 13.8,

$$\varsigma(z^{-1}) = z^{-2n}(z - \varsigma_1)\cdots(z - \varsigma_{2n}) \tag{13.11}$$

From Eq. 13.9 and Eq. 13.11, we obtain the useful property that if ς_i is a zero of $\varsigma(z^{-1})$, $1/\varsigma_i$ also is a zero. It is also clear that if ς_i is inside the unit circle, $1/\varsigma_i$ is outside the unit circle. It follows from the above discussion that such a factorization exists for the self-reciprocal polynomial ς:

$$\varsigma(z^{-1}) = \rho A(z^{-1})A_r(z^{-1}) + z^{-k}B(z^{-1})B_r(z^{-1}) \tag{13.12}$$

We illustrate this in the next example.

Example 13.2 For

$$A(z^{-1}) = 0.9 - 1.9z^{-1} + z^{-2}$$
$$B(z^{-1}) = 0.08 + 0.1z^{-1}$$

carry out the above indicated factorization for $\rho = 1$.

We have

$$\begin{aligned} A(z^{-1})A_r(z^{-1}) &= 0.9 - 3.61z^{-1} + 5.42z^{-2} - 3.61z^{-3} + 0.9z^{-4} \\ z^{-1}B(z^{-1})B_r(z^{-1}) &= 0.008z^{-1} + 0.0164z^{-2} + 0.008z^{-3} \\ \varsigma(z^{-1}) &= A(z^{-1})A_r(z^{-1}) + z^{-1}B(z^{-1})B_r(z^{-1}) \\ &= 0.09 - 0.361z^{-1} + 0.542z^{-2} - 0.361z^{-3} + 0.09z^{-4} \end{aligned}$$

$\varsigma(z^{-1})$ has the following zeros:

$$\begin{aligned} &1 : 1.4668 + 0.9031j \\ &2 : 1.4668 - 0.9031j \\ &3 : 0.4944 + 0.3044j \\ &4 : 0.4944 - 0.3044 \end{aligned}$$

Observe the following:

1. The third zero is the reciprocal of the second and the fourth zero is the reciprocal of the first.
2. The first two zeros are outside the unit circle, while the last two are inside the unit circle.
3. As the coefficients of ς are real, the zeros occur in conjugate pairs, in case any of the zeros is a complex number. Accordingly, there are two pairs of complex conjugate pairs.

∎

If ς is self-reciprocal with no zeros on the unit circle, it will have half the number of zeros inside and an equal number outside the unit circle. Let the product of the factors with zeros inside the unit circle be β and that of the rest β_r. To make this factorization unique, let β be monic, *i.e.*, the constant in the polynomial β is one. Thus, we arrive at the relation

$$\varsigma = \rho AA_r + z^{-k}BB_r = r\beta\beta_r \tag{13.13}$$

where r is the scale factor. This is known as *spectral factorization*. We now illustrate this approach through an example.

Example 13.3 Determine the spectral factorization for the system presented in Example 11.4 on page 413.

We have

$$A = (1 - 0.5z^{-1})(1 - 0.9z^{-1}) = 1 - 1.4z^{-1} + 0.45z^{-2}$$
$$B = 0.5(1 - 0.9z^{-1})$$
$$\rho = 1$$

We obtain

$$\varsigma = 0.45 - 2.255z^{-1} + 3.615z^{-2} - 2.255z^{-3} + 0.45z^{-4}$$

This polynomial has roots 2.618, 1.1111, 0.9 and 0.382. As the last two are inside the unit circle, they make the polynomial β. With the condition that β is monic, we obtain

$$\beta = 1 - 1.282z^{-1} + 0.3438z^{-2}$$

By reversing the coefficients, we obtain

$$\beta_r = 0.3438 - 1.282z^{-1} + z^{-2}$$

The roots of β are 0.9 and 0.382. Using Eq. 13.13, we obtain

$$r = 1.309$$

The commands in M 13.1 show how to do these calculations in Matlab. ∎

In this section, we have achieved spectral factorization through the brute force method of factoring a polynomial into its zeros. Unfortunately, this is not a numerically reliable method, especially for complicated problems. Although there exist reliable methods, they are beyond the scope of this book.

Spectral factorization will play an important role in controller design, the topic of the next section.

13.2 Controller Design

We will now present a transfer function approach to design the LQG controller [54, 26]. Consider a system of the form

$$A(z^{-1})y(n) = B(z^{-1})u(n-k) + \frac{C(z^{-1})}{F(z^{-1})}\xi(n) \tag{13.14}$$

which is in the same form as Eq. 11.1 but for the presence of F in the denominator of the noise term. As before, $\xi(n)$ is the noise, $u(n)$ the input to and $y(n)$ the output from the system. The reason for including $F(z^{-1})$ now is that we allow F to have zeros *on* the unit circle, so that step disturbances can be handled. Thus $F(z^{-1})$ can have zeros inside or on the unit circle.

We would like to design a controller $u(n)$ that minimizes the performance index

$$J = \mathscr{E}\left[\left(V(z^{-1})y(n)\right)^2 + \rho\left(W(z^{-1})F(z^{-1})u(n)\right)^2\right] \tag{13.15}$$

This is different from the performance index given in Eq. 11.61 on page 420 in that it has polynomial matrices $V(z^{-1})$, $W(z^{-1})$ and $F(z^{-1})$, as well. These are included to provide flexibility and to make the problem solvable:

1. Suppose that we wish only the rate at which $u(n)$ changes to be minimized. We can then choose $W(z^{-1})$ to be $1 - z^{-1}$.

2. Suppose that the disturbance has steps in it, say $F = 1 - z^{-1}$. Then it can be controlled only if $u(n)$ also is allowed to drift. In this case, we cannot constrain the absolute value of $u(n)$. The best we can do is to reduce the variations in $u(n)$. The factor $F(z^{-1})$ is included precisely for this reason. Thus in this case, the user is not completely free to choose the weighting matrices. This reasoning is the same as the one used to include Δ in Eq. 11.67 on page 422.

Thus in general, V and W are provided for the flexibility of the problem while F is included to make the problem solvable.

We rewrite Eq. 13.15 in the following form without the z^{-1} argument:

$$J = \mathscr{E}\left(Vy(n)\right)^2 + \rho\mathscr{E}\left(WFu(n)\right)^2 \tag{13.16}$$

We will use the variational method to minimize this objective function. Suppose that there is a controller of the form

$$u(n) = -\frac{S_c(z^{-1})}{R_c(z^{-1})}y(n) \tag{13.17}$$

that minimizes the above performance index. The variational approach is to look for a controller of the form

$$u(n) = -\frac{S_c(z^{-1})}{R_c(z^{-1})}y(n) + T\xi(n) \tag{13.18}$$

which is a variant of Eq. 13.17. We will try to find R_c, S_c and T in such a way that the increase in performance index is zero.

Let us close the loop with this controller. That is, substitute for $u(n)$ from Eq. 13.18 into Eq. 13.14 and simplify. Dropping the argument in z^{-1}, we obtain

$$Ay(n) = z^{-k}B\left[-\frac{S_c}{R_c}y(n) + T\xi(n)\right] + \frac{C}{F}\xi(n) \tag{13.19}$$

This can be simplified as

$$y(n) = \frac{CR_c}{F\alpha}\xi(n) + z^{-k}\frac{BR_c}{\alpha}T\xi(n) \tag{13.20}$$

where

$$\alpha = R_c A + z^{-k} B S_c \tag{13.21}$$

Thus we see that α makes up the characteristic polynomial. Hence we would like to have its zeros inside the unit circle. We write $y(n)$ as

$$y(n) = y_0(n) + \delta y(n) \tag{13.22}$$

where

$$\begin{aligned} y_0(n) &= \frac{CR_c}{F\alpha}\xi(n) \\ \delta y(n) &= z^{-k}\frac{BR_c}{\alpha}T\xi(n) \end{aligned} \tag{13.23}$$

Note that the variational term δy_0 has the variation causing variable, namely T. Substituting Eq. 13.22 and Eq. 13.23 in Eq. 13.18, we obtain

$$u(n) = -\frac{S_c}{R_c}\left[\frac{CR_c}{F\alpha}\xi(n) + z^{-k}\frac{BR_c}{\alpha}T\xi(n)\right] + T\xi(n) \tag{13.24}$$

This can be simplified as

$$u(n) = u_0(n) + \delta u(n) \tag{13.25}$$

where

$$\begin{aligned} u_0(n) &= -\frac{CS_c}{F\alpha}\xi(n) \\ \delta u(n) &= \frac{R_c A}{\alpha}T\xi(n) \end{aligned} \tag{13.26}$$

Let us now write Eq. 13.16 without arguments in z^{-1} and after substituting for $y(n)$ and $u(n)$ from Eq. 13.22 and Eq. 13.25:

$$\begin{aligned} J &= \mathscr{E}\left(V(y_0 + \delta y)\right)^2 + \rho\mathscr{E}\left(WF(u_0 + \delta u)\right)^2 \\ &= \mathscr{E}\left(V^2(y_0^2 + 2y_0\delta y + \delta y^2)\right) + \rho\mathscr{E}\left(W^2F^2(u_0^2 + 2u_0\delta u + \delta u^2)\right) \end{aligned} \tag{13.27}$$

This can be split as

$$J = J_0 + 2J_1 + J_2 \tag{13.28}$$

where

$$\begin{aligned} J_0 &= \mathscr{E}(Vy_0)^2 + \rho\mathscr{E}(WFu_0)^2 \\ J_1 &= \mathscr{E}(Vy_0)(V\delta y) + \rho\mathscr{E}(WFu_0)(WF\delta u) \\ J_2 &= \mathscr{E}(V\delta y)^2 + \rho\mathscr{E}(WF\delta u)^2 \end{aligned} \tag{13.29}$$

As J_0 and J_2 are sums of squares, they cannot be made zero. On the other hand, as it has cross terms, J_1 can be made zero. Thus the condition for minimizing Eq. 13.16 can be expressed as

$$J_1 = \mathscr{E}(Vy_0)(V\delta y) + \rho\mathscr{E}(WFu_0)(WF\delta u) = 0 \tag{13.30}$$

Substituting for y_0, δy_0 and u_0, δu_0 from Eq. 13.23 and Eq. 13.26, we obtain

$$\mathscr{E}\left(\frac{VCR_c}{F\alpha}\xi(n)\right)\left(z^{-k}\frac{VBR_c}{\alpha}T\xi(n)\right) + \rho\mathscr{E}\left(-\frac{WFCS_c}{F\alpha}\xi(n)\right)\left(\frac{WFR_cA}{\alpha}T\xi(n)\right) = 0 \tag{13.31}$$

This can be written as

$$\oint\left[\frac{VCR_c}{F\alpha}\frac{z^kV_*B_*R_{c*}}{\alpha_*}T_* - \rho\frac{WFCS_c}{F\alpha}\frac{W_*F_*R_{c*}A_*}{\alpha_*}T_*\right]\frac{dz}{z} = 0 \tag{13.32}$$

where the starred variables are functions of z, for example $V_* = V_*(z)$, while the unstarred variables are functions of z^{-1}, for example $S_c = S_c(z^{-1})$. The above equation becomes

$$\oint\frac{z^kVCR_cV_*B_* - \rho WFCS_cW_*F_*A_*}{zF\alpha\alpha_*}R_{c*}T_*dz = 0 \tag{13.33}$$

Note that as mentioned earlier, α, a part of the characteristic polynomial, will have zeros inside the unit circle. Thus if $zF\alpha$ divides $z^kVCR_cV_*B_* - \rho WFCS_cW_*F_*A_*$, there will be no residue term within the unit circle, and hence the above integral will vanish. Thus we arrive at the equivalent condition

$$z^kVCR_cV_*B_* - \rho WFCS_cW_*F_*A_* = zF\alpha X_* \tag{13.34}$$

where X_* is an unknown polynomial. Substituting for α from Eq. 13.21, we obtain

$$z^kVCR_cV_*B_* - \rho WFCS_cW_*F_*A_* = zF(R_cA + z^{-k}BS_c)X_* \tag{13.35}$$

Arranging R_c and S_c terms separately, we obtain

$$R_c(z^kCVV_*B_* - zAFX_*) = (\rho WW_*F_*A_*C + z^{-k+1}BX_*)FS_c \tag{13.36}$$

As R_c and S_c are coprime, each of the two sides in the above equation should be equal to $R_cS_cK_*$, where K_* is a polynomial in z, z^{-1}. Thus we obtain

$$(\rho WW_*F_*A_*C + z^{-k+1}BX_*)F = K_*R_c \tag{13.37}$$

$$z^kCVV_*B_* - zAFX_* = K_*S_c \tag{13.38}$$

Multiplying Eq. 13.37 by A and Eq. 13.38 by $z^{-k}B$ and adding, we obtain

$$C(\rho AWFF_*W_*A_* + BVV_*B_*) = K_*\alpha \tag{13.39}$$

Next, we obtain spectral factorization of the left-hand side of the above equation:

$$r\beta\beta_* = \rho AWFF_*W_*A_* + BVV_*B_* \tag{13.40}$$

where r is a positive scalar and $\beta(z^{-1})$ is a stable monic polynomial. When $\rho > 0$, stability of β is assured if BV and AFW have no common factors with zeros on the unit circle. If $\rho = 0$, BV should have no zeros on the unit circle. Using this equation, Eq. 13.39 becomes

$$Cr\beta\beta_* = K_*\alpha \tag{13.41}$$

As α is required to be stable and monic, we let

$$\alpha = C\beta \tag{13.42}$$

$$K_* = r\beta_* \tag{13.43}$$

In Eq. 13.37, F must be a factor of $K_*R_c = r\beta_*R_c$. Since β_* has no zeros on the unit circle, F must be a factor of R_c. Thus, we set

$$R_c = R_1F \tag{13.44}$$

Substituting for K_* from Eq. 13.43 and for R_c from Eq. 13.44 into Eq. 13.37 and Eq. 13.38, we obtain

$$\rho WW_*F_*A_*C + z^{-k+1}BX_* = r\beta_*R_1 \tag{13.45a}$$

$$z^kCVV_*B_* - zAFX_* = r\beta_*S_c \tag{13.45b}$$

We need to solve Eq. 13.40 and Eq. 13.45 for the unknowns R_1, S_c and X. Unfortunately, these relations involve polynomials in powers of z as well as z^{-1}. So we first convert them into polynomials in powers of z^{-1}. Using Eq. 13.3, we can write Eq. 13.40 as

$$z^{\mathrm{d}\beta}r\beta\beta_r = \rho AFW_rF_rA_rz^{\mathrm{d}A+\mathrm{d}F+\mathrm{d}W} + BVV_rB_rz^{\mathrm{d}V+\mathrm{d}B}$$

Generally, $\mathrm{d}A + \mathrm{d}F + \mathrm{d}W \geq \mathrm{d}V + \mathrm{d}B$ and, as a result, we will assume that

$$\mathrm{d}\beta = \mathrm{d}A + \mathrm{d}F + \mathrm{d}W \tag{13.46}$$

Multiplying throughout by $z^{-\mathrm{d}\beta}$, we obtain

$$r\beta\beta_r = \rho AFWW_rF_rA_r + BVV_rB_rz^{-\mathrm{d}\beta+\mathrm{d}V+\mathrm{d}B} \tag{13.47}$$

All the variables in the above relation are polynomials in powers of z^{-1}. In Eq. 13.45, the degrees of polynomials are as follows:

$$\mathrm{d}X = \mathrm{d}\beta + k - 1 \tag{13.48}$$

Using Eq. 13.3 and Eq. 13.46, Eq. 13.45a becomes

$$z^{\mathrm{d}\beta}\rho WW_rF_rA_rC + z^{-k+1}BX_rz^{\mathrm{d}\beta+k-1} = r\beta_rR_1z^{\mathrm{d}\beta}$$

Simplifying and rearranging,

$$r\beta_r R_1 - BX_r = \rho W W_r F_r A_r C \tag{13.49}$$

Similarly, Eq. 13.45b becomes

$$z^k CVV_r B_r z^{\mathrm{d}VB} - zAFX_r z^{\mathrm{d}\beta+k-1} = r\beta_r S_c z^{\mathrm{d}\beta}$$

Simplifying and rearranging,

$$r\beta_r S_c z^{-k} + AFX_r = CVV_r B_r z^{\mathrm{d}VB-\mathrm{d}\beta} \tag{13.50}$$

Eq. 13.49 and Eq. 13.50 can be written in the following form:

$$\begin{bmatrix} R_1 & S_c \end{bmatrix} \begin{bmatrix} r\beta_r & 0 \\ 0 & r\beta_r z^{-k} \end{bmatrix} + X_r \begin{bmatrix} -B & AF \end{bmatrix} = \begin{bmatrix} \rho W W_r F_r A_r C & z^{\mathrm{d}VB-\mathrm{d}\beta} CVV_r B_r \end{bmatrix} \tag{13.51}$$

where $\begin{bmatrix} R_1 & S_c \end{bmatrix}$ and X_r are unknowns. This equation is in the form of the well known Aryabhatta's identity of Eq. 7.119 on page 290. Using the method discussed in Sec. 7.8.3, we can solve this equation for $\begin{bmatrix} R_1 & S_c \end{bmatrix}$ and X_r, although we need only the former for controller design. Using Eq. 13.17 and Eq. 13.44, we obtain the controller as

$$u = -\frac{S_c}{R_1 F} \tag{13.52}$$

M 13.4 implements this controller.

Example 13.4 Let us now design an LQG controller for the system presented in Example 11.4 on page 413. We have

$$\begin{aligned} A &= (1 - 0.5z^{-1})(1 - 0.9z^{-1}) = 1 - 1.4z^{-1} + 0.45z^{-2} \\ B &= 0.5(1 - 0.9z^{-1}) \\ C &= (1 - 0.5z^{-1}) \\ k &= 1, \quad \rho = 1 \\ V &= W = F = 1 \end{aligned}$$

As V, W, F are 1, the spectral factorization is identical to the one obtained in Example 13.3. So we find

$$\begin{aligned} \rho W W_f F_f A_f C &= 0.45 - 1.625z^{-1} + 1.7z^{-3} - 0.5z^{-4} \\ z^{-1} CVV_f B_f &= -0.45z^{-1} + 0.725z^{-2} - 0.25z^{-3} \\ r\beta_f &= 0.45 - 1.6781z^{-1} + 1.309z^{-2} \end{aligned}$$

Substituting these in Eq. 13.51 and solving, we obtain

$$\begin{aligned} R_1(z^{-1}) &= 1 - 0.6439z^{-1} \\ S_c(z^{-1}) &= 0.5239 - 0.2619z^{-1} \\ X_f(z^{-1}) &= 0.7619 - 0.6857z^{-1} \end{aligned}$$

Using Eq. 13.52, we obtain the controller to be

$$u = -\frac{S_c}{R_1 F} y = -\frac{0.5239(1 - 0.5z^{-1})}{1 - 0.6439z^{-1}} y$$

M 13.5 implements this example. ∎

The solution procedure for LQG control remains the same whether the system to be controlled is minimum phase or nonminimum phase. We illustrate this with an example.

Example 13.5 We now design an LQG controller for the viscosity problem [34] presented in Example 11.10 on page 423. Note that in the current notation, ρ is a scalar and, as a result, we use the following values:

$$\begin{aligned} \rho &= 1 \\ F &= 1 - z^{-1} = \Delta \end{aligned}$$

We also have

$$\begin{aligned} A &= 1 - 0.44z^{-1} \\ B &= (0.51 + 1.21z^{-1})(1 - z^{-1}) \\ C &= 1 - 0.44z^{-1} \\ k &= 1 \end{aligned}$$

M 13.6 carries out the control design. We obtain the following results:

$$\begin{aligned} AFW &= 1 - 1.44z^{-1} + 0.44z^{-2} \\ AFWW_fF_fA_f &= 0.44 - 2.0736z^{-1} + 3.2672z^{-2} - 2.0736z^{-3} + 0.44z^{-4} \\ BVV_fB_f &= 0.6171 + 1.7242z^{-1} + 0.6171z^{-2} \\ r\beta\beta_f &= 0.44 - 1.4565z^{-1} + 4.9914z^{-2} - 1.4565z^{-3} + 0.44z^{-4} \end{aligned}$$

This polynomial has roots $1.5095 \pm 2.8443j$ and $0.1456 \pm 0.2743j$. As the last two are inside the unit circle, they make the polynomial β. We obtain

$$\begin{aligned} r &= 4.5622 \\ \beta &= 1 - 0.2912z^{-1} + 0.09645z^{-2} \\ \rho WW_fF_fA_fC &= 0.44 - 1.6336z^{-1} + 1.6336z^{-2} - 0.44z^{-3} \\ z^{-1}CVV_fB_f &= 1.21z^{-1} - 0.0224z^{-2} - 0.2244z^{-3} \\ r\beta_f &= 0.44 - 1.3284z^{-1} + 4.5622z^{-2} \end{aligned}$$

Substituting these in Eq. 13.51 and solving, we obtain

$$\begin{aligned} R_1(z^{-1}) &= 1 + 0.4701z^{-1} \\ S_c(z^{-1}) &= 0.4682 - 0.2060z^{-1} \\ X_f(z^{-1}) &= 2.1359 + 1.004z^{-1} \end{aligned}$$

The first two are obtained in R_c and S_c, respectively, in M 13.6. Note that the variable X_f is not required for controller design. We obtain the controller to be

$$u = -\frac{S_c}{R_c} y = -\frac{S_c}{R_1 F} y$$

Because $F = \Delta = 1 - z^{-1}$, we obtain

$$\Delta u = -\frac{0.4682(1 - 0.4400z^{-1})}{1 + 0.4701z^{-1}} y$$

After closing the loop, we obtain the following relations:

$$\begin{aligned} y(n) &= \frac{1 + 0.4701z^{-1}}{1 - 0.2912z^{-1} + 0.0964z^{-2}} \xi(n) \\ u(n) &= \frac{0.4682 - 0.2060z^{-1}}{1 - 0.2912z^{-1} + 0.0964z^{-2}} \xi(n) \end{aligned}$$

which are stored in `Ny/Dy` and `Nu/Du`, respectively, in M 13.6. The variance of y and u, calculated and stored in `yvar` and `uvar`, respectively, are

$$\begin{aligned} \sigma_y^2 &= 1.5970 \\ \sigma_u^2 &= 0.2285 \end{aligned}$$

It is instructive to compare these results with those of the minimum variance controller, obtained in Example 11.8 on page 419. For a small increase (about 10%) in the variance of y, the variance of u has come down greatly, by about 85%.

It is also instructive to compare these results with those of the generalized minimum variance controller, obtained in Example 11.10 on page 423, for the same value of ρ. The LQG controller results in a smaller output variance, for a larger variance in the input.

∎

In the above example, we have made a comparison of the variance values obtained with different controllers. We will develop this line further in Sec. 13.3.

The LQG control design technique that we developed in this section can easily accommodate polynomial weighting in the sum minimized. We now demonstrate this with an example taken from [54].

Example 13.6 Design the LQG feedback control for the following system:

$$(1 - 0.9z^{-1})y(n) = (0.1 + 0.08z^{-1})u(n-2) + \xi(n)$$

with

$$V = 1, \quad W = 1 - z^{-1}$$

Thus we would like to weight the differential input. We have

$$\begin{aligned}
A &= 1 - 0.9z^{-1} \\
B &= 0.1 + 0.08z^{-1} \\
k &= 2 \\
C &= F = 1 \\
\rho &= 0.1
\end{aligned}$$

Invoking M 13.8, we obtain the following results:

$$\begin{aligned}
AFW &= 1 - 1.9z^{-1} + 0.9z^{-2} \\
AFWW_fF_fA_f &= 0.09 - 0.361z^{-1} + 0.542z^{-2} - 0.361z^{-3} + 0.09z^{-4} \\
BVV_fB_f &= 0.008 + 0.0164z^{-1} + 0.008z^{-2} \\
r\beta\beta_f &= 0.09 - 0.353z^{-1} + 0.5584z^{-2} - 0.353z^{-3} + 0.09z^{-4}
\end{aligned}$$

This polynomial has roots $1.4668 \pm 0.903j$ and $0.4944 \pm 0.3044j$. As the last two are inside the unit circle, they make the polynomial β. We obtain

$$\begin{aligned}
r &= 0.267 \\
\beta &= 1 - 0.9887z^{-1} + 0.337z^{-2} \\
\rho WW_fF_fA_fC &= 0.09 - 0.28z^{-1} + 0.29z^{-2} - 0.1z^{-3} \\
z^{-1}CVV_fB_f &= 0.08z^{-1} + 0.1z^{-2} \\
r\beta_f &= 0.09 - 0.264z^{-1} + 0.267z^{-2}
\end{aligned}$$

Substituting these in Eq. 13.51 and solving, we obtain

$$\begin{aligned}
R_c(z^{-1}) &= 1 - 0.08871z^{-1} + 0.121z^{-2} \\
S_c(z^{-1}) &= 1.3617 \\
X_f(z^{-1}) &= 0.404 + 0.04945z^{-1} + 0.08z^{-2}
\end{aligned}$$

The controller is given by

$$u(n) = -\frac{S_c}{R_c}y(n)$$

Recall once again that X_f is not required in the expression for the controller. ∎

13.3 Simplified LQG Control Design

In this section, we will derive a simpler procedure to design LQG controllers. We will also derive a procedure to design minimum variance controllers when the plant to be controlled is nonminimum phase.

Multiply Eq. 13.45a by AF and Eq. 13.45b by $z^{-k}B$ to arrive at

$$\rho W W_* F_* A_* C A F + z^{-k+1} B X_* A F = r \beta_* R_1 A F$$
$$C V V_* B_* B - z^{-k+1} B A F X_* = r z^{-k} B \beta_* S_c$$

Adding the two equations, we obtain

$$\rho W W_* F_* A_* C A F + C V V_* B_* B = r \beta_* R_1 A F + r z^{-k} B \beta_* S_c$$

The left-hand side of the equation becomes $r\beta\beta_* C$ using Eq. 13.40, and hence the above equation becomes

$$r \beta \beta_* C = r \beta_* R_1 A F + r z^{-k} B \beta_* S_c$$

Cancelling the common factor $r\beta_*$, the above equation becomes

$$\beta C = R_1 A F + z^{-k} B S_c \tag{13.53}$$

We may solve the above equation for R_1 and S_c when A and B don't have a common factor. Note that the solution of the above equation is a lot simpler than that of Eq. 13.51. We now illustrate this simplified approach with an example.

Example 13.7 Solve the viscosity control problem of Example 13.5 using the simplified LQG control design method, derived above.

M 13.6 is once again invoked with the call to `lqg` being replaced with a call to `lqg_simple`. Everything else, including the argument list, remains the same. M 13.7 is now used. We obtain results that are identical to the ones obtained in Example 13.5.

∎

Another utility of this simplified procedure is that it can be used to derive the minimum variance control law when the plant to be controlled is nonminimum phase. When $\rho = 0$, Eq. 13.40 becomes

$$\beta \beta_* = B B_*$$

where we assume that we do not carry out the normalization and hence take r as 1. We have also taken V as 1, as per the convention in minimum variance control. In terms of reciprocal polynomials, we obtain

$$\beta \beta_r = B B_r$$

Splitting B into good and bad factors, namely $B = B^g B^b$, the above equation becomes

$$\beta \beta_r = B^g B^b B^g_r B^b_r \tag{13.54}$$

As β should have only stable factors, we extract the stable factors from the right-hand side and obtain

$$\beta = B^g B^b_r$$

Substituting this expression in Eq. 13.53, we obtain

$$B^g B_r^b C = AFR_1 + z^{-k} BS_c \tag{13.55}$$

and the control law is obtained as

$$u(n) = -\frac{S_c}{R_1 F} \tag{13.56}$$

which is the same as Eq. 13.52. This approach has been used in Sec. 11.2.3.

We have remarked earlier that when A and B do not have a common factor, we may solve Eq. 13.53 for R_1 and S_c. Let us see in the next example what happens when there is a common factor between A and B.

Example 13.8 Try solving the problem presented in Example 13.4 using the simplified LQG design procedure developed in this section.

We have

$$\begin{aligned} A &= (1 - 0.5z^{-1})(1 - 0.9z^{-1}) \\ B &= 0.5(1 - 0.9z^{-1}) \\ C &= (1 - 0.5z^{-1}) \\ k &= F = 1 \end{aligned}$$

Recall that we have calculated β for the same problem in Example 13.3 on page 467:

$$\beta = 1 - 1.282z^{-1} + 0.3438z^{-2} = (1 - 0.9z^{-1})(1 - 0.382z^{-1})$$

Thus, Eq. 13.53 becomes

$$\begin{aligned} &(1 - 0.9z^{-1})(1 - 0.382z^{-1})(1 - 0.5z^{-1}) \\ &\qquad = R_1(1 - 0.5z^{-1})(1 - 0.9z^{-1}) + z^{-1}0.5(1 - 0.9z^{-1})S_c \end{aligned}$$

We obtain the following solution:

$$\begin{aligned} R_1 &= 1 - 0.382z^{-1} \\ S_c &= 0 \end{aligned}$$

From Eq. 13.52, we obtain the control law to be $u = 0$!. ∎

This solution is unacceptable. Recall that we have faced a similar situation in Example 11.7 on page 417.

13.4 Introduction to Performance Analysis of Controllers

Now that we have seen a lot of control design techniques, the natural question that comes to mind is, "Which controller is the best?" We now give a brief answer to this

Table 13.1: Comparison of minimum variance, generalized minimum variance and LQG controllers through variance of input and output signals in viscosity control problem

	MVC	GMVC $\rho = 1$	LQG $\rho = 1$	LQG $\rho = 10$	LQG $\rho = 2$
σ_y^2	1.4070	1.7190	1.5970	2.1107	1.7045
σ_u^2	1.2994	0.1870	0.2285	0.0594	0.1523

Table 13.2: Comparison of minimum variance, generalized minimum variance and LQG controllers through variance of input and output signals in MacGregor's first control problem

	MVC	GMVC $\rho = 1$	LQG $\rho = 1$	LQG $\rho = 2$	LQG $\rho = 1.772$
σ_y^2	1	3.0301	2.3475	3.1765	3.0297
σ_u^2	5.9684	0.5075	1.0250	0.4294	0.5075

question through examples. For details, the reader should consult more specialized books, such as [23].

We designed minimum variance (Example 11.8 on page 419), generalized minimum variance (Example 11.10 on page 423, $\rho = 1$) and LQG (Example 13.5) controllers for the viscosity control problem and obtained the variances listed in Table 13.1. The minimum variance controller may be acceptable if the large variance in u is acceptable. It is not clear whether GMVC or LQG should be chosen. To resolve this, we try to make σ_y^2 of LQG approximately equal to that of GMVC. With $\rho = 10$, σ_y^2 becomes much larger. With $\rho = 2$, we obtain variances that are comparable.

In Table 13.1, we have listed the variances corresponding to controllers obtained with $\rho = 10$ and 2. Comparing GMVC with LQG obtained with $\rho = 2$, we see that the latter achieves a smaller σ_y^2 with a smaller σ_u^2 as well. This shows that LQG achieves a smaller error simultaneously with a smaller control effort as compared to GMVC. In view of this criterion, we would prefer LQG for the problem.

We carry out a similar comparison of these controllers for MacGregor's first control problem, presented in Example 11.4 on page 413, Example 11.9 on page 421 and Example 13.4. The resulting variances are now listed in Table 13.2. From this table, it is clear that the performance of GMVC and LQG are comparable. When the performances of the controllers are comparable for this problem, why is there a difference in the viscosity control problem? One reason we can immediately think of is that the viscosity control problem involves a nonminimum phase system. We discuss this issue further shortly.

We will next discuss how a plot of the output variance *vs.* input variance can be used to assess the performance of a control loop. For example, Fig. 13.1 shows one such plot obtained by varying ρ from 0.1 to 10 for the viscosity control problem. The control designer has the option of running the plant at any point on this curve. The left extreme corresponds to large ρ values, as these weight the input heavily, resulting in small input variance. Similarly, the right extreme of the plot corresponds

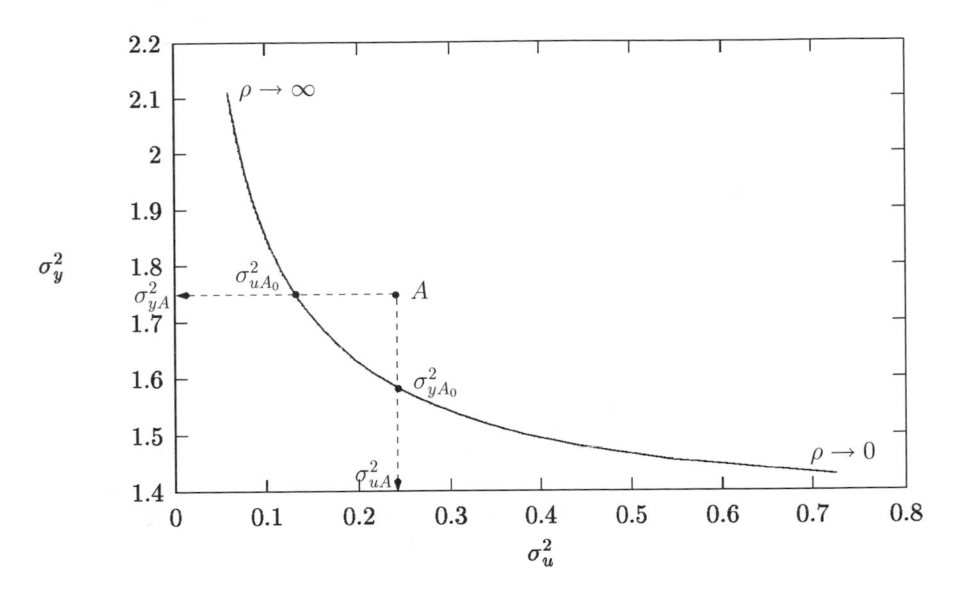

Figure 13.1: Variance of y (σ_y^2) *vs.* variance of Δu (σ_u^2): LQG law for viscosity control problem

to small ρ values, as these result in the input not being constrained, resulting in large σ_u^2. M 13.9 is one way to arrive at this plot, which will be referred to as the *performance curve* [23].

Depending on the current performance level of the plant, one can decide whether it can be improved. Suppose that the actual measured performance of this plant with a particular controller (not necessarily LQG) is given by the point $A = (\sigma_{uA}^2, \sigma_{yA}^2)$. From the plot, it is clear that one of the following is possible, by switching over to an LQG controller:

1. One can obtain the same output variance (σ_{yA}) for a smaller control effort (σ_{uA0}^2).
2. For the same control effort, one can obtain smaller output variance (σ_{yA0}^2).

The performance plots of different controllers can be compared. Fig. 13.2 shows the performance curves of generalized minimum variance and LQG controllers for the viscosity control problem.

In this figure, the performance curve of the GMVC is higher than that of LQG. For large ρ values, the two curves coincide, but as $\rho \to 0$, the curve of GMVC moves upwards. The reason for this is not difficult to see. As $\rho \to 0$, GMVC approaches the standard minimum variance controller, which results in unbounded control effort, because the plant is nonminimum phase. In other words, GMVC does not gracefully degrade to the stable minimum variance controller of Sec. 13.3.

Although not shown, the performance curve of the generalized predictive controller is more or less similar to that of the GMVC. In fact, the former also cannot stabilize the viscosity control problem when $\rho \to 0$. Amongst the controllers that we have seen, the only one that guarantees this for a nonminimum phase is the LQG controller. In view of this, the performance curve of the LQG controller is recommended as a benchmark, against which other controllers can be compared.

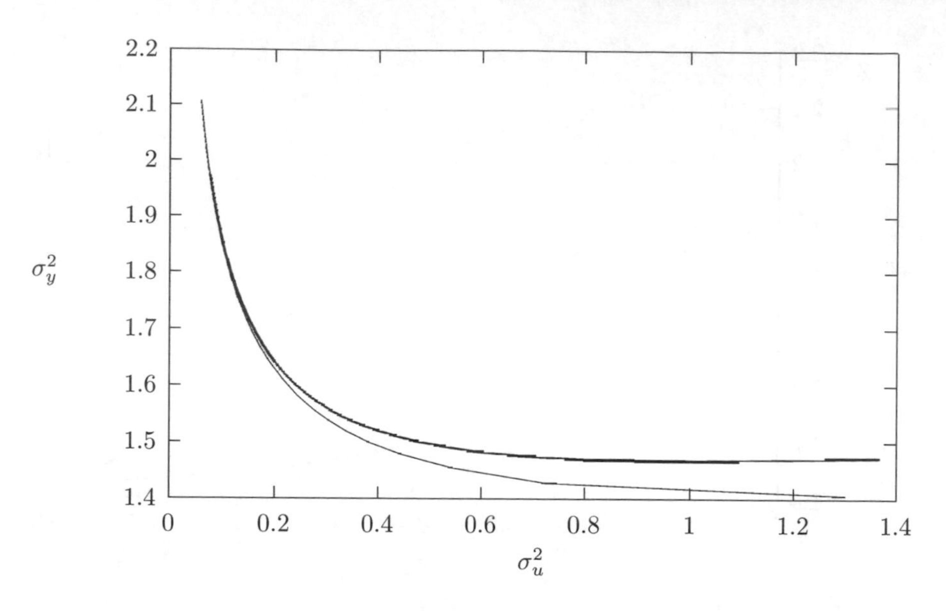

Figure 13.2: Comparison of performance curves for LQG (lower curve) and generalized minimum variance (upper curve) control of the viscosity problem

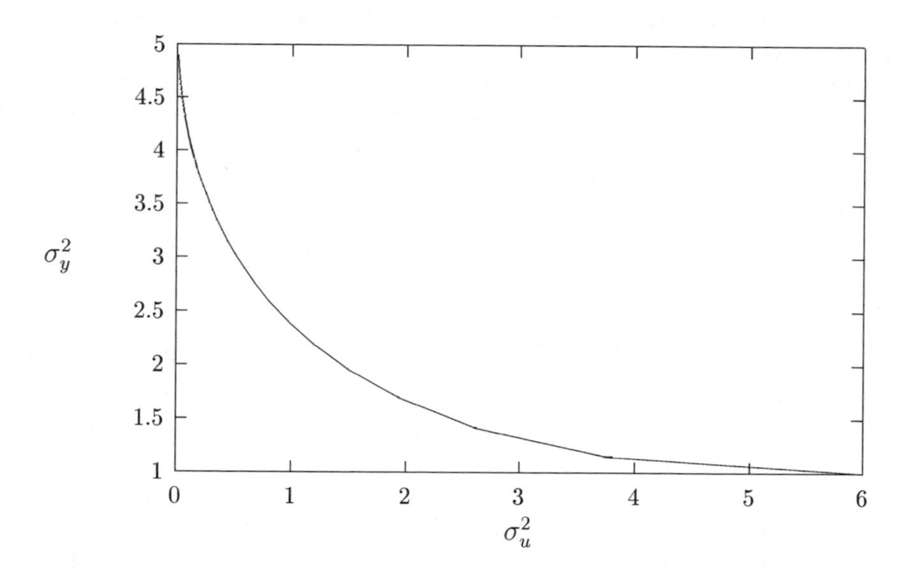

Figure 13.3: Performance curve for GMVC of MacGregor's first example

There is not much difference between the performance curves for MacGregor's first control problem. M 13.10 shows the steps taken to arrive at the performance curve using GMVC. A plot of this curve is shown in Fig. 13.3. Although not shown, the performance curves obtained using GMVC and GPC look similar.

13.5 Matlab Code

Matlab Code 13.1 Spectral factorization, as discussed in Example 13.3 on page 467. This code is available at HOME/lqg/matlab/spec_ex.m[2]

```
1  A = conv([-0.5 1],[-0.9 1]); dA = 2;
2  B = 0.5*[-0.9 1]; dB = 1; rho = 1;
3  [r,beta,sigma] = spec(A,dA,B,dB,rho)
```

Matlab Code 13.2 Function to implement spectral factorization, as discussed in Sec. 13.1. For an example of the usage, see M 13.1. This code is available at HOME/lqg/matlab/spec.m

```
1  function [r,b,rbbr] = spec(A,dA,B,dB,rho)
2  AA = rho * conv(A, flip(A));
3  BB = conv(B, flip(B));
4  diff = dA - dB;
5  dBB = 2*dB;
6  for i = 1:diff
7      [BB,dBB] = polmul(BB,dBB,[0 1],1);
8  end
9  [rbbr,drbbr] = poladd(AA,2*dA,BB,dBB);
10 rts = roots(rbbr); % roots in descending order of magnitude
11 rtsin = rts(dA+1:2*dA);
12 b = 1;
13 for i = 1:dA,
14     b = conv(b,[1 -rtsin(i)]);
15 end
16 br = flip(b);
17 bbr = conv(b,br);
18 r = rbbr(1) / bbr(1);
```

Matlab Code 13.3 Spectral factorization, to solve Eq. 13.47 on page 471. For an example of the usage, see M 13.4. This code is available at HOME/lqg/matlab/specfac.m

```
1  % function [r,b,dAFW] = ...
2  %     specfac(A,dA,B,dB,rho,V,dV,W,dW,F,dF)
3  % Implements the spectral factorization for use with LQG
4  % control by design method of Ahlen and Sternad
5
6  function [r,b,dAFW] = specfac(A,dA,B,dB,rho,V,dV,W,dW,F,dF)
7  AFW = conv(A,conv(W,F));
8  dAFW = dA + dF + dW;
9  AFWWFA = rho * conv(AFW, flip(AFW));
10 BV = conv(B,V);
11 dBV = dB + dV;
```

[2] HOME stands for http://www.moudgalya.org/dc/ – first see the software installation directions, given in Appendix A.2.

```
12 BVVB = conv(BV, flip(BV));
13 diff = dAFW - dBV;
14 dBVVB = 2*dBV;
15 for i = 1:diff
16     [BVVB,dBVVB] = polmul(BVVB,dBVVB,[0 1],1);
17 end
18 [rbb,drbb] = poladd(AFWWFA,2*dAFW,BVVB,dBVVB);
19 rts = roots(rbb);
20 rtsin = rts(dAFW+1:2*dAFW);
21 b = 1;
22 for i = 1:dAFW,
23     b = conv(b,[1 -rtsin(i)]);
24 end
25 b = real(b);
26 br = flip(b);
27 bbr = conv(b,br);
28 r = rbb(1) / bbr(1);
```

Matlab Code 13.4 LQG control design by polynomial method, to solve Eq. 13.51 on page 472. This code is available at `HOME/lqg/matlab/lqg.m`

```
1 % LQG controller design by method of Ahlen and Sternad
2 % function [R1,dR1,Sc,dSc] = ...
3 % lqg(A,dA,B,dB,C,dC,k,rho,V,dV,W,dW,F,dF)
4
5 function [R1,dR1,Sc,dSc] = ...
6 lqg(A,dA,B,dB,C,dC,k,rho,V,dV,W,dW,F,dF)
7 [r,b,db] = specfac(A,dA,B,dB,rho,V,dV,W,dW,F,dF);
8 WFA = flip(conv(A,conv(F,W)));
9 dWFA = dW + dF + dA;
10 [rhs1,drhs1] = polmul(W,dW,WFA,dWFA);
11 [rhs1,drhs1] = polmul(rhs1,drhs1,C,dC);
12 rhs1 = rho * rhs1;
13 rhs2 = conv(C,conv(V,flip(conv(B,V))));
14 drhs2 = dC + 2*dV + dB;
15 for i = 1:db-dB-dV,
16   rhs2 = conv(rhs2,[0,1]);
17 end
18 drhs2 = drhs2 + db-dB-dV;
19 C1 = zeros(1,2);
20 [C1,dC1] = putin(C1,0,rhs1,drhs1,1,1);
21 [C1,dC1] = putin(C1,dC1,rhs2,drhs2,1,2);
22
23 rbf = r * flip(b);
24 D1 = zeros(2);
25 [D1,dD1] = putin(D1,0,rbf,db,1,1);
26 for i = 1:k,
27         rbf = conv(rbf,[0 1]);
28 end
```

```
29 [D1,dD1] = putin(D1,dD1,rbf,db+k,2,2);
30 N = zeros(1,2);
31 [N,dN] = putin(N,0,-B,dB,1,1);
32 [AF,dAF] = polmul(A,dA,F,dF);
33 [N,dN] = putin(N,dN,AF,dAF,1,2);
34 [Y,dY,X,dX] = xdync(N,dN,D1,dD1,C1,dC1);
35 [R1,dR1] = ext(X,dX,1,1);
36 [Sc,dSc] = ext(X,dX,1,2);
37 X = flip(Y);
```

Matlab Code 13.5 LQG design for the problem discussed in Example 13.4 on page 472. This code is available at `HOME/lqg/matlab/lqg_mac1.m`

```
1 % MacGregor's first control problem
2 clear
3 A = [1 -1.4 0.45]; dA = 2; C = [1 -0.5]; dC = 1;
4 B = 0.5*[1 -0.9]; dB = 1; k = 1; int = 0; F = 1; dF = 0;
5 V = 1; W = 1; dV = 0; dW = 0;
6 rho = 1;
7 [R1,dR1,Sc,dSc] = lqg(A,dA,B,dB,C,dC,k,rho,V,dV,W,dW,F,dF)
8 [Nu,dNu,Du,dDu,Ny,dNy,Dy,dDy,yvar,uvar] = ...
9     cl(A,dA,B,dB,C,dC,k,Sc,dSc,R1,dR1,int);
```

Matlab Code 13.6 LQG control design for viscosity control problem discussed in Example 13.5. This code is available at `HOME/transfer/lqg/matlab/lqg_visc.m`

```
1 % Viscosity control problem of MacGregor
2 A = [1 -0.44]; dA = 1; B = [0.51 1.21]; dB = 1;
3 C = [1 -0.44]; dC = 1; k = 1; int = 1; F = [1 -1]; dF = 1;
4 V = 1; W = 1; dV = 0; dW = 0;
5 rho = 1;
6 [R1,dR1,Sc,dSc]=lqg(A,dA,B,dB,C,dC,k,rho,V,dV,W,dW,F,dF)
7 [Nu,dNu,Du,dDu,Ny,dNy,Dy,dDy,yvar,uvar] = ...
8     cl(A,dA,B,dB,C,dC,k,Sc,dSc,R1,dR1,int);
```

Matlab Code 13.7 Simplified LQG control design, obtained by the solution of Eq. 13.53 on page 476. Calling procedure is identical to that of M 13.4. This code is available at `HOME/transfer/lqg/matlab/lqg_simple`

```
1 % LQG controller simple design by method of Ahlen and Sternad
2 % function [R1,dR1,Sc,dSc] = ...
3 % lqg_simple(A,dA,B,dB,C,dC,k,rho,V,dV,W,dW,F,dF)
4
5 function [R1,dR1,Sc,dSc] = ...
6 lqg_simple(A,dA,B,dB,C,dC,k,rho,V,dV,W,dW,F,dF)
7 [r,b,db] = specfac(A,dA,B,dB,rho,V,dV,W,dW,F,dF);
8 [D,dD] = polmul(A,dA,F,dF);
9 [zk,dzk] = zpowk(k);
10 [N,dN] = polmul(zk,dzk,B,dB);
```

```
11 [RHS,dRHS] = polmul(C,dC,b,db);
12 [Sc,dSc,R1,dR1] = xdync(N,dN,D,dD,RHS,dRHS);
```

Matlab Code 13.8 LQG control design for the problem discussed in Example 13.6 on page 474. This code is available at `HOME/lqg/matlab/lqg_as1.m`

```
1 % Solves Example 3.1 of Ahlen and Sternad in Hunt's book
2 A = [1 -0.9]; dA = 1; B = [0.1 0.08]; dB = 1;
3 k = 2; rho = 0.1; C = 1; dC = 0;
4 V = 1; dV = 0; F = 1; dF = 0; W = [1 -1]; dW = 1;
5 [R1,dR1,Sc,dSc] = lqg(A,dA,B,dB,C,dC,k,rho,V,dV,W,dW,F,dF)
```

Matlab Code 13.9 Performance curve for LQG control design of viscosity problem, as discussed in Sec. 13.4. This code is available at `HOME/lqg/matlab/lqg_visc_loop.m`

```
1 % MacGregor's Viscosity control problem
2 A = [1 -0.44]; dA = 1; B = [0.51 1.21]; dB = 1;
3 C = [1 -0.44]; dC = 1; k = 1; int = 1; F = [1 -1]; dF = 1;
4 V = 1; W = 1; dV = 0; dW = 0;
5 u_lqg = []; y_lqg =[]; uy_lqg = [];
6 for rho = 0.001:0.1:3,
7     [R1,dR1,Sc,dSc] = lqg(A,dA,B,dB,C,dC,k,rho,V,dV,W,dW,F,dF
        );
8     [Nu,dNu,Du,dDu,Ny,dNy,Dy,dDy,yvar,uvar] = ...
9         cl(A,dA,B,dB,C,dC,k,Sc,dSc,R1,dR1,int);
10    u_lqg = [u_lqg uvar]; y_lqg = [y_lqg yvar];
11    uy_lqg = [uy_lqg; [rho uvar yvar]];
12 end
13 plot(u_lqg,y_lqg,'g')
14 save -ASCII lqg_visc.dat uy_lqg
```

Matlab Code 13.10 Performance curve for GMVC design of MacGregor's first control problem, as discussed in Sec. 13.4. This code is available at `HOME/minv/matlab/gmv_mac1_loop.m`

```
1 % MacGregor's first control problem
2 clear
3 A = [1 -1.4 0.45]; dA = 2; C = [1 -0.5]; dC = 1;
4 B = 0.5*[1 -0.9]; dB = 1; k = 1; int = 0;
5 u_gmv = []; y_gmv = []; uy_gmv = [];
6 for rho = 0:0.1:10,
7     [S,dS,R,dR] = gmv(A,dA,B,dB,C,dC,k,rho,int);
8     [Nu,dNu,Du,dDu,Ny,dNy,Dy,dDy,yvar,uvar] = ...
9         cl(A,dA,B,dB,C,dC,k,S,dS,R,dR,int);
10    u_gmv = [u_gmv uvar]; y_gmv = [y_gmv yvar];
11    uy_gmv = [uy_gmv; [rho uvar yvar]];
12 end
13 plot(u_gmv,y_gmv,'b')
14 save -ASCII gmv_mac1.dat uy_gmv
```

13.6 Problems

13.1. Using the notation that we used to derive LQG controllers, carry out a spectral factorization for the following set of values:

$$\begin{aligned}
\rho &= 1 \\
V &= 1 \\
W &= 1 \\
F &= 1 \\
A &= 1 + z^{-1} + z^{-2} \\
B &= 1 - 0.5z^{-1}
\end{aligned}$$

Write down the resulting r, β and β^* values.

13.6 Problems

Part IV

State Space Approach to Controller Design

Chapter 14

State Space Techniques in Controller Design

An alternative to the transfer function approach to control system design is the state space approach. The former depends on polynomial algorithms, which are slow to emerge. In contrast, the linear algebraic tools that are required by the state space techniques are a lot more advanced. As a result, many more control design techniques are available in the state space approach. At the current state of development, the state space techniques are preferred over the transfer function approach, especially for multivariable systems. In this chapter, we briefly summarize the state space approach to controller design. For details, we refer the reader to the standard books on this topic, such as [29] and [1].

We begin with the design of pole placement controllers, assuming that all states are measurable. Then we present the design of an observer that helps determine the states, in case all of them are not measured. We combine the two to achieve a pole placement controller when not all states are measured. The design input to pole placement controllers is the location of closed loop poles. An alternative approach is to specify an optimization index to be minimized. We present the concept of linear quadratic regulator (LQR), obtained by minimizing a quadratic index. When the states required for LQR are estimated by a Kalman filter, the estimator–regulator combination is known as the linear quadratic Gaussian (LQG) controller. We conclude this chapter with a brief introduction to LQG controller design by state space approach. Recall the transfer function approach to LQG controller design in Chapter 13.

14.1 Pole Placement

Consider the state space model, given in Eq. 4.28 on page 85, reproduced here for convenience:

$$x(k+1) = Ax(k) + Bu(k) + \delta(k+1)x_0 \tag{14.1}$$

We will now restrict our attention to u being a scalar; that is, there is only one manipulated variable. Thus B becomes a vector. To indicate this, we will use b in place of B. Suppose that all the states are measured. Can we use a state feedback

Digital Control Kannan M. Moudgalya

controller that gives the control law as a function of the state, to obtain a desired closed loop characteristic polynomial? Before we answer this question, let us take the Z-transform of Eq. 14.1 to arrive at

$$\begin{aligned}zX(z) &= AX(z) + bU(z) + x_0 z \\ (zI - A)X(z) &= bU(z) + x_0 z \\ X(z) &= (zI - A)^{-1} bU(z) + x_0 z\end{aligned}$$

Evaluating the inverse, for $x_0 = 0$, we obtain

$$X(z) = \frac{\operatorname{adj}(zI - A)}{|zI - A|} bU(z) \tag{14.2}$$

where adj denotes the adjoint operator. Poles of the transfer function are given by the roots of $|zI - A| = 0$. Let λ denote eigenvalues and v eigenvectors of A. We obtain $Av = \lambda v$ and $|\lambda I - A| = 0$. We see that the poles of the transfer function given in Eq. 14.2 are the same as the eigenvalues of A.

Example 14.1 Discuss the stability of Eq. 14.1, if the system matrix is given by

$$A = \begin{bmatrix} 1 & 2 \\ 0 & 3 \end{bmatrix}$$

Because A is upper triangular, the eigenvalues are given by the diagonal elements. Thus, we see the eigenvalues to be 1 and 3. Because the poles are not inside the unit circle, the system is unstable.

∎

Next, we would like to ask whether with a state feedback controller of the form

$$u(k) = -Kx(k) + v(k) = -\begin{bmatrix} K_1 & K_2 & \cdots & K_n \end{bmatrix} \begin{bmatrix} x_1(k) \\ x_2(k) \\ \vdots \\ x_n(k) \end{bmatrix} + v(k) \tag{14.3}$$

it is possible to change the closed loop characteristic polynomial. Here, we have assumed that the state is an n-dimensional vector. Because all the states are used in the control law, it is known as the *state feedback controller*. Note that K_j, $1 \le j \le n$, are scalars, to be determined. The variable $v(k)$ may be thought of as some kind of offset in the control law.

Let us assume that all the states are measured. Applying this feedback control is equivalent to substituting the expression for u in Eq. 14.1. We obtain

$$x(k+1) = Ax(k) + b[-Kx(k) + v(k)] + x_0 \delta(k+1)$$

Simplifying, we obtain

$$x(k+1) = (A - bK)x(k) + bv(k) + x_0 \delta(k+1) \tag{14.4}$$

It may be possible to find a K such that the above closed loop system is well behaved, *i.e.*, the eigenvalues of $(A - bK)$ are in desirable locations.

Example 14.2 Determine whether the system considered in Example 14.1 can be improved by a suitable K, if $b = \begin{bmatrix} 0 & 1 \end{bmatrix}^T$.
The closed loop system is given by Eq. 14.4. Its system matrix is given by

$$\begin{aligned} A - bK &= \begin{bmatrix} 1 & 2 \\ 0 & 3 \end{bmatrix} - \begin{bmatrix} 0 \\ 1 \end{bmatrix} \begin{bmatrix} K_1 & K_2 \end{bmatrix} = \begin{bmatrix} 1 & 2 \\ 0 & 3 \end{bmatrix} - \begin{bmatrix} 0 & 0 \\ K_1 & K_2 \end{bmatrix} \\ &= \begin{bmatrix} 1 & 2 \\ -K_1 & 3 - K_2 \end{bmatrix} \end{aligned}$$

Suppose that we choose $K_1 = 0.5$, $K_2 = 3.5$. We obtain

$$A - bK = \begin{bmatrix} 1 & 2 \\ -0.5 & -0.5 \end{bmatrix}$$

whose eigenvalues are $0.25 \pm 0.6614j$ with the absolute value as 0.7071. Thus the choice of this K has made the closed loop system stable. ∎

In the above example, we have shown that the introduction of feedback control could change the eigenvalues of the system matrix. We are actually interested in the inverse problem: which K will give desired closed loop pole locations? To answer this, we begin by taking the Z-transform of Eq. 14.4:

$$[zI - (A - bK)]X(z) = bV(z) + x_0 z$$

Simplifying this, we obtain

$$X(z) = \frac{\operatorname{adj}[zI - (A - bK)]}{|zI - (A - bK)|}(bV(z) + x_0 z)$$

$|zI - (A - bK)|$ is the characteristic polynomial of the closed loop. We may want it to be equal to a polynomial $\alpha_c(z)$ of our choice, given by

$$\begin{aligned} \alpha_c(z) &= z^n + \alpha_1 z^{n-1} + \cdots + \alpha_{n-1} z + \alpha_n \\ &= (z - \beta_1)(z - \beta_2) \cdots (z - \beta_n) \end{aligned} \tag{14.5}$$

There is no rigorous theory for the choice of this polynomial. In Sec. 7.7, we have presented a method to select second degree polynomials. We would like to restrict ourselves to the selection of K, given such a polynomial. Can we equate the coefficients and find K? Equating the two expressions results in a system of n nonlinear algebraic equations. This is not an easy problem, especially when n is large. In fact, it may be difficult even to determine whether these equations have a solution. In view of these difficulties, we present an alternative approach now. We will show that if the pair (A, b) is in controller canonical form, to be defined next, K can be easily determined.

The pair (A, b) is said to be in controller canonical form if A and b are as given below:

$$A = \begin{bmatrix} -a_1 & -a_2 & \cdots & -a_{n-2} & -a_{n-1} & -a_n \\ 1 & 0 & \cdots & 0 & 0 & 0 \\ 0 & 1 & \cdots & 0 & 0 & 0 \\ \vdots & & & & & \\ 0 & 0 & \cdots & 1 & 0 & 0 \\ 0 & 0 & \cdots & 0 & 1 & 0 \end{bmatrix}, \quad b = \begin{bmatrix} 1 \\ 0 \\ \vdots \\ 0 \\ 0 \\ 0 \end{bmatrix} \tag{14.6}$$

It can be shown that when A is in this special form, its characteristic polynomial is given by (see Problem 14.1)

$$|zI - A| = z^n + a_1 z^{n-1} + \cdots + a_n \tag{14.7}$$

Note that this characteristic polynomial can be written down, simply by observation. The utility of this canonical form is that with K as in Eq. 14.3, it is possible to determine the characteristic polynomial of $A - bK$, also by observation. By direct calculation, we obtain

$$A - bK$$
$$= \begin{bmatrix} -a_1 - K_1 & -a_2 - K_2 & \cdots & -a_{n-2} - K_{n-2} & -a_{n-1} - K_{n-1} & -a_n - K_n \\ 1 & 0 & \cdots & 0 & 0 & 0 \\ 0 & 1 & \cdots & 0 & 0 & 0 \\ \vdots & & & & & \\ 0 & 0 & \cdots & 1 & 0 & 0 \\ 0 & 0 & \cdots & 0 & 1 & 0 \end{bmatrix}$$

$A - bK$ is also in the same form as A and hence the closed loop characteristic polynomial can be written by observation. Using Eq. 14.7, we obtain

$$|zI - (A - bK)| = z^n + (a_1 + K_1)z^{n-1} + \cdots + (a_{n-1} + K_{n-1})z + (a_n + K_n)$$

Equating the coefficients of powers of z in the above equation and that in Eq. 14.5, we obtain the components of the control vector,

$$K_1 = \alpha_1 - a_1, \ldots, K_n = \alpha_n - a_n \tag{14.8}$$

Thus, we see that the difficult problem of solving a system of n algebraic equations has been greatly simplified by the introduction of the concept of canonical form.

14.1.1 Ackermann's Formula

In this section, we will derive a closed form expression for the control law, given by Eq. 14.8. This will be useful also in the general case of the system not being in controller canonical form. Suppose that for now, A and b are in controller canonical form. Then the characteristic equation of A is given by Eq. 14.7. As per the Cayley–Hamilton theorem [55], a matrix satisfies its own characteristic equation:

$$A^n + a_1 A^{n-1} + \cdots + a_{n-1} A + a_n = 0 \tag{14.9}$$

Replacing z by A in Eq. 14.5, we obtain

$$\alpha_c(A) = A^n + \alpha_1 A^{n-1} + \cdots + \alpha_{n-1} A + \alpha_n I \tag{14.10}$$

Solving for A^n from Eq. 14.9 and substituting in the above equation, we obtain

$$\alpha_c(A) = (\alpha_1 - a_1)A^{n-1} + \cdots + (\alpha_{n-1} - a_{n-1})A + (\alpha_n - a_n)I$$

We will now explore whether we can obtain expressions for powers of A. Note that the last row of A, given in Eq. 14.6, is the transpose of the $(n-1)$th unit vector. That is,

$$e_n^T A = e_{n-1}^T$$

where e_i is the ith unit vector. Continuing in this fashion, we obtain

$$\begin{aligned} e_n^T A^2 &= (e_n^T A)A = e_{n-1}^T A = e_{n-2}^T \\ &\vdots \\ e_n^T A^{n-1} &= e_1^T \end{aligned}$$

Although we want expressions for powers of A, we have obtained e_n times powers of A. In view of this, we premultiply Eq. 14.10 by e_n^T, and use the above expressions to obtain

$$e_n^T \alpha_c(A) = (\alpha_1 - a_1)e_1^T + \cdots + (\alpha_{n-1} - a_{n-1})e_{n-1}^T + (\alpha_n - a_n)e_n^T$$

Using the fact that e_i is the ith unit vector and that $\alpha_i - a_i = K_i$ from Eq. 14.8, we see that the right-hand side is nothing but K. Thus the pole placing controller is given by

$$K = e_n^T \alpha_c(A) = \text{last row of } \alpha_c(A) \tag{14.11}$$

where $\alpha_c(z)$ is the desired characteristic polynomial, given by Eq. 14.5, for the pole placement problem.

We have seen that we can obtain an explicit expression for K when the system is in controller canonical form. We take up the general case in the next section.

14.1.2 Control Law when System is not in Canonical Form

The next natural question that we have to answer is what happens if A and b are not in controller canonical form? We will address this question in this section.

We will first show that the characteristic polynomial is not changed by a similarity transformation. Let the state x be given as

$$x = M\overline{x} \tag{14.12}$$

where M is nonsingular and constant. We say that x and $\overline{x}$ are related by a similarity transformation. We will shortly see that we don't need to know the value of M, as it will be eliminated during the course of these calculations. Substituting in Eq. 14.1, we obtain

$$\begin{aligned} M\overline{x}(k+1) &= AM\overline{x}(k) + bu(k) \\ \overline{x}(k+1) &= M^{-1}AM\overline{x}(k) + M^{-1}bu(k) \end{aligned}$$

Defining new variables

$$\overline{A} \triangleq M^{-1}AM, \quad \overline{b} \triangleq M^{-1}b \tag{14.13}$$

the transformed state space equation becomes

$$\overline{x}(k+1) = \overline{A}\overline{x}(k) + \overline{b}u(k) \tag{14.14}$$

Now we consider using a state feedback controller of the form

$$u(k) = -\overline{K}\overline{x}(k) + v(k) \tag{14.15}$$

using the transformed variable $\overline{x}$. To emphasize the fact that we are working with transformed variables, we have a line over K as well. Using Eq. 14.12, this becomes

$$u(k) = -\overline{K}M^{-1}x(k) + v(k)$$

Comparing this with Eq. 14.3, we obtain

$$K = \overline{K}M^{-1} \tag{14.16}$$

Substituting the expression for the control law from Eq. 14.15 into Eq. 14.14, we obtain

$$\overline{x}(k+1) = \overline{A}\overline{x}(k) - \overline{b}\overline{K}\overline{x} + \overline{b}v(k) = (\overline{A} - \overline{b}\overline{K})\overline{x}(k) + \overline{b}v(k)$$

The characteristic polynomial of the closed loop, in the transformed coordinates, is

$$\overline{\phi}(z) = |zI - (\overline{A} - \overline{b}\overline{K})|$$

Substituting for $\overline{A}$, $\overline{b}$ and $\overline{K}$ from Eq. 14.13 and Eq. 14.16, we obtain

$$\begin{aligned}\overline{\phi}(z) &= \left|zM^{-1}M - \left(M^{-1}AM - M^{-1}bKM\right)\right| \\ &= \det\left\{M^{-1}\left[zI - (A - bK)\right]M\right\}\end{aligned}$$

Using the fact that the determinant of the product of square matrices is the product of the determinant of the corresponding matrices, we obtain

$$\overline{\phi}(z) = |zI - (A - bK)|$$

which is nothing but the characteristic polynomial in the original coordinates. Thus, it is clear that the similarity transformation does not change the characteristic polynomial of the closed loop system. The utility of the transformation matrix M is that $(\overline{A}, \overline{b})$ can be chosen to be in controller canonical form, provided the matrix

$$\mathcal{C} = \begin{bmatrix} b & Ab & \cdots & A^{n-1}b \end{bmatrix} \tag{14.17}$$

is nonsingular, which we will show next. The matrix $\mathcal{C}$ is known as the *controllability matrix* of the system.

There exists a nonsingular matrix M such that

$$M^{-1}AM = \begin{bmatrix} -a_1 & -a_2 & \cdots & -a_{n-2} & -a_{n-1} & -a_n \\ 1 & 0 & \cdots & 0 & 0 & 0 \\ 0 & 1 & \cdots & 0 & 0 & 0 \\ \vdots & & & & & \\ 0 & 0 & \cdots & 1 & 0 & 0 \\ 0 & 0 & \cdots & 0 & 1 & 0 \end{bmatrix}, \quad M^{-1}b = \begin{bmatrix} 1 \\ 0 \\ \vdots \\ 0 \\ 0 \\ 0 \end{bmatrix} \tag{14.18}$$

if and only if Rank $\mathcal{C} = n$. We show this fact as follows. Let the jth column of M be denoted as m_j, *i.e.*,

$$M = \begin{bmatrix} m_1 & m_2 & \cdots & m_n \end{bmatrix} \tag{14.19}$$

Note that the following equations

$$A \begin{bmatrix} m_1 & \cdots & m_n \end{bmatrix} = \begin{bmatrix} m_1 & \cdots & m_n \end{bmatrix} \begin{bmatrix} -a_1 & -a_2 & \cdots & -a_{n-2} & -a_{n-1} & -a_n \\ 1 & 0 & \cdots & 0 & 0 & 0 \\ 0 & 1 & \cdots & 0 & 0 & 0 \\ \vdots & & & & & \\ 0 & 0 & \cdots & 1 & 0 & 0 \\ 0 & 0 & \cdots & 0 & 1 & 0 \end{bmatrix}$$

$$b = \begin{bmatrix} m_1 & \cdots & m_n \end{bmatrix} \begin{bmatrix} 1 \\ 0 \\ \vdots \\ 0 \\ 0 \\ 0 \end{bmatrix}$$

are equivalent to Eq. 14.18 if M is nonsingular. Equating the columns of the first equation, we obtain

$$\begin{aligned} Am_1 &= -a_1 m_1 + m_2 \\ Am_2 &= -a_2 m_1 + m_3 \\ &\vdots \\ Am_{n-1} &= -a_{n-1} m_1 + m_n \end{aligned} \tag{14.20}$$

The reader may find the matrix manipulation techniques described in Sec. 1.4 of [55] useful at this point. On equating the columns of the b equation, we obtain

$$m_1 = b \tag{14.21}$$

On substituting this equation in Eq. 14.20, we obtain

$$\begin{aligned} m_2 &= Am_1 + a_1 m_1 = Ab + a_1 b \\ m_3 &= Am_2 + a_2 m_1 = A^2 b + a_1 Ab + a_2 b \\ &\vdots \\ m_n &= A^{n-1}b + a_1 A^{n-2} b + \cdots + a_{n-1} b \end{aligned} \tag{14.22}$$

Stacking the expressions for m_j from Eq. 14.21 and Eq. 14.22 side by side, we obtain

$$\begin{bmatrix} m_1 & m_2 & \cdots & m_n \end{bmatrix} = \begin{bmatrix} b & Ab & \cdots & A^{n-1}b \end{bmatrix} \begin{bmatrix} 1 & a_1 & a_2 & \cdots & a_{n-1} \\ & 1 & a_1 & \cdots & a_{n-2} \\ & & 1 & \cdots & a_{n-3} \\ & & & \ddots & \\ & & & & 1 \end{bmatrix}$$

which we can write as

$$M = \mathcal{C}U \tag{14.23}$$

where M and $\mathcal{C}$ are given by Eq. 14.17 and Eq. 14.19, respectively. As U is an upper triangular matrix with ones on the diagonal, it is nonsingular. It follows that M is nonsingular if and only if $\mathcal{C}$ is nonsingular.

The control law design involves the following steps. Transform the matrices to arrive at the controller canonical form. Derive the control law $\overline{K}$, in the transformed coordinate system, using Ackermann's formula, Eq. 14.11. Calculate the control law in the original coordinate system, using Eq. 14.16.

This procedure involves a lot of calculations. We will now come up with a simplified approach to address these issues. First, we will arrive at an expression for M. Using Eq. 14.13, we obtain

$$\begin{aligned} \overline{A}\overline{b} &= M^{-1}Ab \\ \overline{A}^2\overline{b} &= M^{-1}A^2 b \\ &\vdots \\ \overline{A}^{n-1}\overline{b} &= M^{-1}A^{n-1}b \end{aligned} \tag{14.24}$$

We now define the controllability matrix in the transformed coordinate system:

$$\overline{\mathcal{C}} = \begin{bmatrix} \overline{b} & \overline{A}\overline{b} & \cdots & \overline{A}^{n-1}\overline{b} \end{bmatrix} \tag{14.25}$$

Substituting for $\overline{b}$ from Eq. 14.13 and $\overline{A}^j\overline{b}$ from Eq. 14.24, we obtain

$$\overline{\mathcal{C}} = M^{-1} \begin{bmatrix} b & Ab & \cdots & A^{n-1}b \end{bmatrix} = M^{-1}\mathcal{C} \tag{14.26}$$

Therefore, $\mathcal{C}$ has full rank, if and only if $\overline{\mathcal{C}}$ is of full rank.

We are now ready to arrive at an explicit expression for K that obviates the need to follow the three step procedure outlined above. Because the transformed system is in controller canonical form, the control law given by Eq. 14.11 becomes $\overline{K} = e_n^T \alpha_c(\overline{A})$. Using Eq. 14.16 in this equation, the controller in the original coordinates becomes $K = e_n^T \alpha_c(\overline{A}) M^{-1}$. By direct calculation, it is easy to verify the relation $\alpha_c(\overline{A}) = M^{-1}\alpha_c(A)M$, using Eq. 14.13. With this, we obtain $K = e_n^T M^{-1} \alpha_c(A)$. Substituting for M^{-1} from Eq. 14.26, this becomes $K = e_n^T \overline{\mathcal{C}} \mathcal{C}^{-1} \alpha_c(A)$. From Problem 14.2, $e_n^T \overline{\mathcal{C}} = e_n^T$, because $\overline{\mathcal{C}}$ refers to the transformed coordinate system. Using this, we finally obtain

$$K = e_n^T \mathcal{C}^{-1} \alpha_c(A) \tag{14.27}$$

This is the explicit formula for computing the control law in the original coordinate system. We now illustrate these ideas with an example.

Example 14.3 Using the controller formulation derived above, move the poles of the system presented in Example 14.2 to $0.25 \pm 0.6614j$.

Let us begin the calculations with the controllability matrix:

$$\mathcal{C} = \begin{bmatrix} b & Ab \end{bmatrix} = \begin{bmatrix} 0 & 2 \\ 1 & 3 \end{bmatrix}$$

The rank of this matrix is 2 and hence it is invertible. As a result, we can proceed to determine the pole placement controller:

$$\mathcal{C}^{-1} = -\frac{1}{2}\begin{bmatrix} 3 & -2 \\ -1 & 0 \end{bmatrix}$$

$$\alpha_c = (z - (0.25 + j0.6614))(z - (0.25 - j0.6614)) = z^2 - 0.5z + 0.5$$

We next evaluate $\alpha_c(A)$ by substituting A in place of z:

$$\begin{aligned}
\alpha_c(A) &= \begin{bmatrix} 1 & 2 \\ 0 & 3 \end{bmatrix}\begin{bmatrix} 1 & 2 \\ 0 & 3 \end{bmatrix} - 0.5\begin{bmatrix} 1 & 2 \\ 0 & 3 \end{bmatrix} + 0.5\begin{bmatrix} 1 & 0 \\ 0 & 1 \end{bmatrix} \\
&= \begin{bmatrix} 1 & 8 \\ 0 & 9 \end{bmatrix} - \begin{bmatrix} 0.5 & 1 \\ 0 & 1.5 \end{bmatrix} + \begin{bmatrix} 0.5 & 0 \\ 0 & 0.5 \end{bmatrix} = \begin{bmatrix} 1 & 7 \\ 0 & 8 \end{bmatrix} \\
K &= e_n^T \mathcal{C}^{-1} \alpha_c(A) \\
&= -\frac{1}{2}\begin{bmatrix} 0 & 1 \end{bmatrix}\begin{bmatrix} 3 & -2 \\ -1 & 0 \end{bmatrix}\begin{bmatrix} 1 & 7 \\ 0 & 8 \end{bmatrix} = -\frac{1}{2}\begin{bmatrix} -1 & 0 \end{bmatrix}\begin{bmatrix} 1 & 7 \\ 0 & 8 \end{bmatrix} \\
&= \frac{1}{2}\begin{bmatrix} 1 & 7 \end{bmatrix} = \begin{bmatrix} 0.5 & 3.5 \end{bmatrix}
\end{aligned}$$

This result is in agreement with the K value used in Example 14.2. ∎

Although Eq. 14.27 uses the inverse of $\mathcal{C}^{-1}$, one does not calculate it in large problems. Instead, let $e_n^T \mathcal{C}^{-1} = w$. By solving $\mathcal{C}w = e_n^T$, w is determined. On substituting for $e_n^T \mathcal{C}^{-1}$ in Eq. 14.27, one gets $K = w\alpha_c(A)$, from which K can be determined.

We have seen that to achieve arbitrary control objectives, the controllability matrix $\mathcal{C}$ has to be nonsingular. We formalize this important concept in the next section.

14.1.3 Controllability

The system given by Eq. 14.1 is said to be *controllable* if it is possible to find an input sequence $\{u(m)\}$ that moves the system from an arbitrary initial state $x(0) = x_0$ to an arbitrary final state $x(j) = x_1$ for some finite j. We have shown in the previous section that this is possible if the controllability matrix $\mathcal{C}$ is nonsingular.

The next question we would like to ask is whether there is a minimum number of control moves that may have to be applied before we succeed in our objective of reaching a desired state. Consider the problem of x_1 being the origin, *i.e.*, we would like to drive the system to the origin in a finite number of moves. Let us start with the control law, given by Eq. 14.4:

$$x(k+1) = (A - bK)x(k) \tag{14.28}$$

where we have omitted v and x_0 for convenience. With the new variable A_c given by $A_c = A - bK$, Eq. 14.28 becomes

$$x(k+1) = A_c x(k) \tag{14.29}$$

which, when solved recursively, becomes

$$x(k) = A_c^k x(0) \tag{14.30}$$

Note that we wish to move the system to the origin as fast as possible. The fastest way to do this is by placing all the closed loop poles at the origin. As all the eigenvalues are zero, the characteristic equation of A_c is

$$\lambda^n = 0 \tag{14.31}$$

Using the Cayley–Hamilton theorem, we obtain $A_c^n = 0$. As we can multiply both sides by A_c without changing the relation, we obtain

$$A_c^k = 0, \quad k \geq n \tag{14.32}$$

As a result, from Eq. 14.30

$$x(k) = 0, \quad k \geq n \tag{14.33}$$

irrespective of the starting value $x(0)$. This control law is known as the *dead-beat control* law, compare with the definition in Sec. 9.1. Assuming that the plant response is not oscillatory in between sampling instants, the settling time of dead-beat control is nT_s.

Now we have an answer to the previous question: how small can j be? From the above discussion, we know that in n steps, we can guarantee that the state can be taken to any desired position.

We now present an example to control the inverted pendulum.

Example 14.4 Stabilize the inverted pendulum modelled by Eq. 2.19 on page 12 through the pole placement technique.

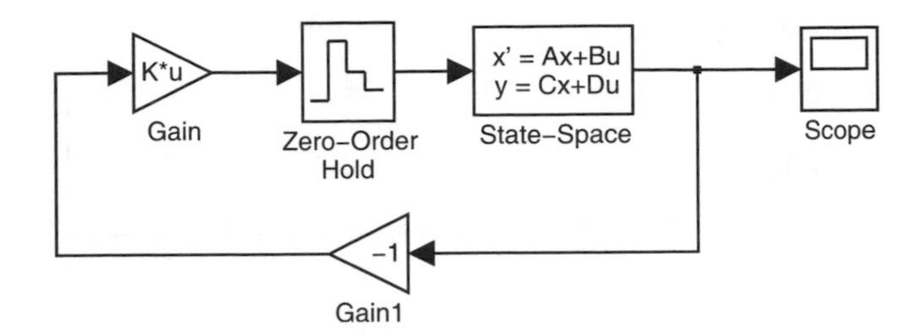

Figure 14.1: Simulink model to simulate inverted pendulum. Code is available at HOME/ss/matlab/pend_ss_c.mdl, see Footnote 1 on page 516.

As the only objective is to stabilize the system, we are free to choose the parameters. Let us design a controller to satisfy the following: $T_s = 0.01$ s, rise time = 0.05 s, $\varepsilon = 0.1$. Using the approach of the desired characteristic polynomial developed in Sec. 7.7, we arrive at

$$N = 0.05/T_s = 5$$
$$\omega = \frac{\pi}{2N}$$
$$r = \varepsilon^{\omega/\pi}$$

The above approach places only two poles. As we have four poles, we select the other two to be placed at $0.9r$.

The Matlab program pend_model.m, given in M 2.1, shows how the model of the inverted pendulum is arrived at. Execution of the program pend.m, given in M 14.1, in sequence, designs the controller. Fig. 14.1 shows the Simulink block diagram used to simulate the system. Fig. 14.2 shows the response of the system for the initial condition $0, 0.1, 0, 0$. ∎

To implement the controller in the above example, all four states are required. In practice, however, only the first two states are measured in such systems. These are the carriage position x_1 and the angle of the pendulum x_2, see Sec. 2.2.3. Calculation of x_3 and x_4 through approximate discretization (see Sec. 2.5.3) gives acceptable performance. We present more sophisticated methods of estimating the unmeasured states in the next section.

14.2 Estimators

In the previous section, we have designed a state feedback controller, assuming that all states are measured. We also know, however, that not all states can be measured, see Sec. 2.4.3 for a detailed discussion on this topic. Does this mean that the state feedback controllers cannot be implemented? The answer is that we can obtain a reasonably equivalent state feedback controller through *estimated* state measurements.

A *state estimator*, or simply an estimator, is a mathematical construct that helps estimate the states of a system. It can also be used to obtain smooth estimates of

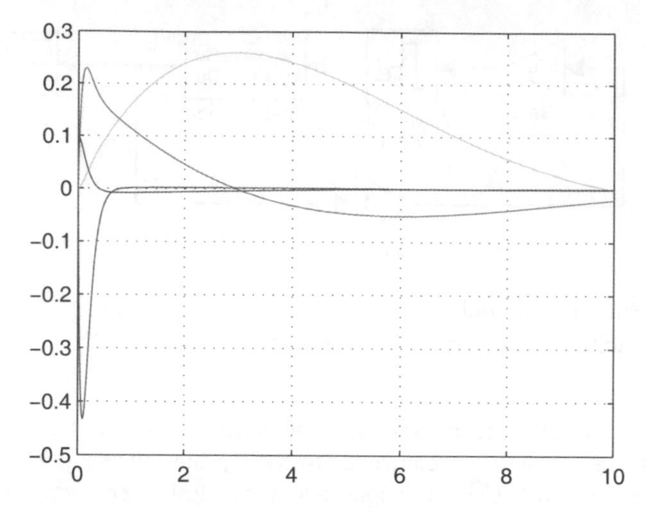

Figure 14.2: Stabilization of inverted pendulum. For initial conditions $(0, 0.1, 0, 0)$, system reaches zero state.

measurements that are corrupted by noise. In this section, we will consider estimators that work with deterministic systems. These are also known as Luenberger observers. Nevertheless, we will refer to them as estimators only, and follow the approach of [17] in this section. We will obtain two types of estimates for the state measurement, $x(k)$: 1. Prediction estimate $\overline{x}(k)$ that uses measurements up to $y(k-1)$. 2. Current estimate $\hat{x}(k)$ that uses measurements $y(k)$ up to and including the kth instant.

14.2.1 Prediction Estimators

As mentioned above, prediction estimators make use of measurements taken up to the sampling instant $k-1$ to estimate the value of $x(k)$. Given

$$\begin{aligned} x(k+1) &= Ax(k) + Bu(k) \\ y(k) &= Cx(k) \end{aligned} \tag{14.34}$$

where $y(k)$ is a scalar, the obvious choice for the estimator is to explore whether the plant model itself can be used for estimation. Thus, in the first instance, we explore whether the following model can be used as an estimator of the states:

$$\begin{aligned} \overline{x}(k+1) &= A\overline{x}(k) + Bu(k) \\ \overline{y}(k) &= C\overline{x}(k) \end{aligned} \tag{14.35}$$

where A, B and $u(k)$ are known. Note that if $x(0)$ and $\overline{x}(0)$ are the same, then $\overline{x}(k)$ will be the same as $x(k)$ for all $m > 0$. Thus, knowing $x(0)$ is equivalent to measuring all the states. Because in this section we assume that the states are not measured, we assume that we do not know $x(0)$. Thus, there is a difference in the values of x and its estimate $\overline{x}$. We define this as the error in estimation, $\widetilde{x}$, as

$$\widetilde{x} = \overline{x} - x \tag{14.36}$$

Subtracting Eq. 14.35 from Eq. 14.34, we obtain

$$\widetilde{x}(k+1) = A\widetilde{x}(k) \tag{14.37}$$

The solution to this equation is given by (see Eq. 3.46 on page 56)

$$\widetilde{x}(k) = A^k \widetilde{x}(0) \tag{14.38}$$

If we assume that we can diagonalize A, as in Sec. A.1.2, we obtain $A = S\Lambda S^{-1}$, $A^k = S\Lambda^k S^{-1}$, where Λ is a diagonal matrix, consisting of the eigenvalues of A. Substituting this in Eq. 14.37, we arrive at the following solution to Eq. 14.37:

$$\widetilde{x}(k) = S\Lambda^k S^{-1}\widetilde{x}(0) \tag{14.39}$$

Recall that we have $\overline{x}(0) \neq 0$. We would like to know whether $\widetilde{x}(k)$ goes to zero, at least, asymptotically. We see from the above equation that this will happen if and only if all eigenvalues of A are inside the unit circle. If even one eigenvalue is on or outside the unit circle, the error will not go to zero. In other words, if $\widetilde{x}(0) \neq 0$, the error will never decrease to zero if the system is unstable or marginally stable. For an asymptotically stable system, an initial error will decrease only because the plant and the estimate will both approach zero.

The main problem with the estimator suggested above is that it is an open loop system, see Eq. 14.38. We now explore what happens if we correct the estimate with the difference between the estimated output and the measured output. This leads us to explore the following:

$$\overline{x}(k+1) = A\overline{x}(k) + Bu(k) + L_p[y(k) - C\overline{x}(k)] \tag{14.40}$$

where L_p is the feedback gain matrix of $n \times 1$. We will now verify whether this is an acceptable estimator. Subtracting Eq. 14.34 from the above, we obtain

$$\widetilde{x}(k+1) = A\widetilde{x}(k) + L_p[y(k) - C\overline{x}(k)]$$

From the definition of y and $\widetilde{x}$, given in Eq. 14.34 and Eq. 14.36, respectively, we obtain

$$\widetilde{x}(k+1) = (A - L_pC)\widetilde{x}(k) \tag{14.41}$$

If this system is asymptotically stable, $\widetilde{x}(k) \rightarrow 0$ as $m \rightarrow \infty$, thus $\overline{x} \rightarrow x$. We will shortly see that it is generally possible to choose L_p so as to make the above error dynamics well behaved: not only will the system be asymptotically stable, but it will also respond with an acceptable speed. Thus, this estimator is better than the previous one.

We will refer to this as the prediction estimator because the measurement at time m results in an estimate of the state at time $m+1$, *i.e.*, the estimate is predicted one cycle in the future.

We will now explore the options for choosing the dynamics of the estimator, decided by the eigenvalues of $A - L_p$, see Eq. 14.41. In other words, we would like to choose L_p in such a way that the roots of the estimator characteristic polynomial

$$\alpha_e(z) = (z - \beta_1)(z - \beta_2)\cdots(z - \beta_n) \tag{14.42}$$

are in desired locations. Here, the β are the desired estimator root locations and represent how fast the estimator state converges towards the plant state. We would want this to be equal to $|zI - A + L_pC|$. Equality results if the coefficient of each power of z is the same in both expressions. Thus, once again, we find ourselves in a situation of solving a system of n equations, where n is the dimension of the state vector. As in the pole placement problem, we use a different strategy to determine L_p.

Note that we want the eigenvalues of $(A-L_pC)$ to be in some desired locations. We know how to assign the eigenvalues of $(A-BK)$ by suitably choosing K. The difference between these two problems is that the positions of L_p and K are interchanged. But this problem is easily solved, because the eigenvalues of a square matrix and its transpose are identical. If we take a transpose of $(A - L_pC)$, we obtain $(A^T - C^T L_p^T)$ which is in the same form as $(A - BK)$, with the unknown variables coming exactly at the same place. Making appropriate changes in Eq. 14.27, we obtain

$$L_p^T = e_n^T \begin{bmatrix} C^T & A^T C^T & \cdots & A^{T^{n-1}} C^T \end{bmatrix}^{-1} \alpha_e(A^T)$$

Taking the transpose of both sides, we arrive at

$$L_p = \alpha_e(A) \begin{bmatrix} C \\ CA \\ \vdots \\ CA^{n-1} \end{bmatrix}^{-1} e_n \tag{14.43}$$

where we have used the fact $(P^{-1})^T = (P^T)^{-1}$ and the transpose rule for the powers of transposed matrices. For example, $\left(A^{T^2}\right)^T = (A^T A^T)^T = AA = A^2$. Next, we will illustrate these ideas with an example.

Example 14.5 Suppose that in the system presented in Example 14.2 and in Example 14.3, only the first state is measured. Determine an estimator with poles at $0.1 \pm 0.1j$.

We begin with the calculation of the observability matrix

$$\mathcal{O} = \begin{bmatrix} c \\ cA \end{bmatrix} = \begin{bmatrix} 1 & 0 \\ 1 & 2 \end{bmatrix}$$

As this matrix is invertible, we can proceed to determine the estimator

$$\mathcal{O}^{-1} = \frac{1}{2} \begin{bmatrix} 2 & 0 \\ -1 & 1 \end{bmatrix}$$

$$\alpha_e(z) = (z - (0.1 + j0.1))(z - (0.1 - j0.1)) = z^2 - 0.2z + 0.02$$

We next evaluate $\alpha_e(A)$ by substituting A in place of z:

$$\alpha_e(A) = \begin{bmatrix} 1 & 2 \\ 0 & 3 \end{bmatrix} \begin{bmatrix} 1 & 2 \\ 0 & 3 \end{bmatrix} - 0.2 \begin{bmatrix} 1 & 2 \\ 0 & 3 \end{bmatrix} + 0.02 \begin{bmatrix} 1 & 0 \\ 0 & 1 \end{bmatrix} = \begin{bmatrix} 0.82 & 7.6 \\ 0 & 8.42 \end{bmatrix}$$

Using the expression given in Eq. 14.43, we calculate L_p:

$$L_p = \begin{bmatrix} 3.8 \\ 4.21 \end{bmatrix}$$

The estimator is given by Eq. 14.40. ∎

The concept of the matrix $\mathcal{O}$ being nonsingular is so important that we formalize this idea in the next section.

14.2.2 Observability

We have seen in the previous section that if the matrix $\mathcal{O}$ is invertible, we can make the estimate of the states converge to the actual value in a finite number of steps. We will use this idea to formally define the concept of observability. Suppose that not all states of a system are measured. The system is said to be observable if it is possible to reconstruct the states in an arbitrarily finite number of sampling periods. We formalize observability as follows:

$$\begin{aligned} x(k+1) &= Ax(k) + Bu(k) \\ y(k) &= Cx(k) \end{aligned}$$

is said to be *observable* if and only if the *observability matrix*

$$\mathcal{O} = \begin{bmatrix} C \\ CA \\ \vdots \\ CA^{n-1} \end{bmatrix}$$

is invertible.

14.2.3 Current Estimators

The prediction estimator uses the measurements up to the previous time instant to provide an estimate of the state at the current instant. This procedure should be followed only if it takes one sample time to carry out the calculations involved in estimation. Modern devices, however, can complete these computations in a negligibly short time. In view of this, we explore the possibility of using the current measurement as well to obtain an estimate of the state. Following the approach of [17], we define the current estimate, denoted by $\widehat{x}(k)$, as an update of the prediction estimate, $\overline{x}(k)$, as follows:

$$\widehat{x}(k) = \overline{x}(k) + L_c(y(k) - C\overline{x}(k)) \tag{14.44}$$

Note that we update the prediction estimate $\overline{x}(k)$ by a constant (L_c) multiplied by the latest error in output prediction. Shortly, we will present a methodology to compute L_c. We propose that the prediction estimate $\overline{x}(k)$ be obtained from the state equation of the plant, as stated next:

$$\overline{x}(k) = A\widehat{x}(k-1) + Bu(k-1) \tag{14.45}$$

Thus, the current estimate $\widehat{x}(k)$ is obtained in two stages, starting from the previous time instant, namely $\widehat{x}(k-1)$. Substituting Eq. 14.44 into Eq. 14.45, we obtain

$$\overline{x}(k+1) = A[\overline{x}(k) + L_c(y(k) - C\overline{x}(k))] + Bu(k)$$

Using the measurement equation $y(k) = Cx(k)$, we obtain

$$\overline{x}(k+1) = A\overline{x}(k) + Bu(k) + AL_cC(x(k) - \overline{x}(k)) \tag{14.46}$$

Subtracting the state equation $x(k+1) = Ax(k) + Bu(k)$ from the above and defining

$$\widetilde{x}(k) = \overline{x}(k) - x(k) \tag{14.47}$$

as earlier, we obtain

$$\widetilde{x}(k+1) = (A - AL_cC)\widetilde{x}(k) \tag{14.48}$$

Comparing this with Eq. 14.41, we find that

$$L_p = AL_c \tag{14.49}$$

Because it is equal to e^{FT_s} in sampled data systems, A is always invertible. We obtain

$$L_c = A^{-1}L_p \tag{14.50}$$

As mentioned earlier, the current estimate of x is obtained in two stages, using Eq. 14.44–14.45. In order to start the calculations, we need $\widehat{x}(0)$, which has to be guessed as the initial state also is not available, as discussed in Sec. 14.2.1.

14.3 Regulator Design – Combined Control Law and Estimator

In this section, we will combine the state feedback controller and the state estimator to arrive at the overall controller design. First we state the separation principle. Using this, we design the controller and the estimator, separately. We next combine them and produce the overall controller, which we will refer to as the *compensator*. We conclude this section with an example.

First we state the important result known as the *separation principle*. The state feedback controller can work only if all the states are available. Thus, while designing the controller, we assume that the states are estimated by an estimator. Next, we look at the assumptions involved in the use of the estimator. It is possible to estimate the states only if the states are bounded. Thus, we assume the stabilizing influence of the controller to keep the states bounded. Thus, the controller assumes the existence of the estimator and vice versa. This separation principle allows us to design the controller and the estimator separately, obviating the need to design both of them simultaneously. Without this result, state feedback controllers would have become a lot more difficult to design.

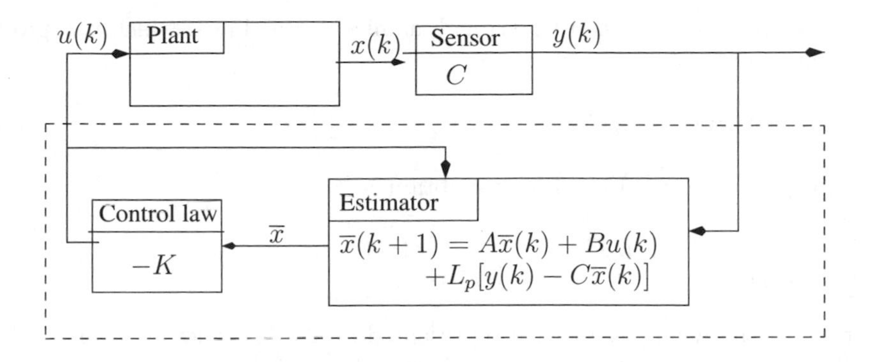

Figure 14.3: Block diagram of the state feedback controller and estimator combination. The blocks within in the dashed lines constitute the compensator, which is similar to the conventional output feedback controller.

The closed loop control configuration is as given in Fig. 14.3. The plant is described by A and b, but not all the states are measured; only the output y is available. We assume that a prediction estimator is used to estimate x. The input u that goes into the plant is sent also to the estimator. The plant output becomes another input to the estimator. Using these, the prediction estimator gives an estimate of the state vector, denoted by $\overline{x}$. The control law works on the basis of $\overline{x}$ and produces the control effort u. Fig. 14.3 shows a schematic of this sequence.

In this figure, we have shown the blocks that implement the controller and the estimator inside dashed lines. As mentioned above, we will refer to this combined block as the compensator. Plant output y is an input to and control effort u is the output from the compensator, which, from the input–output viewpoint, is similar to the traditional output feedback controller. We now proceed to derive the transfer function of this compensator.

Recall that in Eq. 14.3 we have proposed the use of the control law in the form $u = -Kx + v$. Taking $v = 0$ and using $\overline{x}$ in place of x, it becomes

$$u(k) = -K\overline{x}(k) \tag{14.51}$$

Closing the feedback loop is equivalent to substituting for u in the state space equation

$$x(k+1) = Ax(k) + Bu(k) \tag{14.52}$$

Substituting for u as $-K\overline{x}$, we obtain

$$x(k+1) = Ax(k) - BK\overline{x}(k) \tag{14.53}$$

Recall that we have defined the error between the estimated state and the actual state as $\widetilde{x}$:

$$\widetilde{x} = \overline{x} - x \tag{14.54}$$

Substituting the expression for $\overline{x}$ from this equation into Eq. 14.53, we obtain

$$x(k+1) = Ax(k) - BK[x(k) + \widetilde{x}(k)] \tag{14.55}$$

We have derived an expression for error dynamics in Eq. 14.41, which is reproduced here for convenience:

$$\widetilde{x}(k+1) = (A - L_pC)\widetilde{x}(k) \tag{14.56}$$

Stacking Eq. 14.55 below Eq. 14.56, we obtain

$$\begin{bmatrix} \widetilde{x}(k+1) \\ x(k+1) \end{bmatrix} = \begin{bmatrix} A - L_pC & 0 \\ -BK & A - BK \end{bmatrix} \begin{bmatrix} \widetilde{x}(k) \\ x(k) \end{bmatrix} \tag{14.57}$$

Thus, the system dynamics are now represented by $2n$ difference equations, where n is the dimension of the state vector. This is clear from Fig. 14.3 as well. If we measure all the states, however, we need to use only n difference equations to describe the dynamics. Thus, one of the tradeoffs in not measuring the states of the system is that we have to deal with a larger dimensional system.

Next, we will verify that the poles of this augmented system are indeed at the locations we want them. The characteristic equation of Eq. 14.57 is

$$\begin{vmatrix} zI - A + L_pC & 0 \\ BK & zI - A + BK \end{vmatrix} = 0$$

which, because of the zero matrix in the upper right, can be written as

$$|zI - A + L_pC||zI - A + BK| = 0 \tag{14.58}$$

In Eq. 14.42 and in Eq. 14.4, we have defined these as $\alpha_e(z)$ and $\alpha_c(z)$, respectively. We obtain

$$\alpha_e(z)\alpha_c(z) = 0 \tag{14.59}$$

This shows that the poles of the $2n$-dimensional dynamical equations that describe the feedback system are at precisely the locations where we have placed our state feedback controller and the estimator poles.

We now proceed to determine the transfer function of the compensator, given within the dashed lines in Fig. 14.3. The states of the estimator are described by

$$\overline{x}(k+1) = A\overline{x}(k) + Bu(k) + L_p[y(k) - C\overline{x}(k)]$$

Substituting for the control law, from Eq. 14.51, we obtain

$$\overline{x}(k+1) = A\overline{x}(k) - BK\overline{x}(k) + L_p[y(k) - C\overline{x}(k)]$$

Simplifying this, we obtain the state equation for the controller as

$$\overline{x}(k+1) = (A - BK - L_pC)\overline{x}(k) + L_py(k) \tag{14.60}$$

The input to this compensator is y and the output from it is the control effort, u. The output equation is given by Eq. 14.51, reproduced here:

$$u(k) = -K\overline{x}(k)$$

The transfer function of the compensator is given by

$$G_c(z) = \frac{U(z)}{Y(z)} = -K[zI - A + BK + L_pC]^{-1}L_p \tag{14.61}$$

Next, we will discuss the stability of this compensator. The poles of the compensator satisfy the equation

$$|zI - A + BK + L_pC| = 0 \tag{14.62}$$

Recall that we have designed the state feedback controller such that the roots of $|zI - A + BK|$ and $|zI - A + L_pC|$ are at the required locations. But now, we have a different characteristic polynomial altogether: $|zI - A + BK + L_pC|$. Where will its roots lie? The answer is that there is no guarantee that these will lie inside the unit circle, let alone the locations chosen for controller and estimator poles. In other words, the compensator, by itself, could be unstable. This is another price we pay for not measuring all the states. Note that the original state feedback controller is of constant gain, K, which is stable. It should be emphasized that although the compensator, by itself, could be unstable, the closed loop system is stable, as indicated by its characteristic polynomial, given in Eq. 14.59.

We will now revert to the question of where we should choose the poles of the regulator and the estimator. When measurement noise is not an issue, it is convenient to pick the control roots to satisfy the performance specifications and actuator limitations, and then to pick the estimator roots somewhat faster, say, by a factor of 2 to 4, indicated by the rise time, so that the total response is dominated by the response due to the slower control poles. We will illustrate these ideas with an example.

Example 14.6 Determine the compensator for the system presented in Example 14.5. Use the control law calculated in Example 14.3.

$$A = \begin{bmatrix} 1 & 2 \\ 0 & 3 \end{bmatrix}, \quad b = \begin{bmatrix} 0 \\ 1 \end{bmatrix}, \quad c = \begin{bmatrix} 1 \\ 0 \end{bmatrix}$$

$$K = \begin{bmatrix} 0.5 & 3.5 \end{bmatrix}, \quad L_p = \begin{bmatrix} 3.8 \\ 4.21 \end{bmatrix}$$

Substituting in the controller expression

$$G_c = -K[zI - (A - bK - L_pc)]^{-1}L_p$$

$$A - bK - L_pc = \begin{bmatrix} -2.8 & 2 \\ -4.71 & -0.5 \end{bmatrix}$$

$$[zI - (A - bK - L_pc)]^{-1} = \begin{bmatrix} z + 2.8 & -2 \\ 4.71 & z + 0.5 \end{bmatrix}^{-1}$$

$$= -\frac{1}{z^2 + 3.3z + 10.82} \begin{bmatrix} z + 0.5 & 2 \\ -4.71 & z + 2.8 \end{bmatrix}$$

$$G_c = -\frac{16.635z - 16.225}{z^2 + 3.3z + 10.82}$$

What we have derived above is the transfer function of the compensator. Note that its poles are outside the unit circle. M 14.2 shows some of the steps in the calculation of the above.

∎

14.4 Linear Quadratic Regulator

One of the difficulties in designing the pole placement controllers is the selection of closed loop pole locations, especially in a large dimensional state space system. An alternative approach is to use an optimization based strategy to design the controller. Minimization of the square of an error for a linear system results in a linear quadratic regulator, abbreviated as LQR.

The LQR design approach involves the minimization of a weighted sum of quadratic functions of the state and the control effort. Although it can handle tracking problems, the LQR approach is generally applied mainly for disturbance rejection. In view of this, the final objective of an LQR is to take the state vector to zero state. Because the states are usually deviation variables with respect to a steady state, driving the state to zero is equivalent to guiding the system to its steady state.

If we minimize a quadratic function of only the states, the control effort could become unbounded. For example, we could come up with control strategies, such as, *input 1 million watts for 1 picosecond to raise the temperature of a beaker by a small amount.* Because there is no restriction on the control effort, the optimization of states alone would result in such comic solutions. To overcome these difficulties, we minimize a quadratic function of the control effort as well. We will discuss this approach in detail in the next section.

14.4.1 Formulation of Optimal Control Problem

In this section, we formulate the optimal control problem. There are many approaches to this problem. The most popular ones use dynamic programming [2] and the Lagrange multiplier method [17]. We will make use of the latter in this section. Given a discrete time plant

$$x(k+1) = Ax(k) + Bu(k) \tag{14.63}$$
$$x(0) = x_0 \tag{14.64}$$

we wish to find a control law $u(k)$ so that a cost function

$$J = \frac{1}{2}\sum_{k=0}^{N}\left[x^T(k)Q_1x(k) + u^T(k)Q_2u(k)\right] \tag{14.65}$$

is minimized. Q_1 and Q_2 are symmetric weighting matrices to be selected by the designer. As some of the states may be allowed to be zero, we obtain the condition

$$x^TQ_1x \geq 0, \ \forall x \tag{14.66}$$

We obtain a different condition for Q_2 as all control efforts have a cost associated with them:

$$u^TQ_2u > 0, \ \forall u \neq 0 \tag{14.67}$$

We can restate the problem as: minimize J given by Eq. 14.65, subject to

$$-x(k+1) + Ax(k) + Bu(k) = 0, \quad k = 0, 1, \ldots, N \tag{14.68}$$

which is the same as Eq. 14.63. We can solve this problem using the Lagrange multiplier method. There will be one Lagrange multiplier vector, denoted by $\lambda(k+1)$, for every m. We arrive at the new optimal index:

$$J' = \frac{1}{2}\sum_{k=0}^{N}\left[x^T(k)Q_1x(k) + u^T(k)Q_2u(k) + \lambda^T(k+1)[-x(k+1) + Ax(k) + Bu(k)]\right] \tag{14.69}$$

As J has to reach a minimum with respect to $x(k)$, $u(k)$ and $\lambda(k)$, we obtain

$$\frac{\partial J'}{\partial u(k)} = u^T(k)Q_2 + \lambda^T(k+1)B = 0 \quad \text{(control equation)} \tag{14.70}$$

$$\frac{\partial J'}{\partial \lambda(k+1)} = -x(k+1) + Ax(k) + Bu(k) = 0 \quad \text{(state equation)} \tag{14.71}$$

$$\frac{\partial J'}{\partial x(k)} = x^T(k)Q_1 - \lambda^T(k) + \lambda^T(k+1)A = 0 \quad \text{(adjoint equation)} \tag{14.72}$$

Notice that although $x(0)$ is given, $\lambda(0)$ is not. As a result, it is not possible to recursively solve the above equations forward. This forces us to look for alternative means. Notice that we require

$$u(N) = 0 \tag{14.73}$$

because, otherwise, optimal index J will not be zero in Eq. 14.65, as $u(N)$ affects only $x(N+1)$ but not $x(N)$, see Eq. 14.63. If $u(N)$ were to affect $x(N)$, perhaps there could be a way to cancel a nonzero $u(N)$ with an appropriate $x(N)$. Thus to make J small, $u(N)$ has to be zero. The following value for λ satisfies Eq. 14.70 when m takes the value of N:

$$\lambda(N+1) = 0 \tag{14.74}$$

Substituting this in the adjoint equation, Eq. 14.72, we obtain the condition

$$\lambda(N) = Q_1x(N) \tag{14.75}$$

This gives a condition for λ, except that it is available only at $m = N$, but not at $m = 0$. This results in a two point boundary value problem, the solution of which is presented next.

14.4.2 Solution to Optimal Control Problem

In the previous section, we posed the optimal control problem as a two point boundary value problem. We present a solution to this in this section. We conclude this section by stating the steady state version of this solution.

Motivated by Eq. 14.75, where λ appears as a linear function of x at N, we explore the possibility of linearity at all sampling instants [6]:

$$\lambda(k) = S(k)x(k) \tag{14.76}$$

so that the control equation, Eq. 14.70, becomes

$$Q_2u(k) + B^T\lambda(k+1) = 0 \tag{14.77}$$

Substituting for λ from Eq. 14.76, we obtain

$$Q_2u(k) = -B^TS(k+1)x(k+1)$$

Substituting for $x(k+1)$ from Eq. 14.63 and simplifying, we obtain the following equations:

$$\begin{aligned} Q_2u(k) &= -B^TS(k+1)(Ax(k) + Bu(k)) \\ [Q_2 + B^TS(k+1)B]u(k) &= -B^TS(k+1)Ax(k) \end{aligned}$$

Thus, we arrive at the following expression for the control law:

$$u(k) = -[Q_2 + B^TS(k+1)B]^{-1}B^TS(k+1)Ax(k) \tag{14.78}$$

Defining

$$R = Q_2 + B^TS(k+1)B \tag{14.79}$$

the above equation becomes

$$u(k) = -R^{-1}B^TS(k+1)Ax(k) \tag{14.80}$$

The adjoint equation, Eq. 14.72, can be rewritten as

$$\lambda(k) = A^T\lambda(k+1) + Q_1x(k) \tag{14.81}$$

which, on substitution of λ from Eq. 14.76 and simplification, becomes

$$S(k)x(k) = A^TS(k+1)x(k+1) + Q_1x(k)$$

Substituting for $x(k+1)$ from the state equation, we obtain

$$S(k)x(k) = A^TS(k+1)[Ax(k) + Bu(k)] + Q_1x(k)$$

which, on substitution of u from Eq. 14.80, becomes

$$S(k)x(k) = A^TS(k+1)[Ax(k) - BR^{-1}B^TS(k+1)Ax(k)] + Q_1x(k)$$

Collecting terms to one side, we obtain

$$\left[S(k) - A^TS(k+1)A + A^TS(k+1)BR^{-1}B^TS(k+1)A - Q_1\right]x(k) = 0$$

As this must hold for all $x(k)$, we obtain

$$S(k) - A^TS(k+1)A + A^TS(k+1)BR^{-1}B^TS(k+1)A - Q_1 = 0$$

This can be rewritten as

$$S(k) = A^T[S(k+1) - S(k+1)BR^{-1}B^T S(k+1)]A + Q_1 \tag{14.82}$$

This is known as the discrete time Riccati equation. Defining

$$M(k+1) = [S(k+1) - S(k+1)BR^{-1}B^T S(k+1)] \tag{14.83}$$

the above equation becomes

$$S(k) = A^T M(k+1)A + Q_1 \tag{14.84}$$

Using the expression for R from Eq. 14.79, the expression for M given in Eq. 14.83 becomes

$$M(k+1) = [S(k+1) - S(k+1)B[Q_2 + B^T S(k+1)B]^{-1}B^T S(k+1)] \tag{14.85}$$

From Eq. 14.75 and 14.76, we obtain

$$S(N) = Q_1 \tag{14.86}$$

We can calculate the controller backward in time:

$$u(k) = -K(k)x(k) \tag{14.87}$$

where from Eq. 14.78, we obtain the following relation for $K(k)$

$$K(k) = -[Q_2 + B^T S(k+1)B]^{-1}B^T S(k+1)A \tag{14.88}$$

We now summarize the control design procedure.

1: $S(N) = Q_1$, $K(N) = 0$, $m = N$
2: **repeat**
3: $\quad M(k) \leftarrow [S(k) - S(k)B[Q_2 + B^T S(k)B]^{-1}B^T S(k)]$
4: $\quad K(k-1) \leftarrow -[Q_2 + B^T S(k)B]^{-1}B^T S(k)A$
5: $\quad$ Store $K(k-1)$
6: $\quad S(k-1) \leftarrow A^T M(k)A + Q_1$
7: $\quad m \leftarrow k-1$
8: **until** $m = 1$

This algorithm calculates controller gains $K(k)$ for all m and stores them once and for all. Using these previously calculated control moves, one recursively calculates forward the state vector starting from Eq. 14.64, 14.63 and 14.87.

The optimal value of the objective function can be calculated easily. First substitute in Eq. 14.69 the expression for $\lambda^T(k+1)A$ from the adjoint equation,

Eq. 14.72, and for $\lambda^T(k+1)B$ from Eq. 14.70, to arrive at

$$\begin{aligned} J' &= \frac{1}{2}\sum_{k=0}^{N}\left[x^T(k)Q_1x(k) + u^T(k)Q_2u(k)\right. \\ &\quad \left. - \lambda^T(k+1)x(k+1) + (\lambda^T(k) - x^T(k)Q_1)x(k) - u^T(k)Q_2u(k)\right] \\ &= \frac{1}{2}\sum_{k=0}^{N}[\lambda^T(k)x(k) - \lambda^T(k+1)x(k+1)] \\ &= \frac{1}{2}\lambda^T(0)x(0) - \lambda^T(N+1)x(N+1) \end{aligned}$$

But as $\lambda(N+1) = 0$ (see Eq. 14.74), we arrive at

$$J' = J = \frac{1}{2}\lambda^T(0)x(0) = \frac{1}{2}x^T(0)S(0)x(0) \tag{14.89}$$

This equation states that the best value for the optimization index to be minimized is a quadratic function of the initial state vector.

14.4.3 Infinite Horizon Solution to LQR Design

The control law K presented in the last section is time varying. We will be interested in a steady state solution, however. The reasons are that a constant solution is easier to implement. Moreover, even in the time varying solution, the control law could be constant during most of the time. When the objective function is a sum of an infinite number of terms, known as the infinite time problem, the steady state solution is the optimal solution.

Because we are interested in a steady state solution to the optimal control problem, we look for a steady state solution to the Riccati equation, given in Eq. 14.82, and reproduced here for convenience:

$$S(k) = A^T[S(k+1) - S(k+1)BR^{-1}B^TS(k+1)]A + Q_1 \tag{14.90}$$

Because we are interested in a steady state solution, let

$$S(k) = S(k+1) = S_\infty \tag{14.91}$$

Substituting this in Eq. 14.90, we obtain

$$S_\infty = A^T[S_\infty - S_\infty BR^{-1}B^TS_\infty]A + Q_1 \tag{14.92}$$

Although we can solve this by iteration, it is not an easy problem. But there is an alternative, easier, approach to solve this equation. We will only state the solution here. The interested reader can refer to [17] for details.

Construct the control Hamiltonian matrix, H_c, given by

$$H_c = \begin{bmatrix} A + BQ_2^{-1}B^TA^{-T}Q_1 & -BQ_2^{-1}B^TA^{-T} \\ -A^{-T}Q_1 & A^{-T} \end{bmatrix} \tag{14.93}$$

Then, the steady state solution to the Riccati equation is given by

$$S_\infty = \Lambda_I X_I^{-1} \tag{14.94}$$

where $\begin{bmatrix} X_I & \Lambda_I \end{bmatrix}^T$ is the eigenvector of H_c corresponding to the stable eigenvalues. The steady state control law is given by

$$u(k) = -K_\infty x(k) \tag{14.95}$$

where

$$K_\infty = (Q_2 + B^T S_\infty B)^{-1} B^T S_\infty A \tag{14.96}$$

The performance index for the steady state control problem is

$$J_\infty = \frac{1}{2} x^T(0) S_\infty x(0) \tag{14.97}$$

Notice that this control law K_∞ can be calculated once and for all at the very beginning.

14.5 Kalman Filter

In Sec. 14.2, we presented deterministic estimation techniques. In this section, we will assume that the state and measurement equations are corrupted by noise. That is, the model of the plant is now described by

$$\begin{aligned} x(k+1) &= Ax(k) + Bu(k) + B_1 w(k) \\ y(k) &= Cx(k) + v(k) \end{aligned} \tag{14.98}$$

where $w(k)$ and $v(k)$ denote noise sequences that affect the states and the measurements, respectively. We will assume that w and v are zero mean white noise sequences, satisfying the following:

$$\begin{aligned} \mathscr{E}\left[w(k)\right] &= 0 \\ \mathscr{E}\left[v(k)\right] &= 0 \\ \mathscr{E}\left[w(j)w(k)^T\right] &= 0, \quad \text{if } j \neq k \\ \mathscr{E}\left[v(j)v(k)^T\right] &= 0, \quad \text{if } j \neq k \end{aligned} \tag{14.99}$$

Their covariances are assumed as follows:

$$\begin{aligned} \mathscr{E}\left[w(k)w(k)^T\right] &= R_w \\ \mathscr{E}\left[v(k)v(k)^T\right] &= R_v \end{aligned} \tag{14.100}$$

Based on the reasoning given while explaining deterministic estimators, we look for a current estimator of the following form (compare with Eq. 14.44–14.45):

$$\begin{aligned} \widehat{x}(k) &= \overline{x}(k) + L(y(k) - C\overline{x}(k)) \\ \overline{x}(k) &= A\widehat{x}(k-1) + Bu(k-1) \end{aligned} \tag{14.101}$$

Note that we have used L to denote the estimator gain, instead of L_c. This is because we would like choose L so as to make $\widehat{x}$ optimal, in the sense that the error covariance,

$$J_{\widehat{x}} = \mathscr{E}\left[(x(k) - \widehat{x}(k))^T(x(k) - \widehat{x}(k))\right] \tag{14.102}$$

is minimized. The optimal estimator that satisfies these conditions is known as the Kalman filter.

We will now state the procedure to compute the Kalman filter. We define the covariance matrix $P(k)$ as follows:

$$P(k) = \mathscr{E}\left[(x(k) - \widehat{x}(k))(x(k) - \widehat{x}(k))^T\right] \tag{14.103}$$

First, we have to guess the value of $\widehat{x}(0)$ and $P(0)$ while designing the Kalman filter. Next, we denote by M the propagated value of P, but before making use of the next measurement, satisfying the following relation:

$$M(k+1) = AP(k)A^T + B_1R_wB_1^T \tag{14.104}$$

Note that we can calculate $M(1)$ using this equation. Next, we calculate the prediction estimate at the sampling instant $m+1$, as follows:

$$\overline{x}(k+1) = A\widehat{x}(k) + Bu(k) \tag{14.105}$$

Given $\widehat{x}(0)$, we can calculate $\overline{x}(1)$ using this relation. Next, we update the error covariance at the current instant:

$$P(k) = M(k) - M(k)C^T\left[CM(k)C^T + R_v\right]^{-1}CM(k) \tag{14.106}$$

Using this, we can calculate $P(1)$. Finally, we determine the current estimator, the measurement update, as follows:

$$\widehat{x}(k) = \overline{x}(k) + P(k)C^TR_v^{-1}(y(k) - C\overline{x}(k)) \tag{14.107}$$

Using this relation, we can calculate $\widehat{x}(1)$. Note that we have used $P(k)C^TR_v^{-1}$ in place of L in Eq. 14.101, as these are equal. We have indicated how to update all the values to the current time instant, given the values at the previous instant. Thus, it is possible to march forward and calculate all the future values. This procedure is summarized in Fig. 14.4.

As in the deterministic case, the steady state version of the Kalman filter is popular. Just like the deterministic estimators presented in Sec. 14.2, the Kalman filter also takes the plant input $u(k)$ and the output $y(k)$ to give an estimate of $x(k)$. In the former, because there is no noise, there is no question of filtering it. In the stochastic case, however, the Kalman filter gives a filtered estimate of the state vector.

We will end the discussion on the Kalman filter by presenting a simple example, taken from [19].

Example 14.7 (Estimating a constant) Using a Kalman filter, obtain a smooth estimate of a constant, whose measurement is corrupted by noise.

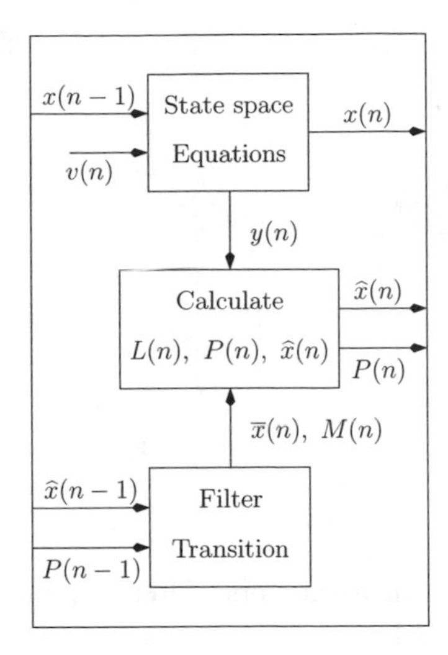

Figure 14.4: Kalman filter calculation at time instant n

This system can be represented by the following state space equations:

$$\begin{aligned} x(n+1) &= x(n) \\ y(n) &= x(n) + v(n) \end{aligned}$$

We have $A = 1$, $B = 0$, $C = 1$, $R_w = 0$ in the notation of this section. We arrive at the following equations:

$$\begin{aligned} \overline{x}(n) &= \widehat{x}(n-1) \\ M(n+1) &= P(n) \\ \overline{x}(n+1) &= \widehat{x}(n) \\ L(n) &= \frac{P(n)}{R_v} \\ \widehat{x}(n) &= \overline{x}(n) + L(n)(y(k) - C\overline{x}(n)) \end{aligned}$$

Execution of M 14.3 generates Fig. 14.5, in which the left hand diagram contains the profiles of the response of the filter, the noisy measurement and the constant. It can be seen that the estimate provided by the filter converges to the constant value of 5, in the presence of a large noise. In the right hand diagram of Fig. 14.5, it can be seen that the profile of $\mathscr{E}\left[(\widehat{x}(n) - x(n))^2\right]$ converges to zero, confirming the fact that the estimate is converging to the actual value of the state. ∎

Recall from Sec. 14.3 the procedure for combining the state feedback controller and the estimator to build the compensator. In a similar way, we can build a compensator combining the LQR and the Kalman filter. This compensator is known as the linear quadratic Gaussian (LQG) controller.

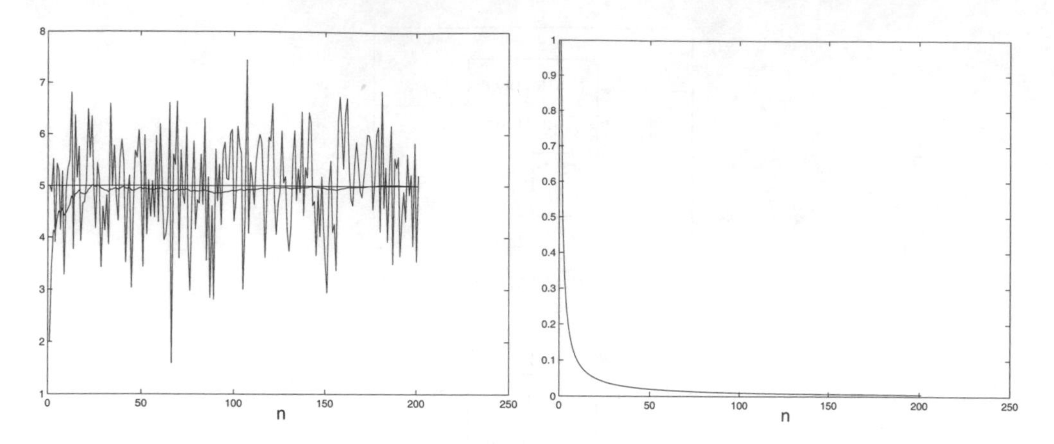

Figure 14.5: Profiles of noisy measurement, filtered value and constant (left), and a profile of error covariance $P(n)$

14.6 Matlab Code

Matlab Code 14.1 Pole placement controller for inverted pendulum, discussed in Example 14.1 on page 490. M 2.1 should be executed before starting this code. This code is available at HOME/ss/matlab/pend.m[1]

```
C = eye(4);
D = zeros(4,1);
Ts = 0.01;
G = ss(A,B,C,D);
H = c2d(G,Ts,'zoh');
[a,b,c,d] = ssdata(H);
rise = 5; epsilon = 0.1;
N = rise/Ts;
omega = pi/2/N;
r = epsilon^(omega/pi);
r1 = r; r2 = 0.9*r;
[x1,y1] = pol2cart(omega,r1);
[x2,y2] = pol2cart(omega,r2);
p1 = x1+j*y1;
p2 = x2+j*y2;
p3 = x1-j*y1;
p4 = x2-j*y2;
P = [p1;p2;p3;p4];
K = place(a,b,P)
```

Matlab Code 14.2 Compensator calculation for Example 14.6 on page 507. This code is available at HOME/ss/matlab/ex_comp.m

```
A = [1 2; 0 3]; c = [1 0];
```

[1]HOME stands for http://www.moudgalya.org/dc/ – first see the software installation directions, given in Appendix A.2.

```
2  p = roots([1 -0.5 0.5]);
3  b = [0; 1];
4  K = place(A,b,p);
5
6  p1=0.1+0.1*j; p2=0.1-0.1*j;
7  phi = real(conv([1 -p1],[1 -p2]));
8  Obs = [c; c*A];
9  alphae = A^2-0.2*A+0.02*eye(2);
10 Lp = alphae*inv(Obs)*[0; 1];
11 Lp = place([1 0;2 3], ...
12 [1; 0],[0.1+0.1*j 0.1-0.1j]);
13 Lp = Lp';
14
15 C = [1 0 0.5 2;0 1 -4.71 2.8];
16 dC = 1;
17 [HD,dHD] = polmul(K,0,C,dC);
18 [HD,dHD] = polmul(HD,dHD,Lp,0);
```

Matlab Code 14.3 Kalman filter example of estimating a constant, discussed in Example 14.7. This code is available at `HOME/ss/matlab/kalrun.m`

```
1  x = 5; xhat = 2; P = 1; xvec = x;
2  xhat_vec = xhat; Pvec = P; yvec = x;
3  for i = 1:200,
4      xline = xhat; M = P;
5      [xhat,P,y] = kal_ex(x,xline,M);
6      xvec = [xvec;x];
7      xhat_vec = [xhat_vec;xhat];
8      Pvec = [Pvec;P]; yvec = [yvec;y];
9  end
10 n = 1:201;
11 plot(Pvec);
12 xlabel('n');
13 pause
14 plot(n,xhat_vec,n,yvec,n,xvec);
15 xlabel('n');
```

Matlab Code 14.4 Kalman filter example of estimating a constant. For a sample call, see M 14.3. This code is available at `HOME/ss/matlab/kal_ex.m`

```
1  function [xhat,P,y] = kal_ex(x,xline,M)
2  y = x + randn;
3  Q = 0; R = 1;
4  xhat_ = xline;
5  P_ = M + Q;
6  K = P_/(P_+R);
7  P = (1-K)*P_;
8  xhat = xhat_ + K*(y-xhat_);
```

14.7 Problems

14.1. This problem addresses the calculation of the determinant of $(zI - A)$ when A is in controller canonical form. One of the ways of calculating the determinant of the matrix

$$|zI - A| = \begin{bmatrix} \lambda + a_1 & a_2 & \cdots & a_{n-2} & a_{n-1} & a_n \\ -1 & \lambda & \cdots & 0 & 0 & 0 \\ 0 & -1 & \cdots & 0 & 0 & 0 \\ \vdots & & & & & \\ 0 & 0 & \cdots & -1 & \lambda & 0 \\ 0 & 0 & \cdots & 0 & -1 & \lambda \end{bmatrix}$$

is given below.

(a) Divide the nth column by λ and add it to the $(n-1)$th. The $(n-1)$th will then become $(a_{n-1} + a_n/\lambda, 0, \ldots, 0, \lambda, 0)$. Note that the subdiagonal entry is made zero by this process.

(b) Repeat this procedure, *i.e.*, divide the $(n-1)$th column by λ and add to the $(n-2)$nd and so on. At the end of this procedure, the (-1) term in the first column will be zeroed. Show that the matrix is upper triangular with the diagonal vector being $(\lambda + a_1/\lambda + a_2/\lambda^2 + \cdots + a_n/\lambda^{n-1}, \lambda, \ldots, \lambda)$.

(c) Complete the arguments to show that the characteristic polynomial of the A matrix of Eq. 14.6 is given by $\lambda^n + a_1\lambda^{n-1} + \cdots + a_n$.

14.2. Verify that when (A, B) are in controller canonical form, (a) the controllability matrix is upper triangular with ones on the diagonal and hence nonsingular, (b) $e_n^T \mathcal{C} = e_n^T$.

14.3. In Eq. 3.46 on page 56, let B be a vector and be denoted by b. Show that Eq. 3.46 can be written as

$$x(n) - A^n x(0) = \mathcal{C} \begin{bmatrix} u(n-1) \\ \vdots \\ u(0) \end{bmatrix}$$

where $\mathcal{C}$ is the controllability matrix given by Eq. 14.17. Argue that one can find the control effort $\{u(0), \ldots, u(n-1)\}$ required to drive the system to *any* final state $x(n)$ starting from *any* starting state $x(0)$ if and only if the controllability matrix is nonsingular (equivalently, the system is controllable).

14.4. Place the closed loop poles of the system $x(k+1) = Ax(k) + Bu(k)$ with

$$A = \begin{bmatrix} 1 & 1 \\ 1 & 1 \end{bmatrix}, \quad B = \begin{bmatrix} 1 \\ 0 \end{bmatrix}$$

at 0 and 0.5 using a state feedback controller. Do this computation first by direct matching and then using Ackermann's formula.

14.5. Note that controllability has nothing to do with A being singular or nonsingular. For example, in the above problem, A is singular yet the system is controllable. Try repeating the above problem, but now with the $(2, 1)$ term of A being 0.

14.6. You have to control a plant with the following state space description:

$$\begin{bmatrix} x_1(k+1) \\ x_2(k+1) \\ x_3(k+1) \end{bmatrix} = \begin{bmatrix} 1 & -1 & 1 \\ 0 & 1 & 1 \\ 0 & 0 & 1 \end{bmatrix} \begin{bmatrix} x_1(k) \\ x_2(k) \\ x_3(k) \end{bmatrix} + \begin{bmatrix} 1 & 0 \\ 1 & 0 \\ 0 & 1 \end{bmatrix} \begin{bmatrix} u_1(k) \\ u_2(k) \end{bmatrix}$$

You are given only one control valve using which either u_1 or u_2 (but not both) can be changed. Explain which control variable you would manipulate and why. You do not have to design the controller. You can assume that all the states are available for measurement.

14.7. Consider a state space model given by $x(k+1) = Ax(k) + Bu(k)$ with

$$A = \begin{bmatrix} -1 & -2 & 0.5 \\ 1 & 0 & 0 \\ 0 & 1 & 0 \end{bmatrix}, \quad b = \begin{bmatrix} 1 \\ 0 \\ 0 \end{bmatrix}$$

(a) Determine a state space controller that will place all the poles at the origin, namely at (0,0).

(b) Recursively calculate the states of the closed loop system at $k = 1$, 2 and 3. You may take the initial state to be (1,1,1).

This controller is known as the dead-beat controller.

14.8. Consider the continuous time system given by

$$F = \begin{bmatrix} 0 & 1 & 0 \\ 0 & -1 & 1 \\ 0 & 0 & 4 \end{bmatrix}, \quad G = \begin{bmatrix} 0 \\ 0 \\ 1 \end{bmatrix}, \quad c = \begin{bmatrix} 1 & 0 & 0 \end{bmatrix}, \quad d = 0$$

Design dead-beat controllers for $T_s = 0.1$ s and for $T_s = 2/3$ s. Check if the controllers indeed work dead-beat. Comment on the magnitude of the control effort produced by each of these controllers. What are the pros and cons of each of them?

14.9. This problem is concerned with placement of poles of the system $x(k+1) = Ax(k) + bu(k)$, $y(k) = cx(k)$, with

$$A = \begin{bmatrix} 1 & 2 \\ 0 & 3 \end{bmatrix}, \quad b = \begin{bmatrix} 0 \\ 1 \end{bmatrix}, \quad c = \begin{bmatrix} 1 & 0 \end{bmatrix}$$

(a) Assuming that both the states are measured, determine the pole placement controller K that will make the system dead-beat, *i.e.*, both the closed loop poles are placed at 0. Answer the following:

i. By substituting this control law, *i.e.*, $u = -Kx(k)$, check that the eigenvalues of the closed loop system matrix are at 0.

ii. With the application of this control law, recursively calculate $x(1)$, $x(2)$ and $x(3)$, assuming the initial state to be $x(0) = \begin{bmatrix} x_{10} & x_{20} \end{bmatrix}^T$, where x_{10} and x_{20} are arbitrary. Does $x(k)$ become zero for any k? Explain.

(b) Now assume that not all states are measured. Determine the estimator L_p that also has its eigenvalues placed at 0.

(c) Determine the compensator that is the combination of the controller K obtained in part (a) and estimator L_p obtained in part (b).

14.10. It is desired to place the poles of the system

$$\begin{bmatrix} x_1(k+1) \\ x_2(k+1) \end{bmatrix} = \begin{bmatrix} 1 & T_s \\ 0 & 1 \end{bmatrix} \begin{bmatrix} x_1(k) \\ x_2(k) \end{bmatrix} + \begin{bmatrix} T_s^2/2 \\ 0.1 \end{bmatrix} u(k)$$

$$y(k) = \begin{bmatrix} 1 & 0 \end{bmatrix} \begin{bmatrix} x_1(k) \\ x_2(k) \end{bmatrix}$$

where $T_s = 0.1$.

(a) Assuming that all the states are measurable, design a pole placing controller that will result in a closed loop system with characteristic equation $z^2 - 1.4z + 0.49 = 0$.

(b) Design an estimator that will have the closed loop poles at $0.5 \pm 0.5j$.

(c) Combine the pole placement controller of part (a) with the estimator of part (b) to arrive at an expression for the compensator.

14.11. It is desired to control a system having two states x_1 and x_2 using two control variables u_1 and u_2. It is necessary to control the first state only. The second state could take any value. The first control variable is ten times more expensive than the second control variable. Pose this as an optimal control problem familiar to you. You do not have to solve the problem.

14.12. Consider the unstable discrete time LTI system

$$x(k+1) = Ax(k) + Bu(k)$$

with

$$A = \begin{bmatrix} 2 & 0 \\ 1 & 0 \end{bmatrix}, \quad B = \begin{bmatrix} 1 \\ 0 \end{bmatrix}$$

(a) Design a steady state LQR with weighting matrices

$$Q_1 = \begin{bmatrix} 1 & 0 \\ 0 & 1 \end{bmatrix}, \quad Q_2 = 1$$

[Hints. (1) The steady state matrix equation can be solved directly. (2) The solution is a symmetric positive definite matrix.]

(b) Discuss the behaviour of the closed loop system with this controller.

(c) Repeat the above calculations but now with the following Q_1:

$$Q_1 = \begin{bmatrix} 1 & 0 \\ 0 & 0 \end{bmatrix}$$

How does the closed loop system behave now? Explain.

Appendix A

Supplementary Material

We provide two types of information in this Appendix. In the first section, we derive a few useful mathematical relations. In the second section, we explain the procedure to use the software that comes with this book.

A.1 Mathematical Relations

We need to differentiate quadratic forms in optimization problems. In this book, we use it to solve the least squares estimation (LSE) problem. In this section, we explain how to carry out this differentiation [13]. Diagonalization of square matrices is very useful while solving systems of equations. We explain how we can diagonalize square matrices.

A.1.1 Differentiation of a Quadratic Form

A quadratic form Q in n variables $x_1, x_2, \ldots, x_n$ is a scalar of the following form

$$Q = x^T A x \tag{A.1}$$

where x is the vector of the above given n variables and A is an $n \times n$ matrix of the following form

$$x = \begin{bmatrix} x_1 \\ \vdots \\ x_n \end{bmatrix}, \quad A = \begin{bmatrix} a_{11} & \cdots & a_{1n} \\ \vdots & & \\ a_{n1} & \cdots & a_{nn} \end{bmatrix} \tag{A.2}$$

With this definition, Eq. A.1 becomes

$$Q = \begin{bmatrix} x_1 & \cdots & x_k & \cdots & x_n \end{bmatrix} \begin{bmatrix} a_{11} & \cdots & a_{1k} & \cdots & a_{1n} \\ \vdots & & & & \\ a_{k1} & \cdots & a_{kk} & \cdots & a_{kn} \\ \vdots & & & & \\ a_{n1} & \cdots & a_{nk} & \cdots & a_{nn} \end{bmatrix} \begin{bmatrix} x_1 \\ \vdots \\ x_k \\ \vdots \\ x_n \end{bmatrix} \tag{A.3}$$

On multiplying it, we obtain

$$Q = \begin{bmatrix} x_1 & \cdots & x_k & \cdots & x_n \end{bmatrix} \begin{bmatrix} a_{11}x_1 + \cdots + a_{1k}x_k + \cdots + a_{1n}x_n \\ \vdots \\ a_{k1}x_1 + \cdots + a_{kk}x_k + \cdots + a_{kn}x_n \\ \vdots \\ a_{n1}x_1 + \cdots + a_{nk}x_k + \cdots + a_{nn}x_n \end{bmatrix} \tag{A.4}$$

Suppose that we want to differentiate the above expression by x_k. It is clear that only the terms of the following type will be nonzero:

$$\begin{array}{ccc} & a_{1k}x_kx_1 & \\ & \vdots & \\ (a_{k1}x_1 + \cdots + & a_{kk}x_k & + \cdots + a_{kn}x_n)x_k \\ & \vdots & \\ & a_{nk}x_kx_n & \end{array}$$

All other terms in the matrix are zero. As a result, it is easy to obtain the following result:

$$\frac{\partial Q}{\partial x_k} = (Ax)_k + (A^T x)_k, \quad k = 1, 2, \ldots, n \tag{A.5}$$

where $(Ax)_k$ and $(A^T x)_k$ are the kth rows of vectors Ax and $A^T x$, respectively.

We have seen how to differentiate the quadratic form by a scalar. Now we will explain how to differentiate with a vector. Suppose that we want to differentiate Q given by Eq. A.1 with respect to x. Using Eq. A.5, we obtain

$$\nabla_x Q \triangleq \begin{bmatrix} \partial/\partial x_1 \\ \vdots \\ \partial/\partial x_n \end{bmatrix} Q = \begin{bmatrix} (Ax)_1 + (A^T x)_1 \\ \vdots \\ (Ax)_n + (A^T x)_n \end{bmatrix} = Ax + A^T x \tag{A.6}$$

If A is symmetric,

$$\nabla_x Q = 2Ax \tag{A.7}$$

We also need to differentiate an asymmetric form, such as

$$R = x^T B y = \begin{bmatrix} x_1 & \cdots & x_n \end{bmatrix} \begin{bmatrix} b_{11} & \cdots & b_{1m} \\ \vdots & & \\ b_{n1} & \cdots & b_{nm} \end{bmatrix} \begin{bmatrix} y_1 \\ \vdots \\ y_m \end{bmatrix} \tag{A.8}$$

with respect to vector x. If we write Eq. A.8 in the following form

$$R = x_1 \sum_{j=1}^{m} b_{1j} y_j + \cdots + x_k \sum_{j=1}^{m} b_{kj} y_j + \cdots + x_n \sum_{j=1}^{m} b_{nj} y_j \tag{A.9}$$

it is clear that

$$\frac{\partial R}{\partial x_k} = \sum_{j=1}^{n} b_{kj} y_j = k\text{th row of } By \tag{A.10}$$

Thus we obtain

$$\nabla_x (x^T B y) = \begin{bmatrix} \text{1st row of } By \\ \vdots \\ n\text{th row of } By \end{bmatrix} = By \tag{A.11}$$

Notice that R is a scalar, see Eq. A.8. It is clear that its derivative with respect to a vector will be a vector. Next we would like to differentiate Eq. A.8 with respect to y. As $x^T B y$ is a scalar, it is equal to $y^T B^T x$ and hence we obtain

$$\nabla_y (x^T B y) = \nabla_y (y^T B^T x) = B^T x \tag{A.12}$$

where to arrive at the last term, we have made use of Eq. A.11.

A.1.2 Diagonalization of Square Matrices

In this section, we will briefly describe a procedure to diagonalize square matrices. Let λ_j, $j = 1, \ldots, n$, and x_j, $j = 1, \ldots, n$, respectively, be eigenvalues and eigenvectors of an $n \times n$ matrix A. Let us assume that the eigenvectors are linearly independent. We obtain

$$A x_j = \lambda_j x_j, \; j = 1, \ldots, n$$

Stacking these vector equations side by side, we obtain

$$A \begin{bmatrix} | & & | \\ x_1 & \cdots & x_n \\ | & & | \end{bmatrix} = \begin{bmatrix} | & & | \\ \lambda_1 x_1 & \cdots & \lambda_n x_n \\ | & & | \end{bmatrix}$$

where we have emphasized the fact that x_j are vectors. The above equation can be written as

$$A \begin{bmatrix} | & & | \\ x_1 & \cdots & x_n \\ | & & | \end{bmatrix} = \begin{bmatrix} | & & | \\ x_1 & \cdots & x_n \\ | & & | \end{bmatrix} \begin{bmatrix} \lambda_1 & & 0 \\ & \ddots & \\ 0 & & \lambda_n \end{bmatrix}$$

where the last matrix is diagonal, denoted by the symbol Λ. Let us denote the matrix of eigenvectors as S. We arrive at

$$AS = S\Lambda$$

Because of the assumption that the eigenvectors are linearly independent, it is possible to invert S:

$$A = S \Lambda S^{-1}$$

This approach has several benefits. It is easy to find the powers of A. For example,

$$A^2 = S\Lambda S^{-1} S\Lambda S^{-1} = S\Lambda^2 S^{-1}$$
$$A^3 = S\Lambda^2 S^{-1} S\Lambda S^{-1} = S\Lambda^3 S^{-1}$$

We can generalize this result:

$$A^k = S\Lambda^k S^{-1}, \quad n \geq k \geq 1 \tag{A.13}$$

This property also helps calculate polynomials of a matrix A. Suppose that we want to calculate

$$\alpha(A) = A^n + \alpha_1 A^{n-1} + \cdots + \alpha_n I$$

Using Eq. A.13, we obtain

$$\alpha(A) = S\left[\Lambda^n + \alpha_1 \Lambda^{n-1} + \cdots + \alpha_n I\right] S^{-1}$$

Because of the ease in evaluating Λ^k, it is straightforward to use the above equation.

A.2 Installation and Use of Software

We now explain the procedure to download and install the Matlab code presented in this book. Before executing anything, the programs in `HOME/matlab`[1] have to be downloaded and made available through Matlab's `set path` statement. After this path is set, one can download the required programs and execute them. For example, one can download `ball_basic.m`, described in M 9.1, and execute it. It is easy to see that this code requires `pp_basic.m`. From the *Index of Matlab Code*, it is easy to locate the details of `pp_basic.m` on 369, including the downloading instructions.

Because of space constraints, the following Matlab programs have not been listed in the book: `ext.m`, `l2r.m`, `left_prm.m`, `poladd.m`, `polmul.m`, `putin.m`, `xdync.m`. Nevertheless, these and the dependent programs are available from `HOME`. Some of these programs have been taken from [27].

The above procedure explains how to download the programs as needed. An easier way to access the programs is to download `http://www.moudgalya.org/dc.tgz` which contains all the Matlab code listed in this book, and the routines named in the previous paragraph. On uncompressing this archive, one will get all the programs in the folder `dc`. The procedure to use this code is as follows:

1. Change the directory to `dc`.

2. Set Matlab path to `matlab`, which is a subdirectory of `dc`.

3. Go to a directory where the Matlab code resides and execute it. For example, if one wants to run the *pole placement* programs, one would change the directory to `place/matlab` (`place` is also a subdirectory of `dc`) and run the relevant program, as explained above. Note that the web addresses of all programs required for the pole placement chapter are given in the captions listed in Sec. 9.9.

[1] `HOME` stands for `http://www.moudgalya.org/dc/`.

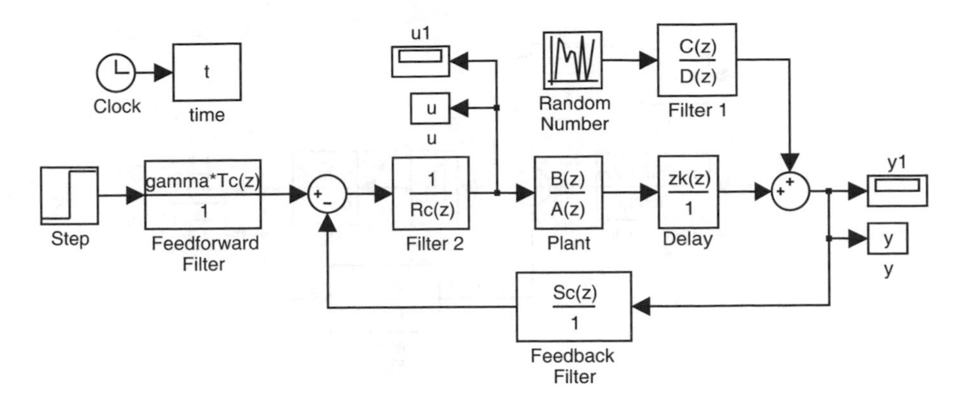

Figure A.1: Simulink block diagram for plants with z domain transfer function with 2-DOF pole placement controller. The code is available at `HOME/matlab/stb_disc.mdl`. Inputs for simulation have to be established by first executing a controller program, such as M 9.1.

One of the reliable and hence popular ways of checking the efficacy of a control system is through simulation. In general, simulation is used to verify that the closed loop system is stable and that the performance requirements are met. In sampled data systems, there are additional issues, such as intrasample behaviour.

In this section, we present a set of Simulink programs that help carry out simulation. The listing of the code `ball_basic.m` says that the Simulink programs `basic.mdl` and `c_ss_cl.mdl` can be used to verify the efficacy of the controller. From the *index of Matlab code*, it is easy to see that details of the Simulink routines are available on pages 526 and 528.

The controller has to be first designed before using the Simulink code. Simulation parameters also have to be set. A common mistake in carrying out this kind of simulation is not initializing the sampling period properly. It is advisable to use a variable, such as `Ts` for this purpose and initialize it in the Matlab workspace. For example, `ball_basic.m` sets up all the variables required to execute either `basic.mdl` or `c_ss_cl.mdl`. The result of simulation can be observed in the scopes. All important variables are also stored in the Matlab workspace, for further processing by any other software.

Simulation using this Simulink code can be used to verify that the design objectives are met. In case there is a discrepancy between what the controller is designed for and what is actually achieved, it could indicate errors in the design procedure, the simulation procedure, etc. More importantly, this could point to a lack of understanding of the design and simulation procedures. It is suggested that the control designer uses this mode to ensure that they understand the theory well before going ahead with implementation on a real plant.

The Simulink program in `basic.mdl` requires the plant to be in Laplace transfer function form, while that in `c_ss_cl.mdl` expects it in state space form. A continuous time version of the plant is used in both of these Simulink routines. Note that the discrete time control action has to be sent through a zero order hold, indicated by `ZOH`, before being sent out to the plant.

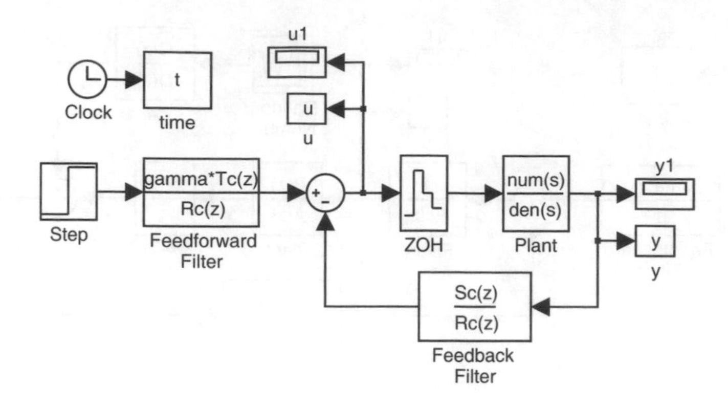

Figure A.2: Simulink block diagram for plants described by s domain transfer function with 2-DOF pole placement controller. The code is available at `HOME/matlab/basic.mdl`

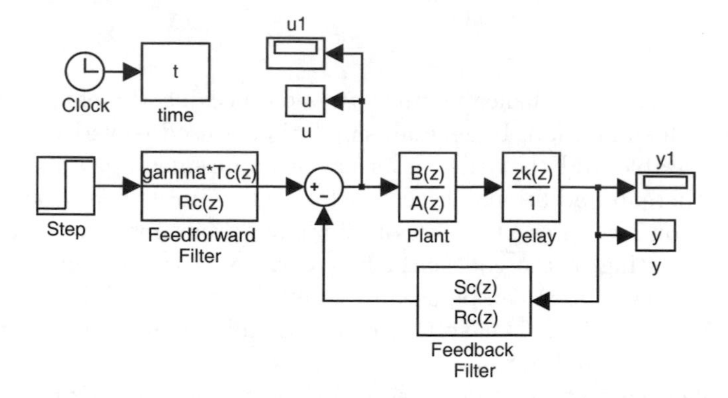

Figure A.3: Simulink block diagram for discrete time plants with 2-DOF pole placement controller. The code is available at `HOME/matlab/basic_disc.mdl`

All digital controllers designed for continuous systems have to be validated through simulation using a program, such as the one being discussed. The reason is that the controller validated using Fig. A.1 does not say anything about the intrasample behaviour. For example, if proper care is not taken, there could be wild oscillations, which cannot be detected through simulations with the discrete time plant. These oscillations are known as intrasample *ripple*. Fortunately, however, such erratic behaviour can be detected through simulations with the continuous plant, as described above.

For those who deal only with discrete time plants, there is no need to use ZOH. They have to pick a suitable Simulink routine, for example `stb_disc.mdl` listed in Fig. A.1 and use it. The utility of the other Simulink programs given in this section is obvious.

All the routines reported in this book have been tested in versions 6.5 and 7 of Matlab.

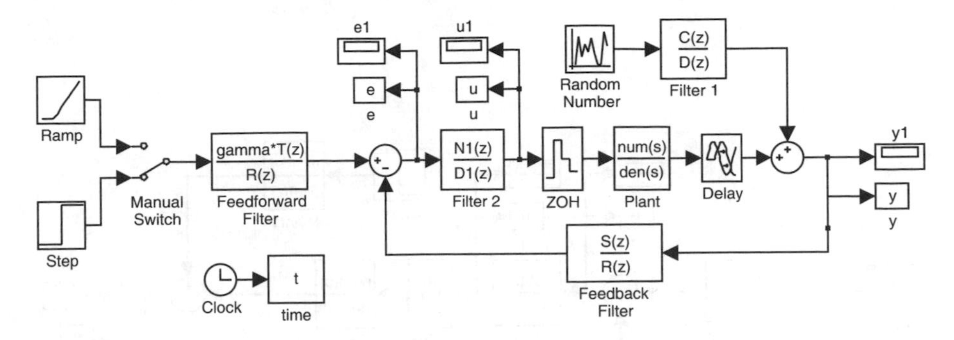

Figure A.4: Simulink block diagram for plants described by s domain transfer function with 2-DOF pole placement controller. The code is available at `HOME/matlab/g_s_cl.mdl`. Inputs for simulation have to be established by first executing a controller program, such as M 9.1.

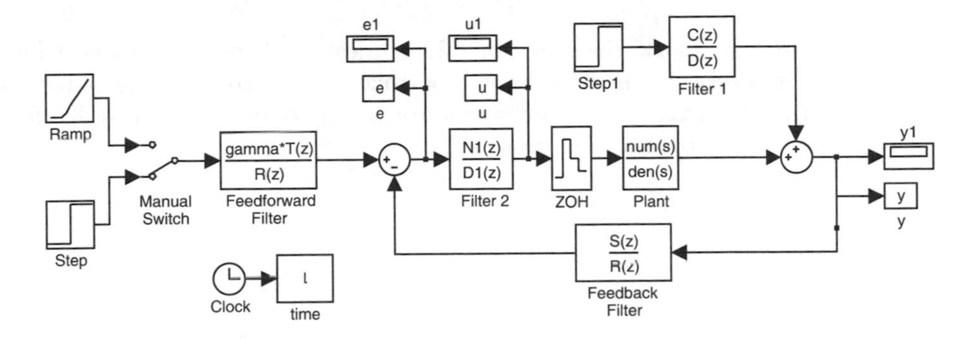

Figure A.5: Simulink block diagram for plants described by s domain transfer function with 2-DOF pole placement controller. The code is available at `HOME/matlab/g_s_cl2.mdl`. Inputs for simulation have to be established by first executing a controller program, such as M 9.21.

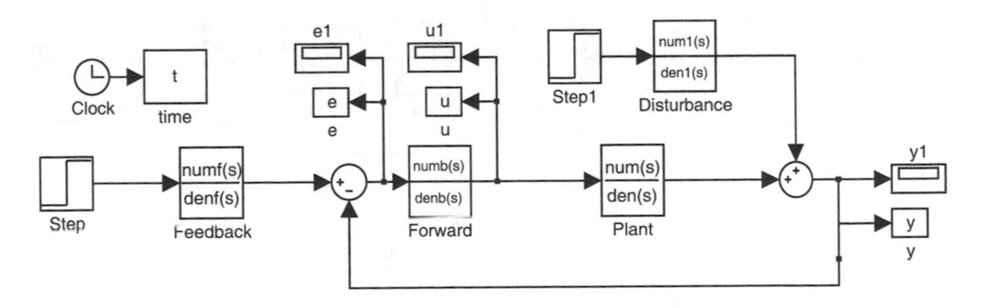

Figure A.6: Simulink block diagram for plants described by s domain transfer function with 2-DOF continuous time controller. The code is available at `HOME/matlab/g_s_cl3.mdl`. Inputs for simulation have to be established by first executing a controller program, such as M 9.21.

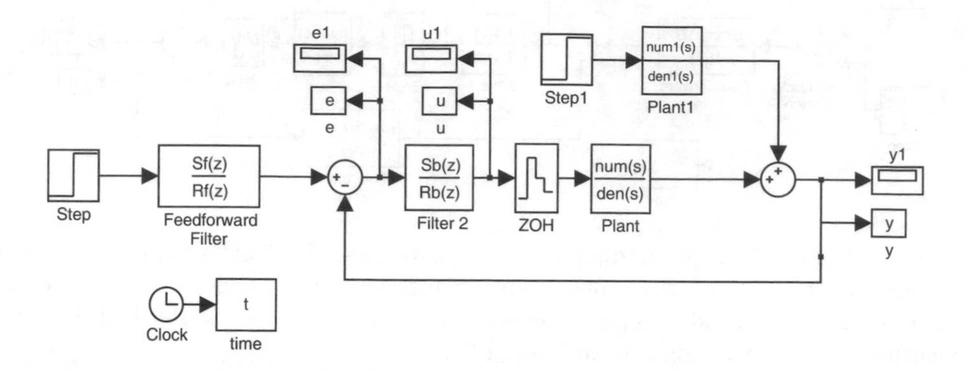

Figure A.7: Simulink block diagram for plants described by s domain transfer function with 2-DOF discrete time controller. The code is available at `HOME/matlab/g_s_cl6.mdl`. Inputs for simulation have to be established by first executing a controller program, such as M 9.15.

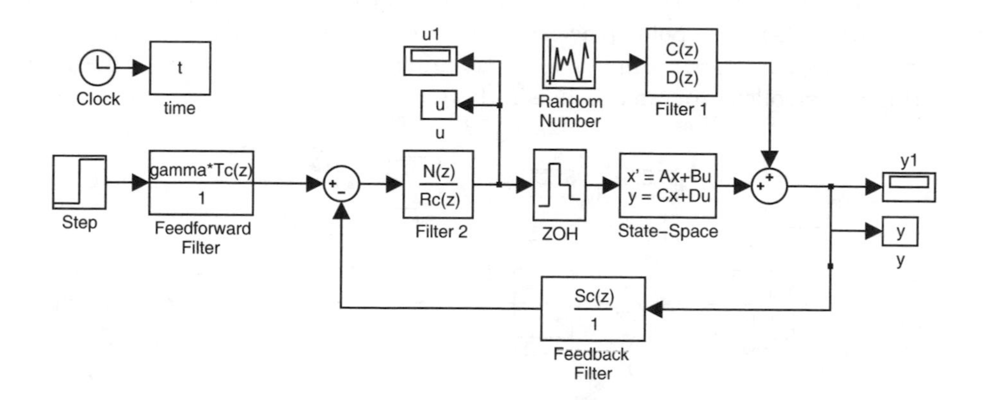

Figure A.8: Simulink block diagram for plants described by state space models with 2-DOF pole placement controller. The code is available at `HOME/matlab/c_ss_cl.mdl`. Inputs for simulation have to be established by first executing a controller program, such as M 9.1.

A.3 Problems

A.1. This problem is concerned with the Jordan canonical form.

(a) Show that the eigenvalues of a matrix A are $-2, 4, 4$, where

$$A = \begin{bmatrix} 5 & 4 & 3 \\ -1 & 0 & -3 \\ 1 & -2 & 1 \end{bmatrix}$$

(b) Calculate the eigenvectors corresponding to these eigen values.

(c) Can you diagonalize this system?

(d) Let q_1 be the eigenvector corresponding to $\lambda_1 = -2$ and q_2 that corresponding to $\lambda_2 = 4$. The third independent vector q_3 is calculated so as to satisfy the following equation:

$$A[q_1\ q_2\ q_3] = [q_1\ q_2\ q_3] \begin{bmatrix} -2 & 0 & 0 \\ 0 & 4 & 1 \\ 0 & 0 & 4 \end{bmatrix}$$

(Notice the "1" in the offdiagonal of the far right matrix. This matrix is known as the Jordan block and the above equation is known as the Jordan canonical form.) Calculate q_3. Notice that q_3 is not unique. Any q_3 that satisfies the equation is OK.

(e) You have a discrete time system

$$x(k+1) = Ax(k)$$

where

$$A = \begin{bmatrix} 3 & -4 \\ 1 & -1 \end{bmatrix}$$

Compute the Jordan canonical form of A satisfying

$$A[q_1\ q_2] = [q_1\ q_2] \begin{bmatrix} 1 & 1 \\ 0 & 1 \end{bmatrix}$$

Using the 100th power of the Jordan block, calculate $x(100)$. The initial value of x is $(1, 1)$.

A.3 Problems

A.1 This problem is concerned with the Jordan canonical form.

(a) Show that [illegible]

[illegible]

References

[1] B. D. O. Anderson and J. B. Moore. *Optimal Control: Linear Quadratic Methods.* Prentice-Hall of India Pvt. Ltd., New Delhi, 1991.

[2] K. J. Åström and B. Wittenmark. *Computer Controlled Systems. Theory and Practice.* Prentice-Hall, Inc., Upper Saddle River, NJ, 3rd edition, 1997.

[3] B. W. Bequette. *Process Control: Modeling, Design and Simulation.* Prentice-Hall, Inc., Upper Saddle River, NJ, 2003.

[4] A. B. Bishop. *Introduction to Discrete Linear Controls: Theory and Application.* Academic Press, New York, 1975.

[5] G. E. P. Box, G. M. Jenkins, and G. C. Reinsel. *Time Series Analysis: Forecasting and Control.* Pearson Education, Delhi, 3rd edition, 2004.

[6] A. E. Bryson and Y. C. Ho. *Applied Optimal Control.* Halsted Press, Washington, DC, 1975.

[7] E. F. Camacho and C. Bordons. *Model Predictive Control.* Springer, London, 1999.

[8] B.-C. Chang and J. B. Pearson. Algorithms for the solution of polynomial equations arising in multivariable control theory. Technical Report 8208, Rice University, Dept. of Electrical Engineering, 1982.

[9] C. T. Chen. *Linear System Theory and Design.* Holt, Rinehart and Winston, Inc., New York, 1970.

[10] T. Chen and B. Francis. *Optimal Sampled-Data Control Systems.* Springer-Verlag, Berlin, 1996.

[11] D. W. Clarke, C. Mohtadi, and P. S. Tuffs. Generalized predictive control - Part I. the basic algorithm, Part II. extensions and interpretations. *Automatica*, 23(2):137–160, 1987.

[12] CTM. State space tutorial. `www.engin.umich.edu/group/ctm/state/state.html`, 2006. Accessed 27 Jan. 2006.

[13] E. P. Cunningham. *Digital Filtering. An Introduction.* Houghton Mifflin, Boston, MA, 1992.

[14] S. M. Disney and D. R. Towill. A discrete transfer function model to determine the dynamic stability of a vendor managed inventory supply chain. *Int. J. Prod. Res.*, 40(1):179–204, 2002.

[15] P. Dorato. *Analytic Feedback System Design. An Interpolation Approach.* Brooks/Cole, Pacific Grove, CA, 2000.

[16] G. F. Franklin, J. D. Powell, and A. Emami-Naeini. *Feedback Control of Dynamic Systems.* Prentice-Hall, Inc., Upper Saddle River, NJ, 4th edition, 2002.

[17] G. F. Franklin, J. D. Powell, and M. Workman. *Digital Control of Dynamic Systems.* Addison Wesley Longman, Menlo Park, CA, 3rd edition, 1998.

[18] P. J. Gawthrop. Some interpretations of the self-tuning controller. *Proc. IEE*, 124(10):889–894, 1977.

[19] A. Gelb, editor. *Applied Optimal Estimation.* MIT Press, Cambridge, MA, 1974.

[20] G. C. Goodwin, S. F. Graebe, and M. E. Salgado. *Control System Design.* Pearson Education, Delhi, 2003.

[21] P. Gupta. Development of PID equivalent control algorithms. B. Tech. Project Report. IIT Bombay, April 2005.

[22] J. L. Hellerstein, Y. Diao, S. Parekh, and D. M. Tilbury. *Feedback Control of Computing Systems.* Wiley–IEEE Press, New York, 2004.

[23] B. Huang and S. L. Shah. *Performance Assessment of Control Loops: Theory and Applications.* Springer-Verlag, Berlin, 1999.

[24] T. Kailath. *Linear Systems.* Prentice-Hall, Inc., Englewood Cliffs, NJ, 1980.

[25] E. Kreyszig. *Advanced Engineering Mathematics.* John Wiley & Sons, Inc., New York, 4th edition, 1979.

[26] V. Kucera. *Discrete Linear Control: The Polynomial Equations Approach.* John Wiley & Sons, Ltd, Chichester, 1979.

[27] H. Kwakernaak. MATLAB macros for polynomial H_∞ control system optimization. Technical Report No. 881, University of Twente, 1990.

[28] H. Kwakernaak and M. Sebek. Polynomial toolbox. `www.polyx.com`, 2000.

[29] H. Kwakernaak and R. Sivan. *Linear Optimal Control Systems.* John Wiley & Sons, Inc., New York, 1972.

[30] I. D. Landau. *System Identification and Control Design.* Prentice-Hall, Inc., Englewood Cliffs, NJ, 1990.

[31] I. D. Landau. Robust digital control of systems with time delay (the Smith predictor revisited). *Int. J. Control*, 62(2):325–347, 1995.

[32] L. Ljung. *System Identification: Theory for the User.* Prentice-Hall, Inc., Upper Saddle River, NJ, 2nd edition, 1999.

[33] W. L. Luyben. *Process Modeling, Simulation and Control for Chemical Engineers.* McGraw-Hill Book Company, New York, 1973, 1989.

[34] J. F. MacGregor. Discrete stochastic control with input constraints. *Proc. IEE*, 124:732–734, 1977.

[35] R. K. Majumdar, K. M. Moudgalya, and K. Ramamritham. Adaptive coherency maintenance techniques for time-varying data. In *Proceedings of 24th IEEE RTSS*, pages 98–107, Cancun, 2003.

[36] J. M. Mendel. *Lessons in Estimation Theory for Signal Processing, Communications, and Control.* Prentice-Hall, Inc., Englewood Cliffs, NJ, 1995.

[37] R. M. Miller, S. L. Shah, R. K. Wood, and E. K. Kwok. Predictive PID. *ISA Trans.*, 38:11–23, 1999.

[38] S. K. Mitra. *Digital Signal Processing.* Tata McGraw Hill, New Delhi, 2001.

[39] K. M. Moudgalya. Polynomial computations in Matlab. Technical report, IIT Bombay, Dept. of Chemical Engineering, 1996. `http://www.moudgalya.org/dc/res/left_prm.pdf`.

[40] K. M. Moudgalya. Discrete time control. A first control course? In *Advances in Control Education*, Madrid, 21–23 June 2006. IFAC.

[41] K. M. Moudgalya and S. L. Shah. A polynomial based first course in digital control. In *Proceedings of IEEE CACSD*, pages 190–195, Taipei, 2–4 Sept. 2004.

[42] B. Noble and J. W. Daniel. *Applied Linear Algebra.* Prentice-Hall, Inc., Englewood Cliffs, NJ, 2nd edition, 1977.

[43] J. P. Norton. *An Introduction to Identification.* Academic Press, London, 1986.

[44] K. Ogata. *Discrete-Time Control Systems.* Prentice-Hall, Inc., Upper Saddle River, NJ, 2nd edition, 1995.

[45] K. Ogata. *Modern Control Engineering.* Pearson Education, Delhi, 2002.

[46] A. V. Oppenheim and R. W. Schafer. *Digital Signal Processing.* Prentice-Hall, Inc., Englewood Cliffs, NJ, 1975.

[47] H. Parthasarathy. *Textbook of Signals and Systems.* I. K. International Pvt. Ltd., New Delhi, 2006.

[48] J. B. Pearson. Course notes for linear system. Rice University, Autumn 1984.

[49] J. C. Proakis and D. G. Manolakis. *Digital Signal Processing: Principles, Algorithms, and Applications.* Prentice-Hall, Inc., Upper Saddle River, NJ, 3rd edition, 1996.

[50] S. Rönnbäck and J. Sternby. A design algorithm for anti-windup compensators. Polynomial approach. In *Proceedings of 12th Triennial World Congress*, pages 619–624, Sydney, Australia, 1993. IFAC.

[51] S. L. Shah. Class notes for CHE 662: System identification. University of Alberta, Autumn 2004.

[52] S. Skogestad and I. Postlewaite. *Multivariable Feedback Control - Analysis and Design.* John Wiley & Sons, Ltd, Chichester, 2nd edition, 2005.

[53] T. Söderström and P. Stoica. *System Identification.* Prentice-Hall, Int. (UK) Ltd., Hertfordshire, 1989.

[54] M. Sternad and A. Ahlen. LQ Controller Design and Self-tuning Control, chapter 3 in *Polynomial Methods in Optimal Control and Filtering*, edited by K. Hunt. Peter Peregrinus, London, 1993.

[55] G. Strang. *Linear Algebra and its Applications.* Thomson Learning, Inc., Singapore, 3rd edition, 1998.

[56] A. K. Tangirala. Course notes for system identification. IIT Madras, Autumn 2005.

[57] P. P. J. van den Bosch and A. C. van der Klauw. *Modeling, Identification and Simulation of Dynamical Systems.* CRC Press, Boca Raton, FL, 1994.

[58] J. Venkateswaran and Y. J. Son. Effect of information update frequency on the stability of production – inventory control systems. *Int. J. Prod. Econ.*, 2006. Available online: 28 July 2006.

[59] M. Vidyasagar. *Control System Synthesis.* MIT Press, Cambridge, MA, 1985.

[60] T. Yamamoto, A. Inoue, and S. L. Shah. Generalized minimum variance self-tuning pole-assignment controller with a PID setting. In *Proceedings of the 1999 IEEE International Conference on Control Applications*, pages 125–130, 1999.

Index

Index of Matlab Code

RECEIVED
NOV 8 2007